Coral Reef Restoration Handbook

Coral Reef Restoration Handbook

edited by
William F. Precht

CRC is an imprint of the Taylor & Francis Group,
an informa business

Published in 2006 by
CRC Press
Taylor & Francis Group
6000 Broken Sound Parkway NW, Suite 300
Boca Raton, FL 33487-2742

Printed in the United States of America on acid-free paper
10 9 8 7 6 5 4 3 2 1

International Standard Book Number-10: 0-8493-2073-9 (Hardcover)
International Standard Book Number-13: 978-0-8493-2073-6 (Hardcover)

Taylor & Francis Group
is the Academic Division of Informa plc.

Visit the Taylor & Francis Web site at
http://www.taylorandfrancis.com

and the CRC Press Web site at
http://www.crcpress.com

Dedication

To Joni, Lindsey, Chandler, and Madison. For making my world complete and for giving me four reasons to make a difference.

Foreword

When I was a child growing up in south Florida, I loved the environmental treasures at my doorstep. With time the lure of the ocean, its blue waters and breathtaking scenery brought me to the Florida Keys. As an adult I learned how to scuba dive and a new underwater world, much of it hidden at first glance, was realized. Beneath the water's surface is a natural wonder comprised of living animals and plants. Diving in this underwater landscape you understand why this resource needs to be preserved and protected for all time. The world of the coral reef is spectacular yet incredibly fragile. As we all know, the coral reefs of the Florida Keys have been an important destination for explorers, scientists, and tourists for centuries. However, their popularity has led to pollution of the marine ecosystem and overuse of resources. Signs of anthropogenic degradation in the Keys became apparent several decades ago. Corals were being damaged and water quality was suffering. Many began to recognize that the Keys' environment and resources needed protection before they were damaged beyond repair.

My deep and abiding love for the reefs of the Florida Keys made me a strong advocate for their protection. Unfortunately, threats to the coral reef ecosystem continued. Proposed oil drilling in the mid- to late-1980s and reports of deteriorating water quality throughout the region surfaced as scientists were assessing the impacts of coral bleaching and the continued spread of coral diseases. The final insult came in the fall of 1989 when three large ships ran aground on the Florida reef tract within an 18-day period, destroying critical reef habitat. This combination of disturbances is why I introduced legislation in November 1989 calling for more protection of the coral reefs. Congress passed the Florida Keys National Marine Sanctuary and Protection Act into law in 1990. The act designated approximately 2800 square nautical miles of state and federal waters in the Keys as the Florida Keys National Marine Sanctuary.

By the designation of this area as a marine sanctuary, the fragile reef habitats were finally afforded the protection and stewardship they required. Today, due to a buy-back of 73 federal oil and gas leases in 1995, exploration for oil and gas is now prohibited off the Florida Keys. As well, the development of an internationally recognized "area to be avoided" provides a 2 1/2-mile buffer zone the length of the reef tract that prevents large tankers and freighters from coming near the reefs. At the same time, however, the Sanctuary permits controlled use of the resource as long as those activities are not injurious to the environment. Unfortunately, accidents do happen and coral reefs are still being undermined from a variety of threats, both natural and anthropogenic. That is why, in some cases, people need to intervene in the process by restoring and rehabilitating the injured resource.

The process of restoring coral reefs is in its infancy when compared to restoration of other ecosystems. It is especially rewarding for me to know that much of what has been performed and learned to date in terms of reef restoration projects has come from experiences within the Florida Keys National Marine Sanctuary. To that end, this book is the first to describe, in detail, the art and science of coral reef restoration. It is to be hoped that the information that can be gleaned within the pages of this book will set a path towards continued preservation of this valuable underwater treasure to be used, appreciated, and experienced for future generations.

Senator Bob Graham (ret.), Miami Lakes, FL

Preface

It is known that coral reefs around the world have changed dramatically over the past two decades. Many types of disturbance separately and in combination are changing the face of reefs. These include hurricanes, coral bleaching, diseases of corals and sea urchins, overfishing, nutrient loading, sedimentation, hyper- and hypothermic stresses, various forms of pollution, harvesting of reef invertebrates, coral mining, trampling by tourists and divers, and the destruction and devastation caused by ship anchors and groundings. It is obvious that this resource needs protection and that many of the cited anthropogenic causes can be reduced or avoided by implementation of science-based management programs.

It seems evident that if we continue the present rate of destruction, reef ecosystems will likely suffer continued significant degradation, possibly to the point of irreversible decline. Accordingly, to continue on the present course is not prudent. It therefore is imperative that we act now to shift this imbalance. The most appropriate course of action is to replace damaged and disturbed reefs with fully functional, restored ecosystems at a rate resulting in no net loss of ecosystem value (i.e., the rate of reef destruction offset by the rate of reef repair). As a practical matter, managers and policymakers also need to understand the effects of human-induced disturbances, be able to properly assess these damages, and develop subsequent restoration efforts on reefs under their stewardship.

To date, most coral reef restoration programs have focused on the physical damage caused by people. Of these, ship groundings are among the most destructive chronic anthropogenic factors causing significant localized damage on coral reefs and have been the focus of many early attempts at reef restoration. In fact, much of what we know about the rehabilitation of coral reef systems stems from our work in trying to repair reefs injured by vessels that have run aground. This is especially true in waters of the United States. To date, however, there is a paucity of published literature regarding the efficacy and/or failure of coral reef restoration techniques. In fact, most of the literature that is available is gray, that is, mostly meeting abstracts, workshops, and technical memoranda. Yet these very papers and reports have forged a scientific framework for future efforts in this field. To hasten our learning curve, it is imperative to understand what works, what does not, and why.

The status of reef restoration has improved a great deal in a very short time. As reef scientists and managers we should glean as much as we can from the work that has gone before. Failure to learn from our past efforts will undoubtedly impede the progress of this enterprise into the future and result in an inferior final restoration product.

Coral reef restoration is both an art and a science if performed well. It is to be hoped that the lessons learned from this synthesis will help to develop successful restoration efforts into the future. As well, because of the infancy of this enterprise, the continued sharing of information will be vital to improving restoration strategies over time.

William F. Precht
Ecological Sciences Program
PBS&J
Miami, Florida

Editor

William Precht is a carbonate sedimentologist by training and has been studying coral reefs since 1978. He was first introduced to coral reefs at Discovery Bay Marine Lab in Jamaica as an undergraduate student and has been working there ever since. His current research areas include the Bahamas, Belize, Florida, Jamaica, Mexico, Puerto Rico, Moorea–French Polynesia, and the Flower Garden Banks in the Gulf of Mexico. His research interests include combining ecological and geological methodologies to decipher "change" in reef communities through time and space. Using this integrated approach, he (with collaborators Richard B. Aronson and Ian Macintyre) has been able to assess the geological and ecological novelty of many of the recent maladies affecting coral reefs. This includes deciphering local anthropogenic signals from overarching global effects. Specific research has included the effects of coral disease and coral bleaching on the trajectories of reef coral communities. Presently, he is developing cutting-edge assessment and restoration strategies for reefs impacted by various anthropogenic sources, including providing expert assistance to a wide array of both national and international clients.

Since completing his graduate degree in marine geology and geophysics from the University of Miami's Rosenstiel School of Marine and Atmospheric Science, Mr. Precht has worked as an environmental scientist specializing in the restoration and rehabilitation of various coastal resources, especially coral reef, seagrass, and mangrove systems. Currently, he is the Ecological Sciences Program Manager for the consulting firm of PBS&J and is located in Miami, Florida. In addition to these duties, Mr. Precht maintains status as a visiting research scientist with the Smithsonian Institution's Caribbean Coral Reef Ecosystem Program in Belize and as adjunct faculty to Northeastern University's Three Seas — East/West Marine Science Program, where he teaches a course in coral reef ecology and geology every winter quarter.

Contributors

Walter H. Adey
National Museum of Natural History
Smithsonian Institution
Washington, D.C.

Andrew W. Bruckner
NOAA/National Marine Fisheries Service
Silver Spring, Maryland

Robin J. Bruckner
NOAA/National Marine Fisheries Service
Silver Spring, Maryland

Greg E. Challenger
Polaris Applied Sciences, Inc.
Seattle, Washington

Mary Gray Davidson
Attorney
Phoenix, Arizona

Donald Deis
PBS&J
Jacksonville, Florida

Gary Fisher
NOAA Center for Coastal Fisheries and Habitat Research
Beaufort, North Carolina

Mark S. Fonseca
NOAA Center for Coastal Fisheries and Habitat Research
Beaufort, North Carolina

Stephen Gittings
NOAA/National Ocean Service
Silver Spring, Maryland

William Goodwin
NOAA/Florida Keys National Marine Sanctuary
Key Largo, Florida

Paul L. Jokiel
Hawaii Institute of Marine Biology
Kaneohe, Hawaii

Brian E. Julius
NOAA/Office of Response and Restoration
Silver Spring, Maryland

Les S. Kaufman
Boston University Marine Program
and
Center for Ecology and Conservation Biology
Boston, Massachusetts

W. Judson Kenworthy
NOAA Center for Coastal Fisheries and Habitat Research
Beaufort, North Carolina

Barbara L. Kojis
Division of Fish and Wildlife
St. Thomas, U.S. Virgin Islands

Steven P. Kolinski
NOAA/National Marine Fisheries Service
Honolulu, Hawaii

Steven J. Lutz
Rosenstiel School of Marine and Atmospheric Science
University of Miami
Virginia Key, Florida

James E. Maragos
U.S. Fish and Wildlife Service
Honolulu, Hawaii

Anne McCarthy
Florida Department of Environmental Protection
Florida Keys National Marine Sanctuary
Key West, Florida

Margaret W. Miller
NOAA/National Marine Fisheries Service
Miami, Florida

John Naughton
NOAA/National Marine Fisheries Service
Honolulu, Hawaii

Tony Penn
NOAA/National Ocean Service
Silver Spring, Maryland

Gregory A. Piniak
NOAA Center for Coastal Fisheries and Habitat Research
Beaufort, North Carolina

William F. Precht
PBS&J
Miami, Florida

Norman J. Quinn
Discovery Bay Marine Laboratory
University of West Indies
St. Ann, Jamaica

Baruch Rinkevich
National Institute of Oceanography
Haifa, Israel

Martha Robbart
PBS&J
Miami, Florida

Joe Schittone
NOAA/National Ocean Service
Silver Spring, Maryland

George P. Schmahl
NOAA/Flower Garden Banks National Marine Sanctuary
Bryan, Texas

Sharon K. Shutler
NOAA/Office of General Counsel for Natural Resources
Silver Spring, Maryland

Alice Stratton
NOAA/National Marine Sanctuaries
Milford, Connecticut

Lisa C. Symons
NOAA/National Marine Sanctuaries
Silver Spring, Maryland

Alina M. Szmant
University of North Carolina at Wilmington
Wilmington, North Carolina

Jessica Tallman
University of Rhode Island
Kingston, Rhode Island

Rebecca L. Vidra
Duke University
Durham, North Carolina

Cheryl Wapnick
PBS&J
Jacksonville, Florida

Paula E. Whitfield
NOAA Center for Coastal Fisheries and Habitat Research
Beaufort, North Carolina

Beth Zimmer
PBS&J
Miami, Florida

Contents

1 Coral Reef Restoration: The Rehabilitation of an Ecosystem under Siege

William F. Precht and Martha Robbart

CONTENTS

1.1 INTRODUCTION

Today, coral reefs are under siege from a number of environmental pressures. Accordingly, the management of the world's coral reef resources is the subject of some controversy.[1] General agreement exists about the value of these ecosystems in terms of ecological, social, and aesthetic benefits.[2] There is also some agreement that an estimated 24% of reefs are in danger of collapse from human pressures[3] and another 26% are under the threat of longer-term degradation and collapse. Admittedly, the numbers and percent devastation may vary regionally, yet no area untouched by humans has gone undisturbed.

Unfortunately, no consensus presently exists on how coral reef protection is to be accomplished. Coral reefs around the world have changed dramatically over the past two decades, particularly in the Caribbean and western Atlantic region.[4–11] Humans can impact reefs directly through vessel groundings, dynamite blasting for fishing and limestone construction materials, and anchor damage, to name but a few relevant activities, and indirectly through pollution, sedimentation associated with coastal activities such as dredging, and river runoff. Humans also are implicated in global warming through the emission of greenhouse gases. Although these anthropogenic impacts have affected coral reefs globally, other natural factors impact reefs as well. It is obvious that the resource needs protection and that many of the cited anthropogenic causes can be reduced, minimized, or avoided by implementing scientifically based management programs.[1,12,13]

An appropriate course of action is to repair or replace damaged and disturbed reefs at a rate resulting in no net loss of ecosystem value; the rate of reef destruction should be offset by the rate of reef repair. Because of financial considerations and logistical problems, this may not always be possible. As a practical matter, however, managers and policymakers need to understand the effects of human-induced disturbances; assess these damages properly; and develop subsequent, appropriate restoration efforts on reefs under their stewardship.[14–17] Most coral reef restoration programs have been focused on the physical damage caused by human activities. Of these, ship groundings are among the most destructive anthropogenic factors on coral reefs and form the basis for much of our present understanding of reef restoration. Some, however, view ship groundings to be only locally significant, implying that groundings do not pose a great threat to coral reef ecosystems or may even be beneficial.[18,19] The recent, staggering history of reported groundings by the Florida Marine Patrol in the Florida Keys (>600 yr^{-1}), however, reveals the significant threat to the health of the reef tract as a whole.[20] Boats of all sizes cause significant destruction.[21] In the case of large vessel groundings, destruction is usually complete and includes the direct loss of corals by dislodgment and pulverization, as well as the crushing, fracturing, and removal of three-dimensional reef structure (Figure 1.1). Secondary impacts include the scarring and abrading of previously undamaged resources as hydrodynamic forces move rubble produced in the initial disturbance. In some cases, increased sedimentation associated with the fracturing and erosion of the underlying exposed reef framework smothers living creatures. Furthermore, collateral damage caused by salvage and towing operations in removing a vessel run hard aground often increases the footprint of the initial damage scar.[15] Careless salvage efforts can destroy vast areas of coral reef unaffected by the initial accident. Fortunately, much of the physical damage caused by vessel groundings can be repaired. Using examples of reefs injured by

FIGURE 1.1 Complete devastation from the impact of a ship-grounding. Note total loss of reef structure, exposed limestone pavement, loose rubble, and residual paint from the hull.

catastrophic vessel groundings in the Florida Keys National Marine Sanctuary (FKNMS),[22–24] Precht et al.[25] developed a process-based scientific approach to coral reef restoration.

Environmental impacts, including hurricanes, tsunamis, global climate change, coral disease, and severe El niño southern oscillation (ENSO) events, have also impacted coral reefs. Though hurricanes and tsunamis are clearly naturally occurring events, the causes of global climate change, coral disease, and increasing severity of ENSO-related warming events are not known and may be related to human activities. Approximately 40% of coral reefs were seriously degraded by the 1998 ENSO-related warming event.[3] Coral disease has devastated Caribbean reefs and was responsible for the almost complete demise of *Acropora cervicornis* and *A. palmata* in the late 1970s and 1980s.[10,26] If we are not able to control or ameliorate the source of coral reef degradation, no matter what the source, we cannot expect to effectively restore these ecosystems.

Coral reef restoration will only be effective in addressing impacts that can be ameliorated and removed from affected coral reefs. Decision makers should decide on a case-by-case basis whether or not to restore a particular coral reef. While destructive and important to address, coral reef degradation through global warming, human-induced climate change, or pandemic coral diseases can only be addressed at the highest levels of government. Restoration of reefs impacted by hurricanes, climate change, coral disease, or other natural agents may be futile because there is no way to prevent the return of these agents.[27] In addition, it would be prohibitively expensive to repair reefs crippled by the negative synergistic effects of multiple types of stressors such as severe storms, global warming, and emergent diseases.

1.2 CORAL REEF RESTORATION — A GUIDE

The most widely accepted definition of ecosystem restoration is "the return of an ecosystem to a close approximation of its condition prior to disturbance."[28] This includes placing all restoration efforts in a landscape context, in which the restored patch is integrated into the ecosystem as a whole.[29] An implicit assumption is that managers and scientists understand the ecological dynamics of the restoration process itself, but most coral reef restoration efforts performed to date have fallen short of these goals.[22,30–36] Rather than being "true" restoration efforts, most of these are rehabilitation projects, with the goal of accelerating natural reef recovery to an endpoint that may or may not resemble predisturbance conditions. Moreover, efforts to evaluate the success of reef restoration projects have been complicated by a lack of scientific goal setting and by a general lack of agreement on what constitutes project success.

The goal of restoration is to restore the structure and function of a degraded ecosystem, habitat area, or site.[29,37–39] As previously mentioned, the word restoration means that you have returned something, in this case a coral reef, back to its original condition. Why then should we not expect restored reefs to look like and provide the same functions as preimpact reefs? We can, but only if we carefully select reefs for restoration. Successful reef restoration requires, first and foremost, an end to impact and/or degradation. This means the agent of destruction is removed from the impacted area. A reef impacted by environmental factors may not be a candidate for restoration in this scenario because the agent cannot be permanently removed (i.e., excessive sedimentation, nutrient pollution, repeated vessel groundings in the same spot). In the case of vessel groundings secondary mitigation such as signage may be implemented to prevent future groundings. Information on the preimpact area and its species composition/structure, financial resources, and guidance on design and construction for the restored coral reef are all necessary for a successful reef restoration project (Figure 1.2). The matrix in Figure 1.2 was developed to serve as a template for coral reef managers faced with the possibility of performing restoration projects on reefs under their stewardship.

Although each restoration project is unique, they all have common elements that can be addressed before action is taken. When faced with a potential restoration site, managers can follow a general set of guidelines, provided here, to decide on whether or not an area should be restored. The following is a narrative of the coral reef restoration decision matrix presented in Figure 1.2.

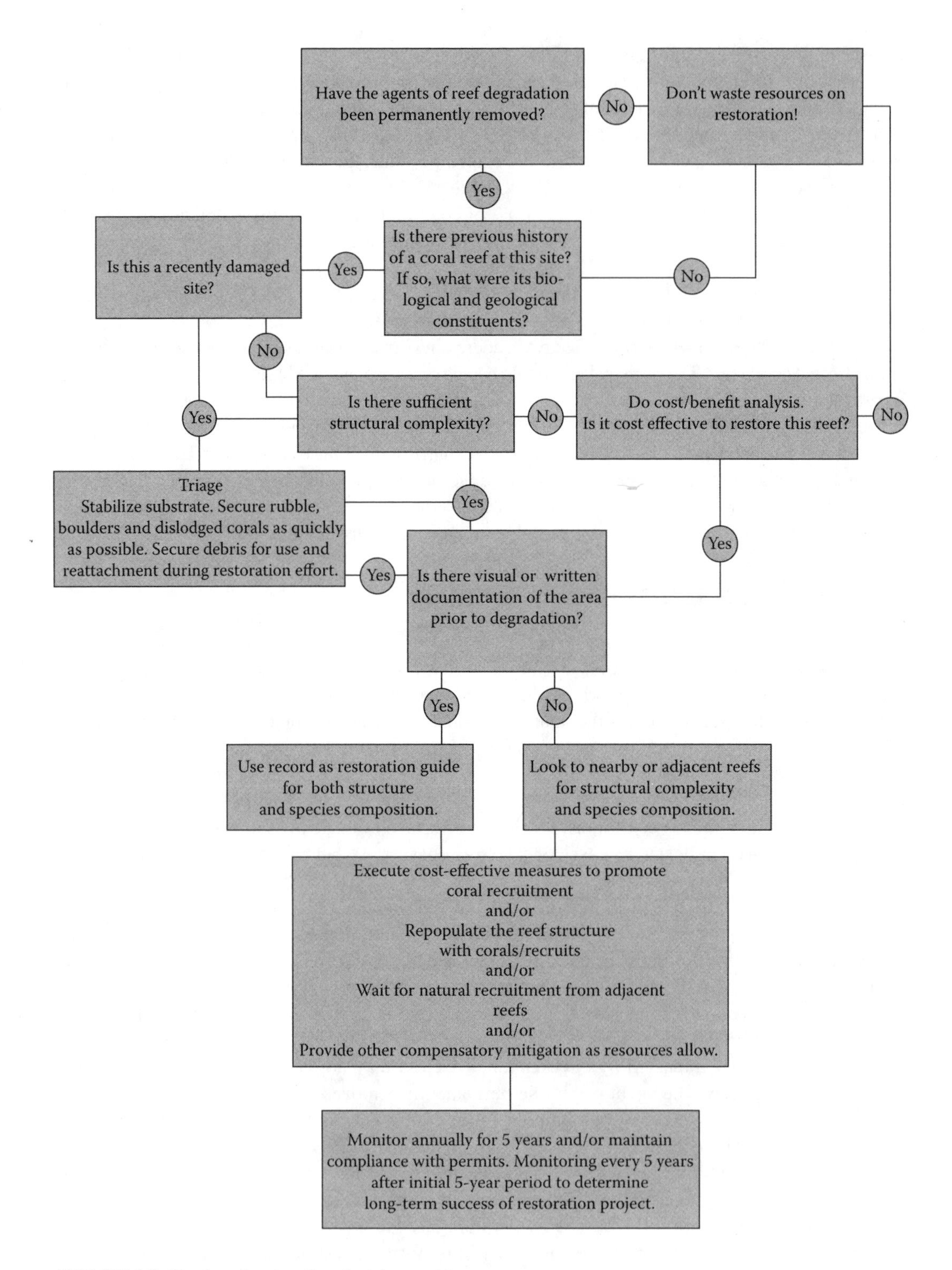

FIGURE 1.2 Coral reef restoration decision matrix.

The primary consideration for managers when confronted with an injured resource of any kind is to ensure that the agent of destruction has been removed. If not, don't even consider restoration; spend your money doing something else. It is only with the removal of the source of degradation that a restoration project has a chance of success. In some cases some additional action is needed to ensure that the cause of injury is not repeated. For example, the installation of mooring buoys can prevent anchor damage and small boat groundings on shallow reefs. Although this may mean spending additional resources it is imperative to include these added measures, otherwise dollars spent on restoration are potentially wasted. It is reasonable to consider not restoring a resource based on these considerations and instead focus resources elsewhere.

When the causative agent of reef destruction has been permanently removed and deemed not to return, the question then is: Was there a coral reef at the site that was injured and what were its geological and biological constituents? If there was never a reef there it is probably best to not create one as the environment (abiotic conditions) may not support a coral reef community. This is an important consideration even though it may seem obvious. To effectively restore a coral reef we must be working toward an achievable goal; therefore, site selection is vitally important. The second portion of the question should be addressed if true restoration is to be achieved. Among other things to consider are the geological constituents (the building blocks of reef framework) that are lost in addition to the biological constituents (scleractinian species, gorgonians, sponges, algae, epifauna, infauna, as well as mobile fauna). In cases where the structural complexity has been reduced or eliminated due to time or severe injury a cost/benefit analysis needs to be completed to determine the appropriate course of action. It is obvious that the more information a manager has about a site preinjury the better. Understandably, these types of records are often unavailable after an injury has occurred. However, adjacent reefs as well as their geological counterparts can be used as a guide in these circumstances. These will be discussed in more detail later in this chapter.

Timing is also critical in terms of restoration of organisms injured at the site. In cases of a recent injury, coral reef triage can be an effective tool. Triage in the form of uprighting and reattaching of corals to the substrate is only possible soon after the injury. Triage can also include large-scale stabilization of loose rubble and/or sediment left by the injury. The goal of triage is to effectively eliminate further damage and degradation to corals that were dislodged and other intact corals in the surrounding and adjacent areas. In many cases it is the most immediate and cost-effective way to begin the restoration process. Triage is a starting point in reef restoration and does not constitute restoration in and of itself. Reestablished corals may also serve as a source of recruitment in recolonizing the surrounding substrate.

Once the decision has been made to move forward with a restoration project, historical photographs and/or other descriptions of the site may be the best (and only) guide to accurately recreating the original reef structure and composition. Since such records may not be available, the adjacent reef or reefs can be used as a template for restoring the type and amount of structural complexity and species composition appropriate for the site. A cost/benefit analysis for the different restoration alternatives needs to be performed, with the knowledge that the combination of possibilities for restoration is only as limited as one's imagination and financial resources. For true reef restoration, that is replacing the reef community structure and function, biodiversity, aesthetics, and socioeconomic value, creative as well as scientific approaches are necessary. Different projects will require varied approaches and may include various techniques including triage, adding structural complexity with reef modules or limestone boulders, stabilizing substrate with mats or cement, and importing corals from nearby donor sites or nurseries, to name but a few. The solutions will depend upon the location of the site and the resources available. Once it is built, will they come? The only way to know is through long-term monitoring.

Monitoring of restored reefs should be treated as part of the reef restoration project itself. Too often this important component is left out of restoration plans. Annual scientifically based monitoring carried out after the completion of restoration can provide critical lessons learned, documenting successes and failures. It is only through these lessons learned that we can improve upon past technologies, techniques, and methods, bringing us closer to the restoration of complex, fully functional reef ecosystems.

1.2.1 Why Restore?

Coral reefs are some of the most productive ecosystems, providing habitat for numerous species and serving important ecologic functions. A coral reef and its specific functions may become degraded when these larger-scale processes are altered or removed. To successfully restore a degraded reef one must examine the important processes that exist within and outside of the spatial and temporal boundaries of a specific reef area. Moreover, numerous spatial and temporal scales must be examined at all stages of coral reef restoration, including identification of degrading agents, selection of processes to be restored, analysis of restoration impacts on the seascape, and long-term, hypothesis-driven monitoring.

1.2.2 Identification of the Degrading Agents

The identification of the agents or actions that caused the degradation of a coral reef is the first step in conducting a restoration effort. If the causes of the reef's degradation (the stressors) are not removed or accounted for than the probability that the site will continue to be negatively impacted is high.[20,40] One must expand the scale of examination to determine whether stressors occur outside of the site's boundaries and are impacting important large-scale processes. For example, if dredging activities are occurring up current from a coral reef, then that reef system may be negatively impacted because of high sediment stress. Additionally, the loss of functioning reef systems is often the result of cumulative disturbances to the ecosystems. Coral reefs, especially those near urban settings, are subject to ongoing large-scale stresses from human activities. A multitude of stressors cumulatively influence a system at different scales and intensities, making the identification of the important degrading agents a very complex process. Many ecosystem restoration projects have failed because they have not accounted for all of the stressors that influence the system.[41–43] To properly identify the important stressors negatively affecting a coral reef ecosystem one must methodically examine the individual reef and its landscape setting at numerous spatial scales.

Coral reef stressors are imbedded at various temporal scales as well. Accordingly, the temporal scale must be considered when identifying the specific factors that caused the degradation. A stressor can occur over a short or long time period and often result in disruptions to a coral reef's function.[42,44] A historical event that occurred over a short time period that resulted in long-term impacts to a coral reef's function will not be identified if the temporal scale is not expanded beyond the current timeframe. A stressor that occurs over a long time period and has long-term impacts may be seen as a constant feature of the landscape and may not be classified as a stressor if the temporal scale of analysis is not increased. In addition, not all degrading events occur within the same time period. The severity of the impact of an existing stressor on a coral reef may not be obvious if it occurred after the site was already injured or disturbed. For instance, before the late 1970s, two species of acroporid corals were the primary builders of coral reefs along the Florida reef tract. These corals have since undergone a regional decline, with losses of 95% or more in some areas during the past few decades. Therefore, it may be hard to determine the true extent of an anthropogenic injury on a shallow reef area that was previously dominated by acroporid corals. These confounding factors require diligence on the part of the restoration scientists.

1.2.3 A Legal Basis for Restoration

When a coral reef is injured, several federal statutes provide the United States government with the authority to recover resource damages.[45] The Comprehensive Environmental Response,

Compensation, and Liability Act (CERCLA) of 1980 and the Oil Pollution Act of 1990 (OPA 90) are the principal federal statutes that authorize trustees to assess damages for trust resources that are lost or destroyed as a result of the discharge of oil or release of hazardous substances. The Department of Commerce, National Oceanic and Atmospheric Administration (NOAA) was charged with developing regulations for OPA 90. The NOAA rule (15 CFR Part 990) was finalized on January 5, 1996. Many of the procedures and techniques developed for assessing natural resource damages under OPA 90 have been applied to the damages caused by the grounding of vessels on coral reefs and other significant natural resources.[46]

1.2.4 Natural Resource Damage Assessment

Natural Resource Damage Assessment (NRDA) is a process for making the public "whole" for direct injury to natural resources and/or the services of natural resources. The primary objectives of the NRDA process are to identify and quantify natural resource injury, determine the damages resulting from the injury, and develop and implement appropriate restoration actions. The primary goal of NRDA is to provide for the restoration of injured natural resources and/or services to preincident conditions.[47] This goal is achieved by implementing a plan for the restoration, rehabilitation, replacement, and/or acquisition of equivalent natural resources. In NRDA, restoring the environment after injury has two basic components. These are "primary restoration," which is the restoration of the injured resources to baseline (i.e., preimpact, unimpaired) conditions, and "compensatory restoration," which is the compensation for interim losses of resources from the time of the injury until the resources recover to the predetermined baseline. Compensation is in the form of additional restoration, replacement, rehabilitation, or acquisition of equivalent natural resources.

NOAA's NRDA rule is intended to promote expeditious and cost-effective recovery of natural resources and the services of these natural resources. Responsible trustees (e.g., authorized federal, state, Indian tribe, and foreign officials) can use the rule to recover losses of natural resources and their services. In addition, companies and/or individuals responsible for natural resource damage (i.e., responsible parties) can use the rule as guidance for determining natural resource damage and shaping proposals to the trustee to repair the damaged resource and compensate for lost services during the recovery of the resource.

The result of the NRDA process is a restoration plan that is developed by the trustee and/or responsible party with input from the public. The process has three phases:

1. Preassessment
2. Restoration planning
3. Restoration implementation

The preassessment phase involves a preliminary determination by the trustees as to whether natural resources and/or services have been injured. The result of the initial preassessment phase is a Notice of Intent to Conduct Restoration Planning, which contains:

- The facts of the incident resulting in ecosystem injury
- Trustee authority to proceed with an assessment
- Natural resources and services that are, or are likely to have been, injured as a result of the incident
- Potential restoration actions relevant to the expected injuries
- Potential assessment procedures to evaluate the injuries and define the appropriate type and scale of restoration for injured natural resources and services

The restoration planning phase includes injury assessment, restoration selection, and selection of preferred restoration alternatives. This document is often referred to as a Damage Assessment and Restoration Plan (DARP).

1.2.5 Crime Scene Investigation

Essentially, the trustees and the responsible party can proceed into an NRDA as partners, cooperatively developing a jointly agreeable restoration plan, or as potential opponents in a legal battle. In either case, however, any preliminary investigations should be treated as the equivalent of a crime scene investigation until the course of the NRDA is determined. Accordingly, as much physical evidence from the site as possible needs to be documented, collected, and quantified as soon as possible after the incident.[48] The evaluating scientific divers are essentially underwater detectives and must use forensic methods and protocols to accurately assess the injured resources. For a ship-grounding, these include measurements that detail both the inbound and outbound paths of the responsible vessel (e.g., hull paint scrapes, scarification and directional striations on the reef surface, keel and chine scars, direction of movement of overturned or toppled corals and/or reef rubble, etc.). There is no substitute for good scientific methodology at any time during the investigative or assessment portions of the NRDA, as they are the building blocks upon which restoration plans are based. In addition, proper chain-of-custody should be maintained at all times for samples, photographs, and other forms of data.

The decision by the responsible party to proceed cooperatively in an NRDA with the trustees will result in the development of a Memorandum of Agreement (MOA), a legally binding document, jointly developed and signed by all parties. This situation can certainly expedite the resolution of an NRDA for the responsible party and may result in a considerable reduction in the cost of assessments and restoration because interim losses can be significantly reduced.

1.2.6 Injury Assessment

Under the NOAA rule, injury is defined as an observable or measurable adverse change in a natural resource or impairment of natural resource service. The trustee or responsible party must quantify the degree and spatial and temporal extent of injuries. Immediately after an injury occurs, a detailed injury assessment should be prepared. As previously mentioned, since many of the injury actions will result in either a settlement or litigation between the trustee and the responsible party, the assessment must also substantiate or refute the description of events that caused the injury.[14] The degree of injury may be expressed in such terms as percent mortality; proportion of a species, community, or habitat affected; extent of injury or damage; and/or availability of substitute resources. Spatial extent may include quantification of the total area or volume of the injury. For a comprehensive review of the protocol detailing the field methodology for coral reef injury assessments, the readers are directed to Hudson and Goodwin[48] and Symons et al.[49]

Temporal extent or duration of the injury may be expressed as the total length of time that the natural resource and/or service is adversely affected, starting at the time of the incident and continuing until the natural resources and services return to baseline. In minor incidents this length of time is usually measured in decades. In some large disturbances, however, impacts to the "nonliving" resource may include the loss of three-dimensional reef structure and removal of vast quantities of reef substrate and sediment, eliminating thousands of years of reef development in one fell swoop. In these cases, without major intervention by humans, it is evident that the resource would not recover to its preinjury baseline for millennia.[25]

1.2.7 Emergency Restoration

Emergency restoration includes those actions that can be taken immediately after an incident that may reduce the overall extent of the injury to the resource. These actions take the form of "triage," which is defined as "the sorting of and allocation of treatment to patients … especially disaster victims according to a system of priorities designed to maximize the numbers of survivors."[50] In the case of groundings on coral reefs, triage can be righting and reattachment of displaced or broken corals, the removal and/or stabilization of loose rubble and sediment, and the stabilization of structural fractures.[51]

In some cases there may be a conflict between the evidence collection portion of the investigation and initial remedial action or reef triage efforts. Reef triage efforts must be implemented in concert with the initial damage assessment to attain maximum success in salvaging the damaged resource. However, these efforts should not be performed in a vacuum, and collaboration among all team members is essential so as not to compromise the integrity of any of the ongoing investigative operations.

1.2.8 Economic Assessment of Damages

The following discussion is meant as a guide for evaluating the economic criteria for determining damages to injured reef resources. This discussion avoids legal analysis and liability and jurisdictional issues, and readers are directed to seek specific regulations pertaining to the complexities of individual cases or areas.[52,53] For example, in south Florida a variety of regulations pertain to the protection of reefs and corals.[14,51] In Federal waters, the National Marine Sanctuaries Program Amendments of 1988 provide that any person who destroys, causes the loss of, or injures living or nonliving resources of a National Marine Sanctuary may be liable to the United States for damages, including the cost of replacing or restoring the resource and the value of the lost use pending the replacement or restoration. The Park Service Resource Protection Act also authorizes the U.S. Secretary of the Interior to recover damages for injuries to National Park System resources.

In assessing the extent of damage from an economic standpoint, the purpose is to estimate the amount of money to be sought as compensation by the trustee from the responsible party for the injury resulting in the damage to the resource. Damages based on restoration costs may include any diminution of use and nonuse values occurring until the recovery is complete (i.e., functional success criteria are attained).

After a detailed DARP is performed, a monetary assessment of damages based on restoration costs should be prepared and a demand for these damages presented to the responsible party. The restoration methodology should be based on the costs of the actions to restore or replace the damaged reef to its predisturbance, baseline condition. Replacement costs are the costs of substitution of the resource that provides the same or substantially similar services as the damaged resource. The restoration or replacement alternatives should be evaluated according to the DARP. The damage amount should be the amount to cover all costs related to the injury and not just limited to an amount used to restore the damaged resources, including:

- All emergency response and/or salvage efforts
- Environmental assessment and mapping of the injured resource (damage assessment)
- Implementation of emergency rehabilitation methodologies (reef triage)
- Preparation of the DARP report
- Implementation and completion of restoration through project success
- Long-term scientific monitoring studies (both functional and compliance)
- Compensatory restoration for interim loss of services

The assessment of natural resource damages requires close interaction between law enforcement officers, scientists, lawyers, resource managers, regulators, and economists. Since many damage cases result in litigation, it is imperative to get the science, law, and economics correct. Damages recovered by the trustee should then be made available to restore, replace, or create equivalent resources.

1.2.9 Selection of Processes to Be Restored

Degradation is a complicated process involving numerous changes to the function of an ecosystem; therefore, the restoration process will be at least as complex.[39] Although the identification and, if possible, removal of stressors is the first step in the restoration process, it is not the only step (see Figure 1.2). Stressors impact the function of a coral reef by altering or removing structural components and ecological processes; therefore, even if the stressors are eliminated, some components may still be absent from the restored ecosystem.[37,54,55] To develop a successful long-term solution, restoration must include the reintroduction or creation of three-dimensional structure and critical small- and large-scale processes that generate the function of a coral reef ecosystem.[54–56]

To restore the necessary small- and large-scale processes, one must first identify the function of the specific coral reef area (usually the unimpaired resource adjacent to the injured site) and by extension, the goal of the restoration project to be performed. There is currently a large debate regarding the appropriateness of restoring ecosystems to their historical functions and engineering these systems to mimic the function of reference sites.[40,42] Historical functions and reference sites can greatly assist with the restoration process but may not always be the most appropriate end goals for restoration. The landscape in which the restored coral reef exists differs from what was historically present. This is especially apparent with the ongoing global coral reef crisis, with stressors being related to coral disease and bleaching. Some stressors may not be removable, and not all of the processes that were present historically can be reestablished because the surrounding landscape has changed.[57,58] When restoring a site one must expand the spatial scale at which the reef is examined to determine which ecological process can be established and will function appropriately given the specific landscape setting of that site.[40,59]

The use of reference ecosystems is a vital component of developing success criteria in restoration programs. In coral reef systems many of these reference sites are heavily disturbed, rendering them useless as templates for the reconstruction of lost ecological services. It is possible, however, to use the paleoecologic information stored in Quaternary reefs as an appropriate analogue for placing current site conditions in context. It has been shown that, almost without exception, Quaternary fossil-reef sections exhibit species composition and zonation similar to those of modern reefs at the same location. Thus, Quaternary reef-coral communities within the same environment are more distinct between reefs of the same age from different places than between reefs formed at different times at the same location. Often, the subsurface Holocene reef history exposed by the injury itself serves as the best reference ecosystem. These Quaternary examples provide a baseline of community composition that predates the impact of humans. Most importantly, these paleoecological examples emphasize the importance of history — succession, assembly rules, and natural system variability — in structuring reef ecosystems through time and space. These fossil and subfossil reference ecosystems also form the basis for identifying desired future conditions for which the resulting restoration should aim. By identifying the ecological processes that generated a site's historical function as well as what processes are influencing other similar reef systems, restoration ecologists can begin to identify the particular large- and small-scale processes that should be established. Thus, the past should be used as a model to reconstruct the future. Because historical science is largely inductive, and interpretation of the fossil record can be highly subjective, the challenge to restoration ecologists is to combine paleoecologic data and reconstructions with reference sites, field experiments, model simulations, and long-term monitoring.

Zedler[60] has suggested that, before any project begins, those performing ecological restoration must have very clear goals for their work. Specific decisions on what aspects of the restoration will be emphasized (structure and/or function) and how those goals will be achieved must be made absolutely clear in order to promote success.[42] Specifically, restoration scientists have a series of "theoretical" decisions to make:

- Whether to use self design or engineered design (i.e., rebuild structure, actively transplant corals and other benthic attributes)
- Whether to create in-kind or out-of-kind restoration projects
- Whether to restore onsite or offsite
- How to use reference sites both as a template and as a means for evaluating restoration success
- How to evaluate/conceptualize coral reefs using hypothesis-driven monitoring programs

Two controversial views in ecosystem restoration are ideas regarding the self design and engineered design of the injured resource. These two views have evolved from Clementsian and Gleasonian succession dynamics:

- In the Clementsian view, the community was interpreted as a superorganism. The component species were highly interactive and their distributions were strongly associated along environmental gradients.[61,62] Clementsian succession claims that the community changes as a whole through different life stages and ends up ultimately in a climax ecosystem. Species are interlinked with one another, and disturbance to the ecosystem interrupts this natural progression to the climax stage of development.[10]
- The Gleasonian model rejected the idea of tight community integration. Instead, the community was seen as a collection of independently distributed species.[63] The Gleasonian model does not exclude the possibility of succession, competition, niche partitioning, assembly rules, and other interspecific interactions. Rather, it denies interspecific interdependence as the cause of species distributions.[64,65] Gleasoninan succession claims that community change can be reduced to the responses of individual species to the environment based on the constraints of their unique life histories.[10]

The controversy of engineered design versus self design centers on the question of whether to rebuild reef structure and transplant corals at a restoration site to jump-start the recovery process or to allow the restoration site to recolonize naturally over time with little or no human intervention. The two concepts differ as follows:

- The main hypothesis of the self-design concept is that over time, a coral reef will restructure itself. The environmental condition determines what organisms will be able to colonize the site. This concept views recolonization as an ecosystem-level process. Proponents of the self-design view believe that intervention in the recovery process is not warranted.
- The main hypothesis of the engineered-design concept is that it is not a matter of time, but intervention, that determines the positive outcome of a restoration project. The most important factors in the success of the restoration project are the life histories of each organism present. The importance of the natural reproductive process (brooders vs. broadcasters) of the corals is often stressed.[66] This concept views recolonization as a population-level process.[67] It seems apparent that for coral reefs, due to the slow rates of natural recovery, intervention is not just warranted but required.[25]

Also, by comparing the restored site to an approximate reference site, restoration scientists can determine how well the restored ecosystem is mimicking the original.[41,42,60,68] However, White and Walker [68] and Grayson et al.[42] have contended that the picking of reference sites for comparison is more complicated than just looking at comparable, adjacent unimpaired settings. Specifically, Grayson et al.[42] suggest that restored sites must be compared to both nondegraded sites and unrestored degraded sites. Thus, if the restored project shows signs of success, more knowledgeable conclusions can be drawn as to whether the success has come from the act of the restoration or whether it is merely a natural response of the ecosystem (which may be evidenced by comparison to the response of the degraded unrestored site).

1.2.10 Analysis of Restoration Impacts on the Landscape Scale

Various spatial and temporal scales need to be examined to determine how the restoration of a coral reef may impact landscape-scale processes and adjacent habitats. Structural complexity has a large influence on what types of habitats are present in a landscape. Most coral reef restoration projects have generally focused on reestablishing coral cover and not structural complexity at the landscape scale. We caution that if restoration is performed on a site-by-site basis without consideration of the structure, we risk a reduction in overall ecosystem function. The restoration process is a series of alterations to the current processes and patch interactions within a landscape. By altering these ecological processes, we may positively or negatively impact other ecosystem patches within a landscape.

By expanding our scale of view to the landscape or regional perspective, a restoration project can be designed so that it adds to the value or function of the entire landscape.[43,60] With the increased need for coral restoration for mitigation purposes such as in ship groundings or dredging projects, there is a danger that restoration projects will be treated as a cookbook-like process in which the same type of reef system is restored to an area regardless of its landscape context.[57] For instance, the placing of a dozen prefabricated reef modules without regard to landscape setting is hardly in-kind restoration. If all reef restoration projects are designed to be of the same type, the diversity of reef functions and habitats as well as the diversity of species within a landscape will be greatly reduced.[43] By including large-scale considerations in restoration activities, restoration projects can be designed to enhance both local and regional ecosystem functions and preserve the diversity of coral reefs present in a landscape.[37,43,54]

1.2.11 Restoration Design

In designing a coral reef restoration project, a reasonable range of restoration alternatives needs to be considered. Evaluation of the alternatives needs to be based at minimum on:

- The cost to carry out the alternative
- The extent to which each alternative is expected to meet the goals and objectives of returning the injured natural resource and services to baseline and/or compensate for interim losses
- The likelihood of success of each alternative
- The extent to which each alternative will prevent future injury as a result of the incident and avoid collateral injury as a result of its own implementation

Determining the benefits of restoration to the affected environment requires an analysis of the ability of the injured natural resources and services to recover naturally. In general, factors to consider include:

- The sensitivity and vulnerability of the injured natural resources and/or services
- The reproductive and recruitment potential of the natural resources and/or services
- The resistance and resilience (stability) of the affected environment

In the case of coral reefs, many things affect the ability of this resource to recover within a measurable time period.[25] The corals themselves are affected by human-induced and natural disturbances (e.g., near-shore pollution, hurricanes, coral diseases, bleaching due to global warming and/or ENSO events, etc.). The growth rates of most coral species are relatively slow. In addition, the distribution of gametes and larvae may affect the potential for recovery of coral species. For instance, in Florida reefs have been shown to be recruitment limited. All of these factors need to be considered during restoration planning.

Restoration ecologists also face the ethical question of whether or not it is actually possible to restore natural habitats such as coral reefs back to their predisturbed state.[69] One of the main goals of restoration ecology is to predict the results of specific restoration actions.[39] The demand for restoration guidelines has often exceeded scientific knowledge on the effects of certain restoration methods.[39] Therefore, published case studies are desperately needed to further understanding of how certain restoration practices affect coral reef ecosystems. Short- and long-term assessments of restoration projects are needed to determine the success (or failure) and function of a particular restoration method or practice.

1.2.12 Success Criteria

The word "success" has a number of meanings as it relates to restoration programs. The success of restoration projects is often evaluated as compliance success: whether environmental permit conditions were met or simply whether the stated projects were implemented or monitored. Quammen[70] distinguished functional from compliance success, noting that functional success is determined by whether the ecological functions of the system have been restored. For example, in evaluating the success of wetland restoration/mitigation projects in Florida, Redmond[71] showed that disconnected decision-making resulted in an abundance of restoration projects but failure in the sense of compliance and function: more than 80% were in noncompliance with permit conditions and/or not achieving expected ecological functions. In the past, many restoration assessments have emphasized structural rather than functional attributes. In fact, many structural attributes, such as species diversity, become indicators of function when monitored over time. The success of restoration efforts, therefore, must be determined by our ability to meet technically feasible and scientifically valid goals and focus our monitoring efforts on both structural and functional attributes. This establishment of realistic, quantifiable, ecologically based criteria is basic to the planning process for all habitat restoration and creation projects. As we have discussed, if the stated goal of reef restoration is to return the ecosystem nearly to predisturbance, baseline conditions and functions, assessment and monitoring programs must be used to evaluate and compare natural, undisturbed reference sites with disturbed and restored sites. For most ecosystem restoration programs, including reef restoration programs, functional analysis has lagged behind project compliance, with the results that goals and success criteria have generally been set ad hoc. To date, it seems as if coral reef restoration ecologists have not learned from one another, and thus the same issues are readdressed and the same problems are confronted over and over again.

1.2.13 Goal Setting

The degree of reef damage by a ship-grounding for instance may set practical limits on the viewpoint and goals of restoration. For example, radical reconstruction is required where large volumes of material have been removed, gouged, fractured, or flattened. Lesser damage may require only partial rehabilitation, such as the reattachment of damaged and overturned corals[15] and coral transplantation or reintroduction.[72–75]

Historically, successful restoration projects have been evaluated primarily by the establishment of certain attributes such as coral cover and/or the abundance of fish species. It is necessary to move beyond this tradition and focus not only on charismatic organisms but on ecosystem function.[39,76] Essentially, all definitions of success are dependent upon the likeness of the restored ecosystem (both in terms of structure and function) to comparable reference sites. However, many would still argue that no restored coral reef (or any ecosystem for that matter) will ever be as successful as the original; therefore some minor relaxations in criteria should be considered.[60] Nevertheless, without using standardized criteria, coral reef success will continue to go unassessed, which in turn may lead to continued mistakes and failures.

Compared to terrestrial and wetland restorations, which range in the thousands of implemented projects, coral reef restoration is in its infancy, with only tens of projects performed. In addition, few of these have been published or described. Therefore, at present there is little basis for understanding what works, what does not, and why. Three of the most important questions that need to be addressed in all restoration programs are:[25]

1. How long will it take for natural recovery to occur at any given site without manipulation?
2. Will natural recovery converge on a community state that is different from its predisturbance state?
3. Will reefs disturbed by humans respond differently than those damaged by natural processes?

Hypothesis-driven ecological studies and quantitative, long-term monitoring programs are the only means of answering these critical questions. Formulating and testing hypotheses about the response of reefs to anthropogenic disturbances allows us to establish the scientific protocol necessary to design and implement restoration strategies, a baseline for developing quantifiable success criteria, and the efficacy of the restoration effort.[25]

1.2.14 A Scientific Basis for Restoration

Understanding whether reefs will heal through self design or need to be actively restored through manipulation and intervention (engineered design) requires a thorough scientific understanding of the recovery process. The basic principles of coral reef restoration are essentially the same as the basic principles of ecological succession. Inasmuch, we are interested in what determines the development of coral reef ecosystems from very early beginnings through senility and what may cause variation in them at points in time and space.

The essential quality of restoration, therefore, is that it is an attempt to test the factors that may alter this ecosystem development through time and space. This gives restoration scientists a powerful opportunity to test in practice their understanding of coral reef ecosystem development and functions. The actual restoration operations that are performed are often dominated by logistical or financial considerations (and possibly by government regulations), but their underlying logic must be driven by ecological hypotheses. Therefore, hypothesis-driven restoration programs are truly an "acid test" for ecological theory and practice.

Formulating and testing hypotheses about the responses of communities and whole ecosystems to disturbances and about the process of recovery will establish:

1. The degree to which the ecosystem in question has the capacity to naturally recover (self design)
2. How intervention (engineered-design) in recovery can retard or enhance the process (or have no effect)
3. The scientific protocols necessary to design and implement restoration strategies
4. A scientific baseline for developing quantifiable success criteria and the efficacy of the restoration effort

Using ship-grounding sites in the Florida Keys, Aronson and Swanson[16,17] and Precht et al.[25] developed and tested hypotheses that take advantage of some simple facts about major reef injuries: when ships contact reefs they break and crush coral rock, kill corals and other sessile organisms, open bare space for colonization, and eliminate topographic (habitat) complexity. Following a ship-grounding, recruitment and growth of sessile organisms can take the community in three possible directions. The first is toward the community structure of the preimpact community, usually judged from the current state of the adjacent undamaged area. The second is toward some other community structure or alternate community state.[77] The third possibility is no change at all from the initially damaged, primary substratum. The probability of the latter, "null" alternative is vanishingly small, given the inevitability of bacterial and algal colonization of primary substratum in the sea. The second alternative leads to an interesting prediction. If a ship-grounding flattens the topographic complexity of a highly structured reef habitat, and if complexity does not recover through coral growth (self design), then community structure could develop so as to converge on that found in natural hardground habitats. Hardground communities typically have low topographic complexity, consisting of flat limestone pavements with crustose coralline algae, gorgonians, and isolated coral colonies. Where ship-groundings occur in hardground habitats, recovery should be back to a hardground community structure.[25]

Aronson and Swanson[16,17] conducted a study in the FKNMS during a 2-year period (1995–1996) that evaluated the 1984 *Wellwood* grounding site. Replicate sampling sites were established within areas of the *Wellwood* grounding site that were formerly spur-and-groove habitat. Benthic assemblages at these sites were surveyed using video techniques.[78] Two types of undamaged reference sites were also surveyed: spur-and-groove sites adjacent to the *Wellwood* site and hardground sites at Conch and Pickles Reefs. Univariate parameters of community structure and biotic composition of the ship-grounding site resembled the natural hardground habitat more closely than they resembled the adjacent spur-and-groove area. When comparing the reference sites to the impacted sites among the sampling years 1995 and 1996, hard coral cover was uniformly low in the ship-grounding and hardground surveys and higher but variable in the spur-and-groove surveys. The spur-and-groove reference sites were significantly more complex topographically than either the grounding sites or the hardground reference sites, which were not significantly different from each other. Interestingly, the Pickles Reef site, which was originally thought to represent a natural hardground, turned out to be the site of two earlier ship groundings; one of the groundings occurred circa 1800 and the other was in 1894. Debris from the two nineteenth-century groundings was still visible, but the Pickles Reef reference site was otherwise indistinguishable from the Conch Reef hardground reference site. The fact that the Pickles Reef site was similar, both visually and quantitatively, to the Conch hardground reference site is strong evidence that ship groundings do indeed produce hardgrounds.[25]

In a companion study evaluating coral recruitment success, Smith et al.[79] showed essentially no increases in juvenile coral abundance and diversity within the *Wellwood* site since 1989 and the relative absence of juveniles of major frame-building corals at all study sites. These results are an indication of recruitment limitation. Overall, this grounding study suggests that the damaged spur-and-groove habitat will not recover to its former state on a time scale of decades without substantial restoration efforts (engineered design). Multivariate analysis indicates that those restoration efforts must include reestablishment of the topographic complexity to enhance the recruitment and growth of coral species that naturally occur in spur-and-groove habitats.[16,25]

In contrast, a study of the 1989 *MV Elpis* grounding site in the FKNMS in 1995 to 1996 revealed that the damaged hardground community was statistically indistinguishable, in univariate and multivariate comparisons, from adjacent hardground reference sites. The *Elpis* site, after a decade of recovery, could not be distinguished from the surrounding hardground habitat. These results suggest that when a ship-grounding occurs in a hardground habitat, it is likely that the community will recover within a decadal time frame. Rehabilitation measures, especially substrate stabilization and coral transplantation, will likely accelerate this natural recovery.[25]

The loss of topographic complexity as a result of vessel groundings in high-relief, spur-and-groove habitats has serious implications for reef recovery. When complexity is reduced, the hydrodynamic forces change and populations of reef fish and sea urchins decrease. Both of these factors influence the trajectories of colonizing reef communities.[15,80] In addition to the lack of recovery of coral fauna mentioned above, Ebersole[81] noted striking differences between fish assemblages on undamaged spur-and-groove sites and both natural hardground and damaged sites, which were themselves indistinguishable. Restoration of habitat complexity may be the vital ingredient in the overall recovery of damaged reefs.[25]

1.2.15 Compensatory Restoration

For any given injury or disturbance on a coral reef, an interim loss of both natural resources and ecological services occurs. Ecological services refer to activities of ecosystems that benefit humans and not the ecosystem itself.[82] Even assuming successful recovery of the damaged resource through restoration, the repair of the damaged area alone is not sufficient to compensate for the total losses incurred due to the incident. Since restoration takes time and the resource will take years (possibly decades) to recover to a functional equivalency after restoration is implemented,[25] compensation for these interim losses must be incorporated into the estimate of the total damages. Accordingly, these interim losses of resource use are often sought as compensatory restoration. The manners in which interim ecosystem losses have been computed have been very inconsistent and have often been driven by financial and not scientific protocol.[57,83] In order to quantify the loss of use, one of the most commonly applied techniques has been a habitat equivalency analysis (HEA). HEA is a method for determining the appropriate compensation for interim loss of natural resources.[46,84,85] The HEA is an appropriate method for quantifying compensation in resource situations where substantial human use is not present (i.e., an adequate measure of human use cannot be calculated for the particular habitat). The concept behind HEA is to provide an equivalency between the ecological functions ("services" that the ecological system provides to humankind and the ecosystem) lost due to the injury and the ecological functions provided by the replacement project. The equivalency allows for the calculation of the size of a habitat replacement project necessary to compensate for the interim loss in habitat services. The HEA methodology combines elements in all components of an NRDA including the quantification of injury, the analysis of restoration projects, and the valuation of lost services.[46] For cases involving injuries to coral reefs, the ecological functions lost due to the injury and those provided by the replacement project are often calculated in terms of coral growth over time on a replacement habitat.

The injury assessment strategy to calculate interim loss on coral reefs should be based on six logical steps:

- Documentation and quantification of the injury
- Intrinsic value of damaged resource
- Identification and evaluation of restoration options
- Estimate (in years) of rates for "natural" reef recovery (self design)
- Determination of the most appropriate means of restoration (engineered design)
- Economic scaling of the restoration project over time until functional success is obtained

Interim loss-of-services is then calculated as an integral of service lost from some reference point or baseline level over time (Figure 1.3). Thus, the HEA is an economically based "model" that provides a means of standardizing computations of interim loss. Recently, Banks et al.[86,87] used a similar habitat equivalency model (HEM) to assess the resource loss when the *USS Memphis* submarine ran aground on a reef off Fort Lauderdale, Florida. Similarly, Deis[46] reported on the use of these methods to determine adequate compensation for impacts from fiber-optic cables on coral bearing hard-bottom in the Fort Lauderdale, Florida area. When coupled with long-term scientific monitoring, these methods also provide a reasonable basis upon which to gauge compensatory restoration success (actual time to establish reef recovery and/or functional success).

In some cases, the most appropriate means of compensatory restoration is a monetary settlement, where the funds are earmarked for specific programs. These might include antigrounding campaigns; coral reef education and outreach programs; interpretive exhibits; boat pilot training; installation of mooring buoys at designated sites; increases in navigational markers; and long-term, scientifically based (not compliance mandated), monitoring studies that empirically gauge the functional success and/or failure of past restoration efforts. Other additional off-site compensatory restoration could include development of coral aquaculture programs (nurseries) and identification

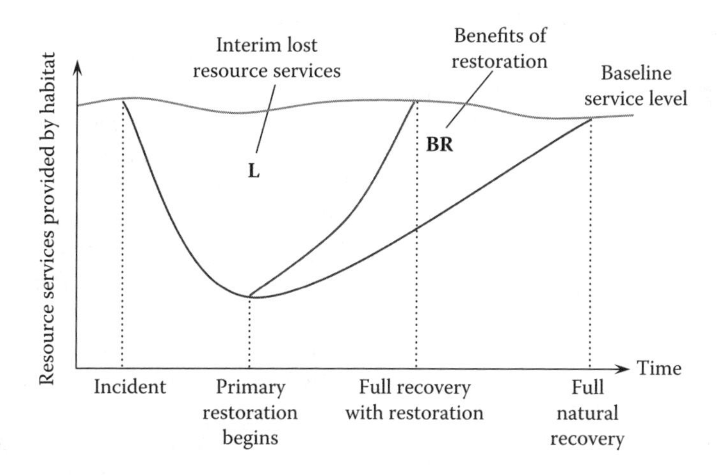

FIGURE 1.3 Graphic depiction of lost ecosystem services due to resource injury and the benefit of performing primary restoration.

of coral donor sites; artificial reef creation projects; establishment of baseline monitoring surveys of undisturbed reef resources; and restoration of damaged "orphan" sites where no responsible party had been identified, yet site rehabilitation/restoration is necessary to repair resource loss.

1.2.16 Long-Term Monitoring and Adaptive Management

After success is defined, the next step becomes working toward the realization of these goals. Specifically, restoration scientists must:[39]

1. Use preexisting ecological theory to maximize their potential for success
2. Periodically evaluate the project(s) via hypothesis-driven monitoring

Although often overlooked, postrestoration monitoring is very important (Figure 1.4). Pickett and Parker[88] noted that one of the pitfalls of restoration is to think of it as a discrete event when

FIGURE 1.4 Scientific diver performing long-term, hypothesis-based monitoring of reef function.

restoration is actually "an ongoing process." When required, monitoring periods typically range from 3 to 5 years, during which the site's structure and function are expected to have become fully established.[41] Because of the slow growth rate of corals, this 3- to 5-year time period is often inadequate for an ecosystem to become established or to determine whether all of the reestablished ecological processes are properly functioning.[41,58] To ensure that ecological processes, especially those that function on larger spatial and temporal scales, have been properly reestablished to a system, restored sites should be monitored and managed for longer periods of time. Moreover, by monitoring restoration activities for longer periods of time, restoration scientists can assess the ability of different restoration activities to achieve desired goals and focus future research efforts where needed.[40,55,58,88]

It is easy to define a coral reef restoration project as successful merely on the establishment of coral cover. While these projects may initially seem to be successful, long-term monitoring has proven that it takes other components (and efforts) than coral cover alone to guarantee the long-term perpetuation of the coral reef ecosystem. In many cases coral reef restoration projects are not monitored at all for success or failure. Others are only monitored for a short period of time after restoration efforts are completed because (1) there is insufficient funding to support continued assessment, and/or (2) legislative regulations do not require monitoring.

Adaptive restoration begins by recognizing what we do not know about restoring a specific site.[89] The unknowns might be what ecologic targets are appropriate, how to achieve desired targets, or how to monitor the site to determine when (or if) these targets are met. For restoration to be truly adaptive, the decision-making structure must include scientists who can best explain how the knowledge can be obtained and what research can be incorporated into the restoration project, and funds need to be earmarked and made available for this strategic, applied research and monitoring effort.

Therefore, all coral reef restoration programs should be based on the following philosophy:

> ...management decisions should be treated as hypotheses of ecosystem response, and restoration programs should be designed as experiments to test them. This approach to ecosystem restoration allows management decisions to be revised (adapted) to meet project goals.

Because coral reef restoration programs are "hypotheses of ecosystem response" based on incomplete information, uncertainty has long been a hallmark of these programs. An adaptive approach to ecosystem management, as described above, must be undertaken to ensure project success.

This progressive view of management recognizes three important principles:

- Management decisions should adapt to the results of the scientific studies and monitoring efforts.
- A multidisciplinary team of competent specialists should direct and guide all scientific studies.
- An independent Quality Assurance/Quality Control (QA/QC) team of highly qualified experts should oversee all projects.

A number of monitoring methodologies have been developed that are diverse in their application as well as their goals. These methods are used to obtain biological and ecological information for effective resource management decision-making. The synthesis of this collected information has four main objectives:

1. To prepare baseline information used in developing a restoration plan for the area being assessed
2. To study patterns and to describe trends through time
3. To determine compliance with all environmental permits
4. To determine whether the project goals have been attained

Successful adaptive monitoring programs dictate that field data must be collected in a manner that accommodates not only traditional methods of characterizing abiotic and biotic associations, but also new developments in spatial statistics. The combination of implementing proven sampling methods with cutting edge geographical positioning systems/geographic information system methods of collecting spatially referenced data in the field meets this objective.

This monitoring approach will allow a straightforward analysis of data and will test all the criteria stated in the restoration design plans and/or permit(s). It will also reveal biologically and statistically significant trends and patterns that could then become the focus of corrective actions in cases where restoration projects are not meeting design or permit criteria. Restoration results may vary significantly with methods and at different locations. If restoration designs are not meeting the desired objectives, modifications should be considered.

For adaptive management to succeed there also needs to be consensus among scientists, managers, and other stakeholders involved in the process, and they all must be willing to change actions in response to knowledge gained. One of the keys in this process is the input of a variety of multidisciplinary experts including biologists and ecologists, geologists, engineers, physical scientists, resource managers and economists, and others dedicated to a common vision — project success. While different experts often have divergent opinions, the Delphi technique has proven to be a successful method for developing consensus among experts. The Delphi technique is based on the following general principles:

1. Opinions of experts are justified as inputs to decision-making where absolute answers are unknown.
2. A consensus of experts will provide a more accurate response to questions or problems than a single expert would.

Part of the ability to run a successful adaptive management strategy on all environmental restoration projects is to have a QA/QC team that functions independently of all elements of the project from design through implementation and evaluation. Specifically, this QA/QC team does not overtly participate in the actual project. This independence assures unbiased oversight and reviews for the benefit of the overall goals of the project and accordingly, the resource.

1.3 CONCLUSIONS

Restoration is a relatively new and rapidly expanding discipline that combines many fields of science including ecology, geology, socioeconomics, and engineering. Although the specific goal of restoration is to restore the ecological function of a particular ecosystem, a multiscale approach is needed to ensure the successful restoration of a site, especially in the case of coral reef restoration. Those conducting restoration activities must examine how ecological processes that vary in spatial and temporal scales have influenced the function of a reef system and determine which processes need to be reestablished to restore critical coral reef functions. A multiscale approach can ensure that stressors to the reef ecosystem are removed or accounted for, that critical ecological processes have been successfully introduced, and that the restoration itself is not negatively impacting the function of the landscape. Additionally, a multiscale approach to restoration may result in greater ecological and environmental benefits because it allows for enhancement to occur at more than one scale.

Restoration is an attempt to overcome, through manipulation, the factors that impede the natural recovery of an impaired resource. For instance, when vessels run aground, they kill coral and reduce topographic complexity, thus dramatically altering the local ecosystem and its services. In these cases the ultimate goal is to restore damaged reefs that are functionally equivalent to their uninjured counterparts. To properly undertake the damage assessment and restoration strategy as outlined above requires a multidisciplinary team of individuals dedicated toward a common goal. Careful documentation of the resultant injury is critical to this planning process.[48] This approach to impact assessment and restoration planning will provide an ecologically defensible basis upon which to document the injury, set restoration goals, implement the appropriate restoration plan, and gauge overall project success. Reef restoration also challenges our understanding of reef ecosystems. Therefore, the logic underlying successful restoration must be rooted in an integrated, multidisciplinary approach that includes engineering, geologic, biologic, aesthetic, and socioeconomic considerations. The outcomes of such efforts will tell us what we know, what we do not know, and what will work in practice. While there is no cookbook for reef restoration, there is a general recipe.

Finally, we must glean as much as we can from the few restoration projects completed to date,[90–92] and we can profit from the vast knowledge gained in performing terrestrial, wetland, and coastal restoration.[85,93–102] Better reef restoration efforts can be achieved by setting goals based on the structure and function of local, unimpaired reefs of similar habitat type and by incorporating what we have learned from the successes and failures of earlier projects. We will learn more from our failures, because failure reveals the inadequacies in our designs.[103] Developing successful restoration efforts in the future will depend upon acquiring and applying a scientific base to this emerging discipline. In addition, because of the infancy of this enterprise, the continued sharing of information will be vital to improving restoration strategies over time. The status of coral reef ecosystem restoration has advanced a great deal in a short time. As restoration scientists and managers, we should be excited with the opportunities that lie ahead.

It is hoped that the protocol established in this document will assist resource managers in developing and guiding coral reef assessment and restoration strategies under their stewardship into the future. Conversely, better quality restoration will in turn lead to better management and more secure protection of the resource for future generations.

ACKNOWLEDGMENTS

We thank PBS&J for their continued support of our coral reef assessment and restoration program and for permission to publish. Rich Aronson, Andy and Robin Bruckner, Billy Causey, Don Deis, Ken Deslarzes, Dick Dodge, Adam Gelber, Dave Gilliam, Steve Gittings, Bob Kaplan, Les Kaufman, Brian Keller, Steven Miller, G.P. Schmahl, Sharon Shutler, and Dione Swanson are thanked for many discussions over many years that helped in formulating the ideas discussed in this chapter. We would also like to thank two anonymous reviewers for their careful critique of an earlier version of this manuscript.

REFERENCES

1. Wilkinson, C.R. 1992. Coral reefs of the world are facing widespread devastation: can we prevent this through sustainable management practices? *Proc. 7th Int'l. Coral Reef Symp.* 1:11–21.
2. Woodley, J.D. and J.R. Clark. 1989. Rehabilitation of degraded coral reefs. Pages 3059–3075 in O.T. Morgan, ed. *Coastal Zone '89*, Charleston, S.C. American Society Coastal Engr.
3. Wilkinson, C. 2004. New initiatives in coral reef monitoring, research, management and conservation. Pages 93–113 in *Status of Coral Reefs of the World: 2004* Volume 1. Australian Institute of Marine Science, 2004.
4. Rogers, C.S. 1985. Degradation of Caribbean and western Atlantic coral reefs and decline of associated fisheries. *Proc. 5th Int'l. Coral Reef Cong.* 6:491–496.
5. Porter, J.W. and O.W. Meier, 1992. Quantification of loss and change in Floridian reef coral populations. *Amer. Zool.* 32:625–640.
6. Ginsburg, R.N., ed. 1994. *Proceedings of the Colloquium on Global Aspects of Coral Reefs: Health Hazards and History, 1993.* Rosenstiel School of Marine and Atmospheric Science, University of Miami, Miami, FL.
7. Hughes, T.P., 1994. Catastrophes, phase shifts and large-scale degradation of a Caribbean coral reef. *Science* 265:1547–1551.

8. Brown, B.E. 1997. Disturbances to reefs in recent times. Pages 354–379 in C. Birkeland, ed. *Life and Death of Coral Reefs*. Chapman and Hall, New York.
9. Connell, J.H., 1997. Disturbance and recovery of coral assemblages. *Coral Reefs* 16: S101–S113.
10. Aronson, R.B. and W. F. Precht. 2001. Evolutionary paleoecology of Caribbean coral reefs. Pages 171–233 in W.D. Allmon and D.J. Bottjer, eds. *Evolutionary Paleoecology: The Ecological Context of Macroevolutionary Change*. Columbia University Press, New York.
11. Gardner, T.A., I.M. Côté, J.A. Gill, A. Grant, and A.R. Watkinson. 2003. Long-term region-wide declines in Caribbean corals. *Science* 301:958–960.
12. Salvat, B. 1987. *Human Impacts on Coral Reefs: Facts and Recommendations*. Antenne Mussee, Ecole Pratique des Hautes Etudes, French Polynesia.
13. Roberts, C.M. 1997. Connectivity and management of Caribbean coral reefs. *Science* 278:1454–1457.
14. Causey, B.D. 1990. Biological assessments of damage to coral reefs following physical impacts resulting from various sources, including boat and ship groundings. Pages 49–57 in W.C. Jaap, ed. *Diving for Science — 1990. Proc. Amer. Acad. Underwater Sci. 10th Ann. Sci. Diving Symp.*
15. Miller, S.L., G.B. McFall, and A.W. Hulbert. 1993. *Guidelines and Recommendations for Coral Reef Restoration in the Florida Keys National Marine Sanctuary.* National Undersea Research Center, University of North Carolina at Wilmington.
16. Aronson, R.B. and D.W. Swanson. 1997. Video surveys of coral reefs: uni- and multivariate applications. *Proc. 8th Int'l. Coral Reef Symp.* 2: 1441–1446.
17. Aronson, R.B. and D.W. Swanson. 1997. Disturbance and recovery from ship groundings in the Florida Keys National Marine Sanctuary. Dauphin Island Sea Lab Tech. Rpt. 97–002.
18. Hatcher, B.G. 1996. Ship and boat groundings on coral reefs: what do they teach us about community responses to disturbance? *Abstracts 8th Int'l. Coral Reef Symp, Panama.* Page 84.
19. Challenger, G.E. 1999. Questions regarding the biological significance of vessel groundings and appropriateness of restoration effort. Abstracts — International Conference on Scientific Aspects of Coral Reef Assessment, Monitoring, and Restoration, Ft. Lauderdale, FL. Page 66.
20. NOAA. 2001. Careless drivers damaging marine habitats in Florida Sanctuary. Coastal Services. September/October, Volume 5 (http://www.csc.noaa.gov/magazine/2001/05/florida.html).
21. Lutz, this volume.
22. Hudson, J.H. and R. Diaz. 1988. Damage survey and restoration of *M/V Wellwood* grounding site, Molasses Reef, Key Largo National Marine Sanctuary, Florida. *Proc. 6th Int'l Coral Reef Symp.* 2:231–236.
23. Gittings, S.R. and T.J. Bright. 1988. The *M/V Wellwood* grounding: a sanctuary case study. *Oceanus* 31:35–41.
24. Gittings, S.R. 1991. Coral reef destruction at the *M/V Elpis* grounding site, Key Largo National Marine Sanctuary. Submitted to U.S. Department of Justice Torts Branch, Civil Division. Texas A&M Research Fdtn. Project 6795.
25. Precht, W.F., R.B. Aronson, and D.W. Swanson. 2001. Improving scientific decision-making in the restoration of ship-grounding sites on coral reefs. *Bull. Mar. Sci.* 69:1001–1012.
26. Gladfelter, W. B. 1982. White band disease in *Acropora palmata*: implications for the structure and growth of shallow reefs. *Bull. Mar. Sci.* 32:639–643.
27. Precht, W.F., R.B. Aronson, S. Miller, B. Keller, and B. Causey. 2005. The folly of coral restoration following natural disturbances in the Florida Keys National Marine Sanctuary. *Restoration Ecol.* 23:24–28.
28. Cairns, J. Jr. 1995. *Rehabilitating Damaged Ecosystems.* Lewis Publishers, Boca Raton, FL.
29. National Research Council. 1992. *Restoration of Aquatic Ecosystems: Science, Technology, and Public Policy.* National Academy Press, Washington, D.C.
30. Maragos, J.E. 1974. Coral transplantation: a method to create, preserve, and manage coral reefs. Sea Grant Advisory Report SEA-GRANT-AR-74-03-COR-MAR-14. University of Hawaii, Honolulu.
31. Maragos, J.E. 1992. Restoring coral reefs with emphasis on Pacific reefs. Pages 141–221 in G.W. Thayer, ed. *Restoring the Nation's Marine Environment*, Maryland Sea Grant, Pub. UM-SG-TS-92-06, College Park, MD.
32. Auberson, B. 1982. Coral transplantation: an approach to the re-establishment of damaged reefs. *Kalikasan, Philippines J. Biol.* 11:158–172.
33. Harriot, V.J. and D.A. Fisk. 1988. Coral transplantation as a reef management option. *Proc. 6th Int'l. Coral Reef Symp.* 2:375–379.

34. Guzman, H.M. 1991. Restoration of coral reefs in the Pacific Costa Rica. *Conserv. Biol.* 5:189–195.
35. Wheaton, J.L., W.C. Jaap, B.L. Kojis, G.P. Schmahl, D.L. Ballantine, and J.E. McKenna. 1994. Transplanting organisms on a damaged reef at Pulaski Shoal, Ft. Jefferson National Monument, Dry Tortugas, Florida. *Bull. Mar. Sci.* 54:1087.
36. Jaap, W.C., B. Grahm, and G. Mauseth. 1996. Reattaching corals using epoxy cement. *Abstracts 8th Int'l. Coral Reef Symp., Panama*. Page 98.
37. Hobbs, R.J., and D. A. Norton. 1996. Towards a conceptual framework for restoration ecology. *Restoration Ecol.* 4:93–110.
38. Palmer, M. A., R.F. Ambrose, and N.L. Poff. 1997. Ecological theory and community restoration ecology. *Restoration Ecol.* 5:291–300.
39. Zedler, J.B. 2000. Progress in restoration ecology. *Trends Ecol. Evol.* 15:402–407.
40. Pastorok, R.A., A. MacDonald, J.R. Sampson, P. Wilber, D.J. Yozzo, and J.P. Titre. 1997. An ecological decision framework for environmental restoration projects. *Ecol. Eng.* 9:89–107.
41. Mitsch, W.J. and R.F. Wilson. 1996. Improving the success of wetland creation and restoration with know-how, time and self-design. *Ecol. Appl.* 6:77–83.
42. Grayson, J. E., M.G. Chapman, and A.J. Underwood. 1999. The assessment of restoration of habitat in urban wetlands. *Landscape Urban Plan.* 43:227–236.
43. Bedford, B.L. 1999. Cumulative effects on wetland landscapes: links to wetland restoration in the United States and Southern Canada. *Wetlands* 19:775–788.
44. Glasby, T.M. and A.J. Underwood. 1995. Sampling to differentiate between pulse and press perturbations. *Environ. Monit. Assess.* 42:241–252.
45. Deis, D.R., and D.P. French. 1998. The use of methods for injury determination and quantification from Natural Resource Damage Assessment in ecological risk assessment. *Human Ecol. Risk Assess.* 4: 887–903.
46. Deis, D.R. 2000. The use of natural resource damage assessment techniques in the assessment of impacts of telecommunication cable installation on hard corals off Hollywood, Florida. In *Overcoming Barriers to Environmental Improvement*, Proceedings of the 25th Annual National Association of Environmental Professionals Conference, Portland, ME.
47. Mauseth, G.S. and D.A. Kane. 1995. The use and misuse of science. In *Natural Resource Damage Assessment.* Prepared for the 1995 International Oil Spill Conference. American Petroleum Institute, Washington, D.C.
48. Hudson, J.H. and W.B. Goodwin. 2001. Assessment of vessel grounding injury to coral reef and seagrass habitats in the Florida Keys National Marine Sanctuary, Florida: protocol and methods. *Bull. Mar. Sci.* 69:509–516.
49. Symons et al., this volume.
50. Jaap, W.C. 2000. Coral reef restoration. *Ecol. Eng.* 15:345–364.
51. Precht, W.F. 1998. The art and science of reef restoration. *Geotimes* 43:16–20.
52. Davidson, this volume.
53. Shutler et al., this volume.
54. Naveh, Z. 1994. From biodiversity to ecodiversity: a landscape-ecology approach to conservation and restoration. *Restoration Ecol.* 2:180–189.
55. Ehrenfeld, J.G. and L.A. Toth. 1997. Restoration ecology and the ecosystem perspective. *Restoration Ecol.* 5:307–317.
56. Aronson, J. and E. Le Floc'h. 1996. Hierarchies and landscape history: dialoguing with Hobbs and Norton. *Restoration Ecol.* 4:327–333.
57. Race, M. and M. Fonseca. 1996. Fixing compensatory mitigation: what will it take? *Ecol. Appl.* 6:94–101.
58. Parker, V.T. 1997. The scale of successional models and restoration objectives. *Restoration Ecol.* 5:301–306.
59. Bell, S.S., M.S. Fonseca, and L.B. Motten. 1997. Linking restoration and landscape ecology. *Restoration Ecol.* 5:318–323.
60. Zedler, J.B. 1996. Ecological issues in wetland mitigation: an introduction to the forum. *Ecol. Appl.* 6:33–37.
61. Clements, F.E. 1916. Plant succession, an analysis of the development of vegetation. *Carnegie Institution of Washington Publication* 242:1–512.

62. Clements, F.E. 1936. Nature and structure of the climax. *J. Ecol.* 24:252–284.
63. Gleason, H.A. 1926. The individualistic concept of plant association. *Bull. Torrey Botanical Club* 53:7–26.
64. Allen, T.F.H. and T.W. Hoekstra. 1992. *Toward a Unified Ecology.* Columbia University Press, New York.
65. McIntosh, R.P. 1995. H.A. Gleason's "individualistic concept" and theory of animal communities: a continuing controversy. *Biol. Rev. Cambridge Phil. Soc.* 70:317–357.
66. Kojis, B.L., and N.J. Quinn. 2001. The importance of regional differences in hard coral recruitment rates for determining the need for coral restoration. *Bull. Mar. Sci.* 69:967–974.
67. Middleton, B. 1999. *Wetland Restoration; Flood Pulsing and Disturbance Dynamics.* John Wiley and Sons, New York.
68. White, P.S. and J.L. Walker. 1997. Approximating nature's variation: selecting and using reference information in restoration ecology. *Restoration Ecol.* 5:338–349.
69. Vidra, this volume.
70. Quammen, M.L. 1986. Measuring the success of wetlands mitigation. *Natl. Wetlands Newslett.* 8(5): 6–8.
71. Redmond, A.M. 1995. Mitigation examples from Florida: what have we learned and where are we going. Pages 259–262 in J.A. Kusler, D.E. Willard, and H.C. Hull, Jr., eds. *Wetlands and Watershed Management – Science Applications and Public Policy.* Inst. Wetland Science and Public Policy — Assoc. State Wetland Managers, Inc., Berne, NY.
72. Kojis, B.L. and N.J. Quinn. 1981. Factors to consider when transplanting hermatypic corals to accelerate regeneration of damaged coral reefs. Pages 183–187 in *Conf. Environmental Engineering*, Townsville, Australia.
73. Clark, S. and A.J. Edwards. 1995. Coral transplantation as an aid to reef rehabilitation: evaluation of a case study in the Maldive Islands. *Coral Reefs* 14:201–213.
74. Rinkevich, B. 1995. Restoration strategies for coral reefs damaged by recreational activities: the use of sexual and asexual recruits. *Restoration Ecol.* 3:241–251.
75. Muñoz-Chagin, R.F. 1997. Coral transplantations program in the Paraiso Coral Reef, Cozumel Island, Mexico. *Proc. 8th Int'l. Coral Reef Symp.* 2: 2075–2078.
76. Kentula, M.E. 2000. Perspectives on setting success criteria for wetland restoration. *Ecol. Eng.* 15:199–209.
77. Hatcher, B.G. 1984. A maritime accident provides evidence for alternate stable states in benthic communities on coral reefs. *Coral Reefs* 3:199–204.
78. Aronson, R.B., P.J. Edmunds, W.F. Precht, D.W. Swanson, and D.R. Levitan. 1994. Large-scale, long-term monitoring of Caribbean coral reefs: simple, quick, inexpensive techniques. *Atoll Res. Bull.* 421:1–19.
79. Smith, S.R., D.C. Hellin, and S.A. McKenna. 1998. Patterns of juvenile coral abundance, mortality, and recruitment at the *M/V Wellwood* and *M/V Elpis* grounding sites and their comparison to undisturbed reefs in the Florida Keys. Final Report to NOAA Sanctuary and Reserves Division and the National Undersea Research Program/Univ. North Carolina at Wilmington.
80. Szmant, A.M. 1997. Nutrient effects on coral reefs: a hypothesis on the importance of topographic and trophic complexity to reef nutrient dynamics. *Proc. 8th Int'l. Coral Reef Symp.* 2:1527–1532.
81. Ebersole, J.P. 2001. Recovery of fish assemblages from ship groundings on coral reefs in the Florida Keys National Marine Sanctuary. *Bull. Mar. Sci.* 69:655–672.
82. Cairns, J., Jr. 1995. *Rehabilitating Damaged Ecosystems.* Lewis Publishers, Boca Raton, FL.
83. Fonseca, M.S., B.E. Julius, and W.J. Kenworthy. 2000. Integrating biology and economics in seagrass restoration: How much is enough and why? *Ecol. Eng.* 15:227–237.
84. NOAA. 1995. Habitat Equivalency Analysis: An Overview. Policy and Technical Paper Series, Number 95–1. Damage Assessment and Restoration Program, National Oceanic and Atmospheric Administration, Department of Commerce.
85. Fonseca, M.S., W.J. Kenworthy, and G.W. Thayer. 1998. *Guidelines for the Conservation and Restoration of Seagrasses in the United States and Adjacent Waters.* NOAA's Coastal Ocean Program, Decision Analysis Series No. 12.
86. Banks, K., R.E. Dodge, L. Fisher, D. Stout, and W. Jaap. 1998. Florida Coral Reef Damage from Nuclear Submarine Grounding and Proposed Restoration. *J. Coastal Res.* Special Issue 26:64–71.

87. Florida Department of Environmental Protection. 1994. *A Natural Resource Damage Assessment for the Grounding of the USS Memphis on the Second Reef in Broward County Florida.* Tech. Economic Rept. DEP-TER: 94-2, May 3, 1994, State of Florida: Department of Environmental Protection, Office of General Council. 22 pp.
88. Pickett, S.T.A., and V.T. Parker. 1994. Avoiding the old pitfalls: opportunities in a new discipline. *Restoration Ecol.* 2:75–79.
89. Zedler, J.B. and J.C. Callaway. 2003. Adaptive restoration: a strategic approach for integrating research into restoration projects. Pages 167–174 in *Managing for Healthy Ecosystems,* D.J. Rapport, et al., eds. Lewis Publishers, Boca Raton, FL.
90. Jaap, W.C. 1999. An historical review of coral reef restoration in Florida. *Abstracts Int'l. Conf. Scientific Aspects of Coral Reef Assessment, Monitoring, and Restoration.* Page 111.
91. NOAA. 1999. *R/V Columbus Iselin* restoration home page. http://www.sanctuaries.nos.noaa.gov/special/columbus/project.html.
92. NOAA. 1999. Damage assessment and restoration program: restoration case histories. http://www.darp.noaa.gov.
93. Lewis, R.R. 1982. *Creation and Restoration of Coastal Plant Communities.* CRC Press, Boca Raton, FL.
94. Lewis, R.R. 1990. Creation and restoration of coastal plain wetlands in Florida. Pages 73–101 in J.A. Kusler and M.E. Kentula, eds. *Wetland Creation and Restoration — The Status of Science.* Island Press, Washington, D.C.
95. Lewis, R.R. 1994. Enhancement, restoration and creation of coastal wetlands. Pages 167–191 in D.M. Kent, ed. *Applied Wetlands Science and Technology.* Lewis Publishers, Boca Raton, FL.
96. Kusler, J.A. and M.E. Kentula. 1990. *Wetland Creation and Restoration — The Status of the Science.* Island Press, Washington, D.C.
97. Thayer, G.W. 1992. Restoring the nation's marine environment, Maryland Sea Grant, Pub. UM-SG-TS-92-06, College Park, MD.
98. Cooke, G.D., E. Welch, S.A. Peterson, and P.R. Newroth. 1993. *Restoration and Management of Lakes and Reservoirs,* 2nd ed. Lewis Publishers, Boca Raton, FL.
99. Moshiri, G.A. 1993. *Constructed Wetlands for Water Quality Improvement.* Lewis Publishers, Boca Raton, FL.
100. Cairns, J., Jr. 1995b. Restoration ecology: protecting our national and global life support systems. Pages 1–12 in J. Cairns, Jr., ed. *Rehabilitating Damaged Ecosystems.* Lewis Publishers, Boca Raton, FL.
101. Snedaker, S.C. and P.D. Biber. 1996. Restoration of mangroves in the United States of America — a case study in Florida. Pages 170–188 in C.D. Field, ed. *Restoration of Mangrove Ecosystems.* Int'l. Soc. Mangrove Ecosystems, Okinawa, Japan.
102. Dennison, M.S. and J.A. Schmid. 1997. Wetland Mitigation. Government Institutions, Rockville, MD.
103. Malakoff, D. 1998. Restored wetlands flunk real world test. *Science* 280:371–372.

2 A Thousand Cuts? An Assessment of Small-Boat Grounding Damage to Shallow Corals of the Florida Keys

Steven J. Lutz

CONTENTS

2.1 INTRODUCTION

For thousands of years coral reefs have survived natural impacts, such as storms, diseases, and predation. What they cannot withstand is the combination of these natural impacts with severe or repeated anthropogenic damage, such as overfishing, sedimentation, and excess nutrients. Reefs around Jamaica and San Andres have been devastated by this combination,[1,2] and Florida reefs are widely reported to decline.[3,4] Indeed, according to Wilkinson (1992),[5] South Florida's reefs are so "threatened" that they may disappear in 20 to 40 years.

Anthropogenic impacts to corals can be divided into direct and indirect effects.[6] Indirect anthropogenic impacts throughout the Florida Keys, which include poor water quality and high sedimentation rates, have received great attention from the scientific community.[4,7–11] However, there is comparatively little information on direct anthropogenic damage, such as broken or overturned corals, on Florida coral reefs. Much of this research has been related to the damage and rehabilitation of larger vessel groundings, which are highly visible and well documented.[12–14]

In contrast, little or no information on direct physical damage to corals caused by smaller vessels is available. Previous studies and reports have noted this form of damage,[12,15–20] also referred to as "orphan groundings" by Florida Keys National Marine Sanctuary staff. However, the amount of damage caused by small vessels that are able to leave grounding incidents under their own power is unreported and may be vast; certainly, such incidents are much more numerous than large vessel groundings. In the Florida Keys small-vessel grounding damage may be particularly widespread because many of the reefs that attract visitors have shallow-water corals. Assessing the extent, amount, and impact of this form of anthropogenic damage to coral is essential for reef management.

This report is the first estimate of the geographic distribution and severity of small-vessel grounding damage on shallow-water massive corals of patch reefs throughout the Florida reef tract. In this assay 315 shallow-water massive coral colonies from 49 reef sites within the Florida reef tract were examined for signs of boat grounding damage.

2.2 MATERIALS AND METHODS

This study was conducted from August 1996 to January 1997 on 49 reef areas with high-profile shallow-water coral heads or clusters of heads in the Florida Keys reef tract (Figure 2.1 and Figure 2.2). All but one of the reef sites surveyed were patch reefs; the exception was Carysfort Reef, a bank-barrier reef. Patch reefs occur throughout the Florida reef tract. They are particularly abundant in the waters off northern Elliot Key and south Key Largo, which include over 5000 patch reefs.[21] Patch reefs typically occur in water 2 to 9 m deep and vary from 30 to 700 m in diameter.[22] In the Florida Keys, the framework builder coral species of patch reefs include *Siderastrea siderea*, *Diploria strigosa*, *D. labyrinthiformis*, *Colpophyllia natans*, *Montastraea annularis*, and *M. faveolata*. These corals have been termed boulder or massive corals.[23] *Montastraea annularis* (*senso lato*) is particularly important as it has been described as a "keystone" species[24] and can exhibit lateral growth as it approaches sea level. This massive coral can be found growing in individual colonies, or heads, and in groups of amalgamated colonies, or clusters, growing together. They can grow to be up to 100 m^2 in area and have up to 5 m of relief.[25] Shallow-water massive coral heads and clusters of shallow massive coral heads are termed head/clusters for the purposes of this study. The geographic location of each reef site was recorded with a hand-held global positioning system.

The exact depth and diameter of each coral head/cluster found within 2 m of the surface was recorded for each reef site. The survey depth of 2 m was chosen to accommodate for tidal range (~1.5 m) and the maximum depth of typical hulls and/or propellers for small vessels (~1 m). The Northern Florida Keys tidal range was determined by inspection of tide tables.[26,27] To account for tidal variation, all *in situ* depth measurements were standardized to depth below spring mean low water tide level. Standardized depths ranged from 0.1 to 1.0 m. According to vessel registration records, the majority of registered vessels in Miami-Dade and Monroe Counties are pleasure craft from 16 to 26 ft in length. In the two counties, 36,312 such vessels were registered in 1994, accounting for 56% of all registered vessels.[28] Miami-Dade and Monroe Counties are the closest counties to the northern Florida reef tract. All of the corals in this survey were potentially susceptible to small-vessel grounding damage.

Reefs surveyed contained from one to 28 shallow-water head/clusters with the majority, 75%, containing from one to five head/clusters. In total, 315 coral head/clusters were measured. Head/clusters ranged in size from less than 1 m (a singular head) to 18 m in diameter (a large cluster of amalgamated heads). The majority, 79%, were less than 5 m in diameter. Tidal range corrected depth of the top surfaces of head/clusters ranged from 25 cm to 1 m in depth; 39% from 0 to 0.25 m deep, 51% from 0.25 to 0.75 m deep, and 10% from 0.75 to 1 m deep.

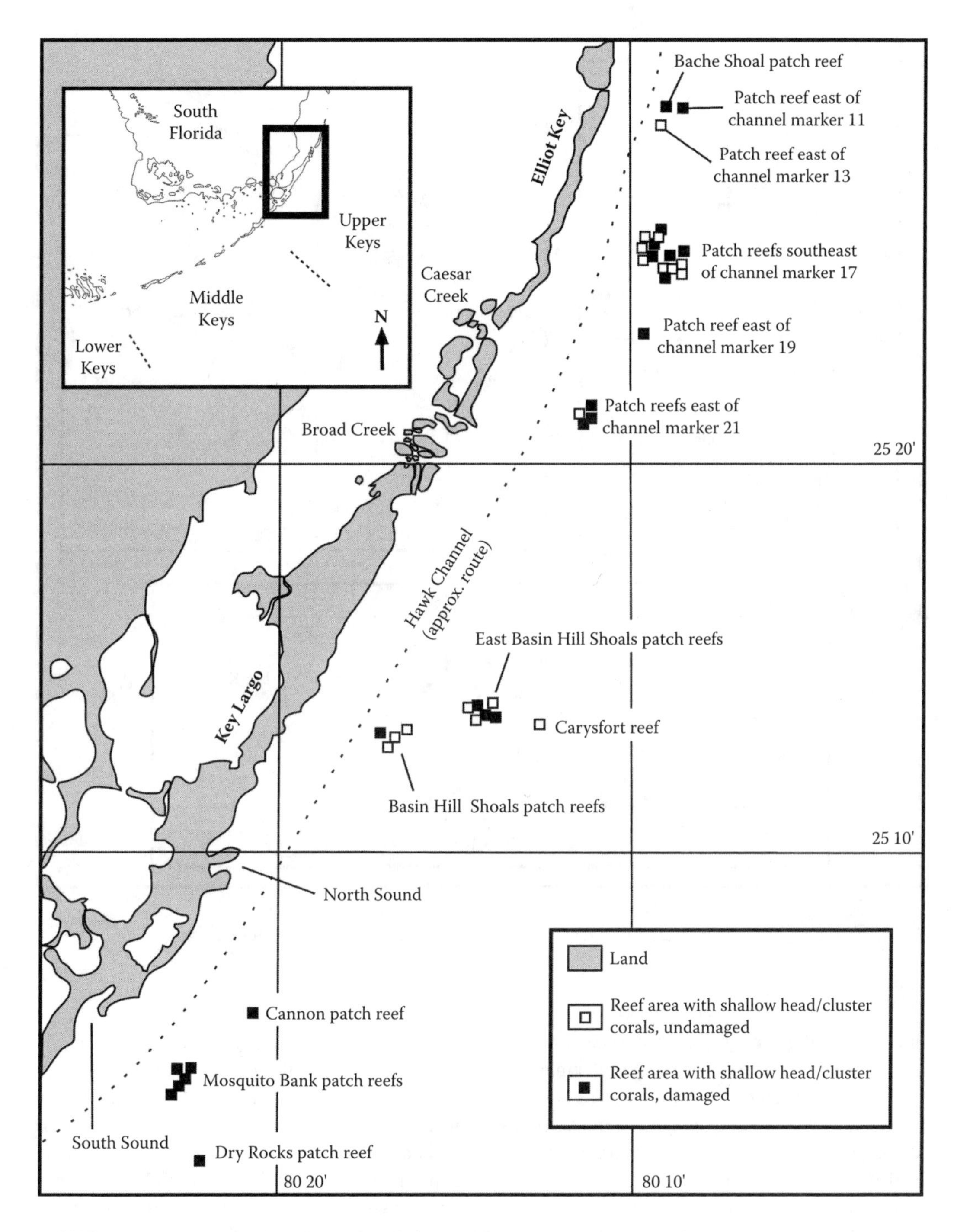

FIGURE 2.1 Approximate locations of reef sites surveyed (North Florida reef tract).

Of the 315 shallow-water head/clusters surveyed, 312 were *Montastraea* spp. and three were *S. siderea. Montastraea* spp. were identified according to the classifications of Weil and Knowlton (1994).[29] *Montastraea annularis* and *M. faveolata* were the only *Montastraea* spp. recorded in the survey. These two coral species commonly co-occur.[30]

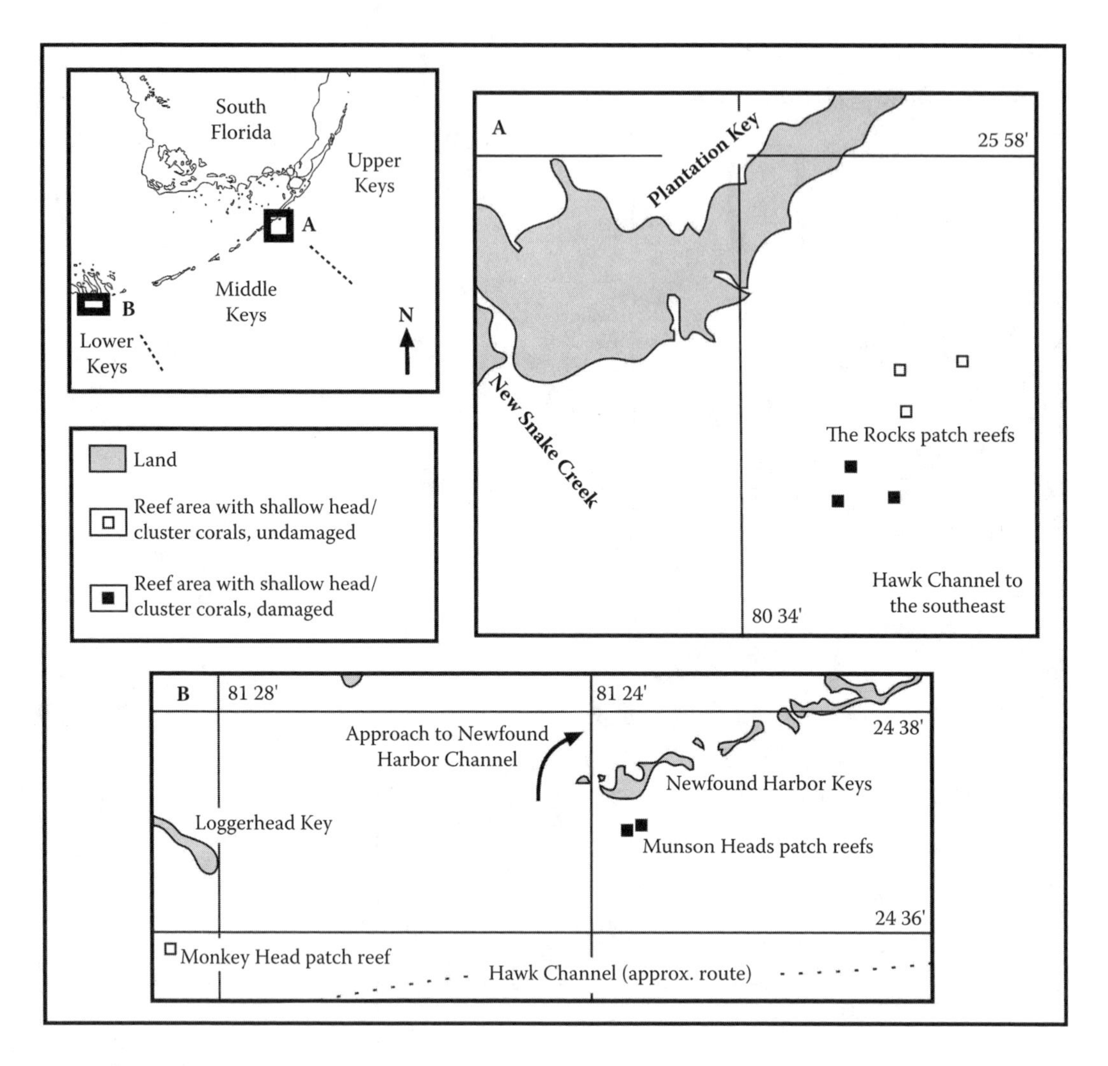

FIGURE 2.2 Approximate locations of reef sites surveyed (Middle and South Florida reef tract).

Although *Acropora palmata* is commonly found growing close to the surface, this coral was not included in the survey. This coral species is particularly vulnerable to natural fragmentation during storms, which renders it difficult to distinguish between natural and anthropogenic damage.[31,32]

For underwater observations of direct physical damage a meter rule marked in 2- and 10-cm increments was used. Damage was recorded in square centimeters and as the extent of surface area destroyed. Two forms of physical damage were identified, collision damage and scarring damage. Collision damage occurs when a coral is crushed and split by a vessel's hull into multiple fragments. Hull paint is often driven into the coral skeleton (Figure 2.3A and Figure 2.3G). Scarring damage, from boat propellers, tears off live coral, exposing the skeleton. In propeller scarring, typical scarlike striations are seen (Figure 2.3B, Figure 2.3C, Figure 2.3D, Figure 2.3E, Figure 2.3F, and Figure 2.3I), and large fragments of coral can be chipped off (Figure 2.3E, Figure 2.3G, Figure 2.3H, and Figure 2.3I). Any damage whose source was not readily identifiable, for example when the surfaces were completely overgrown by turf algae and the corallites were not exposed or identifiable, was not included in the survey.

Statistical analysis was performed with the *t*-test and analysis of variance (ANOVA) where applicable.

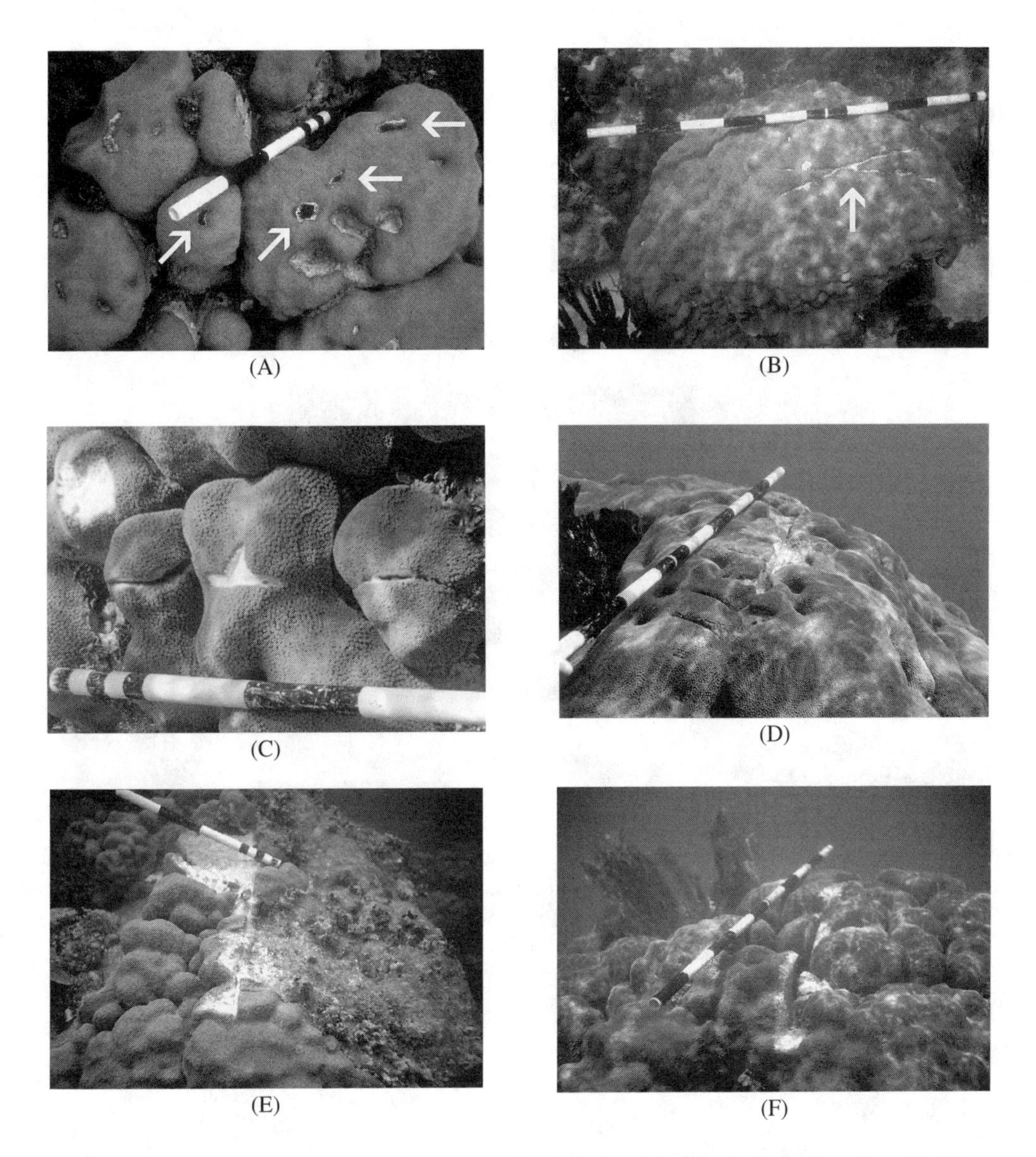

FIGURE 2.3 Damage to various head/clusters. A. Patch reef southeast of channel marker 17, Biscayne National Park (BNP). Arrows indicate boat hull paint embedded in coral. B. Patch reef east of channel marker 21, BNP. Arrow indicates small propeller scar. C. Mosquito Bank patch reef, John Pennekamp Coral Reef State Park (JPCRSP). D. East Basin Hill Shoals patch reef, Florida Keys National Marine Sanctuary (FKNMS). E. Basin Hill Shoals patch reef, JPCRSP. F. Patch reef area southeast of channel marker 17, BNP. G. Bache Shoal patch reef, BNP. Arrows indicate crushed coral and boat hull paint. H. Munson Heads patch reef, FKNMS. I. East Basin Hill Shoals patch reef, FKNMS.

2.3 RESULTS

2.3.1 Geographic Distribution

The results indicate that boat damage was widespread. Most (57.1%) of the shallow-water reef sites surveyed showed signs of damage. Of the 315 coral head/clusters found on those reefs, 79 (25%) had been damaged. The total estimated area of destroyed coral found was 37,675 cm^2. The area of damage to individual head/clusters ranged from 25 to 5800 cm^2. Most damage found on

(G) (H) (I)

FIGURE 2.3 (Continued.)

individual head/clusters was under 250 cm² (illustrated in Table 2.1). Two reefs, Bache Shoal and Mosquito Bank (see Figure 2.1), had much more severe extent of damage than all other reef sites (3366 +/– 1570 cm² (n = 6) on Bache Shoal and Mosquito Bank compared to 775 +/– 109 cm² (n = 22) on all other reef sites, P = 0.0017). These two reefs accounted for 60.2% of all damage found (20,200 cm²). However the occurrence of damage incidents to head/clusters was not statistically significantly higher than at other reef sites (48.5 +/– 12.3% of head/clusters damaged on Bache Shoal and Mosquito Bank compared to 28.9 +/– 5.98% damaged on all other reef sites).

2.3.2 Reef Sites

Reef sites surveyed contained from one to 28 shallow-water massive coral head/clusters. The total amount of damage found on head/clusters per each reef site ranged from 25 to 10,925 cm² coral destroyed.

For a comparative assessment of reef size damage, reef sites were divided into three size categories: small (zero to five head/clusters per reef); medium (six to 15 head/clusters per reef);

TABLE 2.1
Percent of Damaged Head/Clusters by Area of Coral Destroyed

	Area of Coral Destroyed (cm²)				
	≤250	**251 to 500**	**501 to 1500**	**1501 to 3000**	**>3000**
Percent of damaged head/clusters	49.3 (n = 39)	24 (n = 19)	20.3 (n = 16)	5 (n = 4)	1.3 (n = 1)

TABLE 2.2
Reef Size and Damage

	Reef Size Class (Number of Head/Clusters)		
	Small (1 to 5)	**Medium (6 to 15)**	**Large (Over 15)**
Number of head/clusters per reef size class	36	6	7
Percent of reefs damaged	42.2 (n = 17)	66.5 (n = 4)	100 (n = 7)
Mean area (cm^2) of damage	421 +/– 117 (n = 36)	700 +/– 375 (n = 6)	2557 +/– 1414 (n = 7)
Mean number of damaged head/clusters	0.806 +/– 0.19 (n = 36)	2.5 +/– 1.147 (n = 6)	5.0 +/– 0.976 (n = 7)

and large (>15 head/clusters per reef) (illustrated by Table 2.2). Damage among these reef size classes was distributed in the following proportions: 17 of the 36 (47.2%) small reefs, four of the six (66.6%) medium reefs, and seven of the seven (100%) large reefs had signs of damage.

A significant correlation was found between the number of shallow-water head/clusters per reef site and the amount of damage (mean total area in square centimeters per reef site): large reefs = 2557 +/– 1414 cm^2 (n = 7); medium reefs = 700 +/– 375 cm^2 (n = 6); small reefs = 421 +/– 117 cm^2 (n = 36), P = 0.0055. A significant correlation was also found between the number of shallow-water head/clusters per reef site and the mean number of damaged head/clusters per reef site: (large reefs = 5.0 +/– 0.976 (n = 7); medium reefs = 2.5 +/– 1.147 (n = 6); small reefs = 0.806 +/– 0.19 (n = 36), P = 0.0001 (illustrated by Table 2.2).

However, the number of shallow-water head/clusters per reef site did not appear to influence mean damage incidence or wound size: large reefs = 431 +/– 158 cm^2 (n = 17); medium reefs = 218 +/– 68 cm^2 (n = 4); small reefs = 528 +/– 125 cm^2 (n = 7) (illustrated in Table 2.2).

2.3.3 Head/Cluster Size

In order to determine whether head/cluster size influenced damage incidence, the 315 shallow-water massive coral head/clusters were divided into three size categories: small (< 5 m diameter); medium (5 to 10 m diameter); and large (>10 m diameter) (illustrated in Table 2.3). No connection was found concerning damage incidence; 54 of the 240 (22.5%) small head/clusters, 17 of the 47 (36.1%) medium head/clusters, and eight of the 28 (28.5%) large head/clusters were damaged.

TABLE 2.3
Head/Cluster Size and Damage

	Head/Cluster Size		
	Small (<5 m Diameter)	**Medium (5 to 10 m Diameter)**	**Large (>10 m Diameter)**
Number of head/clusters per head/cluster size class	240	47	28
Percent of damaged head/clusters	22.5 (n = 54)	36.1 (n = 17)	28.5 (n = 8)
Mean area (cm^2) of damage per head/cluster size class	77 +/– 15 (n = 240)	282 +/– 131 (n = 47)	194 +/– 109 (n = 28)

TABLE 2.4
Depth of Head/Clusters and Damage

	Head/Cluster Depth		
	0 to 0.3 m	**0.4 to 0.6 m**	**0.7 to 1.0 m**
Number of head/clusters	123	161	31
Percent damaged head/clusters	30 (n = 37)	22.9 (n = 37)	16.1 (n = 5)
Total area (cm^2) of damage per head/cluster	15,200	21,775	700

However, it was found that head/cluster diameter did influence the extent of damage (mean area in square centimeters per head/cluster size class). Medium and large head/clusters had more damage than did those in the small size class: small head/clusters = 77 +/– 15 cm^2 (n = 240); medium head/clusters = 282 +/– 131 cm^2 (n = 47); large head/clusters = 194 +/– 109 cm^2 (n = 28), P = 0.0087.

2.3.4 Depth of Head/Clusters

The depth below mean low-water level of the top surfaces of the shallow-water head/clusters ranged from 0 to 1.0 m. In order to investigate the effect of depth on damage, the head/clusters were divided into three depth categories: 0 to 0.3, 0.4 to 0.6, and 0.7 to 1.0 m depth (illustrated in Table 2.4). Damage incidence among the depth classes was distributed in the following proportions: 37 of the 123 (30%) 0- to 0.3-m deep head/clusters had signs of damage, as did 37 of the 161 (22.9%) 0.4- to 0.6-m deep head/clusters and five of the 31 (16.1%) 0.7- to 1.0-m deep head/clusters. Damage extent (total square centimeters of coral damaged per head/cluster) among the depth classes was distributed in the following proportions: 15,200 cm^2 of the 0- to 0.3-m deep head/clusters coral was destroyed, 21,775 cm^2 of the 0.4- to 0.7-m deep head/clusters coral was destroyed, and 700 cm^2 of the 0.7- to 1.0-m deep head/clusters coral was destroyed. The three depth categories do not significantly differ from each other in either damage incidence or extent. Neither, when damage occurs, does the depth of the top surfaces of shallow-water head/clusters affect the area of coral destroyed (mean area in square centimeters per damaged head/clusters) (0 to 0.3 m depth = 578 +/– 171 cm^2 (n = 37); 0.4 to 0.6 m depth = 410 +/– 76 cm^2 (n = 37); 0.7 to 1 m depth = 140 +/– 67 cm^2 (n = 5).

2.3.5 Mooring Buoys

Of the seven reef sites with mooring buoys that were surveyed, all but one had signs of damage. Of the 42 reefs without mooring buoys surveyed, 22 had signs of damage. However upon statistical evaluation, it was found that whether or not a reef had a mooring buoy did not affect the frequency of damage incidence (37.3 +/– 16.5% for reef sites with buoys [n = 7] compared to 30.3 +/– 5.9% for reef sites without buoys [n = 42]). Similarly, the extent of damage (mean area in square centimeters) found on reef sites is not affected by the presence or absence of mooring buoys (1415 +/– 485 cm^2 [n = 22] on reefs without buoys compared to 1021 +/– 485 cm^2 on reef sites with buoys [n = 6]). The presence or absence of mooring buoys on a reef did not significantly affect the degree of damage caused by small-boat groundings.

2.4 DISCUSSION AND CONCLUSIONS

Most damage found on individual head/clusters was under 250 cm^2. Although this category of damage appears widespread throughout the study range, it does not suggest that it is a cause of any specific decline in the health of corals throughout the Florida Keys reef tract. Additionally, the Florida Keys reef tract is a vast natural structure, most of which remains submerged out of smaller-vessel impact range during tidal fluctuations. While the direct damage from small-boat contact does not pose a serious threat to its overall survival, the accumulated damage can degrade and destroy the structure of localized areas of shallow-water corals and coral clusters, demonstrating this impact's importance to the health of localized head/clusters and contributing to the stresses these corals already experience.

2.4.1 Geographic Distribution

The total amount of damage found at Bache Shoal and Mosqutio Bank was substantial, 60.2% of all damage found. Indeed, these reefs show impact levels significantly higher than those of all other reefs. Bache Shoal is one of the closest shallow reefs with mooring buoys to metropolitan Miami and is directly adjacent to a major boating channel, Hawk Channel (see Figure 2.1). To prevent vessel impacts, it is marked by a triangular reef warning tower and channel marker at its north tip. It is significant to note that all shallow head/clusters surveyed at Bache Shoal were damaged, suggesting that its level of use or boat traffic (and related impacts) exceeds the safety methods used. Mosquito Bank, located adjacent to Hawk Channel and directly in the line of boat traffic coming from slips on Key Largo and South Sound, also has a high percentage (42.6%) of head/clusters damaged, indicating a high level of use or traffic and the need for additional protection. Results also suggest that these reef areas may be experiencing collisions by vessels that are larger and/or going much faster than on other reef sites. Mosquito Bank's high percentage of damage supports the Florida Department of Environmental Protection's findings.[33,34] Farther south, navigation channels and boater access may also play important roles in boat grounding damage, as The Rocks and Munson Heads are both adjacent to boating routes (see Figure 2.2).

2.4.2 Reef Size

It appeared that reefs with five or more shallow-water head/clusters were more susceptible to small-boating damage than were reefs with fewer than five shallow-water head/clusters. The more shallow-water head/clusters that a reef has, the more damage incidents or wounds, but the mean wound size remained the same, regardless of reef size. Larger reefs may receive more damage because the likelihood of collision with a larger reef area is greater, even though smaller reefs are more numerous. However, smaller reefs may also not be as attractive to small-boat traffic from tourists because they have less relief, smaller associated fish populations, and a smaller amount of live coral.

2.4.3 Head/Cluster Size

It appeared that the larger, in diameter, a shallow-water head/cluster, the more damage, but the frequency of damage remains the same, regardless of diameter. It is possible that small-vessel impact damage is infrequent overall and occurs at random. However, when such damage occurs, the larger in diameter a head/cluster, the greater the chance that a single damage incident will result in substantial damage.

2.4.4 Depth of Head/Clusters

It was interesting to find that within the 1-m depth range from spring low tide, the depth of the top surfaces of shallow-water head/clusters did not significantly influence either the degree or extent of damage caused by small-boat groundings. Therefore, all corals within a 1-m depth range from spring low tide are susceptible to small-vessel grounding damage. If the sample depth range of this survey had been extended to 2 or 3 m, frequency and extent of damage might have significantly correlated with depth; this, however, would have greatly lengthened survey time.

2.4.5 Mooring Buoys

One would expect to find higher levels of damage to shallow-water massive corals at reef sites with mooring buoys since mooring buoys tend to attract more recreational boaters. However, the presence or absence of mooring buoys on a reef did not significantly alter the frequency or extent of damage caused by small-boat groundings. More recreational boaters may be drawn to reefs with mooring buoys, but they appear to avoid any significant additional damage to shallow-water massive corals.

2.4.6 Impacts to Individual Coral Heads

It might be expected that small-vessel groundings are an important cause of damage on localized cluster-heads. Because boating damage tends to occur on the top surfaces of coral colonies, their detrimental effects may be more substantial than those of other types of lesions. Damage caused by storm rubble, in contrast, tends to occur more often on the sides of large colonies, rather than the tops. Large lesions may not completely heal, although partial regeneration may occur at the edges.[35,36] It has been found that within a week of a scarring event, filamentous algae colonize exposed skeleton and inhibit coral regeneration. Turf algae or other reef organisms may be well established by the time the healing margin of live coral reaches them. The encrustation of some organisms (e.g., boring sponges, encrusting zoanthids) can lead to further bioerosion of the colony. Meesters (1995),[36] in a study regarding damage and regeneration on scleractinian corals, showed that many lesions on the top surfaces of bleached coral colonies enlarged to numerous times their initial size, occasionally resulting in the death of the entire colony. Indeed, it appears that *M. annularis* may be very sensitive to bleaching.[36–38] Herbivorous fish pecking at the edge of a scar can consume turf algae and coral at the same time.[39] In addition, coral scarring may affect the total health of the colony by forcing the coral to reallocate resources to regeneration, and away from growth, reproduction, and combating disease.

Additionally, the cumulative effect of this form of damage to individual coral heads may have negative tourism consequences. As impacts are to the shallowest and most accessible area of reefs, they are easily within snorkeling range. Figure 2.3 clearly illustrates the diminished aesthetic value of damaged coral heads.

2.4.7 Trend in High User Pressure

The increasing trend of recreational use of South Florida marine habitat is evidenced by the 40.8% increase in registered vessels in 10 years in Miami-Dade and Monroe Counties (from 62,274 in 1993 to 87,699 in 2003).[40,41] Indeed, it appears almost certain that continued high user pressure on the most frequented reefs will, in a short time, degrade the aesthetic and recreational qualities of the reefs. Additionally, the continued high and relentless incidence of damage to these colonies will result in loss of the larger and older massive coral colonies. For these reasons it is imperative that management deal with the small-boat problem as a priority.

2.4.8 Management Considerations

This study indicates that the cumulative effect of small-vessel groundings presents a serious threat to localized coral eco-health and contributes significantly to other reef stresses. Marine parks and management in the Florida Keys are charged with the protection of the natural resources, especially coral reefs, under their jurisdictions. Table 2.5 illustrates coral damage on shallow water reef sites by management authority for reefs surveyed. A comprehensive management plan is needed in order to reduce the number of small-vessel groundings.

Management's options for minimizing this type of anthropogenic damage would vary according to available manpower and funds. In order to present a scientifically based management plan, the author suggests that, first, localized shallow-water reef areas with high levels of user impact must be identified. For "real time" observations, this type of survey should be carried out on an annual basis,

TABLE 2.5
Reef Sites Surveyed and Management Authority

Management Authority	Percentage of Shallow Reef Sites Surveyed	Percentage of Shallow Reef Sites Surveyed with Damage
Biscayne National Park	22	54.5
John Pennekamp Coral Reef State Park	10	70
Florida Keys National Marine Sanctuary	17	52.9

to gauge the direction of user pressure and to determine the effects of preservation and restoration actions. The particular criteria for identifying such reefs should include an estimate of the percentage of impacted shallow-water coral colonies. Special attention should be placed on reefs with over 15 shallow-water head/clusters, reefs with head/clusters over 5 m in diameter, and reefs close to navigation channels and popular marinas. These reefs are especially prone to this type of impact.

Severity of damage incidence can be estimated from measurements of the area of coral destroyed, as laid out in this chapter. Reefs with a high percentage of shallow-water corals damaged or colonies with severe damage should be designated for immediate prevention and restoration action.

As a result of such surveys, the management options for damage prevention would depend upon the severity of impact and could include the following:

1. The placement of additional "shallow reef" markers or other navigational beacons highlighting shallow corals prone to this type of impact
2. The targeted placement of additional mooring buoys (There are currently approximately 40 reef sites with mooring buoys throughout the Florida Keys.)
3. The establishment of small targeted preservation zones, which would restrict a certain use (i.e., boating, diving, or fishing activities) and thereby lessen user pressure on a particularly stressed and sensitive ecosystem
4. The establishment of critical zones, where all recreational and commercial access is prohibited (Currently, approximately 6% of the Florida Keys National Marine Sanctuary is set aside as fully protected zones known as ecological reserves, sanctuary preservation areas, and special use areas.)

Education would also play an important part in reef preservation. Marinas and boat rental shops close to damaged reefs could be targeted for educational materials, and boat pilot training programs highlighting this type of problem could be planned. Advertising fines could help make boaters more cautious while boating in shallow reef areas.

Individual coral head/cluster restoration options would depend upon the severity of impact and could include the following:

1. No action taken: the wound size is so minimal that the coral's natural healing process will suffice to restore the damage, or restoration action may lead to further direct physical damage to corals and surrounding benthos.
2. Stabilization and restructuring of unconsolidated coral fragments in a wound area, as required, in order to mimic the look and function (biological and aesthetic) of the original ecosystem.

The removal of bioeroding and competing organisms and/or the possible transplantation of coral in an unnatural wound area, giving the natural healing process of damaged coral colonies a "boost" (only feasible on the most severe of damage incidents).

2.5 CONCLUSION

In conclusion, this study shows that small-boat groundings on reef areas present a serious widespread negative impact to localized coral eco-health, especially on larger reef areas and to massive corals, and make a significant contribution to the stresses and pressures that corals already endure throughout the Florida Keys reef tract (including bleaching, disease, pollution, large-vessel groundings, high use, etc.).

It is imperative that small-vessel grounding damage be minimized. This distinct category of anthropogenic damage is a major insufficiently recognized negative impact to highly visited coral reef areas that must be dealt with in any scientifically based management plan.

ACKNOWLEDGMENTS

The author thanks the Florida Keys National Marine Sanctuary, Biscayne National Park, and John Pennekamp Coral Reef State Park for their assistance, advice, and permission to conduct studies under their jurisdictions and R.N. Ginsburg for project guidance, comments, and review. This project was funded by the Filipacchi Hachette Foundation as part of the International Year of the Reef Program (IYOR) and produced a masters thesis for the Division of Marine Affairs and Policy, University of Miami, Rosenstiel School of Marine and Atmospheric Science (RSMAS).

REFERENCES

1. Hughes, T., et al. 1985. Mass mortality of the echinoid *Diadema antilarum phlippi* in Jamaica. *Bull. Mar. Sci.* 36:377–384.
2. Hallock, P., et al. 1993. Coral reef decline. *National Geographic Research & Exploration* 9:385–387.
3. Porter, J.W., Meier, O.W. 1992. Quantification of loss and change in Florida reef coral populations. *Amer. Zool.* 32:625–640.
4. La Pointe, B. 1994. Phosphorus inputs and eutrophication on the Florida reef tract. In: Ginsburg, R.N. (ed.). *Proc. Colloq. Global Aspects of Coral Reefs; Health, Hazards, and History,* University of Miami, Florida pp. 106–112.
5. Wilkinson, C. 1992. Coral reefs of the world are facing widespread devastation; can we prevent this through sustainable management practices? In: *Proc. 7th Int. Coral Reef Symp.* Guam, Micronesia, 22–27 June, 1987. Mangilao, University of Guam Marine Lab., pp. 11–21.
6. Grigg, R., Dollar, S. 1990. Natural and anthropogenic disturbance on coral reefs. In: Goodall, D.W. and Dubinsky, Z. (eds.). *Coral Reefs; Ecosystems of the World.* Elsevier, Amsterdam: pp. 439–452.
7. Glynn, P.W. 1989. Condition of coral reef cnidarians from the Northern Florida reef tract: pesticides, heavy metals, and histopathological examination. *Mar. Pol. Bull.* 20:568–576.
8. Shinn, E., Miller, M. 1999, May. The Florida Keys: What is Happening to the Reef Tract and Why? Poster presented at the South Florida Restoration Science Forum, Boca Raton, FL.
9. Shinn, E.A., Lidz, B.H., Hudson, J.H., Kindinger, J.L., Halley R.B. 1989. *Reefs of Florida and the Dry Tortugas: IGC Field Trip Guide T176.* Washington, DC: American Geophysical Union, p. 53.
10. Cole, J. 1990. The state of our seas. *Florida Keys Magazine* 13(6):20–24.
11. Dustan, P., Halas, J.C. 1987. Changes to the reef-coral community of Carysfort Reef, Key Largo, Florida: 1974 to 1982. *Coral Reefs* 6:91–106.
12. Causey, B.D. 1990. Biological assessments of damage to coral reefs following physical impacts resulting from various sources, including boat and ship groundings. In: Jaap, W.C. (ed.). *Proc. Am. Acad. Underwater Sci., 10th Annual Diving Symposium,* St. Petersburg, Florida, pp. 49–57.
13. Wheaton, J.L., Jaap, W.C., Kojis, B.L., Schmahl, G.P., Ballantine, D.L., McKenna, J.E. 1992. Transplanting organisms on a damaged reef at Pulaski Shoal, Ft. Jefferson National Monument, Dry Tortugas, FL. (abstract). *Bull. Mar. Sci.* 54:1087–1088.
14. Miller, S.L., McFall, G.B., Hulbert, A.W. 1993. Guidelines and Recommendations for Coral Reef Restoration in the Florida Keys National Marine Sanctuary, workshop report. NOAA, p. 38.
15. Dustan, P. 1977. Besieged reefs of the Florida Keys. *Nat. His.* 86:73–76.

16. Hudson, J.H., Goodwin, W.B. 2001. Assessment of vessel grounding injury to coral reef and seagrass habitats in the Florida Keys National Marine Sanctuary, Florida: protocols and methods. *Bull. Mar. Sci.* 69:509–516.
17. Jaap, W.C. 1999. Coral Reefs. Presentation to the Florida Keys Carrying Capacity Study Ecosystems Workshop, Marathon, FL, July 7–8, 1999.
18. Tilmant, J.T. 1987. Impacts of recreational activities on coral reefs. In: Salvat, B. (ed.). *Human Impacts of Recreational Activities on Coral Reefs: Facts and Recommendations.* Antenne Museum E.P.H.E. French Polynesia: pp. 195–209.
19. Tilmant, J.T., Schmale, G.P. 1981. A comparative analysis of coral damage on recreationally used reefs within Biscayne National Park, Florida. In: *Proc. 4th Int. Coral Reef Symp.* Manila, 1:187–192.
20. Voss, G. 1983. An Environmental Assessment of the John Pennekamp Coral Reef State Park and the Key Largo Coral Reef Marine Sanctuary (unpublished report).
21. Marszalek, D., Babashoff, G., Noel, M., Worley, P. 1977. Reef distribution in south Florida. *Proc. 3rd Int. Coral Reef Symp.,* Miami, FL. 2:233–230.
22. Jaap, W.C., Hallock, P. 1990. Coral reefs. In Myers, R.N. and Ewel, J.J. (ed.). *Ecosystems of Florida.* University of Central Florida Press: pp. 574–616.
23. Humann, P. 1993. *Reef Coral Identification.* Paramount Miller Graphics, Inc., Jacksonville, FL.
24. Hudson, H. 1981. Growth Rates in *M. annularis*: a record of environmental change in Key Largo Coral Reef Marine Sanctuary, FL, *Bull. Mar. Sci.,* 31:444–459.
25. Ginsburg, R.N., Shinn, E.A. 1993. Preferential distribution of reefs in the Florida reef tract: the past is key to the present. In: Ginsburg, R.N. (ed.). *Global Aspects of Coral Reefs, Health Hazards and History.* University of Miami, FL, pp. 21–26.
26. International Marine. 1996. *Tide Tables: High and Low Water Predictions, East Coast of North and South America.* Camden, ME.
27. International Marine. 1997. *Tide Tables: High and Low Water Predictions, East Coast of North and South America.* Camden, ME.
28. Boating Research Center. 1994. *Atlas of Boats, Florida, 1994.* University of Miami Rosenstiel School of Marine and Atmospheric Science.
29. Weil, E., Knowlton, N., 1994. A multi-character analysis of the Caribbean coral *Montastraea annularis* and its two sibling species, *M. faveolata* and *M. franksi. Bull. Mar. Sci.* 55:151–175.
30. Knowlton, N., Mate, J.L., Guzman, H.M., Rowan, R., Jara, J. 1997. Direct evidence for reproductive isolation among three species of the *Montastraea annularis* complex in Central America (Panama and Honduras). *Mar. Bio.* 127:705–711.
31. Fong, P., Lirman, D. 1995. Hurricanes cause population expansion of the branching coral *Acropora palmata* (Scleractinia): wound healing and growth patterns of asexual recruits. *Mar. Ecol.* 16:317–335.
32. Highsmith, R.C. 1982. Reproduction by fragmentation in corals. *Mar. Ecol.* (Prog. Ser.); vol. 7, no. 2: pp. 207–266.
33. Deaton, A.S., Duquesnel, J.G. 1996. Marine Research and Resource Monitoring in John Pennekamp Coral Reef State Park, 1995 Update Report, Update to Section I, Part C. Boat Grounding Assessments. Florida Dept. of Environmental Protection, Division of Recreation and Parks, District 5 Administration, pp. 1–12.
34. Skinner, R.H., Deaton, A.S., Duquesenel, J.G. 1993. Marine Research and Resource Monitoring in John Pennekamp Coral Reef State Park. Florida Dept. of Environmental Protection (eds.), Division of Recreation and Parks, Region VII Administration: pp. 9–35.
35. Bak, R., Brouns, J., Heys, F. 1977. Regeneration and aspects of spatial competition in the scleractinian corals *Agaricia agaricites* and *Montastrea annularis.* In: *Proc. 3rd Int. Coral Reef Symp.* Miami, FL, pp. 143–149.
36. Meesters, E.H. 1995. Effects of coral bleaching on tissue regeneration potential and colony survival. In Meesters, E.H.: *The Function of Damage and Regeneration in the Ecology of Reef-Building Corals (Scleractinia).* Netherlands Institute for Sea Research (thesis publication): pp. 27–42.
37. Goreau, T.J., Macfarlane, A.H. 1990. Reduced growth rate of *Montastrea annularis* following the 1987–1988 coral bleaching event. *Coral Reefs* 8:211–215.
38. Szmant, A.M., Gassman, N.J. 1990. The effects of prolonged bleaching on the tissue biomass and reproduction of the reef coral *Montastraea annularis. Coral Reefs* 8:217–224.
39. Glynn, P.W. 1990. Coral mortality and disturbances to coral reefs in the tropical eastern Pacific. In Glynn, P.W. (ed.). *Global Ecological Consequences of the 1982–83 El Nino–Southern Oscillation.*

40. Boating Research Center. 1994. *Atlas of Boats, Dade County, Florida, 1993.* University of Miami Rosenstiel School of Marine and Atmospheric Science.
41. Florida Fish and Wildlife Conservation Commission. 2004. 2003 Boating Accident Statistical Report. Retrieved October 2004. Available online at: http://www.floridaconservation.org/law/boating/2003stats/2003StatBook1.pdf.

3 Coral Reef Restoration: An Overview

Beth Zimmer

CONTENTS

3.1 INTRODUCTION

Disturbances in coral reef environments have occurred since the evolution of corals and throughout geologic time to the present. Prior to human existence, natural impacts such as those resulting from storm events, temperature variations, and ultraviolet light exposure had periodic detrimental impacts on coral reefs. Anthropogenic impacts to corals have occurred throughout human history, ranging from collection in early history to vessel groundings, anchor damage, blast fishing, coral mining, dredging, coastal development, water quality degradation, recreation, and others in recent history. Beginning in the late 1970s, coral reef scientists became alarmed by the rapid, widespread decline in coral reefs around the world resulting from a variety of causes, including coral disease, coral bleaching, and anthropogenic impacts.[1–5] The time required for the natural recovery of a disturbed reef is usually on the order of decades.[6–14] However, after a severe disturbance that significantly injures reef structure, the reef can require centuries to recover[15–17] or never effectively recover on a human time scale.[14,18–22]

The global decline of reefs is of particular concern to the scientific community because the general consensus among researchers is that coral reef accretion is currently outpaced by the rate of human-induced destruction.[11,14,23,24] This consensus has inevitably elicited attempts to shift the balance by restoring damaged coral reefs. Coral reef restoration projects and studies have been implemented to prevent injury to corals that would be adversely impacted by planned activity, to hasten coral reef recovery from anthropogenic or natural impacts, to enhance fisheries habitat, and to enhance the aesthetic appearance of reefs for tourism. Throughout this chapter, the term "restoration" will be used as a general term to encompass the restoration, rehabilitation, and creation of coral reefs.

As a science, coral reef restoration faces a variety of challenges including the dire state of coral reefs around the globe, the wide array of disturbances and subsequent ecological responses, and the complexity of the coral reef ecosystem.[25] The science of coral reef restoration is in its infancy and lags far behind related terrestrial and wetland sciences.[14,26–31] Even seagrass restoration science is more highly developed.[32,33] During the initial research for this chapter, it became quite evident that while research relating to restoration techniques is available, a significant lack of published, peer-reviewed literature on actual restoration projects exists. Many restoration projects have been designed and carried out in an *ad hoc* manner, without long-term monitoring for success. Data from unpublished studies is available in the form of gray literature or merely article abstracts. This chapter outlines the currently available restoration techniques and identifies additional research needs in coral reef restoration science.

3.2 RESTORATION TECHNIQUES

A variety of restoration techniques have been explored thus far, including indirect action, reef repair, transplantation, and the installation of artificial reefs. These techniques may focus on the organismal scale by enhancing or restoring lost corals and associated biota (e.g., transplantation, artificial reefs) or on a broader ecological scale by restoring the structure of the reef itself and/or attempting to maximize natural recruitment (e.g., indirect action, reef repair). Available information will be summarized for each restoration technique, including a brief description of the technique and examples of restoration projects that have implemented it. In a restoration plan, techniques have been used individually or in combination.

3.2.1 Indirect Action

The simplest and most essential technique that can be applied to restore a coral reef is to eliminate the source of anthropogenic disturbance(s) causing detrimental impacts to a reef. Such anthropogenic factors might include nutrient loading, anthropogenically induced runoff and sedimentation, water discharges, or frequent injury from vessel groundings. Reefs that are undergoing chronic disturbances will not recover naturally,[10] and failure to remove the source of the chronic disturbance(s) will render other restoration efforts futile.[34–37] Therefore, when applicable, this restoration technique should be applied to all restoration projects; that is, anthropogenic sources of disturbance must be eliminated or at least reduced to a sustainable level in order to achieve "restoration." The following are examples of restoration projects that have implemented indirect action:

- Sewage outfalls were diverted to reduce nutrient loadings in Hawaii.[38]
- Activities causing land-based erosion were discontinued to reduce sedimentation in Hawaii.[39]
- Thermal power plant effluent was diverted into deeper offshore waters in Hawaii.[40]
- Coastal discharges of silt-laden water and bagasse (fibrous residue from pressing sugar cane) were terminated in Hawaii.[41,42]
- Vessel deterrent devices (radar response transmitters) were installed on navigational aids in the Florida Keys as mitigation for the *Containership Houston* grounding.[43]

Although this technique is critical for a successful restoration project, one disadvantage is that, by itself, indirect action may not result in complete restoration of the damaged reef. If the reef has experienced very severe damage (e.g., large-vessel grounding, coral mining, blast fishing), it may require centuries for natural recovery or never fully recover. For this reason, indirect action may be used in combination with other restoration techniques.[44,45] In addition, the cost of this technique may be prohibitive,[46] depending on the individual circumstances (materials, equipment, labor, etc.).

3.2.2 Reef Repair

Reef repair may consist of emergency triage, restoring the structural integrity of the reef framework, and/or restoring topographic complexity. These techniques attempt to minimize additional damage following a disturbance event and enhance natural recruitment. The vast majority of comprehensive restoration projects have included some component of reef repair.

3.2.2.1 Triage

Triage may involve one or more of the following technique components: careful vessel salvage following a grounding event, stabilization or removal of loose sediment and/or coral rubble, removal of debris (foreign objects), and the recovery, storage and/or reattachment of dislodged corals, sponges, and other reef biota. Triage is frequently used in conjunction with other restoration techniques.

After a vessel grounding, it is vital that the vessel be carefully removed from the injury site. Botched vessel salvage efforts lead to additional reef damage, such as was the case with the *M/V Wellwood*[47] and *Caribe Cay* ferry.[48] Procedures used to avoid collateral injury during vessel salvage include off-loading fuel and cargo to gain buoyancy and utilizing floating lines for towing vessels.[49]

Coral rubble and sediment resulting from a disturbance event can increase secondary damage to the reef from resuspension during storm events.[50–52] In addition, the presence of unstable substrate or a layer of fine sediment may delay reef recovery by inhibiting the settlement and growth of corals.[22,51,53–55] Removal or stabilization of the loose rubble reduces secondary damage, increases substrate stability, enhances recruitment, and increases habitat complexity.[50,54,56] Rubble and/or debris (vessel fragments, foreign objects, etc.) may be removed from the damage site using lift bags, lift vacuums, clam dredges, or suction dredges.[49,51] Rubble can be stabilized in place using adhesive materials such as epoxy or overlay structures such as limestone boulders or concrete mats. The potential exists for sponges to aid in rubble consolidation by temporarily stabilizing rubble until carbonate-secreting organisms permanently bind the rubble to the reef framework.[56]

Following disturbance events such as anchor damage, vessel groundings, and dredging damage, triage may involve the emergency recovery of dislodged corals and surviving fragments.[51] Dislodged corals in shallow, high-energy environments may be subject to mortality from inversion, burial, or displacement.[49,57] Dislodged corals that are recovered can be immediately righted and/or reattached or can be stored in a similar, safe environment until reattachment is feasible.[49,51] Reattachment methods might include the use of epoxy, cement, expansion anchors and threaded rod, wire and nails, bamboo skewers,[58] and plastic wire ties,[43,57,59–61] or any of the methods utilized for transplantation (see Section 3.2.3). The following restoration projects have employed triage methodologies:

- Toppled corals were righted, coral fragments were stabilized, and debris was removed following vessel mooring chain damage in Guam.[62]
- Toppled corals were righted following anchor damage in St. John, U.S. Virgin Islands.[43]

- Dislodged corals were reattached after damage from a fiber optic cable installation in Florida.[63]
- Dislodged corals were salvaged and reattached, and rubble was stabilized with epoxy following a vessel grounding (*Containership Houston*) in the Florida Keys.[45]
- Corals dislodged from anchor and grounding damage were reattached following a vessel grounding (*C/V Hind*) in Florida.[64]
- Dislodged corals were salvaged and reattached following a vessel grounding (*M/V Firat*) in Florida.[65,66]
- Fragmented colonies were reattached following a vessel grounding (*Fortuna Reefer*) in Puerto Rico.[60,61]
- Rubble was removed, topographic relief was reestablished, and corals were reattached following a cruise ship (*Maasdam*) impact in Grand Cayman Island.[67]
- Hull paint, rubble, and debris were removed following a vessel grounding (*M/V Horizon*) in St. Maarten, Netherlands Antilles.[68]
- Metallic vessel debris was removed following a vessel grounding (*Jin Shiang Fa*) in Rose Atoll, American Samoa.[69]
- Rubble and sediment were removed and dislodged corals were reattached following a vessel grounding (*M/V Wellwood*) in the Florida Keys.[70]
- Concrete mats were deployed to stabilize rubble at a coral mining site in the Maldives[20,52] and over the *Containership Houston* grounding site in the Florida Keys.[43,45]
- Corals were salvaged and reattached, and rubble and vessel debris was removed following a vessel grounding (*R/V Columbus Iselin*) in the Florida Keys.[71]
- Debris and radioactive material were removed from Enewetak Atoll following nuclear testing (refer to Chapter 15 for additional details).

3.2.2.2 Restoring Structural Integrity

A catastrophic disturbance can drastically damage the structural integrity of the reef framework, creating fractures, fissures, gouges, or craters in the reef limestone. This type of damage is often the case in large-vessel groundings[51,70] and blast fishing.[72] Repairing structural framework damage can prevent further structural deterioration and will minimize or avoid the potential for secondary damage produced from rubble and sand.[51] Techniques for restoring structural integrity have been used mostly for large-vessel groundings. Examples of projects where the reef framework was repaired include the following:

- Fractured reef framework was grouted with Portland cement and molding plaster for stabilization following a vessel grounding (*M/V Wellwood*) in the Florida Keys.[70]
- Craters that threatened the structural stability of the reef, created by the grounding of the *M/V Alec Owen Maitland* in the Florida Keys, were repaired using gravel fill and concrete armoring units.[73,74]
- Craters created by the grounding of the *M/V Elpis* in the Florida Keys were repaired by filling them with rock and rubble from rubble berms caused by the vessel impact, along with boulders and sand transported to site.[75]
- Gouges and cracks in the reef framework, created by the grounding of the *R/V Columbus Iselin* in the Florida Keys, were repaired using limestone boulders stabilized with a tremie pour of concrete and steel bars.[71]
- Concrete mats were installed to prevent additional loss of reef structure on a grounding site (*Containership Houston*) in Florida.[43,45]

It is important to note that careful consideration should be given to the substrate material chosen for restoration of the structural integrity and complexity of a damaged reef. The function of an artificial substrate depends on:

1. Its structural characteristics (composition, surface, design, and stability)
2. The environmental characteristics (temperature, light, sediment, surrounding biota, hydrodynamics, depth, and temporal effects)

These factors are discussed in detail in Spieler et al.[76]

3.2.2.3 Restoring Topographic Complexity

A loss of topographic complexity is often the case following large-vessel groundings,[51,54,70,74,77] coral mining,[52] blast fishing,[72] and major dredging accidents.[51] The reestablishment of topographic complexity and appropriate substrate on a damaged reef is a major aspect of restoration, as these factors affect both coral recruitment and fish abundance.[51,78–80] Coral larvae require specific substrate and environmental conditions for settlement (see Petersen and Tollrian[81] for references). Surfaces that have a higher spatial complexity and rugosity are more suitable for recruitment and survival of biota.[74,82–87] Moreover, coral cover is directly related to fish abundance,[88–90] and topographic complexity shows a positive correlation with reef fish diversity and abundance (see Spieler et al.[76] for references). This positive relationship occurs because topographic complexity and epifauna (corals, alcyonarians, sponges, etc.) provide shelter and food resources for reef fish.[88,91] Lack of herbivorous fish may inhibit the recovery of a reef because coral recruits depend upon herbivory to reduce algal cover.[11,79,92,93] In addition, failure to restore topographic complexity on injured, exposed reef areas could lead to exacerbation of damages by other disturbances, such as storm events.[71] Without restoration, structural destruction resulting from major disturbance events can lead to shifts in community structure.[14,94–96] Examples of projects where a reef's topographic complexity was reestablished include the following:

- Large boulders were installed atop a cruise ship injury site in Grand Cayman, British Virgin Islands[67]
- Large limestone boulders were installed atop concrete mats to provide stability and topographic complexity on a grounding site (*M/V Houston*) in the Florida Keys.[43,45]
- Concrete modular units were placed in areas where dredging had reduced the topographic complexity of the reef in Miami-Dade County, Florida. The modular units were installed to attract epibenthic and cryptic communities that would support fish and invertebrates.[51]

Selection of a substrate appropriate for coral recruitment is vital when restoring topographic complexity.[97] The most common materials employed to reestablish three-dimensional relief are limestone and concrete. Researchers have established that limestone and concrete are appropriate materials for coral recruitment.[49,51,74,98–100] When concrete is chosen for a restoration project, appropriate surface rugosity may be accomplished by:

- The inclusion of rocks in the concrete surface[87]
- The removal of material from the surface as the concrete sets by applying a high water spray, chiseling irregular forms in the concrete, drilling holes, etc.[87,101]
- The addition of material to the concrete surface by spraying additional concrete, adding coarse sand to cement, or attaching items such as bars, plates, or concrete[101,102]
- The creation of a layered, brick-and-mortar style structure using rocks and cement[87]

The potential use of larval attractants to enhance recruitment to a restoration area has been examined. Such attractants include calcium carbonate,[103] coralline algae,[104,105] bacteria from coralline algae,[106] red algae,[107] and neuropeptides from cnidarians.[108,109]

Although reef repair techniques provide numerous benefits to enhance recovery, the disadvantages of these techniques are the intense labor required[43,46,61] and the cost of materials and equipment.[73,75,110]

3.2.3 Transplantation

Coral transplantation was first employed during growth rate studies in the early 1900s[111] and is currently one of the most widely utilized and researched techniques for coral restoration. The principal goal of this restoration technique is to accelerate the damaged reef's natural recovery rate by bypassing the coral's slow-growth, high-mortality life-cycle stage and rapidly improving the impacted reef's coral cover, biodiversity, and three-dimensional topographic complexity.[20,55,110,112,113] In addition to restoring coral cover on a damaged reef, the additional benefits of transplantation may include:

1. Immediate enhancement of coral cover and, potentially, coral diversity[110]
2. Enhancement of coral recruitment to the reef through:
    a. The introduction of reproductive adult corals,[51,110] although whether planulae from transplanted corals would potentially recruit to the damaged reef has been questioned[114,115]
    b. Asexual reproduction through fragmentation[110,112]
    c. The potential for existing transplants to stimulate settlement[110]
3. Enhancement of survival for locally rare species[110]
4. Addition of corals to areas that are recruitment limited because of poor larval supply or high postsettlement larval mortality[110]
5. Habitat enhancement for other reef-dwelling organisms by providing shelter and increasing habitat complexity[54,116–118]
6. Enhancement of the aesthetic value of a damaged reef area, which is important for tourism[119]

Coral transplantation has been attempted in locales around the globe and for every type of coral reef disturbance. Examples of transplantation projects include:

- Avoiding and minimizing impacts from coastal development projects (i.e., transplanting individual corals or entire portions of a reef) in Hawaii,[120–123] Mexico,[124] Guam,[125] Saipan,[126,127] Singapore,[128] Japan,[129] Tutuila, American Samoa,[130] and Palau[131]
- Avoiding and minimizing coral loss from submarine cable installation and replacement in Guam[132–135]
- Avoiding and minimizing coral loss from outfall pipe repairs in Florida[136]
- Avoiding coral loss in Guam from pollution[137]
- Rehabilitation of reefs following dynamite fishing in the Philippines,[17,138–140] Indonesia,[22] and the Solomon Islands[118]
- Rehabilitation of reefs following coral mining in the Maldive Islands[20] and Solomon Islands[118]
- Rehabilitation of reefs following coral mortality from thermal effluent in Guam[34]
- Rehabilitation of reefs following mortality caused by sewage pollution in Hawaii[44,116]
- Attempting to accelerate reef recovery following a submarine grounding (*USS Memphis*) in Florida[141] and vessel groundings in the Florida Keys[43,54,70,74] and Grand Cayman, British Virgin Islands[77]
- Rehabilitation of reefs damaged by tourism in Eliat, Israel[142]
- Enhancement of aesthetics for tourism by transplanting large coral heads into the Gulf of Aqaba[143]
- Rehabilitation of reefs damaged by thermal stress and algal blooms in Costa Rica, Panama, and Colombia[13,144,145]
- Rehabilitation of reefs damaged by crown-of-thorns starfish[146]

Coral branches, colony fragments, entire colonies, and settled planulae may be transplanted. Various techniques for attachment of transplants have been attempted, including epoxy,[43,124,126,132,140,147] Portland cement,[136] Portland cement mixed with molding plaster,[130,148] Portland mortar mix,[149] terra-cotta tiles,[34,137] plastic wire ties,[87,150] rubber-coated wire,[116] steel stakes/bars,[123,144,150] corrosion-resistant hardware,[43] large concrete mats placed over the substrate,[20] transplants wedged into crevices,[151] and others.

The disadvantages of transplantation include:

1. The intensive labor and cost required (i.e., the extensive time underwater removing and transporting colonies and the expensive materials and equipment)[17,46,110,124,151–153]
2. The impacts on donor colonies and populations[20,55,110,112,154–157]
3. The potential for increased mortality rates in transplants[110,112,137,138,140,146]
4. The potential for decreased growth rates in transplants[20,137,158]
5. The potential for dislodgement from the point of attachment due to wave action[20,34,137,146,150,159]
6. The potential for reduced fecundity of transplants resulting from the stress of removal, transport, and transplantation[160]

3.2.3.1 Alternative Transplantation Techniques

3.2.3.1.1 Transplantation without Attachment

Because the typical methods of affixing corals to the substrate require extensive labor and expense, and because the majority of countries that have the greatest need for coral restoration do not have the resources available for such an endeavor, methods of transplantation that would require less cost and labor have been examined. Some studies have focused on transplantation without attachment of the transplanted corals, a method that mimics asexual fragmentation.[22,34,55,113,118,137,143–146,150,161,162] The advantage of this technique is that it does not require the use of SCUBA diving or expensive materials.[55] However, unattached fragments could be displaced and/or subject to mortality from storm events or wave action,[137,146] and this technique cannot be successfully used for high-energy environments.[55]

3.2.3.1.2 Coral Gardening and Coral Seeding

To minimize impacts to donor corals and populations, alternative sources of transplant material have been examined. Possible sources include the collection of juveniles from high-risk, extremely shallow reef environments,[163] the collection of fast-growing "weedy" corals that are outcompeting massive coral colonies,[118] and the use of coral "gardening" to supply transplants.

Coral "gardening" is the mariculture of corals for use in coral restoration.[142,164,165] The concept of coral gardening is similar to that of silviculture,[165] where coral recruits are raised in nurseries (*in situ* or *ex situ*) and then transplanted to restoration sites.[142] Coral gardening studies have shown promise for rehabilitating denuded reef areas.[142,156,166] The advantage of this technique is that it avoids the adverse impacts to donor populations that occur during direct coral transplantation.[142,156,166,167] In addition, the introduced corals may provide genetic diversity to the damaged reef area.[164] The disadvantages of this technique include the lengthy time frame required to establish a viable nursery capable of supplying transplants[167] and the possibility that corals raised *in situ* may be impacted and/or destroyed by environmental disturbances (e.g., storm events, temperature extremes, disease, etc.).

Coral "seeding" has also been proposed as a method of enhancing recruitment for restoration. This technique involves collecting coral larvae from the field, or spawn that has been collected in the field and cultured in the laboratory, and settling these larvae on reef substrate.[166,168–171] While this technique would avoid the damage and removal of healthy donor corals, an efficient and successful methodology has not been developed at this point.

3.2.3.2 Transplantation Questions and Research Needs

Because transplantation is one of the most widely employed and researched restoration techniques, questions and research needs regarding the technique have been noted in the literature. Researchers clearly have not ascertained which coral reproductive strategies are most appropriate for specific transplantation scenarios. Some researchers suggest that branching coral species are preferable for use in transplantation in low-energy areas because they are rapid growers and can quickly increase coral cover and generate conditions that are favorable for recruits.[55,161] Others theorize that slow-growing mounding corals may be more appropriate for transplantation because they are slower to recruit, have longer life spans, and tend to survive severe storm events better than branching species.[110,118] Some suggest that the best corals for transplantation are the massive broadcasters, which have a high survival rate once a specific size is reached,[172] while others suggest that hermaphroditic brooding corals are appropriate for transplantation techniques.[142,173]

Although a fair number of coral species have been used for transplantation research and projects,[25] additional research is needed to define the suitability of a particular species for transplantation with respect to a variety of environmental conditions (i.e., the effects of depth, wave action, water quality, season, and substrate on the effectiveness of transplanting a coral species).[55] In addition to this, the minimum size of a coral transplant that will allow for a 100% survival rate must be determined.[25] Further research is also needed to ascertain the effects of transplanting corals into habitats that are different from the donor site[55] and to expand knowledge on transplant methodology (e.g., species tolerance for transport and transplantation).

3.2.4 Artificial Reefs

In certain situations, restoring the reef via indirect action, reef repair, and/or transplantation is not viable. In such cases, the installation of artificial reefs may be considered. The general goals of an artificial reef installation in a restoration project are to:[76,174,175]

1. Mitigate for reefs damaged by anthropogenic activity
2. Alter currents
3. Restrain rubble
4. Restore habitat by providing substrate and refuge for fish, coral, and other reef organisms
5. Conserve biodiversity and enhance the reestablishment of damaged reefs
6. Provide aesthetically pleasing structure(s) for tourism

The goal of an artificial reef installation that concentrates on restoring corals is to establish a stable, wave-resistant, fixed substrate that provides refuge, where corals can recruit and/or be transplanted.[76] (See Chapter 7.) The refuge provided by such an artificial reef enhances fish and invertebrate communities.[76] The disadvantages of artificial reefs include:

1. The potential for the loss of corals, fish, and other biota by relocation from natural reefs to the artificial structure[46]
2. The potential for exacerbation of overfishing on artificial reefs, as they concentrate fish[46,176]

A vast array of items has been utilized to fabricate artificial reefs (e.g., tires, plastic, metal, wood, fiberglass, polyvinyl chloride (PVC), boulders), although the most commonly used materials for artificial reef restoration projects are concrete and limestone rock, oftentimes in formed modules.[76] (See Chapter 7.) In addition, there has been recent research on the creation of artificial reefs through electrolytic precipitation of minerals, $Mg(OH)_2$ (brucite) and $CaCO_3$ (aragonite), onto conductive metal.[177–181] Corals can then be transplanted onto the mineralized structures.

Artificial structures have been employed in a wide array of restoration projects. Examples include:

- A derelict vessel was sunk to provide additional habitat and for dive tourism in Rota, Commonwealth of the Northern Mariana Islands. (See Chapter 15.)
- Corals and tridacnid clams were transplanted onto hollow, igloo-shaped, stone and cement "fish houses" placed in tide pools on the reef flats in Fiji to enhance fisheries resources.[118]
- Limestone boulders and various concrete modules (tetrahedrons, Reef Balls™, A-Jacks™, and Warren Modules) were deployed as mitigation following the grounding of a submarine (*USS Memphis*) in Florida.[103,141]
- Department of Environmental Resources Management (DERM) modules (concrete base with embedded limestone boulders) were deployed as mitigation following a fiber-optic cable installation in Florida.[141]
- Limestone boulders were deployed as mitigation following a beach nourishment project in Florida.[141]
- Three different concrete modular unit designs were installed off Miami-Dade County, Florida, as mitigation for injuries caused by dredging. The modules were designed to enhance habitat for fish and motile invertebrates and provide refuge for associated organisms.[51,182]
- Artificial reefs constructed of PVC plates were installed off Eliat, Israel to relieve diving pressure on natural reefs.[183]

It is important to carefully consider both the chosen artificial substrate's structural characteristics and the environmental factors of the restoration area, as both will work together to determine how the substrate functions.[76] The chosen material could affect the benthic organisms that can inhabit the substrate.[102,184] The substrate's structural characteristics would include composition, texture, chemistry, color, design, and stability, while the environmental factors to consider would include temperature, light, sediment, surrounding biota, hydrodynamics, depth, and temporal effects.[76] Other aspects to consider would be cost (construction and labor)[76,185] and aesthetic value.[76]

3.3 COST OF CORAL REEF RESTORATION

One of the most challenging aspects of coral reef restoration is the associated cost, which depends on a variety of factors, including materials and labor. The factors that will affect the cost of restoration include the restoration plan's site location, chosen restoration technique(s), site conditions, and the availability of funds.[186] Detailed costs for restoration projects are not generally available in the literature.[36,186] Spurgeon and Lindahl[36] compared the costs of five coral restoration projects that varied in technique (triage, reef repair, transplantation, and/or artificial reefs) and were located in four different countries (United States, Maldives, Australia, and Tanzania). Costs were found to vary tremendously between projects, ranging from approximately US$13,000 per hectare for low-tech methods with local labor to more than US$100 million per hectare for extensive restoration work. Jokiel and Naughton[187] found that many previous restoration projects may have been more cost-effective by concentrating on prevention, preservation, and protection of the resources.

Cost is a major factor in the selection of a suitable restoration technique. A country's economic resources will dictate the restoration options that are available. Hence, the most expensive and comprehensive research projects have been conducted in developed countries (United States and Japan).

Decision-making tools may aid with the selection of a course of action (whether or not to proceed with restoration) to determine the best use of available funds.[186] Benefit-cost analysis assesses the ratio of benefits to costs for a particular course of action and would assist in the selection of the most cost-effective use of funds and maximize the benefits of the chosen path by developing the details of the selected plan.[186] Least-cost analysis identifies the most inexpensive method for realizing a specific environmental goal.[188] Cost-effectiveness analysis may be used to identify the least and most cost-effective methods for realizing a specific environmental goal while comparing various levels of improvement.[188,189] Multi-criteria analysis assists with the selection of a course of action by assigning scores, weights, and priorities to objective criteria without requiring monetary estimates.[190,191] Habitat-equivalency analysis is used to determine the appropriate compensation for interim loss of natural resources.[192] Refer to Spurgeon and Lindahl[36] for a thorough explanation of the benefits and disadvantages of these tools.

3.4 SUCCESS OF CORAL RESTORATION TECHNIQUES

A definitive definition of coral reef restoration "success" has not yet been developed.[14] The majority of existing restoration projects have been oriented toward mitigative compliance success rather than functional and structural attributes.[14] In addition, the overall effectiveness of coral restoration techniques is not clear, as few studies have carried out thorough monitoring programs over a substantial time span. A quantitative comparison of all reef restoration projects/methods has not yet been conducted. Restoration projects to date have varied widely in habitat structure, environmental conditions, method, and species examined. For these reasons, it is difficult to compare the effectiveness of restoration projects. Nevertheless, we must attempt to collect and utilize the knowledge gained from the projects and research that have been conducted thus far.[14]

As has been shown to be the case with related sciences such as seagrass and wetland restoration, site selection is key for coral reef restoration projects.[110,137,193] Failure to select an appropriate site could lead to an unsuccessful restoration project.[110,137] As is the case with seagrass restoration, if a reef has not existed in a particular site over geologic time, there is an underlying reason,[37] and it is therefore not an appropriate site to establish a reef via transplantation or artificial reefs.

Additionally, a successful restoration technique must be appropriate for the selected coral species,[20,140] the environmental conditions of the site, and the economic resources of the country or region.[112] The following sections review the trends in effectiveness for the individual coral reef restoration techniques.

3.4.1 INDIRECT ACTION

Indirect action may be the most essential, successful, and cost-effective technique for coral reef restoration. In cases where the reef framework and topographic complexity remain intact, removal or prevention of a disturbance should allow for natural recovery and recruitment.[35] It has been suggested that this method would be particularly effective on coral reefs with high recruitment rates, such as those in the central and western South Pacific and the Great Barrier Reef. Such reefs have an abundant supply of coral planulae and should be able to recover naturally once the disturbance is removed.[113,194]

3.4.2 REEF REPAIR

Little published experimental or hypothesis-based research has been conducted for reef repair techniques. It has been suggested that reattachment of surviving, dislodged colonies is particularly important in areas where coral recruitment is limited, such as the Caribbean and western Atlantic.[195] In addition, we know that failure to restore topographic complexity following a major disturbance event can lead to shifts in community structure.[14,94–96]

3.4.3 TRANSPLANTATION

As transplantation is one of the most widely used and researched techniques for coral restoration, several trends have emerged with respect to its effectiveness. With regard to site selection, it appears that transplants (both attached and unattached) are more likely to survive when transplanted into low-energy, sheltered environments with good water quality.[20,34,55,113,144–146,150,193,196–198] Selection of transplantation environments that are similar to the donor site appears to be most suitable.[138,199–201] In addition, this restoration technique would be most appropriate on reefs that are recruitment limited, such as those of the Caribbean and western Atlantic.[194,195] However, transplantation may not a viable restoration technique for scenarios where there is a large area of reef destruction with a donor population that could not support the project.[51]

Selecting the appropriate method of transplantation and the appropriate coral species is vital for a successful transplantation project.[20,140,142] Transplantation is most successful after careful removal, transport, and placement of corals.[25,34,137] The studies evaluating transplantation of unattached corals have had varying degrees of success.[55] These lower cost/labor methods of transplantation may be most appropriate on reefs that were previously healthy (i.e., reefs with sufficient larval supply), but where unsuitable substrate (unconsolidated rubble and sediment) now prevents larval recruitment and survival (e.g., areas impacted by blast fishing, dredging, or coral mining).[22,55] In addition, coral gardening would be particularly valuable in areas with low recruitment rates that lack significant donor populations, such as the Florida Keys.[51]

3.4.4 Artificial Reefs

Artificial reefs at least partially composed of limestone are arguably the most suitable for restoration based on limestone's similarity to the natural reef framework.[76] Concrete has also been widely utilized in artificial reef creation. Both limestone and concrete are suitable for bioerosion.[76]

For a successful artificial reef project, it is important to carefully evaluate the local conditions when selecting an artificial reef design for a specific restoration site location.[202] Sherman et al.[203] showed that a particular artificial reef design will not produce the same results (algae, fish, and invertebrate abundance and species richness) at different locations.

3.5 FUTURE OF CORAL RESTORATION RESEARCH

As coral reef restoration science is in its infancy, a great deal of additional research will be needed to develop restoration protocols and a more thorough understanding of the science. As we know from related restoration fields, restoration-specific research is needed to achieve success,[51,204] and research will help to develop more cost-efficient and effective management strategies. (See Chapter 15.) Future restoration efforts should be hypothesis-driven and anchored on scientific principles[14,51,63,205–207] so that managers can establish scientific restoration protocols, develop appropriate success criteria, and determine the success of a restoration project.[63]

Long-term monitoring of restoration projects is essential to gauging their success.[14,51,63,204,205] Besides assessing the success of the project, monitoring also allows for adaptive management, the observation of status and trends essential for the development of methodologies, and the opportunity to improve future restoration efforts by correcting problems with a restoration technique.[14,204,208,209] Chapter 20 describes the necessary components of a well-designed monitoring program.

Most restoration programs lack predisturbance and postrestoration functional analysis, and thus goals and success parameters have been established *ad hoc* and oriented toward mitigative compliance.[14] It is necessary to evaluate both functional and structural attributes of a restored reef.[14,205] The National Oceanographic and Atmospheric Administration (NOAA) compiled the following baseline list of structural and functional criteria that should be considered during design and monitoring of a coral reef restoration plan: topographic complexity, stable three-dimensional hard substrate, breakwater for oceanic swells, cryptic habitat, accretion of hard substrate, biomass production, availability of shelter, shading, identification of biological community structure, and benthic invertebrate and finfish utilization.[205] Functional and structural valuation (i.e., using a rating index to assign values to structural and functional criteria) is an established protocol in terrestrial and wetland restoration. Therefore, a standardized protocol for gauging coral reef restoration success using functional and structural valuation is greatly needed.

More research is needed to assess the costs and cost-effectiveness of coral restoration projects. Future studies should present comprehensive and detailed cost breakdowns in a universal costing framework.[36,186] In addition, financial decision-making tools such as benefit-cost analysis, least-cost analysis, cost-effectiveness analysis, multi-criteria analysis, and habitat-equivalency analysis should be utilized in restoration programs.

Some examples of needed additional development of improved restoration techniques include:

1. Further development of lower cost/labor techniques, which is particularly important in developing countries that do not have the financial means to support typical restoration techniques[186]
2. Further examination of the potential use of larval attractants for recruitment enhancement
3. Research into the function of artificial substrates
4. Determination of the scenarios under which transplantation and artificial reefs will enhance recruitment of corals, fish, and other biota
5. Researching aspects of specific artificial substrates, such as their ecological function, associated biotic assemblages, and interaction with the environment

3.6 SUMMARY

The goal of coral reef restoration is to overcome the factors inhibiting natural reef recovery following a disturbance event. Coral reef restoration projects and studies have been implemented to prevent injury to corals from planned anthropogenic activities and to enhance reef recovery, fisheries habitat, and aesthetics for tourism. The restoration techniques developed thus far include indirect action, reef repair, transplantation, and artificial reefs. Each of these techniques has unique advantages and limitations. The science of reef restoration is in its early stages, and many of the restoration projects that have been conducted to date were designed and implemented in an *ad hoc*, unscientific manner, without hypothesis-based testing or long-term monitoring for success. The development of such restoration projects is oriented around engineering, fiscal concerns, and/or compliance; however, the design and implementation of a restoration project should take into account a more rigorous set of scientific principles and hypothesis-based research. Precht[63] summarizes this idea when he states, "the underlying logic to successful restoration must be rooted in an integrated, multidisciplinary approach that includes engineering, geologic, biologic, aesthetic, and socioeconomic factors."

REFERENCES

1. Brown, B.E., Worldwide death of corals — natural cyclical events or man-made pollution, *Mar. Poll. Bull.*, 18, 9, 1987.
2. Williams, E.H., Goenaga, C., and Vicente, V., Mass bleachings on Atlantic coral reefs, *Science*, 238, 877, 1987.
3. Dight, I.J. and Scherl, L.M., The International Coral Reef Initiative (ICRI): Global priorities for the conservation and management of coral reefs and the need for partnerships, *Coral Reefs*, 16, S139, 1997.
4. Birkeland, C., Ed., *Life and Death of Coral Reefs*, Chapman and Hall, New York, 1997.
5. Bryant, D., Burke, L., McManus, J., and Spaulding, M., *Reefs at Risk*, World Resources Institute, New York, 1998.
6. Stoddart, D.R., Posthurricane changes on the British Honduras reefs and cays, *Nature*, 207, 589, 1965.
7. Grigg, R.W. and Maragos, J.E., Recolonization of hermatypic corals on submerged lava flows in Hawaii, *Ecology*, 55, 387, 1974.
8. Pearson, R.G., Recovery and recolonization of coral reefs, *Mar. Eco. Prog. Ser.*, 4, 105, 1981.

9. Sheppard, C., Coral populations on reef slopes and their major controls, *Mar. Eco. Prog. Ser.*, 7, 83, 1982.
10. Connell, J.H., Hughes, T.P., and Wallace, C.C., A 30-year study of coral abundance, recruitment, and disturbance at several scales in space and time, *Ecol. Monogr.*, 67, 461, 1997.
11. Hughes, T.P., Catastrophes, phase shifts, and large-scale degradation of a Caribbean coral reef, *Science*, 265, 1547, 1994.
12. Dulvy, N.K., Stanwell-Smith, D., Darwall, W.R.T., and Horill, C.J., Coral mining at Mafia Island, Tanzania: a management dilemma, *Ambio*, 24, 358, 1995.
13. Guzman, H.M., Large-scale restoration of eastern Pacific reefs: the need for understanding regional biological processes, in *Program and Abstracts, International Conference on Scientific Aspects of Coral Reef Assessment, Monitoring, and Restoration*, NCRI, Nova Southeastern University, Florida, 1999, 97.
14. Precht, W.F., Aronson, R.B., and Swanson, D.W., Improving scientific decision making in the restoration of ship-grounding sites on coral reefs, *Bull. Mar. Sci.*, 69, 1001, 2001.
15. Alcala, A.C. and Gomez, E.D., Recolonization and growth of hermatypic corals in dynamite blasted coral reefs in the Central Visayas, Philippines, in *Proceedings of the International Symposium on Marine Biogeography and Evolution in the Southern Hemisphere*, Auckland, New Zealand, 137, 645, 1979.
16. Curtis, C., Investigating reef recovery following a freighter grounding in the Key Largo National Marine Sanctuary, *Proc. 5th Int. Coral Reef Symp.*, Tahiti, 6, 471, 1985.
17. Yap, H.T., Licuanan, W.Y., and Gomez, E.D., Studies on coral recovery and coral transplantation in the northern Philippines: aspects relevant to management and conservation, in *Proceedings of the 1st Association of Southeast Asian Marine Scientists (ASEAMS) Symposium on Southeast Asian Marine Science and Environmental Protection*, Yap, H.T., Ed., UNEP Regional Seas Reports and Studies 116, United Nations Environment Programme, Nairobi, 1990.
18. Hedley, C., The natural destruction of a coral reef, *Transactions of the Royal Geographical Society of Australasia (Queensland). Reports of the Great Barrier Reef Committee*, 1, 35, 1925.
19. Stephenson, W., Endean, R., and Bennett, I., An ecological survey of the marine fauna of Low Isles, Queensland, *Australian Journal of Marine and Freshwater Research*, 9, 261, 1958.
20. Clark, S. and Edwards, A.J., Coral transplantation as an aid to reef rehabilitation: evaluation of a case study in the Maldive Islands, *Coral Reefs*, 14, 201, 1995.
21. Reigl, B. and Luke, K.E., Ecological parameters of dynamited reefs in the northern Red Sea and their relevance to reef rehabilitation, *Mar. Poll. Bull.*, 37, 488, 1998.
22. Fox, H.E., Caldwell, R.L., and Pet, J.S., Enhancing coral reef recovery after destructive fishing practices in Indonesia, in *Program and Abstracts, International Conference on Scientific Aspects of Coral Reef Assessment, Monitoring, and Restoration*, NCRI, Nova Southeastern University, Florida, 1999, 88.
23. Pratt, J.R., Artificial habitats and ecosystem restoration: managing for the future, *Bull. Mar. Sci.*, 55, 268, 1994.
24. Grigg, R.W. and Dollar, S.J., Natural and anthropogenic disturbance on coral reefs, in Dubinsky, Z., Ed., *Coral Reefs*, Elsevier Science Publishers, B.V., Amsterdam, 1990, chap. 17.
25. Omori, M. and Fujiwara, S., Eds., *Manual for Restoration and Remediation of Coral Reefs*, Nature Conservation Bureau, Ministry of Environment, Japan, 2004.
26. Zedler, J.B., The Ecology of Southern California Coastal Salt Marshes: A Community Profile, Report FWS/OBS-81/54, US Fish and Wildlife Service, Washington, D.C., 1984.
27. Kusler, J.A. and Kentula, M.E., Eds., *Wetland Creation and Restoration: the Status of the Science*, Island Press, Washington D.C., 1990.
28. Field, C.D., Ed., *Restoration of Mangrove Ecosystems*, International Society for Mangrove Ecosystems, Okinawa, Japan, 1996.
29. Allison, G.W., Lubchenko, J., and Carr, M.H., Marine reserves are necessary but not sufficient for marine conservation, *Ecol. Appl.*, 8, 79, 1998.
30. Keough, M.J. and Quinn, G.P., Legislative vs. practical protection of an intertidal shoreline in South eastern Australia, *Ecol. Appl.*, 10, 871, 2000.
31. Rose, K.A., Why are quantitative relationships between environmental quality and fish populations so illusive? *Ecol. Appl.*, 10, 367, 2000.
32. Fonseca, M.S., A Guide to Transplanting Seagrasses in the Gulf of Mexico, Texas A&M University Sea Grant College Program, Report TAMU-SG-94-601, College Station, TX, 1994.

33. Fonseca, M.S., Kenworthy, W.J., and Thayer, G.W., Guidelines for the conservation and restoration of seagrasses in the United States and adjacent waters, *NOAA Coastal Ocean Program Decision Analysis Series No. 12.*, NOAA Coastal Ocean Office, Silver Spring, MD, 1998.
34. Birkeland, C., Randall, R.H., and Grim, G., Three methods of coral transplantation for the purpose of reestablishing a coral community in the thermal effluent area of the Tanguisson Power Plant, *Univ. Guam Mar. Lab. Tech. Rep.* 60, 1979.
35. Naughton, J. and Jokiel, P.L., Coral reef mitigation and restoration techniques employed in the Pacific Islands: I. Overview, *Oceans MTS/IEEE Conference and Exhibition* 1, 306, 2001.
36. Spurgeon, J.P. and Lindahl, U., Economics of coral reef restoration, in Cesar, H.S.J., Ed., *Collected Essays on the Economics of Coral Reefs*, CORDIO, 125, 2000.
37. Yap, H.T., Coral reef "restoration" and coral transplantation, *Mar. Poll. Bull.*, 46, 529, 2003.
38. Hunter, C.L. and Evans, C.W., Coral reefs in Kaneohe Bay, Hawaii: two centuries of western influence and two decades of data, *Bull. Mar. Sci.*, 57, 501, 1995.
39. Jokiel, P.L., Cox, E.F., and Crosby, M.P., An evaluation of the nearshore coral reef resources of Kahoolawe, Hawaii, Final Report for Co-operative Agreement NA27OM0327, University of Hawaii, Haw. Inst. Mar. Biol., Honolulu, 1993.
40. Coles, S.L., Colonization of Hawaiian reef corals on new and denuded substrata in the vicinity of a Hawaiian power station, *Coral Reefs*, 3, 123, 1984.
41. Grigg, R.W., Hamakua coast sugar mills revisited: an environmental impact analysis in 1983, University of Hawaii, Sea Grant Pub. No. UNIHI-SEAGRANT-TR-85-02, Honolulu, 1985.
42. Grigg, R.W., Hamakua Sugar Company: Haina factories ocean discharges — a comparison analysis of ocean impact from 1971–1991, unpublished.
43. Jaap, W.C., Coral reef restoration, *Ecol. Eng.*, 15, 345, 2000.
44. Maragos, J.E., Evans, C., and Holthus, P., Reef corals in Kaneohe Bay six years before and after termination of sewage discharges, *Proc. 5th Int. Coral Reef Symp.,* Tahiti, 4, 189, 1985.
45. Waxman, J., Shaul, R., Schmahl, G.P., and Julius, B., Innovative tools for reef restoration: the *Contship Houston* grounding, in *Program and Abstracts, International Conference on Scientific Aspects of Coral Reef Assessment, Monitoring, and Restoration*, NCRI, Nova Southeastern University, Florida, 1999, 200.
46. Maragos, J.E., Restoring coral reefs with emphasis on Pacific reefs, in Thayer, G.W., Ed., *Restoring the Nation's Marine Environment*, Maryland Sea Grant College, College Park, MD, 1992, 141.
47. NOAA, Environmental assessment: *M/V Wellwood* grounding site restoration, Florida Keys National Marine Sanctuary, Monroe County, Florida, April 2002, [Online] Available: http://www.sanctuaries.nos.noaa.gov/library/reef_restoration/wellwoodea.pdf, December 30, 2004.
48. Hernandez-Delgado, E., Ortiz-Prosper, A., and Alicea-Rodriguez, L., Ecological effects of ship groundings on coral reefs: two case studies and a proposal for action, *Taking Action for Coral Reefs Workshop*, Mayaguez, Puerto Rico, November 6–8, 1997.
49. Jaap, W.C., Coral reef restoration — a synthesis of information, goal setting and success criteria for coastal habitat restoration, Charleston, SC, January 13–15, 1998.
50. Endean, R. and Stablum, W., The apparent extent of recovery of reefs of Australia's Great Barrier Reef devastated by the crown-of-thorns starfish (*Acanthaster planci*), *Atoll Res. Bull.*, 168, 1, 1973.
51. Miller, S.L., McFall, G.B., and Hulbert, A.W., *Guidelines and Recommendation for Coral Reef Restoration in the Florida Keys National Marine Sanctuary*, National Undersea Research Center, University of North Carolina at Wilmington, 1993.
52. Clarke, S. and Edwards, A.J., The use of artificial reef structures to rehabilitate reef flats degraded by coral mining in the Maldives, *Bull. Mar. Sci.*, 55, 724, 1994.
53. Brown, B.E. and Dunne, R.P., The environmental impact of coral mining on coral reefs in the Maldives, *Environ. Conserv.*, 15, 159, 1988.
54. Gittings, S.R., Bright, T.J., Choi, A., and Barnett, R.R., The recovery process in a mechanically damaged coral reef community: recruitment and growth, *Proc. 6th Int. Coral Reef Symp.,* Townsville, Australia, 2, 225, 1988.
55. Lindahl, U., Low-tech rehabilitation of degraded coral reefs through transplantation of staghorn corals, *Ambio*, 27, 645, 1998.
56. Wulff, J.L., Sponge-mediated coral reef growth and rejuvenation, *Coral Reefs*, 3, 157, 1984.

57. Graham, B.D. and Fitzgerald, P.S., New technique for hard coral reattachment field-tested following two recent ship groundings, in *Program and Abstracts, International Conference on Scientific Aspects of Coral Reef Assessment, Monitoring, and Restoration*, NCRI, Nova Southeastern University, Florida, 1999, 95.
58. Nishihira, M., Transplantation of hermatypic coral using fragments of colonies — brief method using bamboo stick, *Biological Magazine of Okinawa*, 32, 49, 1994.
59. Iliff, J.W., Goodwin, W.B., Hudson, J.H., Miller, M.W., and Timber, J., Emergency stabilization of *Acropora palmata* with stainless steel wire and nails: impressions, lessons learned, and recommendations from Mona Island, Puerto Rico, National Coral Reef Institute Abstract, 110, 1999.
60. Bruckner, A. and Bruckner, R., Condition of restored *Acropora palmata* fragments off Mona Island, Puerto Rico 2 years after the Fortuna Reefer ship grounding, *Coral Reefs*, 20, 235, 2001.
61. NOAA, Restoration activities case: Fortuna Reefer, [Online] Available: http://www.darp.noaa.gov/southeast/fortuna/index.html, June 29, 2004.
62. Richmond, R.H., Recovering populations and restoring ecosystems: restoration of coral reefs and related marine communities, in *Marine Conservation Biology: The Science of Maintaining the Sea's Biodiversity*, Norse, E. and Crowder, L., Eds., Island Press, Washington, DC, 2005, Chap. 23.
63. Precht, W.F., Coral transplant feasibility study for restoration of the *USS Memphis* grounding site, Broward County, Florida, PBS&J, Miami, FL, July 2000, unpublished report.
64. Gilliam, D.S., Dodge, R.E., Thornton, S.L., Glynn, E.A., Jaap, W., and Wheaton, J., Scleractinian coral reattachment success and recruitment on a shallow-water ship grounding site in southeast Florida, USA, date unavailable, [Online] Available: http://www.nova.edu/ocean/ncri/projects/hind/, June 24, 2004.
65. Graham, B. and Shroeder, M., *M/V Firat* removal, grounding assessment, hard coral reattachment, and monitoring — a case study, *Oceans MTS/IEEE Conference and Exhibition*, 3, 1451, 1996.
66. Continental Shelf Associates, Baseline survey: monitoring reattached hard corals at the *Firat* grounding site, Draft report, Jupiter, FL, 1999, unpublished report.
67. Jaap, W.C. and Morelock, J., Baseline monitoring report, restoration project, Soto's Reef, George Town, Grand Cayman Island, British West Indies, *Technical Report of Holland America-Westours and Cayman Islands Department of the Environment*, Seattle and George Town, 1996.
68. Goldberg, W.M. and Caballero, A., Reef damage by large-vessel impact and its mitigation by site cleanup: methods and results after one, in *Program and Abstracts, International Conference on Scientific Aspects of Coral Reef Assessment, Monitoring, and Restoration*, NCRI, Nova Southeastern University, FL, 1999, 94.
69. Green, A., Burgett, J., Molina, M., Palawski, D., and Gabrielson, P., The impact of a ship grounding and associated fuel spill at Rose Atoll National Wildlife Refuge, American Samoa, U.S. Fish and Wildlife Service Report, Honolulu, HI, 1997.
70. Hudson, J.H. and Diaz, R., Damage survey and restoration of *M/V Wellwood* grounding site, Molasses Reef, Key Largo National Marine Sanctuary, in *Proc. 6th Int. Coral Reef Symp.*, Townsville, Australia, 231, 1988.
71. NOAA, Columbus Iselin Coral Reef Restoration Project, [Online] Available: http://www.sanctuaries.nos.noaa.gov/special/columbus/project.html, July 19, 1999.
72. Alcala, A.C. and Gomez, E.D., Dynamiting coral reefs for fish: a resource-destructive fishing method, in Salvat, B., Ed., *Human Impacts on Coral Reefs: Facts and Recommendations*, Antenne de Tahiti Museum, French Polynesia, 1987.
73. NOAA, Restoration activities case: Maitland coral reef restoration, [Online] Available: http://www.darp.noaa.gov/southeast/maitland/index.html, June 29, 2004.
74. Miller, S.L. and Barimo, J., Assessment of juvenile coral populations at two coral reef restoration sites in the Florida Keys National Marine Sanctuary: indicators of success? *Bull. Mar. Sci.*, 69, 395, 2001.
75. NOAA, Restoration activities case: Elpis coral reef restoration, [Online] Available: http://www.darp.noaa.gov/southeast/elpis/index.html, June 29, 2004.
76. Spieler, R.E., Gilliam, D.S., and Sherman, R.L., Artificial substrate and coral reef restoration: what do we need to know to know what we need, *Bull. Mar. Sci.*, 69, 1013, 2001.

77. Jaap, W.C., Reef restoration and monitoring: Soto's Reef, Georgetown, Grand Cayman Island, British West Indies, in *Program and Abstracts, International Conference on Scientific Aspects of Coral Reef Assessment, Monitoring, and Restoration*, NCRI, Nova Southeastern University, FL, 1999, 112.
78. Dennis, G.D. and Bright, T.J., The impact of a ship grounding on the reef fish assemblage at Molasses Reef, Key Largo National Marine Sanctuary, FL, *Proc. 6th Int. Coral Reef Symp.*, 2, 213, 1988.
79. Szmant, A.M., Nutrient effects on coral reefs: a hypothesis on the importance of topographic and trophic complexity to reef nutrient dynamics, *Proc. 8th Int. Coral Reef Symp.*, Lessios, H.A. and Macintyre, I.G., Eds., Smithsonian Tropical Research Institute, Panama, 2, 1527, 1997.
80. Ebersole, J.P., Recovery of fish assemblages from ship grounding on coral reefs in the Florida Keys National Marine Sanctuary, *Bull. Mar. Sci.*, 69, 655, 2001.
81. Petersen, D. and Tollrian, R., Methods to enhance sexual recruitment for restoration of damaged reefs, *Bull. Mar. Sci.*, 69, 989, 2001.
82. Pianka, E.R., Latitudinal gradients in species diversity, *American Naturalist*, 100, 33, 1966.
83. Dahl, A.L., Surface area in ecological analysis: quantification of benthic coral-reef algae, *Mar. Biol.*, 23, 239, 1973.
84. Luckhurst, B.E. and Luckhurst, K., Analysis of the influence of substrate variables on coral reef fish communities, *Mar. Biol.*, 49, 317, 1978.
85. Hixon, M.A. and Brostoff, W.N., Substrate characteristics, fish grazing, and epibenthic reef assemblages off Hawaii, *Bull. Mar. Sci.*, 37, 200, 1985.
86. Thongtham, N. and Chansang, H., Influence of surface complexity on coral recruitment at Maiton Island, Phuket, Thailand, in *Proceedings of an International Workshop on the Rehabilitation of Degraded Coastal Systems*, Phuket Marine Biological Center Special Publication, 20, 93, 1999.
87. Jaap, W.C., Guidelines for restoring marine epibenthic habitats, Florida Artificial Reef Summit, Ft. Lauderdale, FL, 2001.
88. Bell, J.D. and Galzin, R., Influence of live coral cover on coral-reef fish communities, *Mar. Ecol. Prog. Ser.*, 15, 265, 1984.
89. Jones, G.P., Experimental evaluation of the effects of habitat structure and competitive interaction in the juveniles of two coral reef fishes, *J. Exper. Mar. Biol. Ecol.*, 123, 115, 1988.
90. Sale, P.F., Ed., *The Ecology of Fishes on Coral Reefs*, Academic Press, San Diego, 1991.
91. Gladfelter, W.B. and Gladfelter, E.H., Fish community structure as a function of habitat structure on West Indian patch reefs, *Revista de Biologia Tropical*, 26, 65, 1978.
92. Wittenberg, M. and Hunte, W., Effects of eutrophication and sedimentation on juvenile corals, 1. Abundance, mortality and community structure, *Mar. Biol.*, 116, 131, 1992.
93. Hixon, M., Effect of reef fishes on corals and algae, in *Life and Death of Coral Reefs*, Birkelend, C., Ed., Chapman and Hall, New York, 1997, 230.
94. Done, T.J., Phase shifts in coral reef communities and their ecological significance, *Hydrobiologia*, 247, 121, 1992.
95. Ebersole, J.P., Recovery of fish assemblages from ship groundings on coral reefs in the Florida Keys National Marine Sanctuary, in *Program and Abstracts, International Conference on Scientific Aspects of Coral Reef Assessment, Monitoring, and Restoration*, NCRI, Nova Southeastern University, FL, 1999, 82.
96. Swanson, D.W., Aronson, R.B., and Miller, S.L., Ship groundings in the Florida Keys: Implications for reef ecology and management, in *Program and Abstracts, International Conference on Scientific Aspects of Coral Reef Assessment, Monitoring, and Restoration*, NCRI, Nova Southeastern University, FL, 1999, 187.
97. Lirman, D. and Miller, M.W., Modeling and monitoring tools to assess recovery status and convergence rates between restored and undisturbed coral reef habitats, *Restoration Ecol.*, 11, 48, 2003.
98. Reyes, M.Z. and Yap, H.T., Effect of artificial substratum material and resident adults on coral settlement patterns at Danjugan Island, Philippines, *Bull. Mar. Sci.*, 69, 569, 2001.
99. Ikeda, Y. and Iwao, K., Coral transplantation to hardened coal ash, *Abstracts of the 4th Symposium of the Japanese Coral Reef Society*, 2001, 38.
100. Okubo, N., Moderate substratum for coral transplantation, *Midoriishi* [in Japanese], 14, 32, 2003.
101. Harbor and Marine Environment Laboratory, *A Draft Manual on Harbor Construction Harmonized with Coral Reef* [in Japanese], 1999, 99.

102. Fitzhardinge, R.C. and Bailey-Brock, J.H., Colonization of artificial reef materials by corals and other sessile organisms, *Bull. Mar. Sci.*, 44, 567, 1989.
103. Quinn, T.P., Glynn, E.A., Dodge, R.E., Banks, K., Fisher, L., and Spieler, R.E., Hypothesis-based restoration study for mitigation of a damaged SE Florida coral reef: a work in progress, [Online] Available: http://www.nova.edu/ocean/ncri/projects/memphis/index.html, June 24, 2004.
104. Morse, D.E., Morse, A.N.C., Raimondi, P.T., and Hooker, N., Morphogen-based chemical flypaper for *Agaricia humilis* coral larvae, *Biol. Bull.*, 186, 172, 1994.
105. Morse, A.N.C., Iwao, K., Baba, M., Shimoike, K., Hayashibara, T., and Omori, M., An ancient chemosensory mechanism brings new life to coral reefs, *Biol. Bull.*, 191, 149, 1996.
106. Negri, A.P., Webster, N.S., Hill, R.T., and Heyward, A.J., Metamorphosis of broadcast spawning corals in response to bacteria isolated from crustose algae, *Mar. Ecol. Prog. Ser.*, 223, 121, 2001.
107. Iwao, K., Study to find chemical inducer for metamorphosis of scleractinian corals, *Midoriishi* [in Japanese], 8, 20, 1997.
108. Iwao, K., Fujisawa, T., and Hatta, T., A cnidarian neuropeptide of the GLWamide family induces metamorphosis of reef-building corals in the genus *Acropora*, *Coral Reefs*, 21, 127, 2002.
109. Hatta, M. and Iwao, K., Metamorphosis induction and its possible application to coral seedlings production, *Rec. Adv. Mar. Sci. Technol.,* 465, 2003.
110. Edwards, A.J. and Clark, S., Coral transplantation: a useful management tool or misguided meddling? *Mar. Poll. Bull.*, 37, 474, 1998.
111. Vaughan, T.W., Growth rate of the Florida and Bahamian shoal-water corals, *Carnegie Institute of Washington Year Book*, 14, 221, 1916.
112. Harriott, V.J. and Fisk, D.A., Coral transplantation as a reef management option, in *Proc. 6th Int. Coral Reef Symp.,* Townsville, Australia, 1988, 375.
113. Bowden-Kerby, A., Low-tech coral reef restoration methods modeled after natural fragmentation processes, *Bull. Mar. Sci.*, 69, 915, 2001.
114. Babcock, R.C., Growth and mortality in juvenile corals (*Goniastrea*, *Platygyra*, and *Acropora*): the first year, *Proc. 5th Int. Coral Reef Symp.*, Tahiti, 4, 355, 1985.
115. Oliver, J. and Willis, B.L., Coral spawn slicks in the Great Barrier Reef: preliminary observations, *Mar. Biol.*, 94, 521, 1987.
116. Maragos, J.E., Coral transplantation: a method to create, preserve and manage coral reefs, University of Hawaii Sea Grant Program, Report AR-74-03, 1974.
117. Gabrie, C., Porcher, M., and Masson, M., Dredging in French Polynesian coral reefs: towards a general policy of resource exploitation and site development, *Proc. 5th Int. Coral Reef Symp.*, Tahiti, 4, 271, 1985.
118. Bowden-Kerby, A., Coral transplantation and restocking to accelerate the recovery of coral reef habitats and fisheries resources within no-take marine protected areas: hands-on approaches to support community-based coral reef management, Second International Tropical Marine Ecosystems Management Symposium, Manila, Philippines, 2003.
119. Shinn, E.A., Coral reef recovery in Florida and the Persian Gulf, *Environ. Geol.*, 1, 241, 1976.
120. Kolinski, S.P., Harbors and channels as source areas for materials necessary to rehabilitate degraded coral reef ecosystems: a Kaneohe Bay, Oahu, Hawaii case study, *Restoration Ecol.*, in review.
121. Marine Research Consultants, Coral transplantation at box drain project under Bracon P-268T at Marine Corps Base Hawaii (MCBH) Kaneohe Bay, Report submitted to Kiewit Pacific Co., 1998.
122. Marine Research Consultants, Coral transplantation at box drain project under Bracon P-268T at Marine Corps Base Hawaii (MCBH) Kaneohe Bay, baseline B., Report submitted to Kiewit Pacific Co., 1999.
123. Jokiel, P.L., Cox, E.F., Te, F.T., and Irons, D., Mitigation of reef damage at Kawaihae Harbor through transplantation of reef corals, Final Report of Cooperative Agreement 14-48-0001-95801, U.S. Fish and Wildlife Service, Pacific Islands Ecoregion, Honolulu, 1999.
124. Munoz-Chagin, R.F., Coral transplantation program in the Paraiso coral reef, Cozumel Island, Mexico, *Proc. 8th Int. Coral Reef Symp.*, Lessios, H.A. and Macintyre, I.G., Eds., Smithsonian Tropical Research Institute, Panama, 2, 2075, 1997.
125. Pacific Basin Environmental Consultants, Inc., Supplemental Coral Transplanting Methodology, 1995.
126. Cheenis Pacific Company, Coral transplantation at the outer cove of Smiling Cove, Sadog Tase, Saipan, CNMI, Final report submitted to Marine Revitalization Corporation, 1996.

127. Micronesian Environmental Service, Outer Cove Coral Transplantation Project: Supplemental Report, prepared for Marine Revitalization Corporation, 1997.
128. Newman, H. and Chuan, C.S., Transplanting a coral reef: a Singapore community project, *Coastal Management in Tropical Asia*, 3, 11, 1994.
129. Fukunishi, I., Yonaha, K., Morita, S., Yamamoto, H., and Takahashi, Y., A planning method on harbor construction harmonizing with coral reef, *Techno-Ocean '98 International Symposium Proceedings* [in Japanese], 1998, 181.
130. Jeansonne, J., Coral restoration project, Pago Pago, American Samoa, Draft year one monitoring trip report: July 2001, prepared for NOAA Fisheries, 2002.
131. MBA International, Coral transplantation, Palau Pacific Resort, a pilot-demonstration project PODCO No. 2156, Final report prepared for the U. S. Army Corps of Engineers, Honolulu Engineer District, Fort Shafter, HI, 1993.
132. Dueñas and Associates, Inc., Weekly Observations of Transplanted Corals at Gun Beach, North Tumon Bay, Guam, Coral Monitoring Report No. 2, prepared for AT&T Submarine Systems, Inc., 1994.
133. Dueñas and Associates, Inc., Department of the Army permit application: trenching of reef flat, installation of conduits and landing of submarine fiber-optic cables at Tepungan, Piti, Guam, Prepared for TyCom Networks (Guam) LLC, 2000.
134. Dueñas and Associates, Inc., Coral transplant and monitoring plan for Tycom Networks Guam LLC fiber optic cable conduit trench in the Tepungan reef flat Piti, Guam, Prepared for Tycom Networks (Guam) LLC, 2001.
135. Dueñas and Associates, Inc., Coral transplant and follow-up monitoring of transplanted corals at Tepungan, Piti, Guam, 1 June 2001 to 4 September 2001, Final report, prepared for Tycom Networks (Guam) LLC, 2001.
136. Dodge, R.E., Anderegg, D., Fergen, R., and Cooke, P., Coral transplantation following repair of outfall, in *Program and Abstracts, International Conference on Scientific Aspects of Coral Reef Assessment, Monitoring, and Restoration*, NCRI, Nova Southeastern University, FL, 1999, 80.
137. Plucer-Rosario, G.P. and Randall, R.H., Preservation of rare coral species by transplantation: an examination of their recruitment and growth, *Bull. Mar. Sci.*, 41, 585, 1987.
138. Auberson, B., Coral transplantation: an approach to the reestablishment of damaged reefs, *Kalikasan, Philipp. J. Biol.*, 11, 158, 1982.
139. Yap, H.T. and Gomez, E.D., Growth of *Acropora pulchra* II., responses of natural and transplanted colonies to temperature and day length, *Mar. Biol.*, 81, 209, 1984.
140. Yap, H.T., Alino, P.M., and Gomez, E.D., Trends in growth and mortality of three coral species (Anthozoa: Scleractinia), including effects of transplantation, *Mar. Ecol. Prog. Ser.,* 83, 91, 1992.
141. Banks, K. and Fletcher, P., Artificial reefs as mitigation for damage to natural reefs: examples from Broward County, Florida, Florida Artificial Reef Summit, Ft. Lauderdale, FL, October 17–20, 2001. [Online] Available: http://www.broward.org/reefsummit/bri01900.htm, December 21, 2004.
142. Rinkevich, B., Restoration strategies for coral reefs damaged by recreational activities: the use of sexual and asexual recruits, *Restoration Ecol.*, 3, 241, 1995.
143. Bouchon, C., Jauber, J., and Bouchon-Navaro, Y., Evolution of a semiartificial reef built by transplanting coral heads, *Tethys*, 10, 173, 1981.
144. Guzman, H.M., Restoration of coral reefs in Pacific Costa Rica, *Conserv. Biol.*, 5, 189, 1991.
145. Guzman, H.M., Transplanting coral to restore reefs in the eastern Pacific, in *Spirit of Enterprise. The 1993 Rolex Awards*, Reed, D.W., Ed., Bern, Switzerland, 409, 1993.
146. Harriott, V.J. and Fisk, D.A., Accelerated regeneration of hard corals: a manual for coral reef users and managers, *G.B.R.M.P.A. Tech Memorandum*, 16, 1988.
147. Kaly, U.L., Experimental test of the effects of methods of attachment and handling on the rapid transplantation of corals, Technical Report No. 1, CRC Reef Research Center, Townsville, Australia, 1995.
148. Hudson, H., Coral restoration project, Pago Pago, American Samoa. Field trip report, NOAA Fisheries, 2000.
149. Neeley, B.D., Evaluation of concrete mixtures for use in underwater repairs, Tech. Report REMR-18, U.S. Army Corps of Engineers, Vicksberg, MS, 1988.
150. Bowden-Kerby, A., Coral transplantation in sheltered habitats using unattached colonies and cultured colonies, in *Proc. 8th Int. Coral Reef Symp.*, Lessios, H.A. and Macintre, I.G., Eds., Smithsonian Tropical Research Institute, Panama, 2063, 1997.

151. Woodley, J.D. and Clark, J.R., Rehabilitation of degraded coral reefs, in *Coastal Zone '89, Proc. 6th Symposium on Coastal Ocean Management*, 3059, 1989.
152. Kojis, B.L. and Quinn, N.J., Factors to consider when transplanting hermatypic coral to accelerate regeneration of damaged coral reef, in *Conference on Environmental Engineering*, Townsville, 1981, 183.
153. Hatcher, B.G., Johannes, R.E., and Robertson, A.I., Review of research relevant to the conservation of shallow tropical marine ecosystems, *Oceanography and Marine Biology: an Annual Review*, 27, 337, 1989.
154. Bak, R.P.M. and Criens, S.R., Survival after fragmentation of colonies of *Madracis mirabilis*, *Acropora palmata*, and *A. cervicornis* (Scleractinia) and the subsequent impact of a coral disease, *Proc. 4th Int. Coral Reef Symp.*, 2, 221, 1981.
155. Carlson, B.A., Organism response to change: what aquaria tell us about nature, *Am. Zool.*, 39, 44, 1999.
156. Epstein, N., Bak, R.P.M., and Rikevich, B., Strategies for gardening denuded coral reef areas: the applicability of using different types of coral material for reef restoration, *Restoration Ecol.*, 9, 432, 2001.
157. Becker, L.C. and Mueller, E., The culture, transplantation and storage of *Montastraea faveolata, Acropora cervicornis,* and *Acropora palmata*: what we have learned so far, *Bull. Mar. Sci.*, 69, 881, 2001.
158. Yap, H.T. and Gomez, E.D., Growth of *Acropora pulchra* III. Preliminary observations on the effects of transplantation and sediment on the growth and survival of transplants, *Mar. Biol.*, 87, 203, 1985.
159. Alcala, A.C., Gomez, E.D., and Alcala, L.C., Survival and growth of coral transplants in Central Philippines, *Kalikasan, Philipp. J. Biol.*, 11(1), 136, 1982.
160. Rinkevich, B. and Loya, Y., Reproduction in regenerating colonies of the coral *Stylophora pistillata*, in Spanier, E., Steinberger, Y., and Luria, M., Eds., *Environmental Quality and Ecosystem Stability, Vol. IVB, Environmental Quality*, Israel Society for Environmental Quality Sciences Publication, Jerusalem, Israel, 1989, 259.
161. Lindahl, U., Low-tech rehabilitation of coral reefs through transplantation of corals: implications for cost-effective management in developing countries, abstract in *Mar. Poll. Bull.*, 32, 1999.
162. Nagelkerken, S., Bouma, S., van den Akker, S., and Bak, R.P.M., Growth and survival of unattached *Madracis mirabilis* fragments transplanted to different reef sites, and the implication for reef rehabilitation, *Bull. Mar. Sci.*, 66, 497, 2000.
163. Ortiz-Prosper, A.L. and Bowdent-Kerby, W.A., Transformation of artificial concrete "reef ball" structure into living coral heads through the use of implants of juvenile massive corals, in *Program and Abstracts, International Conference on Scientific Aspects of Coral Reef Assessment, Monitoring, and Restoration*, NCRI, Nova Southeastern University, FL, 1999, 148.
164. Borneman, E.H. and Lowrie, J., Advances in captive husbandry and propagation: an easily utilized reef replenishment means from the private sector? *Bull. Mar. Sci.*, 69, 897, 2001.
165. Epstein, N., Bak, R.P.M., and Rinkevich, B., Applying forest restoration principles to coral reef rehabilitation, *Aquatic Conserv.*, 13, 387, 2003.
166. Raymundol, L.J.H., Maypa, A.P., and Luchavez, M.M., Coral seeding as a technology for recovering degraded coral reefs in the Philippines, in *Proceedings of an International Workshop on the Rehabilitation of Degraded Coastal Systems*, Phuket Marine Biological Center Special Publication, 20, 81, 1999.
167. Rinkevich, B., Steps towards the evaluation of coral reef restoration by using small branch fragments, *Mar. Biol.*, 136, 807, 2000.
168. Szmant, A.M., Coral restoration and water quality monitoring with cultured larvae of *Montastraea "annularis"* and *Acropora palmata*, in *Program and Abstracts, International Conference on Scientific Aspects of Coral Reef Assessment, Monitoring, and Restoration*, NCRI, Nova Southeastern University, FL, 1999, 188.
169. Sammarco, P.W., Brazeau, D.A., and Lee, T.N., Enhancement of reef regeneration process: supplementing coral recruitment processes through larval seeding, in *Program and Abstracts, International Conference on Scientific Aspects of Coral Reef Assessment, Monitoring, and Restoration*, NCRI, Nova Southeastern University, FL, 1999, 169.
170. Heyward, A.J., Rees, M., and Smith, L.D., Coral spawning slicks harnessed for large-scale coral culture, in *Program and Abstracts, International Conference on Scientific Aspects of Coral Reef Assessment, Monitoring, and Restoration*, NCRI, Nova Southeastern University, FL, 1999, 104.

171. Heyward, A., Smith, L.D., Rees, M., and Field, S.N., Enhancement of coral recruitment by *in situ* mass culture of coral larvae, *Mar. Ecol. Prog. Ser.*, 230, 113, 2002.
172. Gittings, S.R., Bright, T.J., and Holland, B.S., Five years of coral recovery following a freighter grounding in the Florida Keys, *Proceedings of the American Academy of Underwater Sciences, 10th Annual Symposium*, 1990, 89.
173. Gleason, D.F., Brazeau, D.A., and Munfus, D., Can self-fertilizing coral species be used to enhance restoration of Caribbean reefs?, *Bull. Mar. Sci.*, 69, 933, 2001.
174. Hixon, M.A., Carr, M.A., and Beets, J., Coral reef restoration: potential uses of artificial reefs, in *Program and Abstracts, International Conference on Scientific Aspects of Coral Reef Assessment, Monitoring, and Restoration*, NCRI, Nova Southeastern University, FL, 1999, 104.
175. Seaman, W., International reef technology and research trends, Florida Artificial Reef Summit, Ft. Lauderdale, FL, October 17–20, 2001, [Online] Available: http://www.broward.org/reefsummit/bri01900.htm, December 21, 2004.
176. Polovina, J.J., Artificial reefs: nothing more than benthic fish aggregations, *California Cooperative Oceanographic Fisheries Investigations Reports*, 30, 37, 1989, [Online] Available: http://www.calcofi.org/newhome/publications/CalCOFI_Reports/calcofi_reports.htm, December 21, 2004.
177. Hilbertz, W., Fletcher, D., and Krausse, C., Mineral accretion technology: applications for architecture and aqua-culture, *Indust. Forum*, 8, 75, 1977.
178. Hilbertz, W., Solar-generated building material from seawater as a sink for carbon, *Ambio*, 21, 126, 1992.
179. Schuhmacher, H. and Schillak, L., Integrated electrochemical and biogenic deposition of hard material — a nature-like colonization substrate, *Bull. Mar. Sci.*, 55, 672, 1994.
180. Hilbertz, W. and Goreau, T., Pemuteran coral reef restoration project progress report: May 29, 2001, [Online] Available: globalcoral.org/pemuteran_coral_reef_restoratin.htm, June 24, 2004.
181. Holden, C., Ed., Reef therapy, *Science*, 305, 1398, 2004.
182. Blair, S.M. and Flynn, B.S., Miami-Dade County's Sunny Isles reef restoration: habitat restoration on intermittently impacted hardground reef, in *Program and Abstracts, International Conference on Scientific Aspects of Coral Reef Assessment, Monitoring, and Restoration*, NCRI, Nova Southeastern University, FL, 1999, 56.
183. Oren, U. and Benayahu, Y., Transplantation of juvenile corals: an new approach for enhancing colonization of artificial reefs, *Mar. Biol.*, 127, 499, 1997.
184. Goreau, N.I., Goreau, T.J., and Hayes, R.L., Settling, survivorship, and spatial aggregation in planulae and juveniles of the coral *Porites porites*, *Bull. Mar. Sci.*, 31, 424, 1981.
185. Matthews, H., Comparison of five reef building materials, *Florida Artificial Reef Summit*, Ft. Lauderdale, Florida, October 17–20, 2001, [Online] available: http://www.broward.org/reefsummit/bri01900.htm, December 21, 2004.
186. Spurgeon, J.P., Improving the economic effectiveness of coral reef restoration, *Bull. Mar. Sci.*, 69, 1031, 2001.
187. Jokiel, P.L. and Naughton, J., Coral reef mitigation and restoration techniques employed in the Pacific islands: II guidelines, *Oceans 2001 Conference Proceedings*, Marine Technological Society/Institute of Electrical and Electronics Engineers, Inc., Holland Publications, Escondito, CA, 1, 313, 2001.
188. Dixon, J.A., Carpenter, R.A., Fallon, L.A., Sherman, P.B., and Manipomoke, S., Economic analysis of the environmental impacts of development projects, *Earthscan Publications Ltd.*, United Kingdom, 1988.
189. Ruitenbeek, H.J., Ridgely, M., Dollar, S., and Huber, R., Optimization of economic policies and investment projects using a fuzzy logic based cost-effectiveness model of coral reef quality: empirical results for Montego Bay, Jamaica, *Coral Reefs*, 18, 381, 1999.
190. Korhonen, P.H., Moskowitz, H., and Wallenius, J., Multiple criteria decision support — a review, *Eur. J. Op. Res.*, 63, 61, 1992.
191. Fernandes, L., Ridgley, M.A., and van't Hof, T., Multiple criteria analysis integrates economic, ecological and social objectives for coral reef managers, *Coral Reefs*, 18, 393, 1999.
192. NOAA, Habitat equivalency analysis: an overview, Policy and Technical Paper Series, Number 95-1, Damage Assessment and Restoration Program, National Oceanic and Atmospheric Administration, Department of Commerce, 1995.

193. Clark, T., Tissue regeneration rate of coral transplants in a wave exposed environment, Cape D'Aguilar, Hong Kong, *Proc. 8th Int. Coral Reef Symp.*, Panama City, Panama, 2, 2069, 1996.
194. Kojis, B.L. and Quinn, N.J., The importance of regional differences in hard coral recruitment rates for determining the need for coral restoration, *Bull. Mar. Sci.*, 69, 967, 2001.
195. Kojis, B.L. and Quinn, N.J., Biological limits to Caribbean reef recovery: a comparison with western South Pacific reefs, in Ginsburg, R.N., Ed., *Proceedings of the Colloquium on Global Aspects of Coral Reefs: Health, Hazards and History, 1993*, Rosenstiel School of Marine and Atmospheric Science, University of Miami, FL, 1994, 353.
196. Japan Marine Science and Technology Center, Report of the coral reef project [in Japanese], 1, 1, 1991.
197. Bowden-Kerby, A., The "Johnny coral seed" approach to coral reef restoration: new methodologies appropriate for lower energy reef areas, in *Program and Abstracts, International Conference on Scientific Aspects of Coral Reef Assessment, Monitoring, and Restoration*, NCRI, Nova Southeastern University, FL, 1999, 59.
198. Okinawa General Bureau, Okinawa Development Agency, Research of coral transplantation in the Ishigaki Port [in Japanese], 1997.
199. Marine Parks Center of Japan, *Study on Recovery of Coral Reef Ecological System* [in Japanese], 1993.
200. Marine Parks Center of Japan, *Study on Recovery of Coral Reef Ecological System* [in Japanese], 1994.
201. Marine Parks Center of Japan, *Study on Recovery of Coral Reef Ecological System* [in Japanese], 1995.
202. Beets, J., Carr, M.S., and Hixon, M.A., Artificial reefs may not enhance larval recruitment and juvenile abundance, in *Program and Abstracts, International Conference on Scientific Aspects of Coral Reef Assessment, Monitoring, and Restoration*, NCRI, Nova Southeastern University, FL, 1999, 54.
203. Sherman, R.L., Gilliam, D.S., and Spieler, R.E., Site-dependent differences in artificial reef function: implications for coral reef restoration, *Bull. Mar. Sci.*, 69, 1053, 2001.
204. Mueller, E., Reef restoration: science or technology? In *Program and Abstracts, International Conference on Scientific Aspects of Coral Reef Assessment, Monitoring, and Restoration*, NCRI, Nova Southeastern University, FL, 1999, 142.
205. Pinit, P.T., Bellmer, R.J., and Thayer, G.W., Draft NOAA Technical Guidance Manual for Success Criteria in Restoration Projects, NOAA Restoration Center, Office of Habitat Conservation, no date available, [Online] available: http://www.nmfs.noaa.gov/habitat/restoration/projects_programs/research/SuccessCriteriaGuidance.pdf, December 21, 2004.
206. Aronson, R.B. and Swanson, D.W., Video surveys of coral reefs: uni- and multivariate applications, *Proc. 8th Intl. Coral Reef Symp.*, 2, 1441, 1997.
207. Aronson, R.B. and Swanson, D.W., Disturbance and recovery from ship groundings in the Florida Keys National Marine Sanctuary, Dauphin Island Sea Lab Tech. Rpt. 97-002, 57, 1997.
208. Gomez, E.D. and Yap, H.T., Monitoring reef conditions, in Kenchington, R.A. and Hudson, B.T., Eds., *Coral Reef Management Handbook,* UNESCO/ROSTSEA, Jakarta, 1988, 87.
209. Likens, G.E., *Long-Term Studies in Ecology: Approaches and Alternatives*, Springer Verlag, New York, 1988.

4 Natural Resilience of Coral Reef Ecosystems

Norman J. Quinn and Barbara L. Kojis

CONTENTS

4.1 CORAL REEF COMMUNITIES IN ALTERED STATES

Coral reefs are subject to many forces, both natural and human induced, that can severely damage coral communities. They are susceptible to natural forces such as intense wave action from cyclones and tsunamis, high temperatures, volcanic eruptions, and disease outbreaks, as well as to anthropogenic changes in water quality such as increases in turbidity, suspended sediments, changes in nutrient levels, and toxic chemicals. They are also subject to changes in ecosystem structure as a result of overfishing, the use of destructive fishing methods, and the direct destruction of reefs and associated habitats through dredging and filling. Anthropogenic impacts are often more localized than the impacts of natural forces, but they are increasing in frequency and extent in association with the coastal development and pollution that accompany increasing numbers of people and rising standards of living.[1]

It is now generally accepted that coral cover and the abundance of commercially viable fishes and free-living invertebrates have greatly declined throughout the world.[2] This decline has been well documented in the Caribbean, especially in the Florida Keys, Jamaica, and the Netherlands Antilles.[3]

The reasons for this decline are myriad. Some impacts on reefs occur worldwide. For example, worldwide bleaching events have likely been unprecedented in their global impact on coral reefs.[4] Also, hurricanes occur in most of the tropics and can have localized to devastating impacts on reefs where they strike.[5–7] Other reasons vary by geographical location. From the mid-1970s to the mid-1980s, *Acropora palmata* and *A. cervicornis* were affected by white band disease. By the early 1980s, 95% of the luxuriant shallow-water *A. palmata* reefs and mid-water *A. cervicornis* haystacks in the Caribbean were killed by white band disease.[8] Also, disease killed 90% of the long-spined sea urchin (*Diadema antillarum*), an important herbivore, in the Caribbean in 1983 and 1984.[9,10] These events changed the ecology of Caribbean reefs. In 2002, these species had not recovered, and coral cover of shallow water reefs in the Caribbean in general was still in decline.[2] In the Pacific, the Crown of Thorns (CoT) starfish (*Acanthaster planci*) has devastated coral reefs in many regions.[11,12] However, most Pacific reefs have recovered from the impacts of the CoT.

Reef damage from both natural and human causes is often looked upon with alarm. When the damage is localized or a particular reef is economically important, restoration, especially restoration of hard corals, is often proposed. Restoration can entail a myriad of activities. It most often refers to transplanting corals prior to a construction project to save them. This commonly occurs in areas where commercial resort, marina, or port development may destroy reefs or hard-bottom habitat with scattered hard coral colonies and gorgonians. In such situations agencies/developers may be required by government agencies to transplant threatened corals to mitigate the damage to the reef community in order to obtain development permits. As well, restoration is often required when a reef has been seriously damaged by a shipping accident.[13] The generally high cost of reef restoration or mitigation ensures that the environmental cost is adequately weighed against the economic benefit of a project. It also encourages ship owners to properly maintain their ships and train their captains and crews.

Restoration of reefs that have lost coral cover or species due to naturally occurring events is a much more contentious issue. In most cases it is impractical to enhance natural recovery of reefs damaged by widespread bleaching events, outbreaks of CoT starfish, storms, or disease, or even anthropogenic activity that has global or regional effects. The damage is too severe and widespread for restoration to be economically or even physically feasible. In most cases recovery must be left to nature. However, there may be reef systems or species that have the potential to rapidly recover but which are unable to do so because of lack of recruits. In these cases it may be reasonable to consider methods that may enhance natural restoration of reefs, especially reefs of high economic importance, i.e., those that are heavily used by tourists or are popular sites for local snorkelers and divers.

In this chapter we look at the natural variability of coral reef ecosystems in three distinct geographic seas: the Caribbean (with an emphasis on Jamaica and the U.S. Virgin Islands), the Commonwealth of the Northern Mariana Islands (CNMI) in the northwestern Pacific Ocean, and Fiji in the South Pacific. The consequences of natural variability in coral abundance, diversity, and recruitment among and within these three reef systems will be considered in the context of how this variability is likely to affect the natural recovery of reefs and influence the selection of reefs for restoration.

4.2 ACCELERATED REEF RESTORATION

Coral reef restoration is often promoted as a means of enhancing coral reef recovery, including the recovery of coral species diversity and the structural complexity of reefs. The techniques are controversial because of the expense and uncertainty of the long-term survival of the transplanted colonies.[14,15] Before an expensive reef restoration program is begun, fundamental questions need to be asked about the reef system's natural ability to recover and the time frame in which recovery will occur. Given the high cost of coral restoration, when is coral restoration justified and when should nature be left to take its course?

Coral reef species diversity has two components: the number of species in an area and their patterns of relative abundance. When coral restoration is performed in conjunction with a development project, usually all except very small and thinly encrusting colonies are transplanted. Coral species diversity and the relative abundance of transplanted species at the restoration site are often enhanced.

However, most experimental efforts to enhance coral reef recovery do not focus on recovery of coral species diversity; rather, they concentrate on recovery of reef structural complexity and increase the abundance of fast-growing and/or asexually reproducing species such as *Acropora cervicornis* in the Caribbean, *A. muricata* and *A. vaughani* in the Indian Ocean, *Stylophora pistillata* in the Red Sea and *A. formosa* and *A. nobilis* in Okinawa.[16–20] The efforts are usually monogenetic, with success based on survival and an increase in the weight of live coral.[18] The transplanting of colonies of a single species, often of a single genotype, may increase coral cover in an area, resulting in an increase in food web complexity by attracting more fish and invertebrates associated with these coral species. The result is more attractive areas to view by tourists and possibly increased fisheries production, but the effort is limited in its potential to restore coral biodiversity in the short term.

Two approaches to reef restoration exist: transplanting all or portions of adult corals and releasing planulae raised in the laboratory on damaged reefs. It has long been known that colonial corals have the capacity to survive and grow after breakage.[21] New colony formation by fragmentation occurs in both branching and massive species, but more commonly and successfully in branching corals. This feature of corals has been used to restore reefs damaged by ship groundings.[22] The use of species known to exhibit high survival rates after transplantation is an important factor to consider when selecting species for transplantation.[23] However, transplanting colonies is a laborious, long-term and, therefore expensive means of reef restoration and generally can only be justified when large amounts of commercial funding is available, such as after a ship grounding or when a commercial development will impact a reef.

Restoration by seeding reefs with coral planulae has several advantages. Planulae can be derived directly from brooding species and released onto damaged reefs. Planulae can, with more difficulty, be raised from eggs of broadcast spawning species and released on the reef after they have matured and are ready to settle. However, reseeding reefs using coral planulae requires fairly sophisticated facilities, especially for successfully raising planulae of spawning species. As well, knowledge is required of:

1. The mode and timing of coral reproduction for the species in the local area
2. The requirements for successful fertilization of eggs and subsequent development of larvae for species that spawn
3. The settlement substrate required by the species

A high degree of scientific expertise or technical training is essential. The successful result does produce large quantities of planulae, but the methods used to ensure localized settlement are labor intensive. Of course planulae settlement rates and recruit survival rates would be much lower than fragment transplantation rates, but the much larger number of planulae produced may result in more juveniles and, if planulae from a variety of species and/or individuals of a species are cultured, coral species diversity and/or genetic diversity within a species could be increased.[18,24]

Recruitment rates of hard corals in general and coral families in particular are known to greatly vary within reef systems, e.g., between coastal fringing and mid-shelf reefs and among reefs at different latitudes.[25,26] There is even greater variation between interocean reef systems, with Caribbean recruitment rates generally being lower than those of the Pacific. When coral recruitment rates are naturally high and the damage to a reef has not caused chronic declines in water quality or substrate conditions, then the reef will likely recover rapidly without any need for artificial reef restoration. If coral recruitment rates are low in general or low for the primary reef building species in the geographic area, restoration may significantly enhance recovery. Reefs that resist recovery due to inhibited larval recruitment should become the primary targets of restoration.[17]

Generally, coral reefs chronically degraded by anthropogenic disturbances cannot be effectively restored regardless of the methods used, nor will they recover naturally until the cause of the degradation has been removed. Planting coral fragments or introducing planulae will not restore these reefs until the other problems are fixed.

4.3 GEOGRAPHIC VARIABILITY IN CORAL REEF RESILIENCE AND IMPLICATIONS FOR CORAL REEF RESTORATION

4.3.1 CARIBBEAN

Caribbean coral reefs have been heavily impacted by hurricanes, disease, and other factors, resulting in major changes in coral reef communities. Below is a description of the changes on reefs in two regions, the north coast of Jamaica in the Central Caribbean and the U.S. Virgin Islands (USVI) in the Eastern Caribbean, and the implications for coral reef restoration.

4.3.1.1 Jamaica

The coral reefs around Discovery Bay were the first in the Caribbean to be extensively researched.[27] Until about 25 years ago, the shallow fore-reef slopes of the Jamaican north coast had a high diversity and abundance of coral, dominated by large colonies of *A. palmata, A. cervicornis, Montastraea annularis, Porites* spp. and large towering colonies of *Dendrogyra cylindrus*.[28] Dense, monospecific high-relief thickets of *A. palmata* and *A. cervicornis* were characteristic of shallow and intermediate depth, respectively, coral communities of the north coast of Jamaica.[29,30] After Hurricane Allen in 1980 and Hurricane Gilbert in 1988, coral abundance and diversity decreased.[6,7] Live coral cover of *A. palmata* and *A. cervicornis* dropped from 21% at certain sites to almost 0% around Discovery Bay.[31] During approximately the same period, white band disease swept through the Caribbean, killing up to 95% of acroporids, and an unknown disease killed up to 99% of the long-spined sea urchin (*Diadema antillarum*), an important herbivore.[8–10]

In general, these events changed the coral reef community on the north coast of Jamaica from a community dominated by coral to one dominated by macro algae. This change was repeated throughout the Caribbean.[32]

Large populations of *A. cervicornis* have not reestablished themselves in over 20 years on Discovery Bay reefs, and this species is unlikely to have naturally recovered anywhere else where it declined precipitously in the Caribbean.[33] The only sites near Discovery Bay where *A. cervicornis* cover was high (e.g., >50%) were a small section of reef at 10 m near Dairy Bull and a single spur at East Rio Bueno at 12 m.[34] At most other sites at similar depths, *A. cervicornis* was absent or only small scattered colonies were present. This was true even in areas that had an abundance of dead branches of this species lying on the substrate or forming the underlying reef structure.

While there have been some signs of recovery of *A. palmata* in other areas of the Caribbean, recovery on the north coast of Jamaica has been limited and patchy.[35] In a survey of coral species distribution and abundance at depths of 3 to 30 m on the west fore reef slope of Discovery Bay, *A. palmata* was found only at depths of 3 m, represented only 0.3% of the live coral cover at this depth, and had a density of only 0.01 colonies m^{-2}.[34] The distribution and abundance of *A. palmata* on the West Fore Reef was characterized by single widely separated colonies. In a few locations, small, dense clusters of *A. palmata* colonies were present, e.g., west Rio Bueno reef at 8 m, Braco at 1 to 3 m, and Pear Tree Bottom at 1 to 5 m depth.

Studies of coral recruitment on experimental settlement plates on the fore reef of Discovery Bay have shown recruitment to be low to moderate compared to the South Pacific (Table 4.1). Recruitment was dominated by *Millepora* spp. and scleractinian brooders (Table 4.2) with few broadcast spawning faviids and acroporids recruiting to plates (Table 4.3). This is despite the fact that faviids, especially the *Montastraea* complex, and acroporids were the major structural components of north coast Jamaican reefs. Given the low recruitment rates of *Acropora*, it is believed that the genus had little chance to recover.[36]

However, natural recovery of *A. palmata* may be starting to occur on north coast Jamaican reefs. In contrast to previous Caribbean studies, a few single polyped acroporid recruits did recruit to settlement plates primarily at 3 to 5 m depth with one recruit at 9 m in 2003.[34,37–39] The species identification of these recruits could not be determined. However, juvenile *A. palmata* colonies

TABLE 4.1
Selection of Scleractinian Recruitment Rates in Different Geographic Areas

Site	Mean Number of Recruits m^{-2} (Range of Observed Values) m^{-2}	Duration[a] (months)	Latitude	Ref.
		Caribbean Sea		
USVI				
St. Thomas	139 (89 to 180)	12	18°N	62
St. Croix	6		18°N	37
Jamaica				
Discovery Bay	60		18°N	48
Discovery Bay	378		18°N	77
Discovery Bay	81 (0 to 327)	6 S + W	18°N	34
		Gulf of Mexico		
East Flower Garden	370 (0 to 1254)	3 S	28°N	75
		Atlantic Ocean		
Bermuda	(15 to 160)		29°N	39
		Pacific Ocean		
CNMI, Saipan, Tinian	24 (0 to 112)	6 S[b]	14 to 15°N	62
Guam	(37 to 100)	6 unkn		61
Guam	<1			60
GBR 1980 to 1981	1650 (400 to 3000)	4 S	19°S	66
GBR 1981 to 1982	500 (150 to 750)	4 S	19°S	66
GBR 1982 to 1983	770 (300 to 1300)	4 S	19°S	66
GBR 1983 to 1984	1080 (450 to 2000)	4 S	19°S	66
GBR	4614 (2050 to 7178)	2 S	10 to 23°S	26
GBR	297	5 S and 12	23°S	58
Fiji	734 (322-1812)[c]	12 S + W[d]	17 to 18°S	62

Note: USVI - U.S. Virgin Islands; CNMI - Commonwealth of the Northern Mariana Islands; GBR - Great Barrier Reef.

[a] Number of months settlement tiles were deployed.
[b] Settlement recorded only over summer months
[c] Anomalous inner Suva Harbor site not included.
[d] Two sets of settlement tiles deployed, one set over summer and one over winter, and total settlement pooled from summer and winter tiles.

2 to 10 cm in diameter were also detected in shallow water at depths of 3 to 5 m on the west fore reef of Discovery Bay and at 6 to 14 m on a spur at Braco, 10 km west of Discovery Bay, in 2004.[40] These, primarily encrusting, nonbranching juveniles were found on sections of the reef recently cleared of algae by the long-spined sea urchin, *D. antillarum.* Recruitment of this species was rare on sections of the reef still densely covered by algae.

The variation in the *A. palmata* recruitment between adjacent sections of the same reef at Braco, as well as the widely separated stands of mature colonies of this species, suggest that the chance of reef recovery is quite variable within the same geographic area and may be not only a function of planulae availability but also the availability of cleared space. Small-scale manipulations of grazing echinoids and fish on coral reefs have demonstrated that herbivores significantly reduce

TABLE 4.2
Caribbean and Pacific Ocean Comparison of Settlement of Scleractinian and Milleporan Recruits

Site	Percent Scleractinia	Percent *Millepora*	Ref.
	Caribbean Sea		
USVI, St. Thomas	68	32	59
Jamaica, Discovery Bay	76	24	34
	Pacific Ocean		
CNMI, Saipan, Tinian	89	11	59
Fiji	>99	<1	59

Note: USVI - U.S. Virgin Islands; CNMI - Commonwealth of the Northern Mariana Islands; GBR - Great Barrier Reef).

the quantity of fleshy macroalgae and may help provide the open space necessary for planulae settlement.[41,42] At this time, macroalgae still cover the substrate on most of the north coast reefs, and coral cover and recruitment levels remain low (Table 4.1). However, as *Diadema* abundance increases, macroalgae will be reduced and settlement space will increase. As recruits settle, grow,

TABLE 4.3
Selected Caribbean and Pacific Ocean Settlement of Brooding and Broadcast Spawning Scleractinian Spat

Site	Latitude	Percent Brooding	Percent Spawning	Ref.
		Caribbean Sea		
USVI				
St. Thomas	18°N	96	4	59
Jamaica				
Discovery Bay	18°N	98	2	77
Discovery Bay	18°N	66	34	34
		Atlantic Ocean		
Bermuda	29°N	>99	<1	39
		Pacific Ocean		
CNMI, Saipan, Tinian	14 to 15°N	87	13	59
Fiji	17 to 18°S	30	70	59
GBR (mid-shelf reefs)	16°S	25	75	76
GBR (fringing reefs)	17°S	2	98	76
GBR (mid-shelf reef)	19°S	2	98	66
GBR (Heron Is.)	23°S	80	20	58

Note: USVI - U.S. Virgin Islands; CNMI - Commonwealth of the Northern Mariana Islands; GBR - Great Barrier Reef.

and reproduce, we may see an exponential increase in the abundance of this species on the north coast of Jamaica if no serious setbacks from hurricanes and disease occur.

4.3.1.2 United States Virgin Islands

A similar decline in acroporid and *Diadema* abundance to that recorded in Jamaica occurred in the USVI.[35] Hurricane damage to shallow water *A. palmata* reefs has been extensive since Hurricane David in 1979, resulting in ramparts of dead elkhorn rubble on reef crests. A severe decline in *A. cervicornis* was also noted; these two acroporid species are the Caribbean species most vulnerable to storm damage because of their depth distribution and branching habit and the most susceptible to white band disease.

As in Jamaica, recruitment was dominated by *Millepora* spp. (Table 4.2) and scleractinian brooders with few broadcast spawning faviids. However, unlike in Jamaica no acroporids recruited to plates (Table 4.3). Recruitment of scleractinian corals in the USVI was moderate compared to the rest of the Caribbean (Table 4.1). Although there are limited signs of recovery of *A. palmata* in the USVI at some locations, there is very little sign of recovery of *A. cervicornis*.[35]

4.3.1.3 Implications for Restoration

Many hypotheses have been proposed concerning factors that are currently thwarting the redevelopment of the coral assemblages in the Caribbean. Among these factors are the amount of planulae produced and the amount of space available for planulae settlement.[43] Regarding the latter factor, small-scale manipulations of grazing echinoids and fish on coral reefs have demonstrated that herbivores significantly reduce the quantity of fleshy macroalgae and may help provide the open space necessary for planulae settlement.[41,42] It has also been suggested that predators were preventing the recovery of staghorn populations as predation rates or juvenile corals increased to levels beyond a predicted threshold as the result of a phase shift.[44] While the level of juvenile mortality from predation may be a key factor in the restructuring of some coral communities, we suggest that recruitment failure is more influential than juvenile mortality.[39,45,46]

The failure of a widespread sustained recovery of acroporids in the Caribbean is most likely associated with current production of few planktonic recruits (Figure 4.1). With few adult acroporid colonies, successful production of planulae in the remaining colonies may be thwarted by lack of fertilization success. Existing populations are also under attack by various coral diseases, which are likely to reduce the fecundity of individual colonies.[47] With the additional impediment of

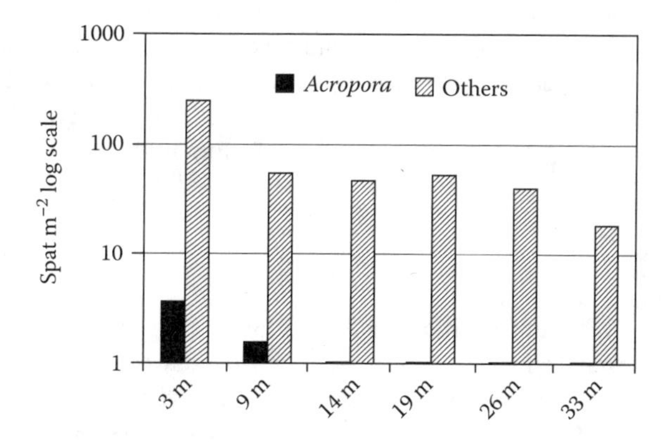

FIGURE 4.1 Total coral settlement by depth in Fiji, 1995 to 1996, normalized to square meters.

macroalgae overgrowing the substrate, the possibility of successful sexual recruitment and survival is further reduced.

In the case of *A. cervicornis*, where sexual recruitment was very infrequent even when colony abundance was high, low fragment survival rate also has affected recovery.[36,48] Because Caribbean acroporid reproductive strategy emphasized asexual fragmentation over sexual recruitment and the existing populations are so small, it is considered unlikely that they will rapidly recover through growth and asexual fragmentation.[36]

With so few acroporid coral planulae there is a greater need to assist in recolonization by transplanting fragments. While acroporid populations are extirpated in some locations, healthy individuals still do exist and several workers tentatively report increased numbers of colonies at a few locations.[49,50] These could be source populations for local transplantation.

Acropora cervicornis is suitable for transplantation because it was once a dominant shallow water coral in the Caribbean. Restoration of acroporids may be necessary on reefs where conditions are not suitable for settlement of spat but are adequate for transplantation of whole colonies or fragments. Reefs absent of *Diadema* and with the substrate covered by algae would be likely sites suitable for restoration if the colonies or fragments are not shaded by algal fronds. Acroporid fragments may rapidly grow and establish themselves in such areas. By increasing the numbers of colonies and density of colonies, gamete production may increase along with fertilization success. Increased local production of planulae may significantly contribute to recruitment at local reefs; especially, *Diadema* populations may recover and open up suitable settlement space and make a small contribution regionally.

In the Indian Ocean transplanted acroporids have been shown to enhance fish abundance and diversity.[18] Transplanting fragments of the fast growing acroporids, such as *A. cervicornis* which has a mean annual growth rate in the Central and Eastern Caribbean ranging from 7.6 to 14.6 cm yr^{-1} in Jamaica and 14.6 cm yr^{-1} in Barbados, may provide the best chance of improving the rate of reef recovery in the immediate future in the Caribbean.[51,52]

4.3.2 Pacific

The Pacific harbors vast expanses of coral reefs, and the scleractinian corals that comprise these reefs are much more diverse than the Caribbean at the family, genus, and especially species level. This diversity is especially apparent in the family Acroporidae, which includes four genera in the Pacific (*Acropora, Anacropora, Astreopora*, and *Montipora*). The genus *Acropora*, which is an important reef-building coral throughout the coral reefs of the world, is represented by 94 species in the Pacific.[53] In contrast, the Acroporidae are represented in the Caribbean by only one genus (*Acropora*) and only three species. While reefs in the Pacific are periodically impacted by hurricanes, disease, crown-of-thorns starfish and corallivorous gastropod predation, many of these reefs, especially in the South Pacific, are highly resilient and recover relatively quickly from perturbations. Below is a description of aspects of reefs from the South Pacific that contribute to this resilience and a discussion of why the reefs of the Commonwealth of the Northern Mariana Islands do not exhibit this same resilience. The implications of these factors in the need for coral reef restoration are discussed.

4.3.2.1 South Pacific

When reefs in the central South Pacific and on the Great Barrier Reef (GBR) are impacted by natural events such as severe tropical storms, bleaching, CoT starfish outbreaks, and disease, human-mediated restoration will usually not be required. High coral recruitment rates in the central South Pacific and GBR result in rapid repopulation of reefs by corals. The coral taxa most susceptible to damage, acroporids and pocilloporids, are the ones with the highest recruitment rates. This likely also holds for human damage to reefs in these areas when the damage is fleeting — e.g., a ship

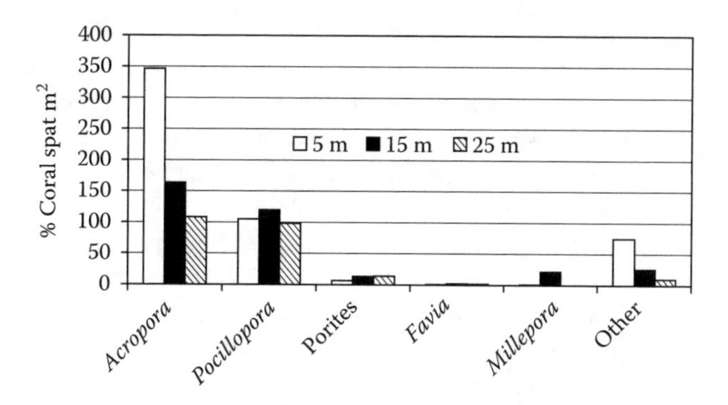

FIGURE 4.2 Total coral family settlement by season from several sites in Fiji normalized to square meters. Y axis is log ($n + 1$)

grounding or anchor damage. The only exception may be in areas that receive frequent human visitation as reefs, which may be thwarted from natural recovery by high usage.

Fijian reefs and certain reefs of the GBR were among the most resilient reefs observed, based on their high coral recruitment rate (Table 4.1). In Fiji, high recruitment rates occurred at all locations except Suva Harbour, a site affected by chronic high turbidity and pollution.[54–56] Acroporids followed by pocilloporids dominated the settlement patterns in Fiji (Figure 4.2), particularly at shallow depths (Figure 4.3). This pattern was true of reefs on the northern and central GBR as well.[26,57] In contrast, in reefs in the Pacific with lower recruitment rates, e.g., Heron Island on the southern end of the GBR and the CNMI, pocilloporids dominated recruitment.[58,59] The predominant taxon on reefs with the highest recruitment rate, the Acroporidae, spawns by releasing gametes (Table 4.3). This mode of reproduction has the greatest dispersal potential. High rates of recruitment and local to long-distance dispersal should enable shallow subtidal reefs of Fiji and the central and northern GBR to rapidly rebound from acute perturbations.

4.3.2.2 Marianas Archipelago, Western Pacific

The reefs of much of the Marianas Archipelago (MA) have low coral cover[60] and low coral recruitment rates.[61,62] Summer scleractinian recruitment to Saipan reefs, MA, from March/May to October, was very low (24 spat m^{-2}) compared with summer settlement (October to March) in Fiji

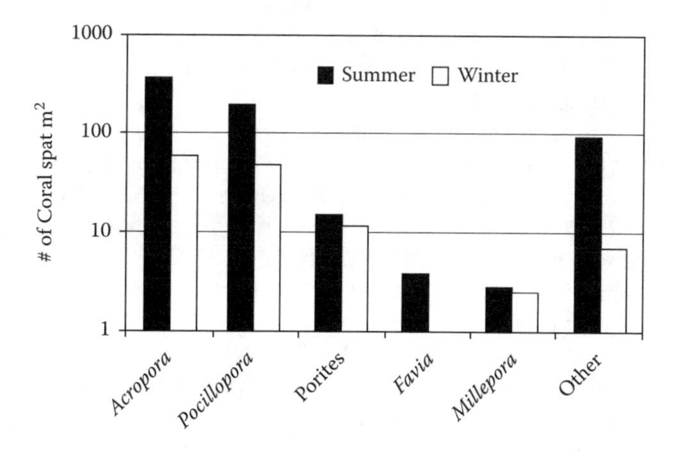

FIGURE 4.3 Total coral settlement by depth pooled from four samples collected between April 2001 and November 2003 at Discovery Bay, Jamaica, normalized to square meters. Y axis is log ($n + 1$).

(500 to 1800 spat m^{-2}). Saipan recruitment rates were also lower than the rates observed in Guam (MA) in 1979, 37 to 100 spat m^{-2}, but higher than those recorded in Guam in 1992, 0.02 to 0.04 spat m^{-2} (Table 4.1).[60,61] More milleporan spat settled than in Fiji but much less than in the Caribbean (Table 4.2), reflecting differences in this taxon's contribution to reef structure. Most of the spat in Saipan were from brooding corals (Table 4.3), which is consistent with the very low percent cover of acroporids in the reef community.[59]

The Marianas Archipelago differs from many of the reef systems in the South Pacific and Great Barrier Reef, where there is considerable "interconnectedness" and possible recruitment from neighboring reefs.[14] The Marianas Archipelago runs in a north-to-south direction, while currents run predominantly in an east-to-west direction with a southeast to northwest component.[63] The closest reef systems to the archipelago are those of Palau and the Caroline Islands. These reef systems lie at least 400 km south of the Marianas Archipelago. It is possible that CNMI reefs occasionally receive larvae from these distant reef systems, but we suggest that most recruits are derived locally.

Acroporid recruitment rates were low in the CNMI, comprising only 7% of recruits, compared to the GBR and Fiji.[14,59,62] This is likely due to the low coral cover of reefs and the low levels of acroporid recruitment.[61] We hypothesize that planulae of broadcast-spawning acroporids, which settle only after a minimum of 4 to 6 days in the plankton, are commonly carried away from local reefs and "lost" to CNMI reefs.[64] If planulae attachment occurs more rapidly, e.g., 2.5 to 2.75 days following fertilization, recruits may not be so readily "lost" and there may be a possibility of higher rates of local recruitment.[65] It has been found that while the attachment stage of settlement occurred rapidly, metamorphosis was delayed and did not commence until nearly 4 days after fertilization in *Platygyra dadaelea* and 6 days in *Goniastrea favulus*.[65] In fact, metamorphosis did not take place in most planula until several days after the first planulae initiated metamorphosis. Planulae that have attached but not yet metamorphosed may readily remobilize and be carried away from their natal reefs.[65]

Guam and the CNMI are also in a "typhoon alley." These coral reefs on these islands are frequently battered by typhoons, which can inflict considerable damage on these coral reefs. However, many coral species in the Pacific, especially digitate acroporids, are adapted to strong storm surge and can survive storm damage if the underlying reef structure does not crumble. These digitate species also recruit rapidly to reefs.[66] Low coral cover, low acroporid recruitment, possible lack of larval interconnectedness between reefs, and residual populations of CoT starfish make recovery of CNMI reefs unlikely in the short term.

4.3.2.3 Implications for Restoration

Coral restoration is not a panacea and will not produce sustainable, healthy reefs if the fundamental cause of the reef's decline has not been addressed. Factors that cause the decline of reefs are likely to differ in various geographical areas.

Low coral cover, low acroporid recruitment, possible lack of larval interconnectedness between reefs, and residual populations of CoT starfish make recovery of CNMI reefs unlikely in the short term. Consequently, reef restoration is worth considering on CNMI reefs if the CoT starfish population is controlled. These reefs were damaged extensively by CoT in the late 1960s/early 1970s. During day dives, patches of small (<25 cm diameter), dead colonies of acroporids and pocilloporids were commonly observed with a CoT usually in the vicinity of the dead colonies and therefore implicated in their demise.[59]

At current recruitment rates it is unlikely that the CNMI reefs will recover to pre–late 1960s coral cover for many decades and would be suitable for an accelerated coral community restoration program in areas important for tourism such as Mañagaha Island, but not at the popular dive site "The Grotto," as the community there is dominated by very slow-growing sclerosponges and a few milleporan *Stylaster* spp.[67] However, before an accelerated coral restoration program commences it is important that non-point source pollution be contained and a water quality standard at which the restored corals are likely to survive be established.[68] It is unwise to commit to a large recolonization of reefs that are still under threat from anthropogenic factors or "natural" factors such as the CoT. Small, initial restoration experiments are suggested to assess the ability of transplanted colonies to tolerate less-than-optimal environmental conditions.

The restoration of whole reef systems is certainly not cost effective. However, the removal of CoT in Tinian at popular dive sites by dive masters seems to be effective.[59] It may be a good business decision by the dive operators to enhance species diversity at these dive sites by scattering acroporid and pocilloporid fragments. Recruitment rates on Green Island reefs on the Great Barrier Reef were higher than at any of the nearby reefs that were less damaged by CoT.[25] As well, the coral taxa most susceptible to predation by CoT, acroporids and pocilloporids, had the highest recruitment rates.[25] Many species of the latter taxon are brooders releasing planulae that are capable of settling immediately after release. This may help increase local recruitment.

In Papua New Guinea, a negative correlation of the fecundity and abundance of *Acropora palifera* with increasing depth, turbidity, and sedimentation was observed.[69] High levels of sedimentation and diminished water transparency limited the depth distribution and reduced the fecundity of this species and probably many other species as well. When water is turbid and sediment laden, reef recovery will be impeded and the success of restoration efforts may be diminished, especially at depth. Reducing the threats to reefs is an important precursor to significant recovery through restoration.

Where human impacts change water quality (e.g., increased nutrient and sediment load) or alter substrate (e.g., bottom characteristics altered from hard, firm substrate to shifting rubble bed), reef recovery will likely be inhibited and/or delayed. Rapid stabilization of damaged reef and the removal of sediment have been demonstrated to be effective in promoting reef restoration.[70] Many reef systems are associated with heavily populated small islands where multiple use of systems is the rule. These islands do not have the coastal area or luxury of separating uses. In Fiji, the Suva Harbor entrance channel is a pass through a reef. A portion of the shipping channel is dredged in a back reef area. The reefs in this area will be periodically subject to the effects of dredging and consequently not suitable for sustained restoration.

Additionally, about four tonnes of hard coral day^{-1} were being removed from Fijian intertidal reefs for international trade in bleached ornamental corals.[71] With such pressure, even high recruitment rates may not be enough. Too often today, people think that they are creating more wealth by developing new enterprises such as coral harvesting. In fact, they may be simply moving wealth from fishers and scuba diving/snorkeling operations to coral harvesters. Loss of the shelter and food resources required by fish that are provided directly and indirectly by corals will reduce not only fishermen's catches but the attractions that bring divers to tourist operations.

Natural replenishment is most likely too slow to support an economical and sustainable coral harvesting industry except for "weedy" species such as *Acropora* and *Pocillopora*.[72] Even these species must only be harvested with a comprehensive "fisheries" management plan and frequent monitoring of the effects of harvest. The culture of fast-growing acroporan and pocilloporid corals in Fiji may permit sustainable production of hard corals and diminish the likelihood of destruction of large areas of reefs. The only exception might be carefully controlled, limited harvest of locally abundant fast-growing species such as the brown-stem *Pocillopora* common in some areas of the South Pacific and GBR.[72]

Pacific workers have proposed that the mainstream of future coral restoration efforts will involve the handling of larvae and/or *in situ* mass culture of juvenile corals.[73,74] This may also be an approach to be considered in the Caribbean, especially for species, such as *A. cervicornis* and *A. palmata*, with relatively small reproductive populations that are susceptible to Allee effects. This, combined with transportation of fragments from healthy populations to areas with very low population levels, may be required in the Caribbean.

4.4 SUMMARY

Specific taxa recruiting to reefs varied between reef systems. Acroporid broadcast spawning species dominated recruitment in Fiji, while recruitment in the CNMI was dominated by the brooding pocilloporans. High recruitment rates by species of the family Acroporidae appear to be pivotal in the ability of Pacific reefs to rapidly recover from major perturbations. The lower coral settlement rates in the Caribbean render these coral reefs more vulnerable to further damage and more likely to benefit for human-assisted restoration efforts.

When reefs in the central and western South Pacific and the GBR are impacted by natural events such as severe tropical storms, bleaching, CoT outbreaks, and disease, human-mediated restoration will usually not be required. This likely also holds for human-induced damage to reefs in the South Pacific and GBR where the damage event is fleeting — e.g., minor ship groundings or anchor damage. High coral recruitment rates result in rapid repopulation of reefs by coral planulae from either the same reef or nearby reefs.

In the Caribbean, recruitment was dominated by brooding species in the families Agariciidae, Poritidae, and Faviidae with few *Acropora* recruits. The decimated *Acropora* populations and the low recruitment levels have created a situation where efforts to transplant *Acropora* in the Caribbean are necessary and will make an interesting long-term experiment challenging our understanding of genetics and coral reef ecology.

ACKNOWLEDGMENTS

We wish to gratefully acknowledge the use of the facilities at the Discovery Bay Marine Laboratory, University of the West Indies. B.L. Kojis contributed to this paper while she was a J.W. Fulbright Senior Fellow at the University of the West Indies. We are also grateful for support through the Tropical Discoveries Fund and the Coastal Water Improvement Project 2/USAID. This is publication #717 from the Discovery Bay Marine Laboratory. We gratefully acknowledge the anonymous reviewers for substantially improving the manuscript.

REFERENCES

1. Birkeland, C., Introduction, in *Life and Death of Coral Reefs*, Birkeland, C., Ed., Chapman and Hall, New York, 1, 1997.
2. Hodgson, G. and Liebeler, J., *The Global Coral Reef Crisis: Trends and Solutions*, Reef Check Foundation, Los Angeles, CA, 79, 2002.
3. Jackson, J., Foreword, in *The Global Coral Reef Crisis: Trends and Solutions*, Hodgson, G. and Liebeler, J., Eds., Reef Check Foundation, Los Angeles, CA, 5, 2002.
4. Schafer, K.L., Coral bleaching event along the Belize Barrier Reef, in *International Conference on Scientific Aspects of Coral Reef Assessment, Monitoring, and Restoration: Program and Abstracts*, The National Coral Reef Institute, Nova Southeastern University, Fort Lauderdale, FL, 172, 1999.
5. Bythell, J.C., Bythell, M., and Gladfelter, E.H., Chronic and catastrophic natural mortality of three common Caribbean reef corals, *Coral Reefs,* 12, 143, 1993.
6. Woodley, J.D., The effects of Hurricane Gilbert on coral reefs at Discovery Bay, in *Assessment of the Economic Impacts of Hurricane Gilbert on Coastal and Marine Resources in Jamaica, CEP Technical Report No. 4*, Bacon, P., Ed., UNEP Caribbean Environment Programme, Kingston, Jamaica, 79, 1989.
7. Woodley, J.D. et al., Hurricane Allen's impact on Jamaican coral reefs, *Science,* 214, 749, 1981.
8. Bythell, J.C. and Sheppard, C.R.C., Mass mortality of Caribbean shallow water corals, *Mar. Pollut. Bull.*, 26, 296, 1993.
9. Carpenter, R., Mass mortality of *Diadema antillarum*: I. Long-term effects on sea urchin population dynamics and coral reef algal communities, *Mar. Biol.,* 104, 67, 1990.
10. Lessios, H.A., Mass mortality of *Diadema antillarum* in the Caribbean: what have we learned? *Annu. Rev. Ecol. Systemat.*, 19, 371, 1988.

11. Endean, R. and Cameron, A.M., *Acanthaster planci* population outbreaks, in *Ecosystems of the World: Coral Reefs,* Vol. 25. Dubinsky, Z., Ed., Elsevier Science Publishing, New York, 419, 1990.
12. Zann, L., Brodie, J., and Vuki, V.C., History and dynamics of the crown-of-thorns starfish *Acanthaster planci* (L.) in the Suva area, Fiji, *Coral Reefs,* 9, 135, 1990.
13. Rogers, C.S. and Garrison, V.H., Ten years after the crime: lasting effects of damage from a cruise ship anchor on a coral reef in St. John, U.S. Virgin Islands, *Bull. Mar. Sci.,* 69, 793, 2001.
14. Harriott, V.J. and Fisk, D.A., Coral transplantation as a reef management option, in *Proc. Sixth Int. Coral Reef Symp.,* 375, 1988.
15. Edwards, A.J. and Clark, S., Coral transplantation: a useful management tool or a misguided meddling? *Mar. Pollut. Bull.*, 37, 474, 1999.
16. Bowden-Kerby, A., Coral transplantation in sheltered habitats using unattached fragments and cultured colonies, in *Proc. 8th Intl. Coral Reef Sym.,* 2063, 1997.
17. Bowden-Kerby, A., Low-tech coral reef restoration methods modeled after natural fragmentation processes, *Bull Mar. Sci.,* 69, 915, 2001.
18. Lindahl, U., Coral reef rehabilitation through transplantation of Staghorn corals: effects of artificial stabilization and mechanical damages, *Coral Reefs*, 22, 217, 2003.
19. Okinawa Development Agency, Research of Coral Transplantation in the Ishigaki Port, 5, 1997.
20. Japan Marine Science and Technology Center, Report of the Coral Reef Project, 1, 1, 1991.
21. Highsmith, R.C., Riggs, A.C., and D'Antonio, C.M., Survival of hurricane-generated coral fragments and a disturbance model of reef calcification/growth rates, *Oecologia*, 46, 322, 1980.
22. Bruckner, A.W. and Bruckner, R.J., Condition of restored *Acropora palmata* fragments off Mona Island, Puerto Rico, 2 years after the *Fortuna Reefer* ship grounding, *Coral Reefs*, 20, 235, 2001.
23. Kojis, B.L. and Quinn, N.J., Factors to consider when transplanting hermatypic coral to accelerate regeneration of damaged coral reefs, in *Conference on Environmental Engineering, Townsville 8–10 July 1981,* 1981, 183.
24. Mundy, C. and Babcock, R., Are vertical distribution patterns of scleractinian corals maintained by pre- or postsettlement processes? A case study of three contrasting species, *Mar. Ecol. Prog. Ser.*, 1998, 109, 2000.
25. Harriott V.J. and Fisk, D.A., Recruitment patterns of Scleractinian corals: a study of three reefs, *Aust. J. Mar. Freshwater Research,* 39, 409, 1988.
26. Hughes, T.P., Baird, A.H., Dinsdale, E.A., Moltschaniwskyj, N.A., Pratchett, M.S., Tanner, J.E., and Willis, B.L., Patterns of recruitment and abundance of corals along the Great Barrier Reef, *Nature,* 397, 59, 1999.
27. Goreau, T.F. and Wells J.W., The shallow–water Scleractinia of Jamaica: revised list of species and their vertical distribution range, *Bull. Mar. Sci.,* 17, 442, 1967.
28. Woodley, J.D. and Robinson, E., Modern reef localities — West Discovery Bay, in *Field Guidebook to the Modern and Ancient Reefs of Jamaica, 3rd International Coral Reef Symposium*, Atlantic Reef Committee, University of Miami, Miami, FL, 1977, 18.
29. Goreau, T.F., The ecology of Jamaican coral reefs. I. Species composition and zonation, *Ecology* 40, 67, 1959.
30. Tunnicliffe, V., Caribbean staghorn coral populations: Pre–Hurricane Allen conditions in Discovery Bay, Jamaica. *Bull. Mar. Sci.,* 33, 132, 1983.
31. Knowlton, N., Lang, J.C., Rooney, M.C., and Clifford, P., When hurricanes kill corals: evidence for delayed mortality in Jamaican staghorns, *Nature,* 294, 251, 1981.
32. Hughes, T.P. and Tanner, J.E., Recruitment failure, life histories, and long-term decline of Caribbean corals, *Ecology,* 81, 2250, 2000.
33. Wapnick, C.M., Precht, W.F., and Aronson, R.B., Millennial-scale dynamics of staghorn coral in Discovery Bay, Jamaica, *Ecol. Lett.*, 7, 1, 2004.
34. Quinn, N.J and Kojis, B.L., *Acropora* recruitment along the West Fore Reef at Discovery Bay, Jamaica, *Int. J. Trop. Biol. Conserv./Revista de Biología Tropica*, 53, 39, 2005.
35. Rogers, C.S. and Beets, J., Degradation of marine ecosystems and decline of fishery resources in marine protected areas in the U.S. Virgin Islands, *Environ. Conserv.*, 28, 312, 2001.
36. Precht, W.F., Bruckner, A.W., Aronson, R.B., and Bruckner, R.J., Endangered acroporid corals of the Caribbean, *Coral Reefs*, 21, 41, 2002.

37. Rogers, C.S., Fitz, H.C. III, Gilnack, M., Beets, J., and Hardin, J., Scleractinian coral recruitment patterns at Salt River Canyon, St. Croix. U.S. Virgin Islands, *Coral Reefs*, 3, 69, 1984.
38. Kojis, B.L., Baseline data on coral recruitment in the northern U.S. Virgin Islands, Caribbean Fisheries Management Council, 1, 1999.
39. Smith, S.R., Patterns of coral recruitment and postsettlement mortality on Bermuda's reefs: comparisons to Caribbean and Pacific reefs, *Am. Zool.*, 32, 663, 1992.
40. Quinn, N.J. and Kojis, B.L., Evaluating the potential of natural reproduction and artificial means to increase *Acropora* populations in Jamaica, *Int. J. Trop. Biol. Conserv./Revista de Biología Tropica*, 55 (in press).
41. Sammarco, P., Echinoid grazing as a structuring force in coral communities: whole reef manipulations, *J. Exp. Mar. Biol. Ecol.*, 6, 31, 1982.
42. Morrison, D., Comparing fish and urchin grazing in shallow and deeper coral reef algal communities, *Ecology*, 69, 1367, 1988.
43. Kojis, B.L. and Quinn, N.J., Biological limits to Caribbean reef recovery. A comparison with western South Pacific reefs, in *Global Aspects of Coral Reefs, Health, Hazards, and History*, Miami, FL, 1993, 353.
44. Knowlton, N., Lang, J.C., and Keller, B.D., Case study of natural population collapse: posthurricane predation on Jamaican staghorn corals, *Smithson. Contr. Mar. Sci.*, 31, 1, 1990.
45. Smith, S.R., Patterns of coral recruitment, recruitment and juvenile mortality with depth at Conch Reef, Florida, in *Proc. 8th International Coral Reef Symp.*, Panama 2, 1197, 1997.
46. Miller, M.W., Weil, E., and Szmant, A.M., Coral recruitment and juvenile mortality as structuring factors for reef benthic communities in Biscayne National Park, USA, *Coral Reefs*, 19, 115, 2000.
47. Aronson, R.B. and Precht, W.F., White-band disease and the changing face of Caribbean coral reefs, *Hydrobiologia*, 460, 25, 2001.
48. Rylaarsdam, K.W., Life histories and abundance patterns of colonial corals on Jamaican reefs, *Mar. Ecol. Prog. Ser.*, 16, 249, 1983.
49. Zubillaga, A.L., Bastidas, C., and Croquer, A., Estandarizacion de un protocolo para la estimacion del estatus poblacional de *Acropora palmata*, in *18th Meeting of the Association of Marine Laboratories of the Caribbean*, Port of Spain, Trinidad (Abstract), 45, 2003.
50. Stennett, M., Quinn, N.J., Gayle, P.M.H., and Bowden-Kerby, A. Enhancing coral habitat for the benefit of the tourist and subsistence fisheries sectors, *18th Annual Scientific Research Council Annual Meeting* Nov 23–26, 2004, Kingston, Jamaica, 54, 2004.
51. Gladfelter, E.H., Monahan, R.K., and Gladfelter, W.B., Growth rates of five reef-building corals in the northeastern Caribbean, *Bull. Mar. Sci.*, 28, 728, 1978.
52. Lewis, J.B., Axelsen, F., Goodbody, I., Page, C., and Chislett, G., Comparative growth rates of some corals in the Caribbean, in *Mar. Sci. Manu. Rep. 10*, McGill Univ., Montreal, 1968.
53. Wallace, C.C., *Staghorn corals of the world: a Revision of the Coral Genus* Acropora *(Scleractinia; Astrocoeniina; Acroporidae) Worldwide, with Emphasis on Morphology, Phylogeny and Biogeography*, CSIRO Publishing, Collingwood, Victoria, Australia, 421, 1999.
54. Naidu, S.D. and Morrison, R.J., Contamination of Suva Harbour, Fiji, *Mar. Poll. Bull.*, 29(1–3), 126, 1994.
55. Quinn, N.J. and Davis, M.T., The productivity and public health considerations of the urban women's daytime subsistence fishery off Suva Peninsula, Fiji, *South Pac. J. Natural Sci.*, 15, 61, 1997.
56. Davis, M.T., Newell, P.F., and Quinn, N.J., TBT contamination of an artisanal subsistence fishery in Suva harbour, Fiji, *Ocean Coastal Manage.*, 42, 591, 1999.
57. Wallace, C.C., Reproduction, recruitment, and fragmentation in nine sympatric species of the coral genus *Acropora, Mar. Biol.*, 88, 217, 1985.
58. Dunstan, P.K. and Johnson, C.R., Spatiotemporal variation in coral recruitment at different scales on Heron Reef, southern Great Barrier Reef, *Coral Reefs*, 17, 71, 1998.
59. Quinn, N.J. and Kojis, B.L., The dynamics of coral reef community structure and recruitment patterns around Rota, Saipan, and Tinian, Western Pacific, *Bull. Mar. Sci.*, 73, 95, 2003.
60. Birkeland, C., Status of coral reefs in the Marianas, in *Status of Coral Reefs in the Pacific*, Gregg, R.W. and Birkeland, C., Eds., University of Hawaii Sea Grant Program, Honolulu, HI, 1997, 91.
61. Birkeland, C., Rowley, D., and Randall, R.H., Coral recruitment patterns at Guam, in *Proc. 4th Int'l. Coral Reef Symp.*, Manila 2, 1982, 339.

62. Kojis, B.L. and Quinn, N.J., The importance of regional differences in hard coral recruitment rates for determining the need for coral restoration, *Bull. Mar. Sci.*, 69, 967, 2001.
63. Pernetta, J., *Atlas of the Oceans*, Sterling, New York, 208, 1994.
64. Harrison, P. and Wallace, C.C., Reproduction, dispersal, and recruitment of scleractinian corals, in *Ecosystems of the World*, Vol. 25, Dubinsky, Z., Ed., Elsevier, New York, 1990, 133.
65. Miller, K. and Mundy, C., Rapid settlement in broadcast spawning corals: implications for larval dispersal, *Coral Reefs*, 22, 99, 2003.
66. Wallace, C.C., Four years of juvenile coral recruitment to five reef front sites, in *Proc 5th Int. Coral Reef Symp.*, Tahiti 4, 385, 1985.
67. Quinn, N.J. and Kojis, B.L., Community structure of the living fossil coralline sponge populations at the Grotto, Saipan, Northern Mariana Islands, *Bull. Mar. Sci.*, 65, 227, 1999.
68. Quinn, N.J., Marine monitoring program for non-point source pollution in the Commonwealth of the Northern Mariana Islands, Division of Environmental Quality Tech. Report, 267, 1995.
69. Kojis, B.L. and Quinn, N.J., Seasonal and depth variation in fecundity of *Acropora palifera* at two reefs in Papua New Guinea, *Coral Reefs,* 3, 165, 1984.
70. Jaap, W.C. and Morelock, J., Soto's reef restoration project, Georgetown, Grand Cayman Island, British West Indies, in Tech. Report Submitted to Holland American Line – Westours and Depart. Environment, Cayman Islands, 16, 1998.
71. Bowden-Kerby, A., Coral aquaculture by Pacific Island communities, Abstract, World Aquaculture Society Meeting, Sydney, Australia, 1999.
72. Harriott, V.J., Can corals be harvested sustainably? *Ambio*, 32, 130, 2003.
73. Omori, M. and Fujiwara, S. (Eds.), *Manual for Restoration and Remediation of Coral Reefs*, Nature Conservation Bureau, Ministry of the Environment, Japan, 2004.
74. Heyward, A.J., Smith, L.C., Rees, M., and Field, S.N., Enhancement of coral recruitment by *in situ* mass culture of juvenile corals, *Mar. Ecol. Prog. Ser.,* 230, 113, 2002.
75. Baggett, L.S. and Bright, T.J., Coral recruitment at the East Flower Garden Reef (Northwestern Gulf of Mexico), in *Proc. 5th International Coral Reef Symp.*, Tahiti, 4, 379, 1985.
76. Fisk, D.A. and Harriott, V.J., Spatial and temporal variation in coral recruitment on the Great Barrier Reef: implications for dispersal hypotheses, *Mar. Biol.*, 107, 485, 1990.
77. Sammarco, P.W., *Diadema* and its relationship to coral spat mortality: Grazing competition and biological disturbance, *J. Exp. Mar. Biol. Ecol.,* 45, 245, 1980.

5 Compensatory Restoration: How Much Is Enough? Legal, Economic, and Ecological Considerations

Sharon K. Shutler, Stephen Gittings, Tony Penn, and Joe Schittone

CONTENTS

5.1 INTRODUCTION

Federal environmental statutes provide liability schemes whereby parties responsible for injuring corals are liable for the costs of restoration. Restoration includes projects designed to return the coral to baseline conditions as well as projects designed to compensate for the coral ecosystem services lost from the time of injury until the time of full recovery. The challenge facing the National Oceanic and Atmospheric Administration (NOAA) and other federal and state trustees is to develop legally defensible methodologies to quantify the amount of compensatory restoration owed by those liable for injuries to coral. This chapter will explore the use of Habitat Equivalency Analysis (HEA) as a legally defensible means to quantify the amount of compensatory restoration owed, examine metrics for appropriate measurement of recovery for western Atlantic coral reefs, recommend coral recovery horizons for use in the HEA, and recommend future research to refine coral recovery horizons.

5.2 NATURAL RESOURCE DAMAGES STATUTORY SCHEME

In the Florida Keys alone, on average over 600 reported boat groundings occur per year. Of these, approximately 20 groundings per year injure coral to an extent significant enough to result in enforcement actions. While these boat groundings cause injuries that range considerably in both size and severity, cumulatively they result in significant destruction of coral reef ecosystems in the Florida Keys National Marine Sanctuary (FKNMS). After the 1984 grounding of the *M/V Wellwood* in the FKNMS, resulting in moderate-to-severe coral destruction in over 2300 m^2 of coral reef,[1] Congress amended the National Marine Sanctuaries Act (NMSA) in 1988 to provide a cause of action to NOAA[2] to redress injuries to natural resources, including coral ecosystems in the nation's National Marine Sanctuaries.[3,4] Two years later, citing the grounding of a freighter in Key Biscayne National Park,[5] Congress passed legislation that provided similar authority to the National Park Service for injuries to natural resources in the nation's National Parks.[6] Prior to the NMSA amendments and 19jj (the analogous statute for the National Parks), both NOAA and the National Park Service (collectively, the "Trustees") had to resort to burdensome common law theories of liability to seek recoveries for injuries to coral or other natural resources within the Sanctuaries and National Parks. These common law theories were subject to a variety of equitable defenses (nonstatutory) that made prosecution challenging. Furthermore, any monetary recoveries by the Trustees reverted to the United States Treasury pursuant to the Miscellaneous Receipts Act and therefore were unavailable for coral restoration.

Recognizing the need for clear statutory authority to prosecute parties responsible for injuries to coral resources and to provide a mechanism to retain monetary recoveries for coral restoration, Congress specifically prohibited activities that cause injuries to living and nonliving resources within the National Marine Sanctuaries and National Parks. While the language in the NMSA and 19jj varies slightly, both establish a liability scheme whereby a person, corporation, or vessel responsible (collectively, Responsible Parties) for causing injury to natural resources, including coral, is liable for damages as defined to include, among other items, the cost of restoration.[7] Unlike common law, defenses available to the Responsible Parties under these statutes are limited. To be entitled to a statutory defense, the Responsible Party must demonstrate that (1) the coral destruction resulted solely from an act of God, war, or the act of a third party; (2) the destruction was authorized by state law; or (3) the destruction was negligible.[8–10]

When Responsible Parties destroy coral resources within a National Marine Sanctuary or a National Park, NOAA or the National Park Service must determine the amount of injury to the coral and the coral reef ecosystems and then determine the appropriate amount of restoration. By statute, restoration includes two separate components. The first component is the cost of replacing, restoring, or acquiring the equivalent of the injured resources.[11,12] The second is the value of the lost use of the injured resources pending their restoration or replacement or the acquisition of the equivalent resource.[13,14]

The NMSA does not define either component of restoration, nor has NOAA promulgated natural resource damage regulations under the NMSA that provide definitions or guidance for calculating the appropriate amounts and measures of restoration. However, NOAA has promulgated regulations governing the development of natural resource damage claims arising from releases of oil pursuant to the Oil Pollution Act (OPA) of 1990.[15] Accordingly, NOAA relies on the OPA natural resource damage regulations in developing coral natural resource damage claims under the NMSA.

The OPA regulations define the two statutory components of restoration. The first component, primary restoration, is any action, including natural recovery, that returns the injured natural resources and services to baseline. The baseline refers to the condition of the natural resources and services that would have existed but for the incident.[16] However, even with primary restoration, resource services are lost from the time of injury until the time when the resources have recovered to baseline or at least to a new level of ecosystem equilibrium. These losses are captured by the second component of restoration under the NMSA, compensatory restoration, which NOAA defines as any action taken to compensate for interim losses of natural resources and services that occur from the date of the incident until recovery.[17]

The concepts of primary and compensatory restoration are derived from tort law and together are intended to make the injured party whole. Take, for example, a claim for personal injury resulting from a car accident. The injured plaintiff can sue for the costs incurred for medical care, including doctors' bills, hospital expenses, prescriptions, and physical therapy. Medical care is necessary for the plaintiff's recovery. Likewise, the plaintiff is entitled to the costs of car repair. Medical care and car repair are similar to primary restoration. They are intended to return the plaintiff to his or her baseline condition. However, from the time of injury until the time the plaintiff is capable of returning to work, the plaintiff has lost wages. The plaintiff is not made fully whole until these wages are compensated for. Lost wages are analogous to compensatory restoration. In the former, the plaintiff is made whole only after payment of his or her interim lost income. For coral reef destruction, the public and the injured resources are made whole only after the ecological services lost during the interim period between injury and full recovery are recaptured through additional restoration projects.

The challenge facing NOAA is how to scale a compensatory restoration project so that it provides the same level of services as the services lost. As will be seen in this chapter, scaling compensatory restoration projects will depend on the ability of the Trustees to quantify service losses; estimate how long it will take for services to return their baseline, either through natural recovery or after primary restoration; and identify projects that will provide the same amount of services as those lost. Employing an appropriate tool to help balance the service losses and benefits sides of the equation is critical to this process. This balancing or scaling process, will be discussed in the next section. Challenges associated with selecting the right service flows to measure and determining the recovery horizons will then be addressed.

5.3 COMPENSATORY RESTORATION SCALING

Scaling determines how much restoration is owed to offset interim ecological service losses until the injured resources and their services return to baseline. NOAA's natural resource damage assessment regulations under OPA define two approaches for determining compensatory restoration scale: the valuation approach and the service-to-service (or resource-to-resource) approach. Under both approaches, the necessary project scale is that which provides value equal to the value of interim losses. The valuation approach is applicable when the lost and restored resources and services are not of the same type and quality, or of comparable value. Then, values are explicitly measured to ensure that the scale of compensatory restoration provides value equal to the value of interim loss.[18] By contrast, the service-to-service approach is appropriate where the restored resources and services are the same type and quality, and of comparable value as the interim losses, and the compensatory restoration project scale is determined by equating the *quantity* of restored resources and services with those lost.[19,20]

By equating quantities under these conditions, it is implicit that the interim loss value and the restoration value are equal. Examples of service-to-service scaling applications can be found in Marine Ecology Progress Series (vol. 264, 2003), which devoted a theme section to research studies that enable service-to-service scaling for different resources and services.[21] In one example, Peterson et al.[22] quantified fish production from restored oyster reef habitat as a way to determine reef restoration scale to compensate for fish biomass loss. French McCay et al.[23] described how fishing restrictions in a lobster fishery can compensate for the loss of lobsters after a major oil spill.

Habitat Equivalency Analysis (HEA) is an example of the service-to-service scaling approach applied to habitat, such as coral, and is a tool for calculating the amount of compensatory restoration owed for injuries to coral. Within the HEA context, interim losses are quantified as lost habitat resources and services. The scaling exercise determines the size or quantity of habitat restoration that provides the equivalent amount of habitat resources and services lost. Unsworth and Bishop[24] first proposed this type of framework and applied it to wetland habitat. Since then, two more HEA applications have been published in the literature. Fonseca et al.[25] and Penn and Tomasi[26] dealt with seagrass and salt marsh injuries, respectively. In addition to HEA's acceptance in the literature, Trustees are increasingly using HEA to quantify the amount of compensatory restoration owed.[27]

To apply HEA in the coral environment or any other environment, a number of parameters are necessary to quantify interim losses, the benefits of restoration, and ultimately the compensatory restoration scale. There are three important injury parameters: (1) the area of habitat impact; (2) the initial degree of impact relative to baseline, which is the condition of the habitat but for the incident; and (3) how the degree of injury changes over time, relative to baseline. The degree of injury change over time is heavily influenced by the choice of primary restoration. It can be determined by evaluating the condition of a resource or service metric, a variable that can be measured that is closely correlated with many if not all of the services provided by a habitat.

Fonseca et al.[25] used short shoot density as a metric to define the level of seagrass function; short shoot density served as a proxy for seagrass functions of providing food, shelter, sediment stabilization, and nutrient cycling. At the Blackbird Mine site, Trustees used the number of naturally spawning chinook salmon as the indicator of the health of Panther Creek, which had been impacted by mining wastes.[28]

The area and functional parameters determine the interim losses by year until the coral habitat returns to its baseline condition. If baseline is never reached, the interim losses continue to accrue in perpetuity. All else being equal, the longer the period of interim losses, the greater amount of compensatory restoration owed.[29] Interim losses are quantified as service square meter-years — the flow of resources and services from 1 m^2 of habitat for 1 year.

The HEA also requires identification of compensatory restoration projects that provide equivalent resources and services and specific parameters to quantify the benefits of restoration. The parameters include the time at which the primary restoration project has been completed, the duration of the project, and the function of the restored habitat through time relative to the injured habitat prior to the incident. The function parameter should be measured using the same metric that defines habitat function in the injury area so that the quantified restoration resources and services are comparable to the quantified interim losses. The restoration benefits defined by these parameters are also quantified as service square meter-years.

Because the services lost due to an incident and the services gained through restoration occur in different time periods, they are not directly comparable. In valuing goods and services, people have a rate of time preference and prefer to use or consume goods and services in the present rather than postpone their use or consumption to some future time. To make the services that occur in different time periods comparable, which is required to determine restoration scale, a discount factor is applied to determine discounted or time-adjusted services.[30]

Once the interim losses and the benefits of restoration are in comparable units, it is possible to determine the compensatory restoration scale that makes them equal. Figure 5.1 graphically demonstrates compensatory restoration scaling; the restoration scale is such that the services represented in Area b just equal the services represented in Area a.[31]

5.4 COMPENSATORY CLAIMS WITHIN THE LEGAL FRAMEWORK

NOAA's compensatory restoration claims, including methodologies to scale restoration, must withstand judicial scrutiny. Accordingly, if NOAA is unable to settle its claims with the Responsible Party, NOAA is authorized to file suit in U.S. District Court and seek damages.[32] NOAA

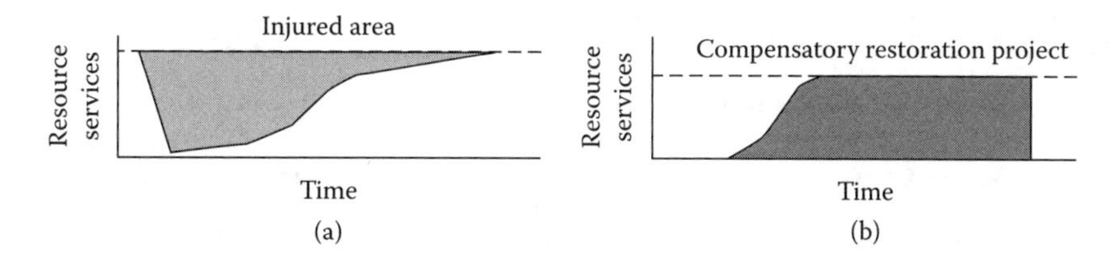

FIGURE 5.1 Graphical depiction of compensatory restoration scaling.

must prove liability and the nature and extent of the injuries, and then demonstrate to the trier of fact (judge or a jury) that the proposed restoration plan is appropriate and reasonable. NOAA has successfully litigated several natural resource damage claims in the FKNMS. *United States v. Great Lakes Dredge & Dock*, 259 F.3d 1300 (11th Cir. 2001); *Great Lakes*, 1999 U.S. Dist. Lexis 17612; *U.S. v. M/V Jacquelyn L.*, 100 F.3d 1520 (11th Cir. 1996); *United States v. Fisher* 22 F.3d 262 (11th Cir. 1994); *Fisher*, 977 F. Supp. at 1201; *U.S. v. M/V Beholden*, 856 F. Supp. 668 (S.D. Fla. 1994).

Two of those cases, *Great Lakes* and *Fisher*, involved substantial injuries to seagrass communities. In both cases, NOAA employed the HEA to determine the appropriate amount of seagrass compensatory restoration. The Defendants/Responsible Parties challenged the HEA, and in both cases, the Courts upheld the use of the HEA to calculate the amount of seagrass compensatory restoration owed. In *Great Lakes,* the Defendant/Responsible Party challenged the very admissibility of the HEA in the trial proceedings. Citing the Supreme Court case *Daubert v. Merrell Dow Pharmaceuticals, Inc.,* the defendants in *Great Lakes* argued that the HEA was not reliable and therefore should be excluded as evidence in the trial proceedings. The defendants argued that the HEA failed to meet the four factors laid out in *Daubert*:

1. Whether a theory can be or has been tested
2. Whether a theory or technique has been peer reviewed
3. The known or potential error rate of a technique
4. Whether the theory or technique is generally accepted

In holding that the HEA was admissible, the court came to the following conclusions. First, the HEA was not a scientific principle subject to testing *per se*, but a mathematical equation subject to the limitations of the input data. Second, the HEA had been peer reviewed and published. Third, the issue of error rate was not applicable as the error rate is determined by errors in the data, not errors in the HEA itself. Finally, while the HEA had not had time to truly gain general acceptance, "the relative youth of a scientific technique does not make it any less valid" (*United States v. Great Lakes Dredge & Dock*, U.S. Dist. Lexis 17612, July 28, 1999, Order at p.4). In both *Great Lakes* and *Fisher*, the Courts found that the HEA methodology would be an appropriate tool to calculate compensatory restoration when, among other things:

- The primary category of lost on-site services pertains to the ecological/biological function of an area.
- Sufficient data to perform the HEA are available or are cost effective to collect.

Accordingly, the HEA can be an appropriate tool for calculating the amount of compensatory restoration owed provided that the judicial factors are met.[33]

The next two sections focus on the application of the two judicial factors necessary to apply the HEA for coral compensatory projects. Section 5.5 addresses whether substantial ecological/biological lost on-site services exist. Section 5.6 addresses whether sufficient coral recovery horizon data to perform the HEA exists and provides reasonable estimates of recovery horizons based on the available data. Section 5.7 discusses, and provides suggestions for, future research needs.

TABLE 5.1
Examples of Services Provided by Coral Ecosystems[a]

	Service Categories	Service Examples
Ecological	Physical structure	Shoreline protection
		Land accretion
		Promotion of mangroves and seagrass beds
		Generation of sand
	Within-ecosystem	Production and maintenance of habitat
		Maintenance of biota (resident or transient)
		Maintenance of biodiversity and genetic diversity
		Regulation of ecosystem functions and processes
		Maintenance of ecosystem resilience
	Between-ecosystem	Connectivity between systems (larval and adult)
		Export of productivity/biomass
		Nitrogen fixation
		Regulation of carbonate budget
		Assimilation/remineralization of waste
Human Use	Information	Documents pollution
		Maintains a geologic, climatologic, and sea level record
	Social/cultural	Recreation/tourism
		Aesthetics (cultural/spiritual/artistic)
		Community livelihoods (jobs/food)
		Biomedicines
		Biotechnology
		Science/education
		Raw materials (constuction)
		Curio/jewelry/aquarium trade
		Minerals (oil/gas)

[a] The services identified in this table are not intended to be comprehensive. Furthermore, several could be classified in multiple categories.

5.5 CORAL ECOLOGICAL SERVICES

Corals provide extensive ecological services, many of which are summarized in Table 5.1. Corals provide essential habitat to a diversity of resident and transient organisms. Animals and plants associated with reef substrates contribute to ecosystem functions and processes such as waste recycling, organic and inorganic carbon production, and trophic dynamics. In combination, these ecological functions and processes create ecosystem resilience, which allows reefs to withstand disturbances without long-lasting changes.[34] One of the most important measures of resilience is the maintenance of corals as dominant species over potentially competitive soft algae.

The literature identifying and attempting to quantify coral reef services is extensive.[35–42] For present purposes we merely highlight the range of ecological services provided by coral ecosystems to demonstrate that primary and compensatory restoration are necessary for injuries caused by vessel groundings.

Services most affected by a vessel grounding are generally those listed under the Ecological Services category in Table 5.1. The ability of reef corals and other species to produce habitat, support the food requirements of associated species, and influence species interactions can be impaired by a ship grounding. The degree of service impairment depends on the nature, extent, and severity of the injury. Generally, the greater the areal extent and severity of the injury, the longer the services will be impaired, resulting in a longer recovery horizon, as compared to an injury of lesser area and severity.[43–47]

5.6 RESTORATION OF ECOLOGICAL SERVICE FLOWS

5.6.1 Injuries and Primary Restoration Options

In the coral context, restoration of ecological service flows to baseline depends on the type of injury. Theoretically, ecological service flows can return to baseline through a combination of primary restoration and natural recovery, though that may take many years to occur. The goal of primary restoration is to return the ecosystem services flows to as close to baseline as possible, in order to reduce the recovery horizon and therefore the amount of compensatory restoration owed. Accordingly, primary restoration is most appropriate when it significantly hastens recovery to baseline when compared to natural recovery. Primary restoration projects undertaken in the FKNMS have involved substrate stabilization, reattachment of dislodged colonies, coral transplantation, and restoration of three-dimensional habitat. Table 5.2 below lists categories of coral injury and types of primary restoration that may be appropriate.

TABLE 5.2
Categories of Coral Injury and Appropriate Primary Restoration

	Injury Category	Characteristics	Examples and Alternative Terms	Primary Restoration Options: Rubble Removal and/or Stabilization	Framework Cementation	Framework Infill	Relief Reconstruction	Coral Reattachment	Biological Enhancement[a]
Reef Framework	Fracturing	Cracks through surface framework, loose reef rock blocks	Cracking Splitting	x	x		x		
	Crushing	Reef rock pulverized by ship hull or other contact	Flattening	x			x		
	Displacement	Displacement of reef rock by hull, anchor, or propwash	Gouging Grooves Hull Scars Blowout Craters	x		x	x		
Coral Tissue and Cover	Toppling	Detached, but otherwise intact colonies	Tumbling					x	
	Abrasion	Soft tissue damage only, no or minimal skeletal damage	Scraping Lesions						
	Gouging	Soft tissue and colony skeleton damaged	Gashes				x		
	Fragmentation	Colony split, or breakage and dismemberment of branches	Fracturing Cracking Splitting					x	
	Destruction	Colony obliterated	Crushing	x	x		x		x

[a] Including, but not limited to coral transplantation, larval seeding, and chemical attractants.

5.6.2 Coral Recovery Horizons

How long recovery will take varies and depends on the nature, extent, and severity of the injury to living and nonliving resources, the requirements for recovery (e.g., reattachment, tissue regrowth, recruitment), and the choice of primary restoration. In addition to these factors, others significantly affect the recovery process and timing, including the local recruitment rate, juvenile mortality, coral growth rates, water quality, competition by macroalgae and other biota, grazing intensity, natural disturbance frequency, disease incidence, and bleaching frequency.[34,48]

The following discussion provides estimates of recovery times for reefs injured by incidents such as ship and boat groundings, anchoring, and other mechanical forms of damage. Recovery times assume the completion of primary restoration (see Table 5.3). The recovery time estimates will be incorporated into HEAs for future incidents to calculate the amount of compensatory restoration owed.[49]

5.6.3 Framework Injury

Some vessel groundings do not affect living assemblages to a significant degree but may affect reef framework (surface and subsurface material) in ways that change the capacity of those habitats and structures to provide ecological or human-use services. Three principal forms of framework injury can occur: fracturing, crushing, and displacement. Each has unique characteristics and poses different threats. Primary restoration techniques are available for responding to some of these injuries. When applied, they affect the rate at which full function can be restored.

5.6.3.1 Framework Fracturing

When a vessel cracks the surface of a coral reef, the framework generally becomes destabilized and prone to subsequent mobilization and destruction. Reef rock can be destroyed and topography altered. Mobile reef rock can cause incidental injury to nearby animals and plants. Repair involves the use of adhesives and other materials to secure broken portions of reef rock together before it is mobilized by subsequent disturbances. While stabilization may fail in cases with extensive fracturing, current technologies enable high success rates for this restoration activity. Recovery from these types of injury is nearly simultaneous with the repair, which recreates the three-dimensional habitat.

5.6.3.2 Framework Crushing

Particularly when large vessels run aground on reefs, surface framework can be obliterated as the hull pulverizes the relatively soft reef rock. This can happen during ingress or when the vessel bounces or twists once aground. Large coral skeletons and high points on the reef can be reduced to fine sediment in such cases, clogging reef crevices and other habitat otherwise suitable for colonization. Small pieces of rubble are also typically generated. Treatment of such areas can involve removal of sediments and rubble, and the construction of structures intended to replace lost three-dimensional surfaces. Current restoration materials and methods do not allow for thorough or exact replacement of the lost structure. Modification through physical and biological processes over many years may be necessary to fully regain all qualities and characteristics affected by the grounding incident. Nevertheless, recovery is considered complete with the conclusion of primary restoration.

5.6.3.3 Framework Displacement

Rubble is often produced during a ship grounding as corals or reef rock are broken free and plowed by a keel or hull moving across the bottom. It is also generated through excavation caused by propeller wash when operators attempt to back off a reef. It is important to remove the rubble either by replacing it into the excavated area or taking it to another location. When it is replaced into the excavated area, it can be stabilized using either a concrete and reef rock cap or pouring concrete into the spaces between the rubble. In either case, recovery consists of the restabilization itself and is considered complete immediately after primary restoration.

TABLE 5.3
Factors Affecting the Recovery Rates of Coral and Coral Reefs, and Predicted Recovery Times Following Various Types of Mechanical Injury

		Primary Restoration Options						Additional Requirements for Recovery							
	Injury Category	Rubble Removal and/or Stabilization	Framework Cementation	Framework Infill	Relief Reconstruction	Coral Reattachment	Biological Enhancement[b]	Retain Stability	Tissue Stabilization	Soft Tissue Replacement	Skeletal Replacement	Recruitment	Reproduction/Fecundity	Limiting Factor(s)	Recovery Time after Primary Restoration[a]
Reef Framework	Fracturing	x	x		x			x						Ensure stability	Immediate
	Crushing	x			x			x						Ensure stability	Immediate
	Displacement	x		x	x			x						Ensure stability	Immediate
Coral Tissue and Cover	Toppling					x		x	x					Ensure tissue stability	>2 yr
	Abrasion								x	x				Growth over abrasion	>6 mo (to max. healing; size dependent)
	Gouging				x				x	x	x			Growth into and over gouge (lateral and accretionary; size and depth dependent)	>5 yr (size dependent)
	Fragmentation					x		x	x	x	x		x	Regaining reproductive capacity	>5 yr (inverse size dependent)
	Destruction	x	x		x		x	x	x	x	x	x		Recruitment and regaining percent cover	>100 yr (area and severity dependent)

[a] Note that minimum recovery time depends on the limiting factor.

[b] Including, but not limited to, coral transplantation, larval seeding, and chemical attractants

5.6.4 Living Coral

5.6.4.1 Toppling

One of the common forms of mechanical injury to reefs involves dislodgement of coral colonies without direct tissue destruction. Detached corals can roll, causing tissue abrasion or movement into unfavorable habitats (sandy[43] or shaded locations, where they may bleach). Most remain unstable in subsequent energetic events and are vulnerable to continued stress or delayed mortality.[50]

Repair often involves reattachment to the substrate using epoxy or cement as soon as possible after the injury. Full recovery requires tissue stabilization followed by the resumption of normal tissue function. Stress in the interim can lead to deterioration of tissues that manifests over several days, weeks, or months (e.g., tissues can bleach in a matter of days or even die over a substantially longer period[1,51]). Recovery of coral tissue requires stabilization of the injured area followed by resumption of normal metabolic function. Evaluation of change in coral cover can be accomplished through repetitive photography or video that monitors tissue necrosis, bleaching, disease incidence, and other changes in tissue condition.[1] When recovery from bleaching occurs, it can take up to a year,[52] and it is presumed that normal function (including gametogenesis) can resume during the subsequent year. Thus, a 2-year recovery horizon might be expected for toppled corals, assuming tissues do not become diseased or die.

5.6.4.2 Abrasion

Contact by a vessel, anchor, or other objects can cause superficial injury to coral, destroying tissue but leaving skeletons intact. For this type of injury, no effective primary restoration exists, and NOAA relies on natural recovery, whereby margins of the tissue grow back over the denuded area. Work by several researchers studying common Caribbean corals suggests corals can regenerate tissue after small abrasions within 6 months of injury (for *Montastraea annularis,*[47,53–56] *Porites astreoides* and *Stephanocoenia michelinii* [57,58]; for *Acropora palmata*[59] and for *Madrais* spp.[57]).

However, abrasions beyond a certain size typically fail to recover, as the lesion area becomes infilled with sediment, or is colonized by algae or excavating sponges.[47,60] This means that the recovery horizon for small abrasions is relatively short, while injuries from larger lesions will be incorporated as losses "in perpetuity." The categorization of what constitutes a small vs. large injury must necessarily be deferred, as it has to be done on a case-by-case basis. This is because it is dependent not only on the size of the abrasion but many other factors, among which are the species, shape of lesion, its depth, colony size, colony reproductive status, bleached (or not) status, depth of colony, colony morphology, lesion location on colony, sedimentation and/or algal colonization rates at site, water quality, and colony growth status.

5.6.4.3 Gouging

Groundings typically cause gouges of varying depths on mechanically injured soft tissues and skeleton. Recovery entails tissue regeneration via lateral growth, as well as accretionary (skeletal) growth, which eventually may fill the gouge. Small gouges are slower to heal than small abrasions, partly because of the need for accretionary growth, but also because the depressions tend to accumulate sediments and other debris inimical to lateral growth.

Gouges on corals tend to remain visible for many years. It is possible, however, that small injuries could heal within a fairly short time without primary restoration. For example, the tissue surrounding a gouge less than 2 cm wide and 2 cm deep could stabilize during the first year following injury, heal over the open space within the next 2 years via lateral growth, then fill the void space over the next 2 years (based on published lateral and accretionary growth rates.[61,62] Gaps as large as 10 cm diameter (core holes closed with cement plugs) have been found to nearly fill in approximately 4 years (Hudson, personal communication), though small central areas typically remain exposed and may contain sediments that inhibit growth. Thus, a reasonable minimum expected recovery time for a small gouge may be 5 years.

Most injuries, however, are substantially larger than this, and recovery times would increase. Theoretical recovery times could be calculated using lateral and accretionary growth rates, if known, for the affected species. Full recovery for large gouges, however, has not been reported, as sediment and debris accumulation and algae growth inevitably preclude it. These are considered injuries in perpetuity for purposes of HEA.

5.6.4.4 Fragmentation

Most incidents of mechanical damage result in either fragmentation of branching corals or fracturing of massive (i.e., nonbranching) corals. Soft tissue injury on affected colonies is comparable to that on abraded or gouged colonies. Skeletal injury is sufficient to dislodge the fragment, making it vulnerable to displacement and later movement, particularly for smaller fragments.[63]

Repair of fractured colonies generally involves reattachment to the substrate or to other colony fragments using epoxy, cement, plastic-covered wire, cable ties, monofilament line, or uncoated wire. For properly secured colonies, lateral and accretionary growth over exposed areas and adhesives could result in recovery in as few as 5 years (by essentially the same mechanisms as those described for gouging).

Though large fragments of broken colonies can remain sexually mature, small ones may lose their ability to reproduce until they grow to a minimal size.[64,65] Lirman[43] found that, 3 years after Hurricane Andrew in South Florida, not only did fragments of *A. palmata* lack gametes, but so did adult colonies from which they detached. The only colonies that contained gametes at that time were those that suffered no breakage. Damaged colonies appeared to reestablish preinjury growth rates and gamete production in approximately 5 years.

5.6.4.5 Crushing

The most severe form of injury to coral reefs is crushing. In extreme cases, like the 1984 *M/V Wellwood* Florida Keys grounding, a vessel may obliterate corals and flatten the coral reef, resulting in large swaths that resemble parking lots.[66,67] The primary difference between this type of injury and others is that recolonization (via recruitment or corals and other species) is required for recovery to occur.

Predicting recovery time for crushed areas is exceedingly difficult, especially for Caribbean reefs, where the recruitment for frame-building corals has been extremely limited in recent years,[68–70] as compared to western Pacific reefs.[71] Recruitment at grounding sites in the Caribbean and western Atlantic have generally involved species that brood larvae (e.g., *P. astreoides* and *Agaricia agaricites*[72,73]). These may be numerically dominant but contribute only minimal cover and framework construction. Gittings[74] found high numbers of recruits of certain species (mostly gorgonians) but negligible cover by hard corals at the *M/V Wellwood* ground site 18 years after the grounding destroyed the reef assemblage. Rogers and Garrison[75] investigated natural recovery of a cruise ship anchor scar in St. John but found no significant increase in cover over 10 years. While coral recruits settled in the scar, survival and growth were poor.

Further, many factors appear to be causing rapid declines in the condition and resilience of coral reefs throughout the world.[76] Accordingly, even with primary restoration, severely injured coral reefs (e.g., crushed corals) may not fully return to pregrounding conditions[77] and may convert to alternative states.[67,78] Instead, recovery can only be defined in terms of a new equilibrium,[67] generally resulting in reduced, or at least different, service flows.

In addition to making recovery estimates difficult, the regulatory definition of recovery, which suggests that it is possible for injured natural resources to return to baseline either through primary restoration or natural recovery, may be inappropriate where living corals are completely destroyed. For these injuries, the natural recovery horizon is extraordinarily long, arguably perpetuity.

In order to avoid protracted litigation regarding the amount of compensatory restoration owed and to settle restoration-based claims, NOAA underestimates the likely recovery horizons for destruction of live coral. NOAA's approach recognizes that full recovery of a destroyed coral assemblage may never occur, and focuses on the reestablishment of the preinjury percentage coral cover to an injured area. It uses coral cover as a measure of reef function — or "metric," in the language of HEA. A metric is a variable that can be measured and represents the condition of many, if not all, habitat resources and services.

Coral cover has been associated with many different aspects of the reef community structure and condition. Among the most important are coral community vitality and resilience, traits reflective of key ecosystem services.[79] It is also critical to creating a high degree of reef architectural complexity, or rugosity[80–82], a principal factor credited with enabling the high diversity that makes reefs able to withstand or recover from disturbance. Recently, it even has been shown to play a significant role in the delivery of organic material to water column bacteria (through mucus secretion and dissolution) and to benthic heterotrophs (through entrapment of planktonic material[83]).

Other metrics, such as rugosity and spatial heterogeneity, also reflect the condition of reef habitat and commensurate services (particularly shelter for resident biota[84,85]). Metrics such as richness and evenness reflect temporal changes in biodiversity,[1] which significantly affect ecosystem resilience.[34] Recruitment[72] and to lesser extent, fecundity,[86] are metrics that also can be measured directly. Process rates, such as calcification, organic production and respiration, photosynthesis, and nitrogen fixation, could also be measured, though their suitability as proxies for overall ecosystem condition has not been demonstrated. Tissue condition, using such measures as zooxanthellae abundance, mucus production, disease prevalence, or biomarkers, has also been suggested for use in reef assessments. The development of coral cover, however, requires recruitment and growth (calcification) by presumably healthy individual colonies, leads to reef structural complexity, and ultimately promotes community richness. Thus, of all these metrics, coral cover may be the best integrator of all requirements for reef ecosystem recovery.

Contrary to many Caribbean studies, work in the northern Red Sea found that cover increased measurably at sites that had experienced ship groundings.[46] It should be noted that the study dealt with recovery in areas without primary restoration, occurred in the Red Sea (where conditions are generally considered more favorable for corals than in the Caribbean), and involved different species. Regardless, predicted recovery times were fairly long, estimated at between 54 and 600 years.[87] The wide range was attributed to the local conditions at each reef studied. For windward upper reef-slope coral reefs (habitats comparable to the Florida Keys Reef Tract), estimates for regrowth to preimpact coverage were in the range of 100 to 150 years.

Piniak et al.[88] used a single-species recovery model and arrived at a similar estimate using data on recruitment, juvenile mortality, and growth of *P. astreoides*, which is common in the western Atlantic region. *Porites astreoides* is a fairly prolific brooding species, found regularly among recruits on natural substrates.[74,89,90] Based on data from this species, Piniak et al.[88] predicted recovery of coral cover in 100 to 150 years. In actuality, full recovery would require the gradual development of an ecologically complex reef with temporally changing structure, biodiversity, species interactions, trophic webs, etc.

For the purpose of calculating the amount of compensatory restoration owed for an injury involving destruction of live corals, NOAA considers the Piniak et al.[88] single species model an acceptable method to estimate recovery from injuries and calculate interim service losses. We recognize that some field observations suggest that much longer recovery periods are more realistic, particularly given the declining state of reefs in the western Atlantic region. Until better data are available for other species and processes, however, *P. astreoides* recruitment and growth rates will serve as the model drivers. NOAA will adopt a recovery horizon of 100 to 150 years for areas destroyed by mechanical injuries, with year-to-year service levels determined by the Piniak et al. single species model.

5.7 RESEARCH NEEDS

Recovery times for most injury categories may vary widely depending on the extent of injury to individual coral colonies or communities. They also vary with environmental factors that differ dramatically on multiple spatial and temporal scales. In some places, small areas of injury may never recover; in others, fairly large injuries may heal fairly fast. Minimal recovery times for most categories of injury are fairly easy to estimate; accurate estimates for more severe injuries are not. For example, estimates of recovery time for severely abraded or gouged corals depend on improved understanding of local differences in such variables as tissue growth rates and potential competitive interactions that may affect space competition during recovery. There appear to be significant differences between reefs with regard to the potential for growth (particularly lateral growth[91]). The potential for recovery following abrasions and other tissue injuries is likely to vary accordingly. It is important to understand local differences and the environmental factors affecting this potential.

Data are surprisingly sparse for current rates of recruitment, juvenile mortality, and growth of Atlantic reef corals. Information is needed on regional differences in these variables, particularly within the western Atlantic, where large discrepancies may exist. In addition, there is clearly a need to improve our understanding of how the declining state of reefs is affecting these processes. One of the most significant mysteries relates to the apparent lack of recruitment of massive frame-building species,[68] particularly the *Montastraea* complex. These corals are known to reproduce prolifically during annual mass spawning events,[92] but they are seldom found on reefs as juveniles. Understanding why could improve not only our ability to improve estimates of recovery times, but also to judge when it might be prudent to enhance recruitment artificially via seeding (capturing gametes and later releasing larvae) or otherwise.[93,94]

Other metrics may serve as better proxies for lost services than coral cover does. If so, then they may also be better measures of recovery and could be used to evaluate both natural rates of recovery and the effectiveness of artificial enhancements in stimulating it. Proxies that reflect changes in resilience on reefs would be especially useful. They might initially include recruitment diversity, then later track community succession through measures of diversity and trophic structure. Such proxies could become important as more is learned about their interrelationships. Research investigating these possibilities would significantly improve the application of HEA to coral reefs.

ACKNOWLEDGMENT

The authors would like to thank the following individuals for input on the chapter and assistance in preparing the manuscript: Harriet Sopher, Scott Donahue, Kim Barry, Brian Julius, and Kathy Dalton.

REFERENCES

1. Gittings, S.R., T.J. Bright, A. Choi, and R.R. Barnett. 1988. The recovery process in a mechanically damaged coral reef community: recruitment and growth. *Proc. 6th Int. Coral Reef Symp.,* 2:225–230.
2. Specifically, the NMSA provides authority to the Secretary of Commerce. The Department of Commerce has, in turn, delegated this authority to NOAA.
3. See S. REP 100-595, #2, 1988. U.S.C.C.A.N. 4387–4388.
4. 16 U.S.C.§ 1441, *et seq.*
5. See S. REP 101-328, 1990 U.S.C.C.A.N. 603.
6. 16 U.S.C. §19jj, *et seq.*
7. Under the NMSA, the damages are defined to include: (1) the costs to restore, replace, or acquire the equivalent of the resources injured; (2) the value of the lost use of the resources pending restoration; (3) the costs of assessing the injuries to those resources; (4) cost of monitoring the restoration; (5) the costs of enforcement. 16 U.S.C. § 1443 (a).
8. 19jj does not include as a defense that the destruction was negligible.
9. 16 U.S.C. § 1443(a)(3).
10. 16 U.S.C. § 19jj-1(c).
11. 16 U.S.C. § 1432(6)(A)(i)(I).
12. 16 U.S.C. § 19jj(b)(A)(i).

13. 16 U.S.C. § 1432(6)(A)(i)(II).
14. 16 U.S.C. § 19jj(b)(A)(ii).
15. 33 U.S.C. § 2701, *et seq.*
16. 15 C.F.R. § 990.30.
17. 15 C.F.R. § 990.
18. In some circumstances, the "value-to-cost" variant of the valuation approach may be employed. Value-to-cost is only appropriate when valuation of the lost services is practicable but valuation of the restored natural resources and services cannot be performed within a reasonable time frame or at a reasonable cost. With this approach, the compensatory restoration is scaled by equating the cost of the restoration to the value (in dollar terms) of losses due to the injury.
19. Even without in-kind equivalency, the service-to-service approach can still be applied as long as the value differences between the restored and lost resources and services are known.
20. Jones, C.A. and K.A. Pease. 1997. Restoration-based compensation measures in natural resource liability statutes. *Cont. Econ. Policy* 15(4):111–122.
21. See the theme section "Restoration scaling in the marine environment." *Mar. Ecol. Prog. Ser.* 264:173–307.
22. Peterson, C.H., J.H. Grabowski, and S.P. Powers. 2003. Estimated enhancement of fish production resulting from restoring oyster reef habitat: quantitative valuation. *Mar. Ecol. Prog. Ser.* 264:249–264.
23. French McCay, D.P., M. Gibson, and J.S. Cobb. 2003. Scaling restoration of American lobsters: combined demographic and discounting model for an exploited species. *Mar. Ecol. Prog. Ser.* 264:177–196.
24. Unsworth, R.E. and R.C. Bishop. 1994. Assessing natural resource damages using environmental annuities. *Ecol. Econ.* 11:35–41.
25. Fonseca, M.S., B.E. Julius, and W.J. Kenworthy. 2000. Integrating biology and economics in seagrass restoration: how much is enough and why? *Ecol. Eng.* 15:227–237.
26. Penn, T. and T. Tomasi. 2002. Calculating resource restoration for an oil discharge in Lake Barre, Louisiana. *Env. Mgt.* 29(5):691–702.
27. See, for example, the Damage Assessment and Restoration Plans Chalk for Lavaca Bay, Kiroshima, and Westchester at www.darp.noaa.gov/publicat.htm and Chalk Point at www.darp.noaa.gov/neregion/chlkptar.htm.
28. Chapman, D., N. Iadanza, and T. Penn. 1997. Calculating resource compensation: an application of the service-to-service approach to the Blackbird mine hazardous waste site. Technical Paper, vol. 97-1. Damage Assessment and Restoration Program, National Oceanic and Atmospheric Administration, Silver Spring, MD. October 16.
29. For any one incident, there may be multiple sets of parameters to capture different types of injury.
30. A discount rate of three percent is commonly used in damage assessment cases. It is a rate that is in accordance with accepted economic principles (Freeman, 1993; and Lind, 1982).
31. For simplicity, discounting is not represented in Figure 5.1.
32. 16 U.S.C. § 1443(c).
33. The Court also stated that for the HEA to be appropriate, feasible compensatory restoration projects should be available that provide services of the same type, quality, and comparable value to those that were lost. This chapter does not focus on the availability of feasible compensatory restoration projects. Rather, it focuses more on the practical issues associated with available data for HEA inputs.
34. McClanahan, T.R., N.V.C. Polunin, and T.J. Done. 2002. Resilience of Coral Reefs. In *Resilience and the Behavior of Large-Scale Systems*, L.H. Gunderson and L. Pritchard Jr. (eds.) Island Press. Washington D.C.
35. Spurgeon, J.P.G. 1992. The economic valuation of coral reefs. *Mar. Poll. Bull.* 24(11):529–536.
36. Costanza, R., R. d'Arge, R. de Groot, S. Farber, M. Grasso, B. Hannon, K. Limburg, S. Naeem, R.V. O'Neill, J. Paruelo, R.G. Rashin, and P. Sutton. 1997. The value of the world's ecosystem services and natural capital. *Nature* 387:253–260.
37. Moberg, F. and C. Folke. 1999. Ecological goods and services of coral reef ecosystems. *Ecol. Econ.* 29:215–233.
38. Cesar, H.S.J. (ed.) 2000. *Collected Essays on the Economics of Coral Reefs*. Kalmar, Sweden: CORDIO.
39. Souter, D.W. and O. Linden. 2000. The health and future of coral reef systems. *Ocean Coastal Mgt.* 43:657–688.

40. Nystroem, M. and C. Folke. 2001. Spatial resilience of coral reefs. *Ecosystems* 4:406–417.
41. Spurgeon, J.P.G. 2001. Valuation of Coral Reefs: The Next Ten Years. Paper presented at workshop organized by ICLARM, in Penang, Malaysia. December 2001.
42. Moberg, F. and P. Rönnbäck. 2003. Ecosystem services of the tropical seascape: interactions, substitutions and restoration. *Ocean Coastal Mgt.* 46:27–46.
43. Lirman, D. 2000. Fragmentation in the branching coral *Acropora palmata* (Lamarck): growth, survivorship, and reproduction of colonies and fragments. *J. Exp. Mar. Biol. Ecol.* 251:41–57.
44. Hall, V. 2001. The response of *Acropora hyacinthus* and *Montipora tuberculosa* to three different types of colony damage: scraping injury, tissue mortality, and breakage. *J. Exp. Mar. Biol. Ecol.* 264: 209–223.
45. Oren, U., Y. Benayahu, H. Lubinevsky, and Y. Loya. 2001. Colony integration during regeneration in the stony coral *Favia favus. Ecology.* 82:802–813.
46. Riegl, B. 2001. Degradation of reef structure, coral and fish communities in the Red Sea by ship groundings and dynamite fisheries. *Bull. Mar. Sci.* 69:595–611.
47. Cróquer, A., E. Villamizar, and N. Noriega. 2002. Environmental factors affecting tissue regeneration of the reef-building coral *Montastraea annularis* (Faviidae) at Los Roques National Park, Venezuela. Revista Biología Tropical 50(3/4):1055–1065.
48. Fox, H.E., J.S. Pet, R. Dahure, and R.L. Caldwell. 2003. Recovery in rubble fields: long-term impacts of blast fishing. *Mar. Poll. Bull.* 46:1024–1031.
49. Recovery is also determined by the rate of change in services provided over time. This rate of change is captured by the selection of an appropriate path over the recovery duration. Absent data on a specific path, NOAA will assume a linear function.
50. Knowlton, N., J.C. Lang, M.C. Rooney, and P. Clifford. 1981. Evidence for delayed mortality in hurricane-damaged Jamaican staghorn corals. *Nature* 294:251–252.
51. Goreau, T.F. 1964. Mass expulsion of zooxanthellae from Jamaican reef communities after Hurricane Flora. *Science* 145:383–386.
52. Wilkinson, C., O. Linden, H. Cesar, G. Hodgson, J. Rubens, and A.E. Strong. 1999. Ecological and socioeconomic impacts of 1998 coral mortality in the Indian Ocean: An ENSO impact and a warning of future change? *Ambio* 28:188–196.
53. Meesters, E.H., M. Noordeloos, and R.P.M. Bak. 1994. Damage and regeneration: Links to growth in the reef-building coral *Montastrea annularis. Mar. Eco. Prog. Ser.* 112:119–128.
54. Van Veghel, M.L.J, and R.P.M. Bak. 1994. Reproductive characteristics of the polymorphic Caribbean reef building coral *Montastrea annularis.* 3. Reproduction in damaged and regeneration colonies. *Mar. Eco. Prog. Ser.* 109:229-233.
55. Meesters, E.H., W. Pauchli, and R.P.M. Bak. 1997. Predicting regeneration of physical damage on a reef-building coral by regeneration capacity and lesion shape. *Mar. Ecol. Prog. Ser.* 146:91–99.
56. Mascarelli, P.E. and L. Bunkley-Williams. 1999. An experimental field evaluation of healing in damaged, unbleached, and artificially bleached star coral, *Montastraea annularis. Bull. Mar. Sci.* 65:577–586.
57. Nagelkerken, I. and R.P.M. Bak. 1998. Differential regeneration of artificial lesions among sympatric morphs of the Caribbean corals *Porites astreoides* and *Stephanocoenia michelinii. Mar. Ecol. Prog. Ser.* 163:279–283.
58. Nagelkerken, I., E.H. Meesters, and R.P.M. Bak. 1999. Depth-related variation in regeneration of artificial lesions in the Caribbean corals *Porites astreoides* and *Stephanocoenia michelinii. J. Exp. Mar. Biol. Ecol.* 234:29–39.
59. Lirman, D. 2000. Lesion regeneration in the branching coral *Acropora palmata*: effects of colonization, colony size, lesion size, and lesion shape. *Mar. Ecol. Prog. Ser.* 197:209–215.
60. Meesters, E.H., I. Wesseling, and R.P.M. Bak. 1996. Partial mortality in three species of reef-building corals and the relation with colony morphology. *Bull. Mar. Sci.* 58:838–852.
61. Hudson, J.H. 1981. Growth rates of *Montastraea annularis*: a record of environmental change in Key Largo Coral Reef Marine Sanctuary, Florida. *Bull. Mar. Sci.* 31(2):444–459.
62. Gittings, S.R., T.J. Bright, and D.K. Hagman. 1993. Protection and monitoring of reefs on the Flower Garden Banks, 1972–1992. In Ginsburg, R.N. (Compiler) *Proc. Colloq. Global Aspects of Coral Reefs: Health, Hazards and History, 1993,* Rosensteil School of Marine and Atmospheric Sciences, University of Miami, pp. 181–187.

63. Bowden-Kirby, A. 2001. Low-tech coral reef restoration methods modeled after natural fragmentation processes. *Bull. Mar. Sci.* 69:915–931.
64. Bak, R.P.M. 1976. The growth of coral colonies and the importance of crustose coralline algae and burrowing sponges in relation with carbonate accumulation. *Neth. J. Sea Res.* 10:285–337.
65. Szmant-Froelich, A. 1985. The effect of colony size on the reproductive ability of the Caribbean coral *Montastrea annularis* (Ellis and Solander). *Proc. 5th Int. Coral Reef Symp.* 4:295–300.
66. Gittings, S.R. and T.J. Bright. 1988. The *M/V Wellwood* grounding: A sanctuary case study. The Science. *Oceanus* 31(1):35–41.
67. Precht, W., R. Aronson, and D. Swanson. 2001. Improving scientific decision-making in the restoration of ship-grounding sites on coral reefs. *Bull. Mar. Sci.* 69:1001–1012.
68. Hughes, T., and J. Tanner. 2000. Recruitment failure, life histories, and long-term decline of Caribbean corals. *Ecology.* 81(8):2250–2263.
69. Miller, M.W., E. Weil, and A. M. Szmant. 2000. Coral recruitment and juvenile mortality as structuring factors for reef benthic communities in Biscayne National Park, U.S.A. *Coral Reefs* 19:115–123.
70. Tougas, J. I. and J. W. Porter. 2002. Differential coral recruitment patterns in the Florida Keys. In J.W. Porter and K.G. Porter (eds.) *The Everglades, Florida Bay, and Coral Reefs of the Florida Keys: An Ecosystem Sourcebook.* CRC Press, Boca Raton, FL, pp. 789–811.
71. Kojis, B.L., and N.J. Quinn. 1994. Biological limits to Caribbean reef recovery. A comparison with western and south Pacific reefs. In Ginsburg, R.N. (compiler) *Proc. Colloq. Global Aspects of Coral Reefs: Health, Hazards and History,* 1993. Rosensteil School of Marine and Atmospheric Sciences, University of Miami, FL, pp. 353–359.
72. Smith, S.R. 1985. Reef damage and recovery after ship groundings on Bermuda. *Proc. 5th Int. Coral Reef Symp.* 6:497–502.
73. Gittings, S.R., T.J. Bright, and B.S. Holland. 1990. Five years of coral recovery following a freighter grounding in the Florida Keys. *Proc. American Academy of Underwater Sciences,* 10th Annual Symposium. pp. 89–105.
74. Gittings, S.R. 2002. Preconstruction coral survey of the *M/V Wellwood* grounding site. Marine Sanctuaries Conservation Series MSD-03-1. National Oceanic and Atmospheric Administration, National Ocean Service, Marine Sanctuaries Division. Silver Spring, MD.
75. Rogers, C.S. and V.H. Garrison. 2001. Ten years after the crime: Lasting effects of damage from a cruise ship anchor on a coral reef in St. John, U.S. Virgin Islands. *Bull. Mar. Sci.* 69:793–803.
76. Wilkinson, C. (ed.) 2000. *Status of Coral Reefs of the World: 2000.* Australian Institute of Marine Science. Cape Ferguson, Queensland, Australia.
77. Woodley, W., P. Alcolado, T. Austin, J. Barnes, R. Claro-Madruga, G. Ebanks-Petrie, R. Estrada, F. Geraldes, A. Glasspool, F. Homer, B. Luckhurst, E. Phillips, D. Shim, R. Smith, K. Sullivan Sealey, M. Vega, J. Ward, and J. Wiener. 2000. Status of coral reefs in the northern Caribbean and western Atlantic. pp. 261–285 in: Wilkinson, C. (ed.). *Status of Coral Reefs of the World: 2000.* Australian Institute of Marine Science. Cape Ferguson, Queensland, Australia. 363 pp.
78. Hatcher, B.G. 1984. A maritime accident provides evidence for alternate stable states in benthic communities on coral reefs. *Coral Reefs* 3:199–204.
79. Gardner, T., I. Côté, J. Gill, A. Grant, and A. Watkinson. 2003. Long-term region-wide declines in Caribbean corals. *Science* 301:958–960.
80. Chiappone, M. and K. Sullivan. 1996. Distribution, abundance, and species composition of juvenile scleractinian corals in the Florida reef tract. *Bull. Mar. Sci.* 58:555–569.
81. Edmunds, P. 2002. Long-term dynamics of coral reefs in St. John, US Virgin Islands. *Coral Reefs* 21:357–367.
82. Miller, M., and C. Gerstner. 2002. Reefs of an uninhabited Caribbean Island: fishes, benthic habitat, and opportunities to discern reef fishery impact. *Biol. Conserv.* 106:37–44.
83. Wild, C., M. Huettel, A. Klueter, S.G. Kremb, M.V.M. Rasheed, and B.B. Jergensen. 2004. Coral mucus functions as an energy carrier and particle trap in the reef ecosystem. *Nature* 428:66–70.
84. Harger, J.R.E., and K. Tustin. 1973. Succession and stability in biological communities. Part I. Diversity. *Int. J. Environ. Stud.* 5:117–130.
85. Carleton, J.H. and P.W. Sammarco 1987. Effects of substratum irregularity on success of coral settlement: quantification by comparative geomorphological techniques. *Bull. Mar. Sci.* 40:85–98.

86. Wyers, S.C. 1985. Sexual reproduction of the coral *Diploria strigosa* (Scleractinia, Faviidae) in Bermuda: research in progress. *Proc. 5th Int. Coral Reef Symp.* 4:301–306.
87. The author reported that most grounding occurred on reefs "populated by faster-growing communities."
88. Piniak, G. et al. Ch. 6. this volume
89. Miller, M.W. and J. Barimo. 2001. Assessment of juvenile coral populations at two reef restoration sites in the Florida Keys National Marine Sanctuary: Indicators of success? *Bull. Mar. Sci.* 69:395–405.
90. Lirman, D. and M.W. Miller. 2003. Modeling and monitoring tools to assess recovery status and convergence rates between restored and undisturbed coral reef habitats. *Restor. Ecol.* 11:448–456.
91. Gittings, S.R. 1998. Reef community stability on the Flower Garden Banks, NW Gulf of Mexico. *Gulf of Mexico Science* 16(2):161–169.
92. Gittings, S.R., G.S. Boland, K.J.P. Deslarzes, C.L. Combs, B.S. Holland, and T.J. Bright. 1992. Mass spawning and reproductive viability of reef corals at the East Flower Garden Bank, northwest Gulf of Mexico. *Bull. Mar. Sci.* 51:420–428.
93. Szmant, A.M., M.W. Miller, and T. Capo. Propagation of Scleractinian Corals from Wild-Captured Gametes: Mass-Culture from Mass-Spawning. Marine Ornamentals 2001: Collection, Culture & Conservation Program and Abstracts. Florida Sea Grant College Program, Gainesville FL.
94. Heyward, A.J. , L.D. Smith, M. Rees, and S. N. Field. 2002. Enhancement of coral recruitment by *in situ* mass culture of coral larvae. *Mar. Ecol. Prog. Ser.* 230:113–118.

6 Applied Modeling of Coral Reef Ecosystem Function and Recovery

Gregory A. Piniak, Mark S. Fonseca, W. Judson Kenworthy, Paula E. Whitfield, Gary Fisher, and Brian E. Julius

CONTENTS

6.1 INTRODUCTION: WHY MODEL CORAL REEF ECOSYSTEMS?

Many coral reefs worldwide are in a state of precipitous decline;[1–3] consequently, a need exists for modeling efforts to understand basic coral ecology and to estimate the recovery of coral reefs from natural physical disturbances (storms, bleaching events, outbreaks of disease, predators) and anthropogenic perturbations (vessel groundings, climate change, land runoff). Models are simplified representations of biological systems and processes, most often expressed as mathematical equations or computer simulations. As such, they are amenable to manipulations that are difficult to test in the field, such as predicting the outcome of different management scenarios, especially those that are expected to occur far in the future. Modeling can therefore be a powerful tool for applied coral research, provided sufficient data to parameterize the model are available.

A large body of coral modeling literature has developed over the last 30 years. While coral models can include interesting applications such as habitat connectivity, food web dynamics, reef

TABLE 6.1
Classification Scheme for Coral Modeling Review

Model Function	Published Models
Demography	4–30
Physiology/growth	5, 31–49
Ecology	8, 9, 11, 14, 17, 18, 25, 30, 50–89
Management	4, 7, 12, 19, 27, 31, 50, 51, 56, 63, 74, 82, 90–93
	Model Structure
Spatial/cellular	25, 28, 30, 52, 53, 76–78, 80, 86, 88
Classic matrix	4, 7–9, 13–15, 17, 19–24, 26, 41, 58, 61, 62, 70, 75, 83, 84
Differential equations	10, 11, 36, 50, 51, 54, 56, 59, 67–69, 73, 74, 89
Simple exponential	5, 6, 12, 27, 29, 32, 34, 37, 43, 45, 47–49, 66, 81, 85, 92, 93
Linear	29, 31, 33, 35, 38–40, 46, 57, 60, 65, 71, 90, 91
Miscellaneous	16, 18, 42, 44, 53, 55, 62–64, 72, 74, 79, 82, 87

Note: See text for definitions.

zonation, hydrodynamic-mediated processes, colony morphology, and reef accretion on geologic time scales, this chapter focuses solely on models that address the population ecology and biology of corals or hypothetical sessile benthic clonal invertebrates[4–93] because these types of models are most easily adapted for management questions on ecological time scales and have the greatest implications for restoration projects. For our purposes, coral models can be classified according to model function and structure (Table 6.1), using the definitions that follow below. These model classifications serve as a guide from which researchers can draw to adapt published models for a particular application, or for lessons to apply to original model development.

The coral models were assigned to four functional groups; in instances where model functions were not mutually exclusive, models were assigned to multiple categories. *Demographic models* are pure population models or space occupancy models that typically treat each species separately. *Physiology and growth models* describe the biological performance of coral colonies or populations, including predictions of skeletal growth, tissue growth, lesion recovery, biomass, von Bertalanffy growth, or colony survival/mortality. *Ecological models* describe how corals interact with other coral species, other reef organisms, or the environment. This category includes such topics as community demographics, competition, disturbance and recovery/succession, diversity, and metapopulations. Finally, *management models* consider direct responses to human activities, including fisheries questions (Beverton-Holt modeling), vessel groundings, and economics.

Structural model categories were defined as follows. *Spatial/cellular models* are cellular automaton models in which the identity of each pixel is tracked over time. *Classic matrix models* are age- or size-based matrix models, including Leslie matrices and Markov chains. *Differential equation models* can have several formats, including Lotka-Volterra models, the spatial colonization framework,[89] space limitation models,[94] diffusion models,[95] and metapopulation models. *Simple exponential models* have exponential or time-step forms, and include von Bertalanffy, Beverton-Holt, and simple polynomial models. *Linear models* are relatively simple mathematical expressions such as additive models, analysis of variance (ANOVA), regression, and log-linear models. *Miscellaneous* models include all other model types; in the coral literature noteworthy examples include fuzzy logic models,[64] simple probabilistic applications, and conceptual models.

Using these basic model structural and functional classifications, some general patterns emerge from an examination of the published literature (Table 6.1). More than half of the models in Table 6.1 describe how corals interact with other species or with the environment, while 30% focus on population demography. Models that address colony-level physiological questions (22%) tend to be linear or simple exponential in structure, while relatively few models (17%) have been directly developed for management questions. Matrix models are the most common format (26%) because this structure is particularly well-suited for demographic simulations and applied ecological questions such as disturbance and competition. Spatial models are relatively rare (13%) and have not been an area of active model development since the late 1990s.

To illustrate an example of how we may use this classification process, the remainder of this chapter is devoted to the theoretical and practical considerations in developing an applied coral model, using as a case study the development of a spatially explicit coral injury recovery model for vessel groundings for the Florida Keys National Marine Sanctuary (FKNMS). Although we discuss this specific example, the basic philosophical and methodological approach is not restricted to the Florida Keys and can be easily modified for management needs in other locations. First, we identify a need and justification for model development. We then address basic model definitions, such as what to model and how to define recovery, and how the model was constructed. We then describe model output and behavior, and how the model may be applied within the context of restoration protocols in the FKNMS.

6.2 THE FLORIDA KEYS NATIONAL MARINE SANCTUARY: MODEL NEEDS AND CONTEXT

Vessel groundings are a major threat to coral reefs in the FKNMS. Over 600 vessel groundings are reported annually in seagrass beds and coral reefs within the FKNMS.[96] Under the National Marine Sanctuaries Act, the National Oceanic and Atmospheric Administration (NOAA) has legal authority to seek monetary damages from the responsible party (RP) for injuries to resources in the FKNMS as well as in other National Marine Sanctuaries. NOAA and the State of Florida act as joint trustees for the FKNMS; response to a vessel grounding can include injury assessment, emergency triage, possible litigation between the trustee and the RP, and a restoration project.[97] Monies recovered from the RP are used to restore the natural resources to as close to their preinjury condition as possible (primary restoration) and to compensate the public for the lost use of those resources until such time as those resources have fully recovered (compensatory restoration).[98] Habitat Equivalency Analysis (HEA) is a method used by NOAA and other trustees to determine the amount of compensatory restoration required for such injuries. HEA combines biological and economic principles to calculate ecosystem services lost in the interim between the time of the injury and return to baseline conditions.[99] If no recovery is expected, the losses are calculated in perpetuity.[99] The calculation of interim losses requires an estimate of the time it takes for the injured services to recover, as well as the shape of that recovery function.[97,99,100] Computer models have been successfully used to generate recovery horizons for HEA in natural resource damage claims.[100]

6.3 BALANCING BIOLOGICAL REALISM AND RESTORATION REALITY

When developing an applied model, it is important to consider what clients the model serves since the specific needs of an end user may modify model assumptions and structure. In developing our spatially explicit coral Injury Recovery Model (IRMc), we worked from two complementary perspectives. The first was a restorationist or "Optimal Path" approach, which required the IRMc structure, assumptions, and data inputs to be as biologically and ecologically accurate as possible. The goal of this nuts-and-bolts process was to determine the ideal biological conditions for effective restoration or fishery management solution. We also considered a management or legal perspective, driven by product as well as process. In many ways, the restorationist and management needs are the same; however, in certain circumstances the needs of the legal client may supersede the restorationist needs, resulting in model simplifications or stipulations that are less ecologically satisfying but legally more conservative. The legal situation in the FKNMS allows authorities to seek compensation for vessel grounding injuries.[98] The IRMc therefore must meet Daubert standards for legal admissibility, which essentially requires expert scientific testimony to be tested and based upon peer-reviewed publication.[102] In addition, the model must be legally defensible (e.g., explainable), which implies a less complicated model structure. In practice this required that when decisions were made regarding model structure or valuation that will influence the outcome, the model default was toward faster recovery times. For example, scientists might predict that the intensity and frequency of storms will increase as a result of global climate change. In a restoration context this implies that frequent disturbance will reduce the recovery rate of an injury, but unless sufficient peer-reviewed scientific data exist for a specific, quantifiable increase in coral mortality caused by the change in storm regime, that effect would not be included in the model.

When an ecological model is used for economic calculations and legal proceedings, the model goals should be twofold: to prevail in litigation and to recover sufficient funds to implement a restoration project(s) that will compensate the public and the environment for the resource injuries. It is important to note that a model used in this way should not be used to design a restoration project, nor is such a model used to determine the ecological "success" of restoration. Instead it is merely a way to determine the appropriate scale of compensation. It is likely that no single metric (or handful of metrics; see below) will adequately characterize the full functionality of the resource services the ecosystem provides, and that any metric chosen will underestimate the value of the system as a whole. The model merely provides the means to an end, rather than the end itself — i.e., the model does not determine how to restore the system effectively; it only helps procure funds to make the restoration possible.

6.4 DECIDING WHAT TO MODEL

A fundamental step in model development is choosing a metric — that is, which aspect(s) of the system the model should describe. At the most basic level, a metric should be biologically meaningful and amenable to modeling. In the case of an applied model, the metric should meet additional practical criteria. For example, the metric should be cost-effective to measure using standardized methodology that can be incorporated into an injury assessment protocol. If the model output is to be used for HEA, the metric should also adequately represent the ecosystem services the resource provides. However, coral reefs are highly complex systems that deliver a large array of ecosystem services[103,104] that might not be easily collapsed into a single metric. We used these criteria to evaluate a suite of possible metrics for our coral vessel grounding model, which specifically applies to physical destruction of coral habitat and not other injury types.[98] However, it is important to note that different modeling needs may require the selection of a different metric.

Community composition and diversity are of particular interest for their ecosystem services such as resistance to disturbance and in providing biomedicines.[103,105] However, the FKNMS supports relatively few coral species. The mean stony coral cover in the FKNMS from 1996 to 2000 was 8.3%, and five scleractinian coral species and one hydrocoral accounted for 94.8% of that total.[106] Therefore, for our current model applications in the FKNMS, we assume that the mere presence of coral is of greater importance than the type of coral present, given the high likelihood it will be one of only a few species. This implies a need for a metric based on space occupancy by stony coral. We note that this assumption will likely not hold for other, more diverse reef systems in the Indo-Pacific or elsewhere in the Caribbean.

Metrics that describe space occupancy may be either three-dimensional (3-D) or 2-dimensional (2-D) (Table 6.2). From a modeling perspective, a 3-D model is more complicated and time-consuming than a 2-D model, but not prohibitively so. Many of the ecosystem services a coral reef provides

TABLE 6.2
Application of Selection Criteria to Four Potential Space-Occupancy Model Metrics

Model Type	Biologically Meaningful	Easy to Model	Easy to Measure	Reflects Ecosystem Services
Three-Dimensional				
Rugosity	X		X	X
Coral volume	X	X		X
Two-Dimensional				
Size-frequency	X	X	X	X
Percent cover	X	X	X	X

are predicated on its 3-D structure, generally composed of an underlying carbonate framework topped with a veneer of live coral. For example, vertical relief is required for wave dissipation and shoreline protection; at smaller scales reef topographic complexity provides sand and sediment storage to minimize abrasion on the reef and serves as important habitat for reef inhabitants such as fish, invertebrates, or microbes. For modeling applications we considered two potential 3-D metrics: rugosity and coral volume. Reef rugosity is commonly measured by carefully laying a known length of small-linked chain along the contour of a reef and measuring the linear distance over which the chain extends. The resulting ratio provides a dimensionless index of topographic complexity, which reflects both the underlying geology and live benthic community.[107] Rugosity often correlates well with disturbance regime, coral cover, diversity, fish community characteristics, and possibly nutrient uptake.[108–112] However, because it is dimensionless, modeling rugosity is not straightforward. Rugosity does not discriminate between reef framework and live coral, and it may be of limited use for groundings on low-relief livebottom areas. Because topographic complexity would recover on geological rather than ecological time scales, and because restoration projects have the technical capacity to restore some 3-D structure of the reef framework,[113] it may be more expedient to measure live coral volume directly for a modeling metric. While not technically demanding, measuring coral volume may be prohibitively time-consuming, not cost effective, and not feasible for a small injury assessment team, particularly in an undisturbed reference site.

If the presence of corals implies the existence of 3-D structure, the ecosystem services for most coral community structures can be approximated by a 2-D metric such as percent cover or size-frequency distributions. These 2-D metrics indirectly reflect the full suite of ecosystem services that a 3-D metric may encompass, but they are easier to measure and their biological effects are easier to describe. For example, coral size-frequency distributions provide information on population structure that is sensitive to environmental conditions[9,17,114–116] and have been successfully used in modeling grounding injuries.[7] Despite the wealth of information it provides about a coral community at a given site, measuring size-frequency distributions is time consuming, particularly for repeated monitoring surveys. Therefore it may not be feasible to use size-frequency as a metric in an area that is as heavily impacted by vessel groundings as the FKNMS. In contrast, percent coral cover is a simple, relatively rapid measurement that is a basic component of monitoring programs in the FKNMS and worldwide.[106,117–120] Coral cover and topographic complexity are correlated,[107,109,110] although this may not hold for encrusting corals or hardbottom habitats. Percent cover also facilitates direct comparisons with other benthic groups (algae, etc.) that are less compatible with coral size-frequency distributions. As a result, percent coral cover provides the most comprehensive metric currently available for IRMc. As an added benefit, vertical growth of live coral cover may be reduced to a simple function of time. Therefore, simply tracking the time that a pixel in a spatial model has been occupied provides a convenient and simple multiplier to convert coral cover at each time step into an approximation of the habitat complexity of the recovering community.[121]

For vessel grounding IRMc application in the FKNMS, injury was defined as the physical impact that a vessel has on a coral reef, as demonstrated by a decrease in coral cover, and recovery occurred when a damaged area provided levels of ecosystem services, ecological function, and resilience (i.e., ability to persist despite natural disturbance) that were indistinguishable from those of similar reef areas not affected by the injury. Functionally, this implied that the recovery in the IRMc is a return to 100% of baseline conditions, where baseline was the percent coral cover in an undisturbed reference population (URP; an adjacent area of reef not affected by the injury, or in the case of a patch reef, an analogous site nearby), as measured during the injury assessment process. By plotting percent recovery against time, the IRMc provided an estimate of the time required for 100% recovery and the shape of that recovery function. The result was a recovery horizon that can be used in an HEA to calculate the discounted value of lost ecosystem services until such time that they return. Discounting is necessary to account for when the delivery of resources and services occurs. The public values current delivery over future delivery at an annual rate of 3%.[122] Functionally, this means that the present value of any services that return after 70 or more years in the future is essentially zero. This is a key point for the development of the IRMc because it means that any recovery that occurs more than ~70 years after the injury does not contribute measurably to the recovering services computed under HEA.

For comparison, three previously published coral recovery models would meet the criteria for HEA (i.e., provide an estimated time for recovery and describe the shape of that curve). A linear model by Riegl[57] predicts coral coverage to return 100 to 160 years after vessel groundings in the Red Sea, while coral cover and diversity in the Philippines did not recover within the 100-year simulation period of Salia et al.'s exponential model.[92] These two recovery curves would be subject to discounting within HEA. In contrast, HEA discounting has a smaller effect on the matrix model of Lirman and Miller[7] for coral size-frequency distribution at a grounding site in the FKNMS, which predicts a minimum of 10 years for recovery. This last example also observed recovery of species richness and abundance of the dominant coral within 6 years and maximum colony size of the dominant species in ~20 years — a good example of how recovery horizons are "greatly dependent on the endpoints being measured."[7]

6.5 CORAL INJURY RECOVERY MODEL (IRMc): STRUCTURE, ASSUMPTIONS, AND APPLICATIONS

We employed two complementary spatial models to predict recovery,[101,121] one in SAS and one in ARCInfo. The models shared similar structures and assumptions and typically produced interchangeable results. When two different programming languages produce comparable results, it helps verify that model predictions are generated by our understanding of the system rather than vagaries of the software. These models were originally developed to model recovery of seagrass communities from vessel grounding injuries and adapted for use on coral reefs.[121] NOAA's current approach to coral grounding injuries is in many ways an extension of the seagrass process, and many of the lessons learned in the seagrass program[101,123,124] are directly transferable to coral injuries. For example, the only major structural difference between the seagrass and coral models was that seagrass recovery is modeled primarily via ingrowth of rhizome apical meristems from undisturbed areas on the injury perimeter with additional new recruits as seeding propagules;[100] in contrast, coral recovers primarily via new recruits,[125] so perimeter ingrowth did not contribute to recovery in the IRMc.[121] Similar modifications are easily performed to adapt the IRMc for other coastal habitats and trust resources, including mangroves, marshes, and rocky intertidal systems.

6.5.1 Model Structure

The coral ARCInfo model used a deterministic approach that combined the Euclidian distance function with a weighted cost distance function to represent biological growth. The injury footprint was input into ARCInfo to serve as a boundary grid, and no lateral growth was allowed from the injury perimeter in the coral model. In each model time step, recruit propagules were added to the grid; those that landed on unoccupied cells became new recruits. The Euclidian Distance function created a grid that measures the distance between unoccupied cells and the nearest occupied cells (i.e., the injury perimeter plus new recruits). The Cost Distance function incorporated both distance and direction to produce a Cost Grid and a Cost Allocation Grid for each species in the model. These grids represented growth by converting the cells with the lowest cost values from vacant to occupied cells. In addition, a mortality rate could be applied to reduce the number of occupied cells. The resulting grid then served as the source grid for the next model iteration. The number of colonized cells at each time step was output as an ASCII file and plotted in SAS to determine a recovery horizon.

The second model was a probabilistic model developed in SAS. An injury footprint was created as a matrix in Proc IML. In each time step, recruit propagules were added to the model, and those that landed on unoccupied cells became new recruits. A second matrix kept track of the number of occupied neighbor cells adjacent to each pixel. The model then examined each pixel, and those with occupied neighbor cells were assigned a random number between 0 and 1. A Boolean test converted the vacant pixel to a filled state if the random number exceeded a preassigned value that represented the growth rate. A mortality rate could also be applied to create newly vacant space. The resulting matrix served as the source grid for the next iteration, and the matrices at each time step copied into SAS data sets to plot a recovery horizon.

6.5.2 Data Inputs and Model Assumptions

Because the seagrass models have been used successfully in the development of natural resource damage claims,[100] we designed the IRMc to be as congruent as possible with the structure and the assumptions made in the seagrass models. Model concepts, assumptions, and input parameters all drew on research published for the wider Caribbean, but we focused the IRMc for application in the FKNMS. In evaluating each assumption or input parameter, we considered three criteria:

1. The quality of the data
2. The expected effect of an assumption on the behavior and output of the coral models (e.g., predicted effect on recovery time)
3. Our ability to defend that assumption in court

When the model structure and parameters were confounded by the absence of peer-reviewed, published data, we elected to make legally conservative assumptions that result in faster recovery times. This model uncertainty can be reduced via expanded injury assessments and independent scientific research. Model assumptions and parameters will continue to evolve as additional data become available but are currently based on the extant data.

The model universe was defined using the area and shape of an injury, measured in the field during the injury assessment process. The FKNMS personnel have developed injury assessment techniques for both seagrass injuries[124] and coral reef injuries.[126] These techniques produce an injury footprint, the size and shape of which define the model boundaries within the IRMc, with the injury represented as a matrix of 10 cm × 10 cm pixels. The injury matrix in the IRMc was assumed to represent the total area of coral lost as a result of the injury, and following emergency triage and primary restoration. That is, toppled corals that have been restabilized and framework damage that has been repaired are assumed to have already recovered and do not contribute to the injury area in the model. If the reattached colonies were spatially mapped within the injury footprint, they could be incorporated into the initial model conditions; alternatively, the area of reattached colonies could be subtracted from the calculated injury area to create a new injury footprint for model input.

Models often keep track of populations by separating corals into size classes[7,8,17,18,26] or by assigning an identity to each model pixel.[30,78,121] We employed the latter approach and tracked the fate of each pixel by applying three standard demographic parameters: growth, recruitment, and mortality. For the IRMc, growth was assumed to be constant, regardless of colony age, size, water depth, water quality, and yearly variability; in addition, colony size had no limit. Because coral growth rates were very slow relative to pixel size, the duration of a time step was scaled so that the growth rate was one pixel per time step. The IRMc made no assumptions about the reproductive characteristics of adult corals (size at maturity, fecundity, etc.) or the relative importance of reef connectivity and self-seeding. The source of the larvae (local or external) was not specified; the larvae were simply assumed to be present in the water column and could recruit to any unoccupied space in the model. Certain types of physical disturbances can produce viable colony fragments that can contribute to reef recovery. In the case of groundings, we consider the stabilization of such fragments to be part of the primary restoration process, so these fragments were not included in the coral models (which could slightly underestimate recovery time). Because the model pixel (10 cm) was substantially larger than the size of a coral recruit (<1 cm), the coral models used net recruitment rates. Net recruitment was determined by multiplying an average recruit density by the duration of a time step, then applying a postsettlement mortality rate so that recruitment occurred at the scale of the pixel. The use of postsettlement mortality in the net recruitment rate calculation was the only appearance of mortality in the base version of the IRMc. This assumption results in a more legally conservative model that predicts faster recovery times. We did not include natural background mortality, competition, bleaching, disease, or mortality from storm disturbance, although the model structure would easily accommodate these factors if quantitative, peer-reviewed data become available. Although size-class models can allow for colony shrinkage,[8,18] partial colony mortality is also not included in IRMc.

The base version of the coral IRMc used only a single hypothetical coral species, with the reproductive characteristics of a brooding species but the age/size characteristics of a framework-building broadcast spawning species. Because the vast majority of coral recruits in the Florida Keys are brooding genera like *Porites* and *Agaricia*,[127–130] we assumed that the hypothetical species in the IRMc had the general characteristics of a brooding species, loosely based on *P. astreoides*. After a grounding injury, coral abundance recovers more quickly than percent cover;[131] we therefore had to convert coral recruit densities to pixel occupancy (i.e., cover). Based on an extensive literature review,[132]* we utilized an average growth rate of 5 mm/yr, which corresponds to a 20-year model time step to fill a 10 cm × 10 cm pixel. Because recruitment is not commonly measured over such long time periods, we extrapolated yearly recruitment rates over a 20-year time step. Coral recruitment rates were calculated from the data of Miller et al.[130] for reefs in Biscayne National Park, in the upper Florida Keys. Their maximum reported recruitment rate of 15.2 new corals/m^2/yr for 20 years gives 304 corals/m^2/20 yr. The lowest annual mortality rate reported (22%), compounded over 20 years, yields a net recruitment of 2.7 corals/m^2/20 yr, which we rounded up to 3.0 corals/m^2/ 20 yr to use for model input. As previously mentioned, there was no other mortality applied once a pixel was occupied. The maximum colony age for *P. astreoides* in the FKNMS is approximately 20 years,[7,133] or essentially one model time step — thus making model change difficult to observe at this temporal resolution. We therefore assumed that our hypothetical species had a life expectancy of a framework-building broadcast spawning species and did not impose age or size restrictions on the species.

In choosing this hypothetical species, we faced the fundamental problem of replacing a coral type most likely to be damaged in a vessel grounding (i.e., framework-building genera like *Montastrea* or *Acropora*), with a different type of coral that may not provide an identical level of certain ecosystem services (e.g., a brooding genera like *Porites* or *Agaricia*). However, the injury compensation calculations heavily discount recovery that occurs after ~70 years, so framework species with negligible recruitment rates will not make substantial contributions to the calculation

* Note: This reference was originally submitted to NOAA as an interim report subject to nondisclosure. It is cited here because the model has developed beyond the scope of the report, which will remain protected by deliberative process.

in any event, while brooding species recruit in rapidly and would provide the greatest amount of recovered services for the HEA. Replacing damaged framework genera with brooding species not only reflects observations of natural recovery[128,131] but results in a more legally conservative model that produces faster recovery times. Framework-building species could be incorporated, and the coral models are flexible enough to incorporate a multispecies approach, if desired.

6.6 IRMc: THEORETICAL EXERCISES

We used the base form of the IRMc with a single hypothetical coral species to address a series of theoretical questions regarding model behavior. The first step was to cross-calibrate the SAS and ARCInfo models to ensure that they produced equivalent output and that we had not inadvertently embedded computational errors. If the output of the two software packages was similar, they could be used interchangeably to test model behavior and the effects of injury size and shape on recovery. We then determined recovery horizons for brooding vs. broadcast species using the ARCInfo model and used the SAS model to determine sensitivity to recruitment and mortality rates and the effect of storm disturbance on injury recovery.

6.6.1 Effect of Spatial Geometry

We used the single-species parameters to calibrate the coral models against each other and to determine the effect of injury geometry of the recovery of coral reef injuries. Because seagrass recovery was driven largely by rhizome ingrowth from undisturbed side populations on the injury perimeter, injury geometry had a large effect on the recovery time of a seagrass system.[101] However, lateral encroachment has little or no role in the recovery of coral injuries, with recolonization occurring as recruitment from the water column.[125] This implies that injury geometry would have a minimal effect on coral recovery. To test this, we created a 900-m^2 grid and computed recovery for four versions of this size grid: a square, and 6-, 4-, and 2-m wide rectangles, using 0.01-m^2 pixels and 0.005 m/yr as a growth rate. The recovery horizons predicted by the SAS and ARCInfo coral models were comparable and did not vary appreciably among shapes (Figure 6.1). In the SAS

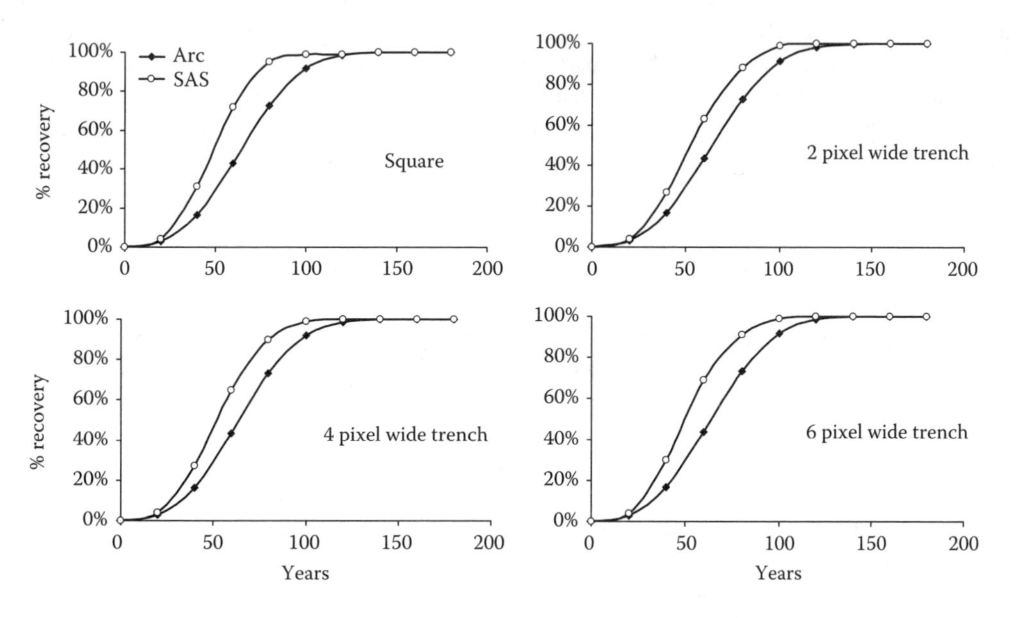

FIGURE 6.1 IRMc recovery horizons of hypothetical injuries of different geometries, using SAS and ARCInfo. Each injury geometry used a grid size of 90,000 pixels (90 m^2 with 10 cm resolution).

model, the number of time steps required for recovery was similar over a 50-fold range of injury sizes,[134] and any small differences between combinations of recruitment and mortality as a function of injury size were attributable only to the small variation that accompanies every run (SAS results are based on probabilistic draws and as such have small variations among individual runs with identical settings).

6.6.2 Effect of Life History Strategy

As stated above, the base form of the IRMc used a hypothetical species with the reproductive characteristics of a brooding coral because most corals that recruit after an injury are brooding species. Because brooders do not generally contribute to the reef framework, we have conducted trial model runs with a hypothetical framework-building broadcast spawning species based on *Montastrea annularis*. To run broadcast spawner models, we made two assumptions. The first was that broadcast spawner growth was equal in all directions. Most of the growth rates measured for *M. annularis* are derived from cores that measure vertical growth; our literature search indicated that the long-term vertical growth rate for this species averages 9 mm/yr across the Caribbean. However, occupation of a pixel in the model requires estimates of lateral growth, not vertical accretion, and lateral growth rate data for framework-building species are rarely reported in the literature. The second assumption was that despite observed recruitment failure of broadcast spawners in the Caribbean,[13] broadcast spawners were allowed to recruit successfully in the IRMc, otherwise recovery obviously would not occur.

We ran trial spawner models using the ARCInfo model version with 10 cm pixels and the following input parameters. High growth (HG) runs assumed that the lateral growth rate and vertical growth rate of colonies were equal at 9 mm/yr, which was rounded to 1.0 cm/yr for modeling convenience (10 cm pixel, 10-year time step). Low growth (LG) trials were conducted at a growth rate of 0.5 cm/yr, which is the only published core data for lateral growth in *M. annularis.*[135] The LG runs thus required a 20-year time step for a 10-cm pixel. Total coral recruitment rates were calculated for 10- and 20-year time steps using the same input data as the brooder models,[132] and the results converted to spawner recruitment rates by applying the following correction factors. High recruitment rates (HR) were generated using the highest reported recruitment rate of spawners in the Caribbean: that 7% of the total recruits in Barbados were *M. annularis.*[136] Low recruitment (LR) rates were meant to reflect the apparent current recruitment failure of *M. annularis* in the FKNMS.[127,129,130,137] The models will not run without recruitment rates, so for the purposes of this theoretical exercise only, we arbitrarily assumed LR to be 20% of HR. The input data for the broadcast spawner trials (10-cm pixel) are summarized in Table 6.3.

TABLE 6.3
Input Parameters for Base IRMc Model with a Hypothetical Brooding Species and Four Broadcast Spawning Scenarios

Model	Growth Rate (cm/yr)	Time Step (yr)	Recruitment Rate (per m^2/20 yr)
Brooder	0.5	20	3.0
HGHR	1.0	10	1.0
HGLR	1.0	10	0.2
LGHR	0.5	20	0.2
LGLR	0.5	20	0.04

Note: HG = high growth, LG = low growth, HR = high recruitment, LR = low recruitment.

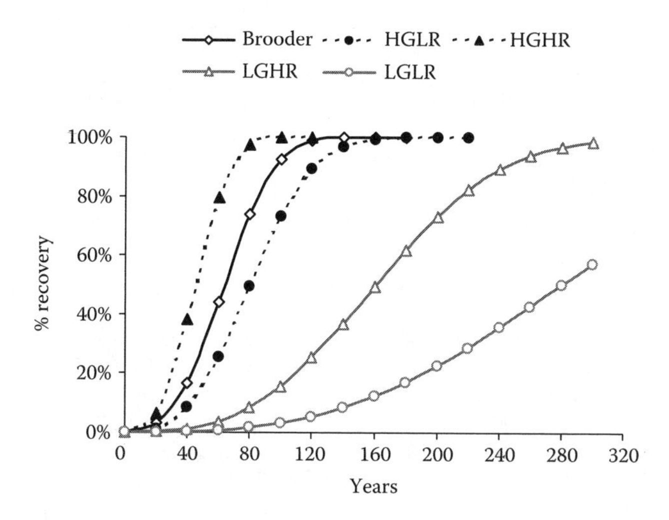

FIGURE 6.2 ARCInfo IRMc recovery horizons for a brooder and for hypothetical broadcast spawner scenarios (HG = high growth, LG = low growth, HR = high recruitment, LR = low recruitment). All models were run with a square, 30 × 30 m injury at 10 cm resolution.

These broadcast spawner model parameters generate very different recovery horizons (Figure 6.2); because the broadcast spawner data require additional refinement, these horizons should be considered upper and lower bounds of possible broadcast spawner models rather than case-ready predictions to be applied to specific grounding injuries. The HGHR model produced a recovery function suggesting very rapid recovery, which was totally inconsistent with observations of recovering grounding sites in the FKNMS within the last 20 years (i.e., almost no broadcast spawner recovery). Therefore, those results were considered spurious and dropped from the analysis. However, reducing recruitment rates to a more realistic level (HGLR) produces broadcast spawner recovery slightly below that of the brooder model. Reducing the growth rate has a more profound effect, so that even with the high recruitment rates (LGHR), 240 years were required to achieve 90% recovery. These predictions could be improved by reevaluating the assumption that *M. annularis* vertical growth rates in the literature are representative of lateral growth required in the model, and by developing more comprehensive data for the relatively rare *M. annularis* recruitment estimates.

6.6.3 Sensitivity to Demography and Disturbance

The SAS model was used to test the sensitivity of the IRMc to variability in recruitment rate, postsettlement intrinsic mortality, and disturbance events (severity and time of onset since injury occurrence), using the parameters in Table 6.4. Previous recovery modeling efforts indicate that recovery time is particularly sensitive to recruitment rate.[7] In the basic brooder IRMc,[132] the deterministic recruitment rate is $3/m^2/20$ yr; this is equivalent to a 3% recruitment probability per pixel per time step for this SAS exercise. To better test uncertainty in the recruitment parameter, the SAS model was run with recruitment probabilities of 1%, 5%, and 20%. Because the base IRMc does not include postrecruitment intrinsic mortality, we selected a range of low (1%) to moderate (20%) mortality rates that are representative of those reported for Caribbean corals.[128,137–140] Disturbance events were arbitrarily set to 5 and 15 time steps to evaluate time-of-onset effects. Similarly, severity of the storm events (percent mortality) of previously colonized sites was varied arbitrarily (50 and 90%) to simulate storm effects. Storm mortality was moderate to severe, as compared to field observations[141–145] and other modeling efforts.[8,18,25] Storm mortality

TABLE 6.4
Factors Used in Sensitivity Analyses

Injury size (pixels)	100 × 100	300 × 300		700 × 700
Postsettlement intrinsic mortality rates	1%	5%		20%
	($p > 0.99$)	($p > 0.95$)		($p > 0.80$)
Time of disturbance (storm onset)	5 iterations		15 iterations	
Disturbance (storm-induced mortality rates)	50%		90%	
	($p > 0.50$)		($p > 0.90$)	

Note: Recruitment percentages indicate the average number of vacant pixels occupied per model iteration. Similarly, mortality rates indicate the percentage of occupied pixels returned to a vacant status per iteration. Numbers in parentheses indicate the p value used as input for the SAS model and how the statistical argument was framed. All scenarios were run at recruitment rates of 1, 5, and 20% ($p = 0.99, 0.95, 0.80$). In addition, 100 × 100 pixels were modeled at 3% recruitment to scale results to previous findings.[132]

rates were applied uniformly to all pixels. All of the aforementioned combinations were run on grids of 100 × 100, 300 × 300, and 700 × 700 pixels. All simulations were run for 30 iterations, and the percent recovery plotted as a function of recruitment and postsettlement intrinsic mortality.

Recruitment rate produced dramatic differences in the final recolonization rate, and with the imposition of mortality, the asymptote of the recovery occurs prior to achieving 100% injury recovery (Figure 6.3). As expected, the postsettlement intrinsic mortality and recruitment rates had similar effects on percent recovery where their ranges overlapped. Similarly, where mortality was run up to 20% without a balanced recruitment rate (which only was run as high as 10%), the effect of mortality was pronounced. This indicates that the model was correctly balancing the occupation and vacating of pixels in response to the mortality and recruitment probabilities.

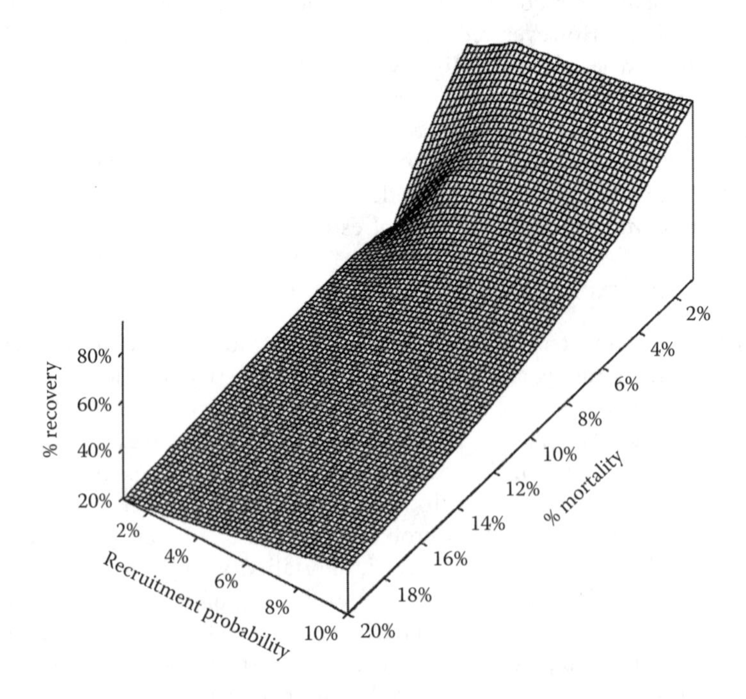

FIGURE 6.3 Krigged surface response of percent recovery as a function of recruitment and postsettlement intrinsic mortality after 30 iterations of the SAS IRMc.

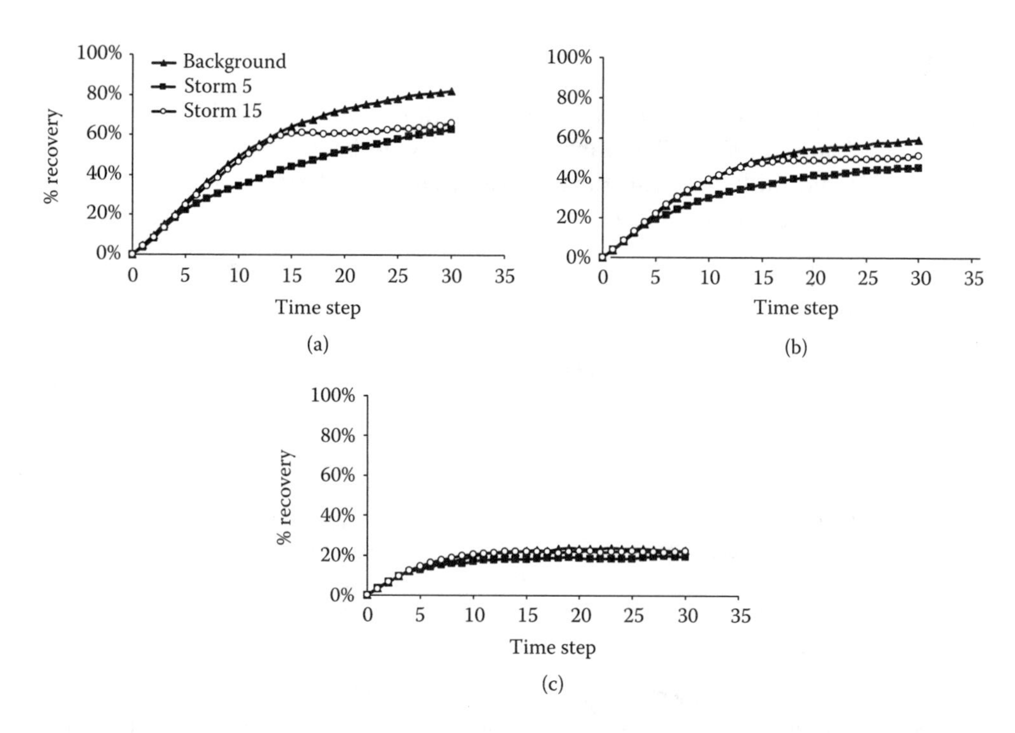

FIGURE 6.4 Storm effects (50% mortality at time step 5 or 15) on the recovery trajectory (3% recruitment, 100 × 100 pixel injury) with background mortality of 1% (a), 5% (b), and 20% (c).

The basic effect of storm-induced mortality was to slow the recovery horizon. Storm events reduced the overall recovery horizon (Figure 6.4), with early storms having a greater impact on the initial slope of the horizon (i.e., the area of the curve more heavily weighted in HEA). As background mortality levels increased, storm effects were less important (Figure 6.4). Very severe storms (90% mortality) reduced the overall recovery achieved even further (data not shown). Although storm mortality was set to 50 and 90% of the landscape, the model did not produce a corresponding 50 or 90% drop in cover with storm onset. This was because recruitment and spreading all occur within each model iteration. Recruitment had occurred (along with growth of recruits and surviving corals) prior to the tally of cover within the model iteration. Therefore, the full decline in cover was never output for graphical representation. High storm-induced mortality, combined with an early storm onset, produced the lowest recovery. In general, storms occurring earlier in the recovery process had a pronounced, negative effect on the eventual degree of overall recovery that could be expected as compared with storms occurring later in the recovery process. Thus, an analysis of the storm history in the geographic area of a given injury may provide useful heuristic information as to the timing and severity of storm effects in the model. Such information would ultimately help guide the timing of restoration projects to minimize the setback of recovery due to storm disturbance.

The combined effects of disturbance (both severity and timing), recruitment, and postsettlement intrinsic mortality were evaluated by utilizing the endpoints of the model results at 30 iterations. The relative contribution of all these factors to percent recovery was compared quantitatively using a stepwise multiple linear regression in SAS (Version 8.0; Table 6.5) on nontransformed data. All four factors — recruitment rate, intrinsic postsettlement mortality, storm-induced mortality and time of storm onset — were highly significant ($p < 0.01$) and combined to explain 92% of the variation in percent recovery. Intrinsic mortality explained the greatest amount of model variation (~42%), closely followed by recruitment (~38%). Storm-induced mortality explained ~7% and time of storm onset ~5% of the variation in recovery.

TABLE 6.5
Stepwise Multiple Linear Regression Indicating the Relative Contribution of Recruitment, Mortality (Individual and Storm-Induced), and Time of Storm Onset to Percent Recovery of an Injury

Source	df	SS	MS	F	Pr > F
Model	4	16,039	4009.78	81.04	<0.0001
Error	31	1533.89	49.48		
Total	35	17,573			

Step	Variable	Cumulative R^2	F	Pr > F
1	Mortality	0.4171	24.33	<0.0001
2	Recruitment	0.7940	60.40	<0.0001
3	Storm mortality	0.8637	16.37	0.0003
4	Storm onset	0.9127	17.39	0.0002

Recruitment rate and postsettlement intrinsic mortality were key drivers of the rate of recovery. Because occupation of space in this model was driven by random recruitment and subsequent spread based on proximity of previously occupied space (but not lateral encroachment from adjacent uninjured areas), the importance of occupying space was an understandable first-order factor. Mortality, which subtracts from occupied space, was a similarly strong determinant of recovery. Recruitment and spread outstrip mortality in this model, however, because spread is contagious. Space occupation can occur in more than one direction from a currently occupied pixel per iteration, meaning that more than one adjacent pixel can be occupied as the result of the proximity of a currently occupied pixel per model iteration. In contrast, mortality was not contagious, meaning that it is induced by a simple binary choice on a per pixel basis and is not dependent on the proximity state of any other pixel.

6.6.4 Intended Application of the IRMc in the Florida Keys National Marine Sanctuary

The FKNMS represents an ideal proving ground for IRMc applications because it is a relatively low-diversity, simple coral reef system with a strong statutory mandate to restore injured natural resources and an established economic process to calculate damages.[146] The recovery time generated by the IRMc is used as input for the HEA to calculate the proper amount of compensatory restoration owed by the Responsible Party. The default position of our base IRMc is to make legally conservative decisions whenever there is uncertainty in any model parameter or assumption, even when such decisions produce recovery times that are faster than are biologically realistic. Accordingly, from the perspective of the model users, it will always generate an amount of compensatory restoration that is legally defensible. As we continue to refine model parameters and assumptions as more data become available, the increasing biological realism would often result in slower recovery times and larger interim losses. Some of these recovery levels for our simulations were quite low, meaning that the area above the curve — the prediscounted interim losses — was much increased. If HEA reveals that changes in these values are rendered only marginally important because they occur too far in the future (i.e., approaching 70 years after the injury), then the amount of compensatory restoration owed may not significantly increase. For example, a preliminary application of the base IRMc to an actual coral grounding injury indicated that the IRMc reduced the compensatory damages to 80% of those required for the same injury if services are lost into perpetuity.[147] That is, a highly legally conservative IRMc provided damages that were not much lower than assuming that the resource never returns — a testament to the slow recovery of damaged coral reefs.

The IRMc is an ongoing process, and we have identified several ways to improve model performance and assumptions. The first and most critical step would be to reduce model uncertainty to provide better input parameters, which is true for any model. In a best-case scenario, the IRMc would be calibrated against a complete set of long-term demographic parameters (growth, recruitment, mortality rates) for injury sites of known age under a variety of ecological conditions (reef type, depth, geographical location within the FKNMS), to provide a more accurate estimate of recovery time and the shape of the recovery curve. The IRMc could also be used to evaluate the efficacy of different restoration options — that is, if available restoration options produce different demographic parameters in different environments, the IRMc could be used to choose the most appropriate or cost-effective option. For example, the sensitivity exercise suggested that with the current model structure and parameters, the model is highly sensitive to recruitment rate. Therefore, selecting restoration options that facilitate recruitment might be the most valid way to improve recovery times. Finally, the IRMc could be refined to accommodate the different reef types (spur-and-groove, patch reefs, hardbottom communities, etc.) that exist in the FKNMS. This could be done by adopting a multispecies approach to apply across all reef types or producing several different IRMs that are tailored to key community components of each reef type.

6.6.5 Application to Other Coral Reef Systems

The IRMc provides a basic modeling approach that could be adapted for virtually any reef system, as long as a few caveats are considered. First, like any management tool, IRMc works best if supported by a strong legal, economic, and political framework. With the possible exception of the Great Barrier Reef Marine Protected Area, the FKNMS is unique because it applies this framework to such a large geographic area. Vessel groundings are by nature a localized injury, so the IRMc would work equally well on a smaller, site-specific scale if the proper sociopolitical framework exists.

Second, adapting the IRMc to different reef systems requires a careful evaluation of that particular reef environment. For example, the model should transfer easily to a location like Bermuda, a relatively low-diversity system where ship groundings are prevalent, damaged framework-building species recruit more slowly than brooding corals, and expected recovery times are 100 to 150 years.[148] Vessel groundings are a relatively ubiquitous threat in which susceptibility to the threat is largely driven by depth distribution; as a result, coral species suffer the same general effect and response, making it easy to simplify the effect by modeling percent cover of a generic coral species. Percent cover may work well as a metric in areas where wave exposure limits coral abundance or on reefs where eutrophication places a heavy emphasis on coral–algal interactions.

However, in certain locations reefs may face a different suite of threats that require different metrics or altered model assumptions. These concerns exist for both the physical and biological aspects of the intended application. For example, modeling applications in which species-specific vulnerability and response is important (bleaching, disease, predation by crown of thorns starfish, etc.) should focus on specific populations and use metrics that incorporate colony condition (such as size-frequency distributions or a percent live tissue index). In highly diverse areas such as the Indo-Pacific, a metric that incorporates coral diversity or the abundance of different functional groups may be more appropriate than the generic percent coral cover we currently use in the relatively species-depauperate FKNMS. These environmental differences would also affect the model assumptions. For example, recruitment levels are much higher in the Indo-Pacific than in the Caribbean,[13,128] so model applications in those areas may not be as sensitive to recruitment rates as the FKNMS IRMc.

6.7 CONCLUDING REMARKS AND FUTURE DIRECTIONS

We feel that the model by itself is conceptually very well situated to begin application in the FKNMS. As the quantity and quality of the data improve, so will the model's predictive capability, resulting in the calculation of the full amount of compensatory restoration owed. Complete demographic data on the spatial and temporal scales needed for landscape-scale representation of the recovery process are limited — few papers provide simultaneous estimates of growth, recruitment, and mortality. However, the IRMc results reported here suggest the potential to streamline the modeling process. For example, the spatial insensitivity of IRMc suggests that at certain spatial scales where patch dynamics and habitat fragmentation are deemed unimportant, the modeling process could be collapsed into a single arithmetic representation, forgoing the spatial modeling process entirely. Work would be required to determine the spatial scales to which the modeling process could be collapsed (a geostatistical evaluation of the modeling inputs and outputs would solve this), as would solving and documenting the mathematical solution. However, it is important to note that an algorithmic version of the IRMc may not be desirable for more complex model applications and is not appropriate for other systems in which vegetative propagation predominates (seagrasses, marshes, etc.).

In an ideal world one might expect that coral modeling, restoration, and monitoring would all be based on a common currency. There is a fundamental disconnect here, however. In a restoration context the main purpose of the IRMc is to scale the appropriate level of compensation needed, rather than to assist in restoration design or determination of success through monitoring. Restoration is largely an engineering solution, while monitoring of an injured site may often be limited to the structural integrity of the project and a visual assessment of coral condition. As an added complication, the economic HEA is based on resource services (which speaks to the ecological functionality of the system), and does not directly incorporate human uses (e.g., recreational uses) or other socioeconomic use. Reef scientists have long recognized that an integrated, multidisciplinary approach is a key to understanding how coral reefs function and the reasons for their decline, but recognizing that fact has not made that job any easier. Coral reef restoration is a far younger science, and the more daunting challenge of combining multidisciplinary natural science, engineering, economics, and law must be met to make restoration most effective.

ACKNOWLEDGMENTS

Model development was funded by the NOAA Damage Assessment Center. We gratefully thank NOAA General Counsel, the Florida Keys National Marine Sanctuary, the NOAA National Marine Sanctuaries office, and the Florida Marine Research Institute for their collaboration. The manuscript was improved by comments from Nicole Fogarty, Shay Viehman, Sharon Shutler, Bill Precht, and two anonymous reviewers.

REFERENCES

1. Turgeon, D.D., The state of coral reef ecosystems of the United States and Pacific Freely Associated States: 2002, National Oceanic and Atmospheric Administration/National Ocean Service/National Centers for Coastal Ocean Science, Silver Spring, MD, 2002.
2. Wilkinson, C. (ed.), *Status of Coral Reefs of the World,* Australian Institute of Marine Science, Townsville, Australia, 2002.
3. Gardner, T.A., Cote, I.M., Gill, J.A., Grant, A., and Watkinson, A.R., Long-term region-wide declines in Caribbean corals, *Science*, 301, 958–960, 2003.

4. Smith, L.D., Devlin, M., Haynes, D., and Gilmour, J.P., A demographic approach to monitoring the health of coral reefs, *Mar. Poll. Bull.*, 51, 399–407, 2005.
5. Goffredo, S., Mattioli, G., and Zaccanti, F., Growth and population dynamics model of the Mediterranean solitary coral *Balanophyllia europaea* (Scleractinia, Dendrophylliidae), *Coral Reefs*, 23, 433–443, 2004.
6. Santangelo, G., Maggi, E., Bramanti, L., and Bongiorni, L., Demography of the overexploited Mediterranean red coral (*Corallium rubrum* L. 1758), *Sci. Mar.*, 68, 199–204, 2004.
7. Lirman, D. and Miller, M.W., Modeling and monitoring tools to assess recovery status and convergence rates between restored and undisturbed coral reef habitats, *Restor. Ecol.*, 11, 448–456, 2003.
8. Lirman, D., A simulation model of the population dynamics of the branching coral *Acropora palmata*: effects of storm intensity and frequency, *Ecol. Model.*, 161, 169–180, 2003.
9. Fong, P. and Glynn, P.W., Population abundance and size-structure of an eastern tropical Pacific reef coral after the 1997–98 ENSO: simulation model predicts field measures, *Bull. Mar. Sci.*, 69, 187–202, 2001.
10. Muko, S., Sakai, K., and Iwasa, Y., Size distribution dynamics of marine sessile organisms with space-limited growth and recruitment: application to a coral population, *J. Animal Ecol.*, 70, 579–589, 2001.
11. Muko, S., Sakai, K., and Iwasa, Y., Dynamics of marine sessile organisms with space-limited growth and recruitment: application to corals, *J. Theor. Biol.*, 219, 67–80, 2001.
12. Chadwick-Furman, N.E., Goffredo, S., and Loya, Y., Growth and population dynamic model of the reef coral *Fungia granulosa* Klunzinger, 1879 at Eilat, northern Red Sea, *J. Exp. Mar. Biol. Ecol.*, 249, 199–218, 2000.
13. Hughes, T.P. and Tanner, J.E., Recruitment failure, life histories, and long-term decline of Caribbean corals, *Ecology*, 81, 2250–2263, 2000.
14. Fong, P. and Glynn, P.W., A regional model to predict coral population dynamics in response to El Nino Southern Oscillation, *Ecol. Appl.*, 10, 842–854, 2000.
15. Tanner, J.E., Density-dependent population dynamics in clonal organisms: a modelling approach, *J. Anim. Ecol.*, 68, 390–399, 1999.
16. Coffroth, M.A. and Lasker, H.R., Population structure of a clonal gorgonian coral: The interplay between clonal reproduction and disturbance, *Evolution*, 52, 379–393, 1998.
17. Fong, P. and Glynn, P.W., A dynamic size-structured population model: Does disturbance control size structure of a population of the massive coral *Gardineroseris planulata* in the eastern Pacific? *Mar. Biol.*, 130, 663–674, 1998.
18. Andres, N.G. and Rodenhouse, N.L., Resilience of corals to hurricanes: a simulation model, *Coral Reefs*, 12, 167–175, 1993.
19. Abbiati, M., Buffoni, G., Caforio, G., DiCola, G., and Santangelo, G., Harvesting, predation and competition effects on a red coral population, *Neth. J. Sea Res.*, 30, 219–228, 1992.
20. Babcock, R.C., Comparative demography of three species of scleractinian corals using age- and size-dependent classifications, *Ecol. Monogr.*, 61, 225–244, 1991.
21. McFadden, C.S., A comparative demographic analysis of clonal reproduction in a temperate soft coral, *Ecology*, 72, 1849–1866, 1991.
22. Gotelli, N.J., Demographic models for *Leptogorgia virgulata*, a shallow-water gorgonian, *Ecology*, 72, 457–467, 1991.
23. Lasker, H.R., Population growth of a gorgonian coral: equilibrium and nonequilibrium sensitivity to changes in life history variables, *Oecologia*, 86, 503–509, 1991.
24. Lasker, H.R., Clonal propagation and population dynamics of a gorgonian coral, *Ecology*, 71, 1578–1589, 1990.
25. Reichelt, R.E., Green, D.G., and Bradbury, R.H., Discrete simulation of cyclone effects on the spatial pattern and community structure of a coral reef, in *Proc. 5th Int. Coral Reef Congr.*, Antenne Museum, Moorea, French Polynesia, 3, 337–342, 1985.
26. Hughes, T.P., Population dynamics based on individual size rather than age: a general model with a reef coral example, *Am. Nat.*, 123, 778–795, 1984.
27. Ross, M.A., A quantitative study of the stony coral fishery in Cebu, Philipplines, *Mar. Ecol.*, 5, 75–91, 1984.
28. Neigel, J.E. and Avise, J.C., Clonal diversity and population structure in a reef-building coral, *Acropora cervicornis*: self-recognition analysis and demographic interpretation, *Evolution*, 37, 437–453, 1983.

29. Gerrodette, T., Dispersal of the solitary coral *Balanophyllia elegans* by demersal planular larvae, *Ecology*, 62, 611–619, 1981.
30. Maguire, L.A. and Porter, J.W., A spatial model of growth and competition strategies in coral communities, *Ecol. Model.*, 3, 249–271, 1977.
31. Goffredo, S. and Chadwick-Furman, N.E., Comparative demography of mushroom corals (Scleractinia: Fungiidae) at Eilat, northern Red Sea, *Mar. Biol.*, 142, 411–418, 2003.
32. Crabbe, M.J.C. and Smith, D.J., Computer modelling and estimation of recruitment patterns of nonbranching coral colonies at three sites in the Wakatobi Marine Park, S.E. Sulawesi, Indonesia; implications for coral reef conservation, *Comp. Biol. Chem.*, 27, 17–27, 2003.
33. Cumming, R.L., Tissue injury predicts colony decline in reef-building corals, *Mar. Ecol. Prog. Ser.*, 242, 131–141, 2002.
34. Crabbe, M.J.C., Mendes, J.M., and Warner, G.F., Lack of recruitment of nonbranching corals in Discovery Bay is linked to severe storms, *Bull. Mar. Sci.*, 70, 939–945, 2002.
35. Anthony, K.R.N., Connolly, S.R., and Willis, B.L., Comparative analysis of energy allocation to tissue and skeletal growth in corals, *Limnol. Oceanogr.*, 47, 1417–1429, 2002.
36. Kizner, Z., Vago, R., and Vaky, L., Growth forms of hermatypic corals: stable states and noise-induced transitions, *Ecol. Model.*, 141, 227–239, 2001.
37. Lirman, D., Lesion regeneration in the branching coral *Acropora palmata*: effects of colonization, colony size, lesion size, and lesion shape, *Mar. Ecol. Prog. Ser.*, 197, 209–215, 2000.
38. Porter, J.W., Lewis, S.K., and Porter, K.G., The effect of multiple stressors on the Florida Keys coral reef ecosystem: a landscape hypothesis and a physiological test, *Limnol. Oceanogr.*, 44, 941–949, 1999.
39. Anthony, K.R.N., A tank system for studying benthic aquatic organisms at predictable levels of turbidity and sedimentation: case study examining coral growth, *Limnol. Oceanogr.*, 44, 1415–1422, 1999.
40. Kim, K. and Lasker, H.R., Allometry of resource capture in colonial cnidarians and constraints on modular growth, *Funct. Ecol.*, 12, 646–654, 1998.
41. Ruesink, J.L., Coral injury and recovery: matrix models link process to pattern, *J. Exp. Mar. Biol. Ecol.*, 210, 187–208, 1997.
42. Riegl, B. and Branch, G.M., Effects of sediment on the energy budgets of four scleractinian (Bourne 1900) and five alcyonacean (Lamouroux 1816) corals, *J. Exp. Mar. Biol. Ecol.*, 186, 259–275, 1995.
43. Meesters, E.H., Noordeloos, M., and Bak, R.P.M., Damage and regeneration: Links to growth in the reef-building coral *Montastrea annularis*, *Mar. Ecol. Prog. Ser.*, 112, 119–128, 1994.
44. Done, T.J. and Potts, D.C., Influences of habitat and natural disturbances on contributions of massive *Porites* corals to reef communities, *Mar. Biol.*, 114, 479–493, 1992.
45. Graus, R.R. and Macintyre, I.G., Variation in growth forms of the reef coral *Montastrea annularis* (Ellis and Solander): a quantitative evaluation of growth response to light distribution using computer simulation, *Smithson. Contrib. Mar. Sci.*, 12, 441–464, 1982.
46. Zou, R.L., A mathematical model of the hermatypic coral community of the Xisha Islands, Guangdong Province, China, in *Proc. 4th Int. Coral Reef Symp.*, Gomez, E.D. et al., (eds.), University of the Philippines Marine Science Center, Quezon City, Philippines, 2, 329–332, 1981.
47. Highsmith, R.C., Riggs, A.C., and D'Antonio, C.M., Survival of hurricane-generated coral fragments and a disturbance model of reef calcification/growth rates, *Oecologia*, 46, 332–329, 1980.
48. Graus, R.R. and Macintyre, I.G., Light control of growth form in colonial reef corals: computer simulation, *Science*, 193, 895–897, 1976.
49. Bak, R.P.M., The growth of coral colonies and the importance of crustose coralline algae and burrowing sponges in relation with carbonate accumulation, *Neth. J. Sea Res.*, 10, 285–337, 1976.
50. Wolanski, E. and De'ath, G., Predicting the impact of present and future human land-use on the Great Barrier Reef, *Est. Coast Shelf Sci.*, 64, 504–508, 2005.
51. Wolanski, E., Richmond, R.H., and McCook, L., A model of the effects of land-based, human activities on the health of coral reefs in the Great Barrier Reef and in Fouha Bay, Guam, Micronesia, *J. Mar. Syst.*, 46, 133–144, 2004.
52. Langmead, O. and Sheppard, C., Coral reef community dynamics and disturbance: a simulation model, *Ecol. Model.*, 175, 271–290, 2004.

53. Garza-Perez, J.R., Lehmann, A., and Arias-Gonzalez, J.E., Spatial prediction of coral reef habitats: integrating ecology with spatial modeling and remote sensing, *Mar. Ecol. Prog. Ser.*, 269, 141–152, 2004.
54. Connolly, S.R. and Muko, S., Space preemption, size-dependent competition, and the coexistence of clonal growth forms, *Ecology*, 84, 2979–2988, 2003.
55. Harriott, V.J. and Banks, S.A., Latitudinal variation in coral communities in eastern Australia: a qualitative biophysical model of factors regulating coral reefs, *Coral Reefs*, 21, 83–94, 2002.
56. McCook, L.J., Wolanski, E., and Spagnol, S., Modelling and visualizing interactions between natural disturbances and eutrophication as causes of coral reef degradation, in Wolanski, E. (ed.), *Oceanographic Processes of Coral Reefs: Physical and Biological Links in the Great Barrier Reef*, CRC Press, Boca Raton, FL, 113–125, 2001.
57. Riegl, B., Degradation of reef structure, coral and fish communities in the Red Sea by ship groundings and dynamite fisheries, *Bull. Mar. Sci.,* 69, 595–611, 2001.
58. Wootton, J.T., Causes of species diversity differences: a comparative analysis of Markov models, *Ecol. Lett.*, 4, 46–56, 2001.
59. van Woesik, R., Modelling processes that generate and maintain coral community diversity, *Biodivers. Conserv.*, 9, 1219–1233, 2000.
60. De'ath, G. and Fabricius, K.E., Classification and regression trees: a powerful yet simple technique for ecological data analysis, *Ecology*, 81, 3178–3198, 2000.
61. Scandol, J.P., CotSim — an interactive *Acanthaster planci* metapopulation model for the central Great Barrier Reef, *Mar. Models Online,* 1, 39–81, 1999.
62. Seymour, R.M. and Bradbury, R.H., Lengthening reef recovery times from crown-of-thorns outbreaks signal systemic degradation of the Great Barrier Reef, *Mar. Ecol. Prog. Ser.,* 176, 1–10, 1999.
63. Ruitenbeek, J., Ridgley, M., Dollar, S., and Huber, R., Optimization of economic policies and investment projects using a fuzzy logic based cost-effectiveness model of coral reef quality: empirical results from Montego Bay, Jamaica, *Coral Reefs*, 18, 381–392, 1999.
64. Meesters, E.H., Bak, R.P.M., Westmacott, S., Ridgley, M., and Dollar, S., A fuzzy logic model to predict coral reef development under nutrient and sediment stress, *Conserv. Biol.,* 12, 957–965, 1998.
65. Karlson, R.H. and Cornell, H.V., Scale-dependent variation in local vs. regional effects on coral species richness, *Ecol. Monogr.*, 68, 259–274, 1998.
66. Bak, R.P.M. and Meesters, E.H., Coral population structure: the hidden information of colony size-frequency distributions, *Mar. Ecol. Prog. Ser.*, 162, 301–306, 1998.
67. Kudo, K. and Yamano, H., Simulation coral reef study: differential equations model (SIMREEF), *Rep. Japan Mar. Sci. Technol. Cent.*, 37, 43–54, 1998.
68. Stone, L., Eliam, E., Abelson, A., and Ilan, M., Modelling coral reef biodiversity and habitat destruction, *Mar. Ecol. Prog. Ser.*, 134, 299–302, 1996.
69. Ware, J.R., Fautin, D.G., and Buddemeier, R.W., Patterns of coral bleaching: modeling the adaptive bleaching hypothesis, *Ecol. Model.*, 84, 199–214, 1996.
70. Tanner, J.E., Hughes, T.P., and Connell, J.H., The role of history in community dynamics: a modelling approach, *Ecology*, 77, 108–117, 1996.
71. Bak, R.P.M. and Nieuwland, G., Long-term change in coral communities along depth gradients over leeward reefs in the Netherlands Antilles, *Bull. Mar. Sci.*, 56, 609–619, 1995.
72. Fong, P. and Lirman, D., Hurricanes cause population expansion of the branching coral *Acropora palmata* (Scleractinia): Wound healing and growth patterns of asexual recruits, *Mar. Ecol.,* 16, 317–335, 1995.
73. Stone, L., Biodiversity and habitat destruction: a comparative study of model forest and coral reef ecosystems, *Proc. R. Soc. Lond. B*, 261, 381–388, 1995.
74. McClanahan, T.R., A coral reef ecosystem–fisheries model: impacts of fishing intensity and catch selection on reef structure and process, *Ecol. Model.*, 80, 1–19, 1995.
75. Tanner, J.E., Hughes, T.P., and Connell, J.H., Species coexistence, keystone species, and succession: a sensitivity analysis, *Ecology*, 75, 2204–2219, 1994.
76. Preece, A.L. and Johnson, C.R., Recovery of model coral communities: complex behaviours from interaction of parameters operating at different spatial scales, in *Complex Systems: From Biology to Computation*, Green, D.G. and Bossomaier, T. (eds.), IOS Press, Amsterdam, 69–81, 1993.

77. Crimp, O.N., and Braddock, R.D., Simulation of coral reefs and crown-of-thorns starfish, *Environmetrics*, 4, 53–74, 1993.
78. Johnson, C.R. and Preece, A.L., Damage, scale and recovery in model coral communities: the importance of system state, in *Proc. 7th Int. Coral Reef Symp.*, Richmond, R.H. (ed.), University of Guam Press, Mangilao, Guam, 1, 606–615, 1993.
79. Done, T.J., Constancy and change in some Great Barrier Reef coral communities: 1980–1990, *Am. Zool.*, 32, 655–662, 1992.
80. Jokiel, P. and Martinelli, F.J., The vortex model of coral reef biogeography, *J. Biogeography*, 19, 449–458, 1992.
81. Reichelt, R.E., Greve, W., Bradbury, R.H., and Moran, P.J., *Acanthaster planci* outbreak initiation: a starfish–coral site model, *Ecol. Model.*, 49, 153–177, 1990.
82. Berwick, N.L. and Faeth, P.E., Simulating the impacts of sewage disposal on coral reefs, in *Proc. 6th Int. Coral Reef Symp.*, Choat, J.H. et al. (eds.), The Symposium Executive Committee, James Cook University, Townsville, Australia, 2, 353–361, 1988.
83. Done, T.J., Simulation of recovery of pre-disturbance size structure in populations of *Porites* spp. damaged by the crown of thorns starfish *Acanthaster planci*, *Mar. Biol.*, 100, 51–61, 1988.
84. Done, T.J., Simulation of the effects of *Acanthaster planci* on the population structure of massive corals in the genus *Porites*: Evidence of population resilience? *Coral Reefs*, 6, 75–90, 1987.
85. Green, D.G., Bradbury, R.H., and Reichelt, R.E., Patterns of predictability in coral reef community structure, *Coral Reefs*, 6, 27–34, 1987.
86. Karlson, R.H. and Buss, L.W., Competition, disturbance and local diversity patterns of substratum-bound clonal organisms: a simulation, *Ecol. Model.*, 23, 243–255, 1984.
87. Abel, D.J., Williams, W.T., Sammarco, P.W., and Bunt, J.S., A new numerical model for coral distribution, *Mar. Ecol. Prog. Ser.*, 12, 257–265, 1983.
88. Karlson, R.H. and Jackson, J.B.C., Competitive networks and community structure: a simulation study, *Ecology*, 62, 670–678, 1981.
89. Hastings, A., Disturbance, coexistence, history, and competition for space, *Theor. Pop. Biol.*, 18, 363–373, 1980.
90. Saphier, A.D. and Hoffman, T.C., Forecasting models to quantify three anthropogenic stresses on coral reefs from marine recreation: Anchor damage, diver contact and copper emission from antifouling paint. *Mar. Poll. Bull.*, 51, 590–598, 2005.
91. McManus, J.W., Reyes, R.B. Jr., and Nanola, C.L. Jr., Effects of some destructive fishing methods on coral cover and potential rates of recovery. *Environ. Manage.*, 21, 69–78, 1997.
92. Salia, S.B., Kocic, V.L., and McManus, J.W., Modelling the effects of destructive fishing practices on tropical coral reefs, *Mar. Ecol. Prog. Ser.*, 94, 51–60, 1993.
93. Grigg, R.W., Resource management of precious corals: a review and application to shallow water reef building corals, *Mar. Ecol.* 5, 57–74, 1984.
94. Roughgarden, J., Iwasa, Y., and Baxter, C., Demographic theory for an open marine population with space-limited recruitment, *Ecology*, 66, 54–67, 1985.
95. Hara, T., A stochastic model and the moment dynamics of the growth and size distribution in plant populations, *J. Theor. Biol.*, 109, 173–190, 1984.
96. NOAA NOS Sanctuaries, *Wellwood* coral reef restoration project, http://www.sanctuaries.nos.noaa.gov/special/wellwood/project.html (accessed March 2004).
97. Precht, W.F., Deis, D.R., and Gelber, A.R., Damage assessment protocol and restoration of coral reefs injured by vessel groundings, in *Proc. 9th Int. Coral Reef Symp.*, Moosa, M.K. et al. (eds.), Ministry of Environment, Indonesian Institute of Sciences, Bali, Indonesia, 2003, 2, 963–968.
98. Shutler, S.K., Gittings, S., Penn, T., and Schittone, J., Coral restoration: how much is enough? In *Coral Reef Restoration Handbook: The Rehabilitation of an Ecosystem under Siege*, Precht, W.F. (ed.), CRC Press, Boca Raton, FL, 2005, chap. 10.
99. National Oceanic and Atmospheric Administration, Damage Assessment and Restoration Program, 2000. Habitat Equivalency Analysis: an overview. Policy and Technical Paper Series. 23p. http://www.darp.noaa.gov/library/pdf/heaoverv.pdf (accessed December 2005).
100. Fonseca, M.S., Julius, B.E., and Kenworthy, W.J., Integrating biology and economics in seagrass restoration: how much is enough and why? *Ecol. Eng.*, 15, 227–237, 2000.

101. Fonseca, M.S., Whitfield, P.E., Kenworthy, W.J., Colby, D.R., and Julius, B.E., Use of two spatially explicit models to determine the effect of injury geometry on natural resource recovery, *Aquat. Conserv. Mar. Freshwat. Ecosyst.*,14, 281–298, 2004.
102. Daubert v. Merrell Dow Pharmaceuticals, Inc., 113 S. Ct 2786 (1993).
103. Moberg, F. and Folke, C., Ecological goods and services of coral reef ecosystems, *Ecol. Econ.,* 29, 215–233, 1999.
104. Moberg, F. and Ronnback, P., Ecosystem services of the tropical seascape: interactions, substitutions, and restoration, *Oce. Coast. Manage.*, 46, 27–46, 2003.
105. Souter, D.W. and Linden, O., The health and future of coral reef systems, *Oce. Coast. Manage.*, 43, 657–688, 2000.
106. Wheaton, J., Jaap, W.C., Porter, J.W., Kosminyn, V., Hackett, K., Lybolt, M., Callaham, M.K., Kidney, J., Kupfner, S., Tsokos, C., and Yanev, G., EPA/FKNMS Coral Reef Monitoring Project, Executive Summary 2001. Project report to the NOAA Coastal Ocean Program for the Florida Keys National Marine Sanctuary, 2001.
107. Miller, M.W. and Gerstner, C.L., Reefs of an uninhabited Caribbean island: fishes, benthic habitat, and opportunities to discern reef fishery impact, *Biol. Conserv.,* 106, 37–44, 2002.
108. Aronson, R.B. and Precht, W.F., Landscape patterns of coral reef diversity: a test of the intermediate disturbance hypothesis, *J. Exp. Mar. Biol. Ecol.*, 121, 1–14, 1995.
109. Chiappone, M. and Sullivan, K.M., Distribution, abundance, and species composition of juvenile scleractinian corals in the Florida reef tract, *Bull. Mar. Sci.*, 58, 555–569, 1996.
110. Edmunds, P.J., Long-term dynamics of coral reefs in St. John, U.S. Virgin Islands, *Coral Reefs*, 21, 357–367, 2002.
111. Luckhurst, B.E. and Luckhurst, K., Analysis of the influence of substrate variables on coral reef fish communities, *Mar. Biol.*, 49, 317–323, 1978.
112. Atkinson, M.J., Topographical relief as a proxy for the friction factors of reefs: estimates of nutrient uptake into coral reef benthos, in *Proc. Hawaii Coral Reef Monitoring Workshop*, Maragos, J.E. and Grober-Dunsmore, R. (eds.), Hawaii Division of Land and Natural Resources, Honolulu, 99–103, 1999.
113. National Oceanic and Atmospheric Administration and Florida Department of Environmental Protection, Coral reef and hardbottom restoration plan, Florida Keys National Marine Sanctuary, Monroe County, FL, 2004.
114. Meesters, E.H., Hillerman, M., Kardinaal, E., Keetman, M., DeVreis, M., and Bak, R.P.M., Colony size-frequency distributions of scleractinian coral populations: spatial and interpsecific variation, *Mar. Ecol. Prog. Ser.*, 209, 43–54, 2001.
115. Bak, R.P.M. and Meesters, E.H., Coral population structure: the hidden information of colony size-frequency distribution, *Mar. Ecol. Prog. Ser.,* 162, 301–306, 1998.
116. Lewis, J.B., Abundance, distribution and partial mortality of the massive coral *Siderastrea siderea* on degrading coral reefs at Barbados, West Indies, *Mar. Poll. Bull.*, 34, 622–627, 1997.
117. Murdoch, T.J.T. and Aronson, R.B., Scale-dependent spatial variability of coral assemblages along the Florida Reef Tract, *Coral Reefs*, 18, 341–351, 1999.
118. Caribbean Coastal Marine Productivity Program (CARICOMP) Data Management Center and Florida Institute of Oceanography, CARICOMP Methods Manual Levels I and II: Methods for mapping and monitoring of physical and biological parameters in the coastal zone of the Caribbean, http://www.ccdc.org/jm/methods_manual.html (accessed December 2005).
119. Ninio, R., Meekan, M., Tone, T., and Sweatman, H., Temporal patterns in coral assemblages on the Great Barrier Reef from local to large spatial scales, *Mar. Ecol. Prog. Ser.,* 194, 65–74, 2000.
120. Brown, E., Cox, E., Jokiel, P., Rodgers, K., Smith, W., Tissot, B., Coles, S.L., and Hultquist, J., Development of benthic sampling methods for the Coral Reef Assessment and Monitoring Program (CRAMP) in Hawai'i, *Pac. Sci.,* 58, 145–158, 2004.
121. Whitfield, P.E., Fonseca, M.S., and Kenworthy, W.J., Coral damage assessment and restoration tools for small vessel groundings (extension of seagrass mini-312 program to coral reefs). Project Report to NOAA Damage Assessment Center, Silver Spring, MD, http://shrimp.ccfhrb.noaa.gov/~mfonseca/coral_modeling_report_4.html (accessed December 2005).
122. Julius, B.E., Expert report submitted in U.S. vs. Melvin A. Fisher et al.

123. Fonseca, M.S., Kenworthy, W.J., Julius, B.E., Shutler, S., and Fluke, S., Seagrasses, in *Handbook of Ecological Restoration*, Davy, A.J. and Perrow, M. (eds.), Cambridge University Press, 2002, chap. 7.
124. Kirsch, K.D., Barry, K.A., Fonseca, M.S., Whitfield, P.E., Meehan, S.R., Kenworthy, W.J., and Julius, B.E., The mini-312 program: an expedited damage assessment and restoration process for seagrasses in the Florida Keys National Marine Sanctuary, *J. Coast. Res.*, 40, 109–119, 2005.
125. Connell, J.H., Hughes, T.P., and Wallace, C.C., A 30-year study of coral abundance, recruitment, and disturbance at several scales in space and time, *Ecol. Monogr.*, 67, 461–488, 1997.
126. Hudson, J.H. and Goodwin, W.B., Assessment of vessel grounding injury to coral reef and seagrass habitats in the Florida Keys National Marine Sanctuary, Florida: protocol and methods, *Bull. Mar. Sci.*, 69, 509–516, 2001.
127. Jaap, W.C., Wheaton, J.L., and Donnelly, K.B., A 3-year evaluation of community dynamics of coral reefs at Fort Jefferson National Monument (Dry Tortugas National Park) Dry Tortugas, Florida, U.S.A., Florida Marine Research Institute manuscript, final report to National Park Service, 1994.
128. Smith, S.R., Patterns of coral recruitment and post-settlement mortality on Bermuda's reefs: comparisons to Caribbean and Pacific reefs, *Am. Zool.*, 32, 663–673, 1992.
129. Smith, S.R., Patterns of coral settlement, recruitment, and juvenile mortality with depth at Conch Reef, Florida, in *Proc. 8th Int. Coral Reef Symp.*, Lessios, H.A. and Macintyre, I.G. (eds.), Smithsonian Tropical Research Institute, Balboa, Panama, 2, 1197–1202, 1997.
130. Miller, M.W., Weil, E., and Szmant, A.M., Coral recruitment and juvenile mortality as structuring factors for reef benthic communities in Biscayne National Park, U.S.A., *Coral Reefs*, 19, 115–123, 2000.
131. Gittings, S.R., Bright, T.J., and Holland, B.S., Five years of coral recovery following a freighter grounding in the Florida Keys, *Proc. American Academy of Underwater Science, 10th Ann. Symp.*, Jaap, W. (ed.), AAUS, Costa Mesa, CA, 1990, 89–105.
132. Piniak, G.A., Whitfield, P.E., Fonseca, M.S., Kenworthy, W.J., Fisher, G., and Fogarty, N., A spatially explicit modeling approach for the recovery of coral reef grounding injuries, Project report to NOAA Damage Assessment Center, Silver Spring, MD, 2003.
133. Precht, B., personal communication, 2004.
134. Fonseca, M.S., Piniak, G.A., Kenworthy, W.J., Whitfield, P.E., and Fisher, G., Report on application and sensitivity of the coral injury recovery model, project report to NOAA Damage Assessment Center, Silver Spring, MD, 2003.
135. Bright T.J., Kraemer, G.P., Minnery, G.A., and Viada, S.T., Hermatypes of the Flower Garden Banks, northwestern Gulf of Mexico: a comparison to other western Atlantic reefs, *Bull. Mar. Sci.*, 34, 461–476, 1984.
136. Tomascik, T., Settlement patterns of Caribbean scleractinian corals on artificial substrata along a eutrophication gradient, Barbados, West Indies, *Mar. Ecol. Prog. Ser.*, 77, 261–269, 1991.
137. Smith, S.R., Hellin, D.C., and McKenna, S.A., Patterns of juvenile coral abundance, mortality and recruitment at the *M/V Wellwood* and *M/V Elpis* grounding sites and their comparison to undisturbed reefs in the Florida Keys, Report to NOAA Sanctuary and Reserves Division, and University of North Carolina at Wilmington, 1998.
138. Hughes, T.P. and Jackson, J.B.C., Population dynamics and life histories of foliaceous corals, *Ecol. Monogr.*, 55, 141–166, 1985.
139. Wittenberg, M. and Hunte, W., Effects of eutrophication and sedimentation on juvenile corals. I. Abundance, mortality, and community structure, *Mar. Biol.*, 112, 131–188, 1992.
140. Bythell, J.C., Gladfelter, E.H., and Bythell, M., Chronic and catastrophic natural mortality of three common Caribbean reef corals, *Coral Reefs*, 12, 143–152, 1993.
141. Bythell, J.C., Hillis-Starr, Z.M., and Rogers, C.S., Local variability but landscape stability in coral reef communities following repeated hurricane impacts, *Mar. Ecol. Prog. Ser.*, 204, 93–100, 2000.
142. Blair, S.M., McIntosh, T.L., and Mostkoff, B.J., Impacts of Hurricane Andrew on the offshore reef systems of central and northern Dade County, Florida, *Bull. Mar. Sci.*, 54, 961–973, 1994.
143. Dollar, S.J. and Tribble, G.W., Recurrent storm disturbance and recovery—a long-term study of coral communities in Hawaii, *Coral Reefs*, 12, 223–233, 1993.
144. Bythell, J.C., Bythell, M., and Gladfelter, E.H., Initial results of a long-term coral reef monitoring program—impact of Hurricane Hugo at Buck Island Reef National Monument, St. Croix, United States Virgin Islands, *J. Exp. Mar. Biol. Ecol.*, 172, 171–183, 1993.

145. Woodley, J.D., Chornesky, E.A., Clifford, P.A., Jackson, J.B.C., Kaufman, L.S., Knowlton, N., Lang, J.C., Pearson, M.P., Porter, J.W., Rooney, M.C., Rylaarsdam, K.W., Tunnicliffe, V.J., Wahle, C.M., Wulff, J.L., Curtis, A.S.G., Dallmeyer, M.D., Jupp, B.P., Koehl, M.A.R., Neigel, J., and Sides, E.M., Hurricane Allen's impact on Jamaican coral reefs, *Science*, 214, 749, 1981.
146. National Marine Sanctuaries Act, 16 U.S.C.§ 1443, *et seq.*
147. Penn, T. and Barry, K., personal communication, 2003.
148. Cook, C.B., Dodge, R.E., and Smith, S.R., Fifty years of impacts on coral reefs in Bermuda, in *Proceedings of the Colloquium on Global Aspects of Coral Reefs: Health, Hazards, and History*, Ginsburg, R.N. (ed.), University of Miami Press, Miami, FL, 1994, 161–166.

7 If You Build It, Will They Come? Toward a Concrete Basis for Coral Reef Gardening

Les S. Kaufman

CONTENTS

7.1 INTRODUCTION

Coral reefs are in precipitous global decline before the eyes of a single generation of biologists. The reason is simple and familiar: a confluence of desperate poverty and demanding affluence, of swelling hunger and rapacious consumption by a mobile, high-tech society. Combined human impacts have corroded the quality of ecological services provided by coral reefs and compromised their capacity to heal after even the normal annoyances of hurricanes and predators. These impacts include climate change, pollution, depressed aragonite saturation levels (due to elevated atmospheric CO_2), coral disease, coral mining, the elimination of keystone species called "ecological engineers," and destructive resource extraction.[1–7] Coral reef decline often manifests itself as a dramatic shift in dominance from hard corals to fleshy algae — very dramatically so in the tropical Atlantic.[2,8,9] Coral reefs are not the only ecosystems around the world that are shifting from desirable to undesirable states due to human activity, a process that has been in play for centuries.[10] However, the drama unfolding on coral reefs is especially stunning because it is happening around us right now. Coral reefs support a massive tourism industry, protect shorelines, and feed a large portion of the world's population dependent upon subsistence fishing. As the highly popular subjects of television nature specials, coral reefs enjoy a broad, if distant, constituency. Even neophytes may be shocked when the see a degraded reef, for they have built up quite reasonable expectations from all the nature films.[11] Consequently, where reefs die a fierce enthusiasm for technological "solutions" brews. Often, these are myopically conceived. Wisdom would put prevention first: reduce coastal eutrophication, eliminate destructive fishing practices, drastically lower CO_2 emissions, establish marine reserves, and enforce laws to protect coral reef species and habitats. Nonetheless, when the next reef to go is the one within *your* village, resort, or national marine sanctuary, you are likely to take a sudden interest in coral reef restoration.

Reef restoration is expensive, whether measured in sweat or dollars or both, but it can work on a small scale.[12] There have been quite successful projects to propagate corals, giant clams, sea urchins, and other reef framework builders and ecological facilitators. Orphaned coral fragments and mass-produced propagules have been placed out in reef habitats and affixed to artificial reeflike structures, even electrically self-healing ones, to good effect (e.g., reference 13). The technologies have the potential to someday rescue some coral reef species and to heal some types of damage, such as the dynamite blast craters that spread like a pox across the Coral Triangle. Restoration is a very tough job in the tropical west Atlantic, and it is definitively *not* the first priority in reef conservation anywhere, but neither is it the very last. Restoration can be emotionally and politically therapeutic and can produce a beautiful small aquatic garden, just so long as this tiny balm does not quell the greater winds of change.

The primary driver for coral reef gardening will always be its value to the local economy. It attracts tourists. It can also attract and possibly boost the productivity of some fishery species. Coral gardens can concentrate marine ornamental fishes for extraction[14] and have been promoted as "arks" that could one day seed and rejuvenate neighboring natural reefs. Isolated reef systems and possibly even most coral reefs are dependent largely upon self-supply of invertebrate and fish larvae.[15,16] To the extent that this is true at any given site, the "ark" idea may not be so farfetched.

Coral gardens can alter the dynamic of a society's relationship with its coral reefs; this is their most unsung value. Cognitive dissonance (e.g., you love the things you suffer for) is writ large on the human psyche. The intensive stewardship required to start and keep a coral garden going is bound to generate greater compassion for the ebb and flow of life on coral shores. The Balinese, a distinctly land-oriented people, have been brought to glance lovingly and frequently out to sea by coral gardening projects in the tourist towns of Seraya, Tulamben, and Permuteran (personal observation). Diving at Tulamben supports tiny village women who famously walk scuba tanks to the beach, balanced on their heads, for overweight tourists. Coral gardens maintained by local citizens have galvanized communities and awakened citizens to attend closely to human impacts on coral reef health. Finally, by focusing human use and extraction on a managed, nearby corner of the sea, the coral reef garden can be a sort of a halfway house or easement between the crush of civilization and true wilderness reserves farther afield. Coral gardens can act as diver aggregation devices or "DADs," drawing divers away from fragile natural reefs into areas managed expressly for this purpose.[17] Like all else in conservation, coral gardening and restoration are about people. People are inspired by small but beautiful works to do much grander and more practical things. Coral gardening has its place. So, how can we make it work really well?

The challenge in coral gardening is to trigger a self-assembly process that supplants its human preparations and culminates in a functional reef community within a practical time frame — say, less than 10 years. Epstein et al.[18,19] have explored in detail the validity of the silvicultural analogy. Success will be measured by what an initial, undesirable state — i.e., bare construction site or fleshy algal-shrouded reef surface — eventually turns into ecologically.[20] This means the benthic community that jackets it and the motile fishes and invertebrates that recruit to it. In this garden, corals and fishes are the wildflowers, shrubs, and trees, and we for the most part are but tilling and doing the initial rockwork … and then waiting. We want to know: if we build it…will they come? And who, exactly, are "they"? And by the way, what creatures would we *not* want to see arriving in droves, and how might these be discouraged? We are gardeners in the truest sense of the word.

7.2 CORAL REEF GARDENING IN THE CONTEXT OF REEF RESTORATION

Gardening can be a valuable adjunct to coral reef restoration when there is need:

1. To accelerate healing on damaged patches of an otherwise healthy reef: e.g., dynamite blast fields and ship groundings
2. To heal relatively isolated target patches in order to create or restore entire reef sections within a reef system
3. For a scheme to produce coral reef goods and services on a sustainable basis in a place where they were absent before, such as in the middle of a sand plain or seagrass bed

What these applications have in common is that all aim to deflect community ontogeny on a patch of sea bottom away from alternative states (such as seagrass, seaweed, or dancing rubble) and toward a physically stable, accretionary state dominated by crustose coralline algae, hard corals, and other reef builders along with their highly valued associated fauna and flora. Reef restoration technologies range from very simple to highly sophisticated and expensive and include the following:

1. Dump bulky solid waste (tires, cars, ships, preformed concrete) on the seabottom in one area.
2. Build a carefully designed and placed wave-resistant structure from mass-produced modules of natural and/or recycled materials.
3. Design modular deployments to kick-start succession toward hard coral dominance by modulating flow fields and recruitment.
4. Dope structure with chemical morphogens to promote the settlement of larvae of hard coral and other desired reef organisms.
5. Engineer structures for self-repair and self-renewal during the establishment phase for a hard coral community; corals eventually subsume these functions.
6. Manually plant structure with framework builders (corals, giant clams) removed from a natural reef.
7. Manually plant structures with framework builders produced in aquaculture.
8. Manipulate the community of motile invertebrates and fishes (facilitators) to enhance the robustness and growth rates of the framework builders.
9. Manipulate the community of motile invertebrates and fishes to maximize levels of sustainable production of extractive resources.

For the most part, these techniques have been employed individually; a comprehensive protocol for coral reef gardening using integrated best practices does not exist. We can get there more quickly by parsing the problem and considering the ecological processes attendent to each critical step toward a self-sustaining patch of coral reef.[21] Here are four candidate steps for such a clinical protocol:

Task 1: Restore or create a wave-resistant structure.
Task 2: Induce circumstances on and around the structure toward a state conducive to the growth of living framework builders.
Task 3: Speed succession toward a community dominated by framework builders that will continually renew and regrow the structure.
Task 4: Craft community ontogeny so as to maximize the value of goods and services produced by this engineered patch of reef.

7.2.1 Task 1: Restore and Create Wave-Resistant Structures

A coral reef begins with topographic relief on the seabottom that resists the destructive forces of waves and sea. The engineering of a wave-resistant structure is what ultimately leads to a coral reef's valued biological features, such as its high local productivity and high species diversity.

Unprepared to wait the thousands of years required for a reef to grow where desired or for a damaged reef to repair large sections of its basic architecture, coastal communities have lobbied hard for the creation of artificial reefs to enhance fishing and diving. There is now an established lore on the best ways to go about that exercise. Coral reef restoration projects usually focus less on massive structural architecture and more on providing an adequate foundation for coral growth. Concrete is clearly better than tires, old cars, or even ships. Simple piles of limestone boulders may be even better, and coating culverts with limestone rocks can improve the outcome as well. Three standardized technologies for fabricating modular structures are now being marketed: ReefBalls (and the conceptually similar Grouper Ghettos), Ecoreefs, and Biorock.

ReefBalls and Grouper Ghettos are cast cement hemispheres (ReefBalls) or angular forms with holes and open spaces in them. ReefBalls in particular can be mass produced and deployed by the hundreds or thousands to construct composite structures as large as breakwaters and sections of reef. The largest such structure to date is an artificial fringing barrier reef in Antigua. Modular concrete structures are comparable to a surgical implant; they replace coral reef framework in the short term, while coral and other invertebrate fragments that recruit or are secured to the balls have a chance to become established. Over time, the implant is obscured and supplanted by desirable epifauna, including native framework-builders. Concrete modules have been used to routinely create about 1 to 3 m of vertical relief. Information on ReefBalls is consolidated on the company's web site, but rigorous, long-term study comparing ReefBalls to alternative structures is still wanting. Eventually, however, we shall discover how long it takes for the concrete hemispheres to be supplanted by a hard coral superstructure under varied conditions and in different parts of the world. One other advantage of the modular concrete approach is that it may provide a scaffolding capable of surviving storm damage, bleaching events, and predator outbreaks. In fairness, an artful pile of boulders can do the much same thing. The remnant, wave-resistant, rough-surfaced habitat of encrusted modules might regenerate hard coral cover more quickly than a low-lying rubble field.

EcoReefs (Figure 7.1a) are ceramic modules similar to a snowflake-shaped staghorn coral colony, which are arranged in clusters over reef rubble left behind from physical damage, such as dynamite explosion fields. An EcoReef is only a few decimeters high at most, but the structure is meant to serve only as a physically stable and hydrodynamically attractive foundation for coral settlement. EcoReef was designed for a very specific purpose: to provide physical stability and create a hydrodynamic current field conducive to hard coral recruitment (Figure 7.1b). Thus, it can jump-start the reestablishment of hard coral communities on low-relief rubble fields otherwise recalcitrant to hard coral regeneration for decades.[22] Careful research is under way to assess the performance of EcoReef clusters in repairing rubble fields on reefs at Bunaken National Park, Indonesia; a new project is starting up in the Phillipines.

Electrochemical precipitation of mineral from seawater, based on a different kind of current, is a patented process invented by Wolf Hilbertz and perfected for use in artificial reef construction by Hilbertz and Thomas Goreau[23–26] under the name "Biorock." Biorock installations are purveyed through the Global Coral Reef Alliance, a small not-for-profit enterprise with a well-illustrated web site. Biorock is one of several experiments in which an electric current has been run through submerged wire frames to build artificial reefs. In Biorock, a hard carbonate material is deposited electrolytically over a form constructed principally from iron reinforcing bars (Figure 7.2). The material that accretes on properly electrified structures is a variable mixture (depending upon how well the array is working) of aragonite (calcium carbonate) and brucite (magnesium hydroxide) (Land, personal communication). Brucite is highly soluble in seawater, and a high proportion of brucite in the electrically deposited material would therefore be undesirable. This author has seen Biorock installations in Jamaica, St. Croix, and Bali (Seraya, Tulamben, and Permuteran). The mineral accreted to the structures in Bali was durable and strongly adherent to the embedded reinforcing bar, both on structures with live current and on those that had experienced a cessation of current within the preceding month or two. I did not have the chance to examine a structure that had been without electricity for a very long time to see if the mineral coating was durable without

(a)

(b)

FIGURE 7.1 (a) Demonstration project of EcoReefs® modules deployed on a coral rubble field at Bunaken National Park, Indonesia. (b) Coral recruit *Stylophora* sp. on a shaded lowering setting plate on an EcoReefs® array.

recent additional precipitation. The accreted mineral can be deliberately sloughed off by reversing the charge, though to make this a useful feature would require some imagination.

Little is known about the submarine diagenesis of Biorock, or about its endolithic community or vulnerability to bioeroders. However, a sliced Biorock sample with accretion of more than 4 cm radius showed little or no boring at all (personal observation). Either it formed very rapidly and thus there had not been time for endoliths to accumulate, or Biorock is resistant to boring. On the one hand, the material may remain quite solid and durable, unlike natural reef rock, which is usually highly tunnelled and friable. Endoliths can play both positive and negative roles in reef development. The endolithic community contains both primary producers and pathogens,[27–29] as well as pore spaces that may be critical to nutrient cycling and the submarine diagenesis of reef rock. Nonetheless, in coral reef restoration as opposed to underwater architecture, the primary function of the accreted mineral in Biorock is to attract coral recruits and serve as an anchoring point for explanted coral colonies. For the former, it may perform more poorly than a rougher, natural surface would. As a system for securing coral fragments, however, electrochemical accretion is quite good.

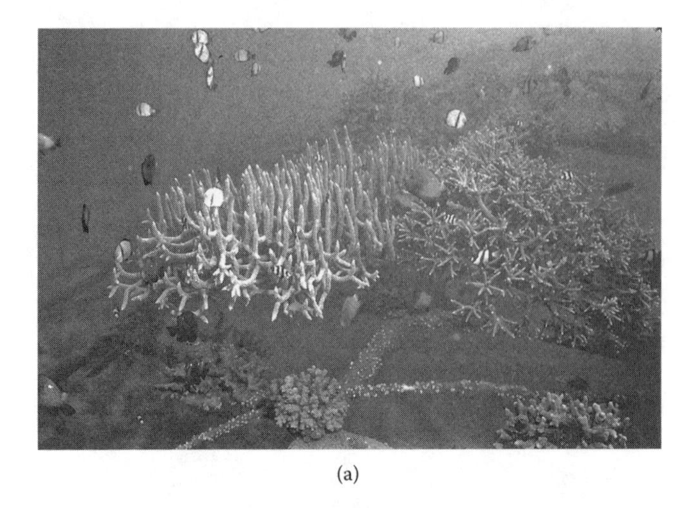

(a)

(b)

FIGURE 7.2 (a) Prolific coral growth and fish on a 3-year-old "Biorock" reef in Bali, Indonesia. (b) Coral recruitment of acroporid coral on "Biorock" mesh.

All three of these methods result in a submarine structure that can resist currents and waves, setting up flow fields that are attractive to both juvenile and adult fishes, and may enhance coral settlement. In all cases there is the possibility that the resulting structures can be large enough to also act as a measure to control beach erosion.

7.2.2 Task 2: Induce Circumstances on and around the Structure That Are Conducive to the Establishment and Growth of Living Framework Builders

Placing a hard substratum on the seabottom triggers the development of a fouling community of sessile organisms and an associated assemblage of motile organisms dependent upon the fouling

(a)

(b)

FIGURE 7.3 (a) Mobile coral rubble produced by blast fishing in Indonesia. (b) Coral recruitment on a pile of large boulder-sized rubble placed on dead reef in Indonesia.

community for food and habitat. The hope in coral reef restoration is of course that this fouling community will, in its exposed portions, quickly become dominated by hard corals, and that these will in turn attract other creatures that help to foster the further development of a healthy coral reef community. Whether or not events actually unfold in this manner depends upon numerous factors that are not fully understood.[30] Frequently, the experiment fails, or at least the desired community takes so long to develop that the attendants give up.

A simple pile of limestone rocks in Indonesia may be obscured by lush growth of living corals in as little as 2 years (Figure 7.3). The same pile of rocks off Negril, Jamaica, would likely be covered by unsightly macroalgae in a period of weeks and then retain this appearance for years. Why such extreme variation in outcome? We know that ambient nutrient levels, the density of coral reef herbivores, and priority effects (i.e., the timing of bare substratum availability relative to larval supply) can be important determinants of fouling community structure. Clearly it is ill-advised to place a coral gardening project just offshore from a sewer outfall or anywhere near a denuded watershed (though even this is context dependent: e.g., reference 31). Even under locally favorable conditions, however, artificial reef structures, and natural reef and hard bottoms as well, have transformed into fleshy algal reefs or pavements. Once we understand fouling community dynamics well enough it may be possible to prevent artificial reef structures from being coated by intransigent sessile invertebrates or macroalgae, which can delay hard coral dominance indefinitely and greatly diminish the return on investment.

Coral planulae exhibit strong preferences as to where they will settle and establish themselves. A good deal of research has been done on the suitability of various substrata to surface artificial reef structures and settlement tiles. Throughout the world, early (and perhaps all) hard coral recruits settle preferentially on crustose coralline algae (CCA). Particular species of CCA exude chemical signatures that are particularly attractive. These settlement-inducing substances can be used directly by doping substrata where coral settlement is desired.[32–34] Many coral planulae exhibit a settling preference for the undersides of surfaces, and settlement can be strongly inhibited by the presence of macroalgae or anything else that physically obstructs the openings to crevices and overhangs. The combination of a limestone or concrete foundation, CCA dominance, a rough surface texture, and many unobstructed overhangs is a killer combination for attracting coral spat.

Dominance of the sessile community by coralline algae is seen within a broad range of light and nutrient levels but often does not occur due to competition from fleshy algae. Indeed, the interactions among macrophytes, crustose coralline algae, and hard corals are at the crux of reef conservation and restoration. Macrophytes probably capitalize on readily accessible nutrient pools faster, or at least in a more competitive manner than corallines do, though it is not clear that oligotrophic conditions are particularly favorable to coralline growth except for their tendency to reduce spatial competition from macrophytes. Corallines enjoy a competitive advantage over turf algae under grazing pressures that are high enough to exclude macrophytes but not so high as to compromise coralline growth and accretion. Intuitively, corallines should do best at the junction of intermediate light level and intermediate grazing pressure. The higher the light and nutrients, the higher the grazing pressure necessary to achieve the same effect. Thus, a second way of increasing the efficacy of coral gardening would be to terminate fishing for herbivorous fishes, particularly on exposed reefs in the Indo-Pacific where fishes are the dominant large algal grazers. In the tropical west Atlantic and on inshore Pacific reefs urchins of the genera *Diadema* and *Echinometra* can be the primary grazers. Their importance is heightened once overfishing has reached the point where large predators are rare and herbivorous fish populations, the next fishery resource in line, have been chewed down.[2,8]

It is worth keeping in mind that the relationship among corals, corallines, and fleshy algae has been broadly oversimplified. Although a fundamental play-off between macrophytes and hermatypic corals or coralline algae (or both) does exist,[35] there are complexities.[36] Macrophyte appearance or disappearance can have cascading effects on the community.[37,38] Fleshy algae can kill adjacent corals in a variety of ways besides overshading and abrasion. They can trigger disease[39] and harbor other organisms, such as fireworms, that can do this themselves, as well as feed directly on the corals.[40] Interactions among sessile benthos are mediated by a diverse host of inquilines and herbivores across a broad size range, as well as a complex web of inducible chemical weapons and herbivore associations.[41] Circumstances may even exist under which fleshy algae are beneficial to hard corals. One example of this arose in an experimental study of coral-algal-herbivore interactions led by the author in the ocean mesocosm of Biosphere II during its tenure as a Columbia University Earth Institute research facility. Biosphere II was a nearly 4-acre, hermetically sealed greenhouse in Oracle, Arizona. Harboring an assortment of ecosystem mesocosms including a 750,000-gallon tropical reef, Biosphere II was conceived initially as a test of technologies and psychology for the long-term human habitation of space and later as an experimental system for biospheric research. Our experiment pitted hard corals against fleshy algae under various experimental treatments: grazed

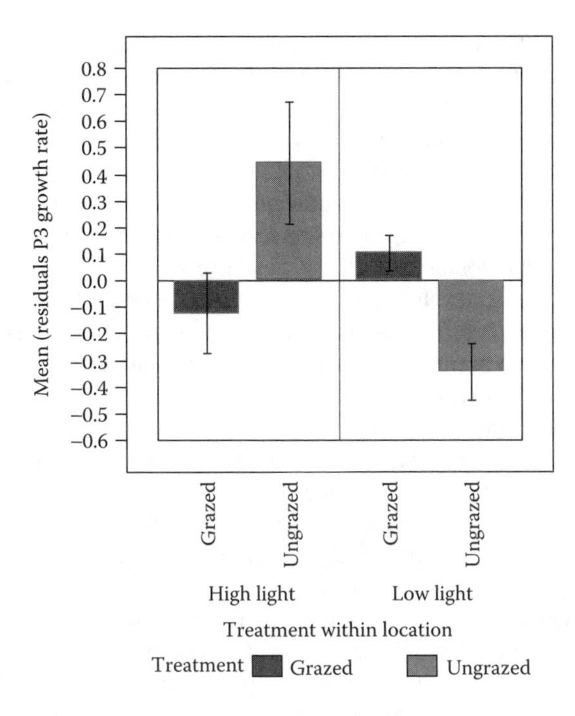

FIGURE 7.4 Results of Biosphere II experiment. Coral growth of *Montipora capitata* under high light and low light conditions. Treatments are grazed and ungrazed.

versus ungrazed, high versus low light, and three different levels of aragonite saturation state, mimicking the effects of varied atmospheric carbon dioxide concentrations. New data on CO_2 effects are still being analyzed. Independent of these effects, however, we noted an odd relationship between axial extension rates of the coral *Montipora capitata* and growths of a red alga, *Polysiphonia* sp. (Figure 7.4). The best coral growth was expected at high light levels (ca. 800 $u^3/m^2/min$), with regular removal of fleshy algal growth by a technician (the "grazer" in this portion of the experiment) due to direct algal–coral interactions.[42] Growth was fastest at the higher light level, but in the ungrazed condition with its lush *Polysiphonia* mat, not the grazed treatments, free of the fleshy algae. At lower light levels, grazed corals did grow faster than ungrazed. Our candidate hypotheses are that so long as light was not a limiting factor, the algae had a positive effect on coral growth either due to reduced CO^2 levels in the immediately adjacent water or to nutrient provisioning via algal exudates. Just because the relationship between reef corals and fleshy algae is complex does not make it inaccessible to reason and experiment, but it does make it a lot more interesting, and coral reef gardening that much more challenging.[43]

It is important to consider how we might construct a reef surface so as to attract the settling juveniles of facilitator species that promote coralline algal growth. Alternatively, for the impatient, it might be fruitful to cultivate and force-recruit crustose corallines where desired. Who are the crustose coralline facilitators? The herbivore guild on coral reefs is a diverse, polyphyletic assemblage of species that vary in their mode of feeding, and hence in the degree to which they are likely to assist in achieving coralline dominance. A full cast of reef herbivores is familiar to "living reef" aquarists, who struggle mightily to achieve the same effects in their living room microcosms that others strive for on damaged coral reefs. The players include microcrustaceans, browsing and grazing hermit crabs, majid brachyurans (e.g., *Mithrax* sp.), a variety of gastropods, regular echinoids, and herbivorous reef fishes, principally members of the families Scaridae and Acanthuridae, along with representatives of the Pomacentridae, Pomacanthidae, Blenniidae, Gobiidae, Kyphosidae, and others.[44,45] Just as intermediate grazing pressures are best for coral settlement and growth, a coral gardener's favored actors in the reef herbivore guild should be species that graze at an intermediate level of intensity. In other words, browsers on algal fronds are worse than useless, while monster mashers of coral rock (variously in search of live coral tissue, cryptic sponges, boring clams, and endolithic algae) may be equally unwelcome. Aquarists lump the monsters, such as triggerfishes, under the term "not reef-safe."

Macrophytic algae are not the only potential competitors capable of displacing corallines and thus inhibiting settlement by hard corals. Fouling communities may be composed of any number of sessile invertebrates, both colonial and solitary, plus diverse algae.[46] The invertebrate assemblage is dominated by sponges, tunicates, bryozoans, and nonscleractinian Cnidaria. Typically we think of invertebrate fouling assemblages as being limited to cryptic environments and undersides of things by spatial competition from photosynthetic scleractinians and macroalgae, but predation (often by supposed herbivores) can be a factor as well.[47] Many sessile invertebrates other than corals host photosynthetic bacterial or protistan symbionts and abound on exposed hard surfaces in shallow waters in the tropics. For example, the emerald-mouthed didemnid tunicate *Didemnon molle* and allied species, colored by their symbiotic strains of the cyanobacterium *Prochloron*, are extremely common recruits to bare limestone (and concrete) in the Indo-Pacific. Even nonsymbiotic sponges and colonial tunicates can monopolize exposed surfaces of both natural and artificial structures, provided they are adequately defended.[47]

In addition to occupying space that might otherwise be available to coral recruits, many fouling organisms are capable of inhibiting coral settlement through allelochemical interactions, or killing and overgrowing young coral recruits. Figure 7.3b shows an unidentified invertebrate that has recently become a pest on Biorock frameworks in Permuteran, Bali. It covers surfaces, making them unavailable to coral settlement, and also attacks and overgrows hard corals already established on these structures. Maida et al.[48] have demonstrated that soft corals produce substances that can inhibit hard coral settlement. These same soft corals are highly opportunistic space occupants on newl[illegible] substratum. For example, hard corals were formerly dominant over a large area of seabottom near Tulamben, Bali. These were killed in November 2002 by sedimentation in an area close to a river mouth. This bottom is currently occupied by a nearly continuous mat of soft corals (personal observation, 2004). The widespread dominance of soft corals on the tops of Indo-Pacific platform reefs — regarded by diving afficianados and scientists alike as the norm — may rather be a reflection of the differential vulnerability of hard corals to extreme low tides, foul storms, and bleaching events on these shallow (<2 m at mean low water) reef and seamount tops.

Despite a healthy literature on invertebrate fouling assemblages on coral reefs,[49–52] we know next to nothing about how to influence their behavior on reef restoration frameworks. Consequently, we also do not know how to control them when they become a problem. In the Permuteran Biorock installation, the human stewards are manually peeling invertebrate mats away in areas where they are overgrowing living coral (Figure 7.3b). This is probably not going to be a viable option on large or remote installations. There is a guild of coral reef fishes, mostly tetraodontiforms, that feed on fouling organisms. Filefishes, puffers, triggerfishes, and Moorish idol (*Zanclus canescens*) are common members of this group. In 2004 all were very conspicuous on Biorock installations at Seraya, Tulamben, and Permuteran, along the northern coast of Bali. It is possible that they were drawn to the structures by the opportunity to forage on these invertebrates; if so, however, they did not succeed in preventing the invertebrate mats from overgrowing hard corals.

Nutrient dynamics are another potentially crucial but puzzling determinant of the ontogeny of benthic communities on artificial reef structures.[53–59] Coral reefs develop best under conditions where ambient waters are highly oligotrophic (low in dissolved nutrients). However, this does not mean that nutrients are bad for reef-building corals. Several experiments have demonstrated that corals grow faster, or are at least not directly compromised, under elevated levels of limiting nutrients (e.g., reference 31). Furthermore, low dissolved nutrient levels in the water column do not automatically mean that nutrient availability is low.[56] It can also be that nutrient turnover rates are high and nutrient cycling is tight — precisely the situation thought to exist on most healthy coral reefs. The expected negative relationship between coral reef growth and eutrophic conditions derives from positive relationships between high water column nutrients and low light (due to phytoplankton blooms or suspended solids), or high nutrients and rapid overgrowth of benthic substrata by macrophytic algae. The secret to lush coral reef growth is not low nutrients *per se* but rather low nutrient levels in the water column and benthic nutrient pools, and high nutrient levels available to coral symbionts. These conditions can be achieved through tight cycling of nutrients with coral zooxanthellae as the primary beneficiaries.

Some of the nutrients available to coral reef organisms are regenerated from sedimentary and reef pore space nutrient pools.[59] Even these, however, were initially imported into the reef system from the water column and adjacent benthic marine habitats by motile reef-associated animals. Recycled nutrients are deposited on the reef in the form of these animals' waste products. Let us break this phenomenon down into its component parts. First comes the fouling community itself, which we can think of as a "rug of mouths." The reef's mat of sessile invertebrates, including all of the corals (both with and without zooxanthellae), is perpetually engaged in filtering the waters that pass over it for organic matter, whether living (i.e., bacteria, phytoplankton, zooplankton) or nonliving (POM). The rug of mouths is best developed on reef surfaces most directly exposed to food-laden currents. Due to the shear gradient and boundary layer formation over the reef–water interface, exposure is enhanced by height in the water column.[60] This "reef effect" is the very essence of why a wave-resistant structure so quickly forms the foundation for a rich benthic community such as a coral reef.[61]

Attracted by and sheltering amid the mat of fouling organisms are three guilds of motile animals, each of which contributes further to nutrient flow into the reef community: the "wall of mouths," "commuters," and "residents." The wall of mouths[62] is composed mostly of zooplanktivorous fishes, plus a few motile filter-feeding crustaceans such as porcellanid crabs, holothurians, and ophiuroids. The wall of mouths extends outward from perches and hidey-holes on the reef surface up into the living, shimmering curtain of zooplanktivorous fishes that is so familiar to divers and followers of underwater photography. There are both a diurnal and a nocturnal wall of mouths, which differ in taxonomic composition and behavior. All of the diverse participants in the wall of mouths void most of their concentrated, nitrogen- and mineral-rich waste products directly over or onto the reef. For example, diurnal wall-of-mouths participants feed until dusk but continue to pass waste products for hours afterwards; it is doubtful (not that we really know) that they would wait until morning to defecate. Thus a good deal of the nutrients in their waste stream is probably released beneath the boundary layer, readily accessible to corals and their dinoflagellate symbionts. On temperate California kelp reefs, a causal link has been demonstrated between wall-of-mouths waste products and growth of the primary structural organisms, the kelps;[63] a similar relationship probably exists for coral reef zooplanktivores and the corals in which they shelter.[64] If this relationship is generalizable, it could help to explain the great success of *Stylophora*, *Pocillopora*, and tabulate acroporids as the premier early colonists of regenerating coral reef in the Indo-Pacific. Such corals are strongly targeted for settlement by numerous species of damselfishes (e.g., *Dascyllus aruanus, D. reticulates, D. trimaculatus, Pomacentrus molluccensis, Chromis viridis, C. atripectoralis*), butterflyfishes (*Chaetodon trifascialis, C. trifasciatus*), acanthurids (*Paracanthurus hepatus*), and serranids (*Pseudanthias* spp.).[65] The zooplanktivores form dense aggregations on these corals, sleeping among the closely spaced branchlets. The butterflyfishes feed on coral mucus and bits of the polyps themselves without damaging the colonies, and numerous reef fishes feed on coral larvae and sperm–egg bundles,[66] both processes furthering nutrient regeneration and recycling.

The commuters are mostly fishes, though some crustaceans and regular echinoids could also function in this manner. Among the major commuters on both Indo-Pacific and Atlantic coral reefs are fishes of the families Haemulidae (grunts), Lutjanidae (snappers), Holocentridae (squirrelfishes), and Mullidae (goatfishes). These species mill about the reef by day, sometimes forming huge aggregations, and fan out across nearby seagrass beds and sand flats by night to hunt for the benthic invertebrates and fishes that comprise their prey base. It has been demonstrated for one Atlantic species, the French grunt (*Haemulon flavolineatum*), that a commuter can be a significant source of nutrient input for a coral (in this case elkhorn coral, *Acropora palmata*), that builds the commuter's resting habitat.[67,68] Large parrotfishes that move between coral reef and seagrass or mangrove habitats may function in a similar way, but on an opposite time cycle.

Residents are motile species that never leave the coral reef environment but can simultaneously mediate spatial competition and nutrient cycling by removing one group of space occupants and fertilizing others with their refuse. Herbivores, invertebrate feeders, and piscivores all contribute to this process, but the outcome can vary widely as a consequence of the overall food web dynamics of the community. For example, large reef piscivores could facilitate nutrient flow into corals by concentrating their nitrogen in daytime resting spots. By the same token, predators could have indirect negative effects on corals by consuming resident herbivores, thus triggering a trophic cascade favoring macrophytic algae.

Nitrogen-fixing cyanobacteria are another taxon whose contribution to nutrient dynamics must be considered. Williams and Carpenter[54] found them to be a surprisingly minor factor. On the other hand, Mazel and the author several years ago discovered what appeared to be cyanobacteria in close association with the Caribbean coralliomorph *Ricordia*. This possibility has since been confirmed through the discovery of symbiotic cyanobacteria in hard corals.[69] Perhaps cyanobacteria are important to nitrogen flux on coral reefs after all, but in a different way than previously suspected. A related but much larger issue is that of microbial community function on coral reefs as a whole. Certainly we are past the level of sophistication where we can acknowledge the uniqueness of host–dinoflagellate symbioses, but we are still not completely comfortable with the reality that functional corals, fishes — indeed any multicellular organism in any environment — would not be itself without its associated microflora. Reef-building corals *are*, in part, their rich associated microflora.[70,71] We must understand corals' prokaryotic camp followers and inquilines if we wish to influence their state of wellness and functionality in a community context.

Rugs of mouths, walls of mouths, commuters, and residents are not limited to coral reefs, nor are corals their only beneficiaries. The reef effect operates conspicuously on any upright hard structure exposed to current, in aquatic ecosystems of all sorts. Hecky et al.[70] described the impact on nutrient flux of reef-like communities in the Laurentian Great Lakes built by two introduced species of dresseinid mussel, the infamous zebra and quagga mussels. Luxuriant growths of freshwater *Cladophora* feed off nutrients cycled from water column to benthos via the mussels' waste products. Hecky et al.[70] saw parallels between nutrient cycling in these lacustrine bivalve bioherms and coastal coral reef communities. Mangrove trees can similarly benefit from nutrients concentrated by their filter-feeding prop-root epibionts.[71] Benthic–pelagic coupling can also cycle the other way, with benthic filter feeders contributing to regeneration of nutrients in the water column, thus fueling phytoplankton growth and eutrophication.[72,73] Sometimes, the switch among alternative nutrient flow channels is on a hair trigger. In heavily exploited estuaries, the switch may be thrown by a change in season.[74] On coral reefs, the switch settings correspond to alternate living fabrics: a rug of mouths on a living coral mound, or a carpet of seaweed on carbonate rock (the corals' dead skeletons). What throws the switch on coral reefs? The amount of nutrients can sometimes matter, but it is probably the routing of these nutrients that matters most. If zooxanthellae are on the receiving end, then symbiotic corals dominate the benthos. When the links to zooxanthellae are broken, phase transition from hard corals to fleshy algae becomes a likely event. In an intact, healthy coral reef system, organisms are passing nutrients to each other quickly and efficiently. Removing key species or functional groups, however, can cause the nutrient pools in reef rock and sediments to grow and ultimately bleed into near-reef waters.

The herbivore guild has always been the prime suspect for head switch-thrower on coral reefs. Herbivores do regulate the transformation of nutrients into algal biomass, so clearly they do help keep the lid on the system. But they may not be the fire under the pot, after all. Normally, the nutrients concentrated on a coral reef are sequestered in the living organisms themselves which, aside from corals, hail from about 30 other phyla and hundreds of thousands of species. If enough of those species that create nutrient closure on the reef are destroyed by some disturbance event, then nutrients from all of the importers will pool and adventitious weeds can prosper. The actual fire under the pot could be the destruction of species critical to nutrient cycling. We do not understand reef nutrient dynamics well enough to know how all the species on a reef function together to achieve such tight recycling. All these species are what Aldo Leopold was referring to when he talked about all the little "cogs and wheels" that you had better not lose. That is, they are the pieces that an intelligent tinkerer would be careful to save because without them the system as a whole will not work. Herbivores are not the only cogs that could make the ultimate difference. Most coral reef animals are at least facultative detritivores. Then there are corallivores, coral mucous-eaters, cleaners, sand sifters, microbes of myriad functions, and so on. Notwithstanding scientists' ongoing obsession with herbivores, there may not be any one key functional group. When a coral reef shifts to fleshy algal pavement, the key may just be whatever species was, by chance, the last one keeping the degraded system functional. We are left with the sage but intensely dissatisfying conclusion that to make coral reef restoration work, we must bring back all the pieces and let them fit themselves together properly. Leopold was right.

In addition to the internal ecological dynamics of a reef restoration site, extrinsic considerations could influence success. The odds of a coral community regenerating on a disturbed reef are enhanced if the restoration site is physically well disposed for the interception of both coral recruits and plankton and if the substratum characteristics are highly attractive to coral larvae ready to settle out of the plankton. The importance of crustose coralline algae has already been mentioned. Several measures can be taken to ensure conditions that favor coral recruitment. Most important of these is to have healthy natural reef close at hand and well positioned to serve as a larval source. For this reason, the major application of reef restoration technology will probably always be the repair of relatively small disturbed patches surrounded by a still-intact natural reef or reef system. In the case of more isolated restoration projects it is crucial that the site be exposed to a heavy larval rain from upstream reefs in at least a fair-to-good condition. It may be possible to maximize the volume of settlers to a structure by attaching buoyed ropes, leads, or attraction devices (e.g., reference 75). Remember that these same attraction devices can efficiently lead waves of undesirables to the reef as well.

The functional profile of the herbivore community is another important factor. Herbivores range from obligate browsers to rock-excavating grazers. The cumulative effect of the herbivore community is thus a product of the density profile of herbivores along this functional spectrum and not just the total number or size of herbivores present. The persistence of fleshy algal pavements, once established, suggests that browsers capable of bringing macrophyte turfs within the range of what grazers will eat may be very important in the phase transition from fleshy algal pavement to crustose coralline dominance. Several papers have examined the synergistic effects of browsers and grazers on coral reefs,[43,76] and hence, once again, the importance of functional diversity to coral reef resiliency (and hence the success of restoration attempts). Nonetheless, it has been shown that moving adult *Diadema antillarum* onto hardbottom ("dead" reef) substrata in Florida can facilitate the settlement and establishment of corals. So, in at least some cases, a single species of hard-grazer (albeit one with rather catholic tastes) can be enough to do the trick … for a while.

Once a coral planula or asexual fragment has arrived, physical stability becomes the next big concern.[77] Settlement on small, unstable bits of coral rubble will lead to high mortality as the rubble moves about. The faster that baby corals and coral fragments can attach themselves and establish a stable base, the sooner the restoration site will be entirely covered with living coral. This process is influenced by the growth rate of sexual propagules and the rate of basal expansion of coral fragments, by which means coral fragments establish a firm attachment to the reef. Several practitioners have found that setting a coral fragment upside down can speed attachment by exposing fast-growing branch tips to anchoring substrata (Erdmann and Moore, personal communication).

The larger a chunk of limestone or other attractive settlement substratum, the greater its stability. On reefs blasted into unstable rubble in Komodo, simply dumping piles of coral rocks into organized structures has been enough to trigger coral reef regeneration (Figure 7.3b), and similar results have been achieved in the Maldives with only slightly more sophisticated structures.

7.2.3 Task 3: Speed Succession toward a Community Dominated by Framework Builders That Will Continually Renew and Grow the Structure

Throughout the world, fast-recruiting, fast-growing coral species tend to be the first hermatypic corals to form a canopy over disturbed reef surfaces in shallow waters (less than 20 m depth). This list is dominated by branching and foliaceous members of the families Acroporidae, Pocilloporidae, and Agariciidae. These are the coral reef equivalents of tropical forest tree families like Cecropiaceae, Moraceae, Simaroubaceae, Ochroma (balsa wood) and Bombacaceae. In tropical rainforest, the pioneers and early builders of high canopy create a wind-resistant structure that pushes the boundary layer upward and creates a seal for climate control in the understory layers. On a coral reef something analogous happens when ramose corals grow upwards to create a structure resistant to waves and current, thus thickening the boundary layer and creating a zone of turbulent but slow-moving water. Any scheme for quick restoration of coral cover must have these species in mind at the outset. Not surprisingly, the acroporids dominate the early recovery phases of damaged reefs and usually comprise the bulk of the colonies explanted onto artificial reef structures.

In the tropical western Atlantic there are only two acroporid species: *A. cervicornis* (staghorn coral), and *A. palmata* (elkhorn coral) plus their hybrid species *A. prolifera.* Over most of the Indo-Pacific, many species of branching and tabulate acroporids fill the bill. The species and ecological diversity of Indo-Pacific acroporids may facilitate regeneration of branching coral cover on Indo-Pacific coral reefs and permit it to occur over a wider range of conditions than in the tropical Atlantic. It may also be significant that none of the Atlantic acroporids habitually produces a tabulate growth form, while several very common Indo-Pacific species do. A tabulate form could be a more formidable spatial competitor than an open-branched colony. Tabulates are able to easily rise above a substratum dominated by either fleshy algae or other sessile invertebrates and stretch an optically and hydrodynamically dense shadow across them, a feat not readily available to Atlantic acroporid species. Furthermore, they are among the most preferred recruitment sites for wall-of-mouths fishes.

The western Atlantic as a whole should be much more vulnerable to the removal of acroporids than is the central Indo-Pacific, where acroporid species diversity and sexual recruitment are both very high. Even within the tropical western Atlantic differences exist, however. In the eastern Caribbean a very high proportion of young acroporids is sexually produced, and here genetic diversity (as indicated by microsatellite heterozygosity) is high. In areas west of Puerto Rico, however, the genetic diversity of acroporid clones is much lower, at least for the staghorn coral *A. cervicornis* (Baums, personal communication). Thus, large-scale disturbances to hermatypic corals, such as bleaching and outbreaks of disease, should be expected to have the most severe and spatially extensive impacts in the western Caribbean, where both species and genetic diversity in acroporids are low. Regions east of Puerto Rico may be slightly less vulnerable due to a higher genetic diversity and prevalence of sexual recruits. Nowhere in the tropical Atlantic, however, do acroporids recruit as aggressively as in much of the Indo-Pacific. Folicaeous and submassive weeds do exist in the western Atlantic in the form of *Agaricia agaricites* and *A. tenuifolia. Agaricia agaracites* recruits heavily to open patches on tropical western Atlantic (TWA) reefs. For example, ribbon-like strips of the Christianstead (St. Croix, U.S. Virgin Islands) fore-reef killed by sediment plumes were covered within a few years by near-total coverage of *A. agaricites* (personal observation, ca. 1981). Most of the time, however, *A. agaricites* recruits heavily but fails to contribute much to reef framework. *Agaricia tenuifolia* filled a lot of space vacated by acroporids on the Meso-American barrier reef in Belize and helped to build strong coral buttresses out of a diverse amalgam of strains, but later the species succumbed to bleaching. Furthermore, *A. tenuifolia* is nearly restricted to continental reefs, leaving most island reefs bereft of its services. Pocilloporids are extinct in the TWA save for the relatively modest contribution of *Madracis* spp., and so these reefs are also missing aggressive early space colonists like *Pocillopora damicornis.* In addition, the TWA lacks large, aggressive soft corals like the *Sarcophyton*, *Sinularia*, and *Xenia* that occupy so much of the surface of platform reefs in the Indo-Pacific; in their place there is only *Erythropodium* and the zoantharian *Palythoa carribea.* Shallow TWA reefs are therefore unusually vulnerable to becoming stuck following phase shifts from cnidarian to algal dominance in the wake of large-scale disturbances like hurricanes ... or people. A useful analogy can be drawn between the TWA and forest regeneration in parts of East Africa such as Queen Elizabeth National Park in Uganda. Here the tree community is deficient in pioneers and early builders, and felled forest shifts to recalcitrant dominance by giant herbs such as *Acanthus*.[78] Regeneration of closed canopy forest at a high enough rate to keep up with disturbance would require the deliberate propagation and planting of large numbers of forest tree species of diverse functional groups, and that is the plan. Uganda has begun aggressive afforestation programs to stave off loss of the last few shreds of its "timber estate," today far more valuable as safe harbor for the chimps, gorillas, and elephants beloved of well-heeled tourists.

It may not be practical in coral reef restoration to wait for the sexual recruits of framework-building corals to arrive and establish themselves on a restoration site. Two technologies hold promise for expediting the development of hard coral dominance on reef restoration sites: the gathering or propagation and guided settlement of coral planulae, and methods for mass propagation, grow-out, and emplacement of coral fragments. Both efforts have been facilitated by the explosive growth of coral propagation techniques, as well as coral gardening in home and public aquariums.[79–82] Since mortality of newly settled recruits is high, it may always be worth the effort of growing out corals to the stage of small colonies and then manually securing them to the restoration site. As in conventional gardening, the spacing and species of nearest neighbors can have important and unexpected effects that must be known and appreciated.[83] There are also key issues concerning the survival of small coral colonies raised under relatively sheltered conditions and then placed out into the harsher, real-world environment where it is hoped they might spread and propagate.

Study of morbidity and mortality of small coral fragments set out on reefs or in grassbeds has yielded a useful lore on the best ways to do this in reef restoration (e.g., reference 84). Large-scale coral propagation and replanting projects have been implemented in the Philippines, Fiji, and Florida, including grow-out and restoration of tridacnid clams in the Philippines, Fiji, Palau, and elsewhere.[85,86] The work on giant clams is particularly important for two reasons: tridacnid clams have themselves been heavily overexploited for their meat and shells, and the growing shells of the larger species are favored places for coral colonies to settle and become established. Thus the mass-propagation and setting out of large tridacnids (*Tridacna gigas, T. maxima*) could in itself provide a basis, or at least an important adjunct strategy, for coral reef restoration. In sum, in many parts of the Indo-Pacific, coral succession on artificial structures can result in a taxonomically and functionally diverse assemblage within a few years or decades at most.[87–90]

7.2.4 Task 4: Craft Community Ontogeny So as to Maximize the Value and Sustainability of Goods and Services Produced by Restored or Created Reef

Once small corals are established and growing on the framework, the next challenge is to guide the development of the reef as a whole to produce a self-sustaining and productive community that is a worthy and capable substitute for the natural reef whose destruction sparked the project in the first place. Several aspects of this task deserve systematic study. One is the forced assemblage of corals and other species established in the garden. It may make sense to allow fast-growing branching corals to dominate in early phases of a project, to occupy space quickly and bring in populations of coral reef facilitators. But as already mentioned, branching corals are also highly vulnerable to a suite of transient disturbances, so the community must be balanced by species with foliaceous, columnar, and massive growth forms. These species can be harvested, propagated, and established just as successfully as branching species,[91] but it may be necessary for the slower-growing corals to be nurtured, supplemented, and stewarded for years or even decades until they achieve a stable position in the community. On the other hand, they are the species most likely to survive bleaching and storms.[92] Indeed, isolated, open-branched acroporid and pocilloporid species are preferentially attacked by algal gardening damselfishes of the genera *Stegastes*, *Plectroglyphidodon*, *Dischistodus*, and *Pomacentrus*, by drupellid and coralliophilid gastropods, and by crown-of-thorns starfish. The early regenerative phase of acroporid haystacks on Jamaican reefs following hurricanes and other disturbances was arrested multiple times by the aggregation of predators on the few, isolated colonies that managed to rise from the rubble.[93–95] The effects of Indo-Pacific corallivorous fishes that normally present no great threat to intact corals can be magnified when corals are subjected to extensive damage by other agents, setting off a feeding frenzy.[96]

Gardened corals are likely to be just as vulnerable to natural predators and disease as their wild counterparts (e.g., reference 97). Sometimes the worst enemies of young corals can be former best friends with catholic tastes. Miller and Hay[76] found parrotfishes to be an important source of mortality on young corals in an experiment. Similarly, a large rainbow parrotfish (*Scarus guacamaia*) in the Biosphere II ocean mesocosm reversed the anticipated outcome of an experiment by eating the uncaged corals exposed to the "benefits" of grazing (Kaufman, Schwartz, and Shank, in preparation). One option for coral gardeners, at least for sessile enemies, is to remove the predators, competitors, and diseased portions of colonies by hand,[98] as is being done lovingly by the caretakers of the Permuteran installation in Bali. Another is to attract or force-recruit resident species that will feed on coral predators (e.g., reference 99). It is crucial to remember that the differences between sets of dynamic interactions that foster versus forbid coral growth may be subtle, surprising, and labile.[43] Trophic plasticity and functional ambiguousness are problems not easily surmounted, and whether a species is friend or foe to the coral gardener is context dependent. In at least a few situations, however, simplicity reigns. At many localities, but most especially in the tropical west Atlantic, the effective control of benthic algae is a major determinant of success in coral gardening. In this regard it came as welcome news when cultivation techniques for *Diadema antillarum*, the Atlantic longspine sea urchin, were first perfected.[100]

Another possible approach to giving planted corals a competitive edge is to enhance their growth artificially, thus giving them greater reserve capacity for dealing with skeletal damage and overgrowth. One of the more outstanding claims made for the Biorock process is that the electric current running through the reinforcing bars (responsible for the deposition of mineral on the bars) also accelerates rates of growth in attached corals. As stated on the Global Coral Reef Alliance (GCRA) web site, "Immediately, these coral pieces (attached to Biorock structures) begin to bond to the accreted mineral substrate and start to grow — typically three to five times faster than normal." Elsewhere it is stated: "Healthy corals grow quickly — up to 10 times faster than normal when exposed to the Biorock process, even in poor water conditions" (http://globalcoral.org/Solution%20for%20Corals%20in%20Peril.htm November 2004). Preliminary experiments have demonstrated a statistically significant boost in zooxanthellae density in some coral colonies on Biorock structures as compared to genetically identical control colonies on natural reef.[101] Further work is needed, however, to demonstrate that these changes are attributable directly to the Biorock process, as opposed to the transplantation of corals to a superior location, and are associated with an increase in rates of either carbonate accretion or axial extension. It is also claimed that the Biorock process can foster coral growth and health under conditions where neighboring corals on natural reef are overgrown by fleshy algae or bleached during warm-water anomalies. This is perhaps the most intriguing and important of the claims made by GCRA and is certainly worth rigorous experimental assessment. Electrolytic mineral accretion has been tried by other investigators as well, with limited but intriguing degrees of success (e.g., reference 102).

Even without careful records, it could be possible to see evidence of accelerated coral growth at Permuteran by examining the size and appearance of coral colonies of known age, some of which had been photographed months earlier. The size of such colonies is inconsistent with the hypothesis that they had been growing at an unusually high rate (personal observation). However, to the extent that any differences could be inferred on inspection, colonies growing high on the rebar structures did appear much larger and more robust than those near the sediment–water interface. It is thus possible that one factor contributing to an elevated growth rate on Biorock structures could be placement higher in the water column than colonies chosen for comparison on nearby natural reef.

Without an abundance of data to corroborate the claims of their supporters, Biorock and other electrolytic methods should be considered a promising but uncertain technology with regard to effects on coral biology. In other regards, mineral does accumulate quickly on the electrified structures, and the technology is obviously of potential benefit in controlling shoreline erosion and in creating an inexpensive framework that attracts resting schools of fishes and can serve as an excellent foundation for coral gardening.

Since the community of fishes and motile invertebrates that accumulate on a reef restoration framework can have a profound influence on corals and other sessile invertebrates (e.g., reference 103), these moving pieces ought to be gardened as thoughtfully as their homebuilders stuck to the seabottom. The first step would be to make the initial sessile community as attractive as possible to the right clientele. In the Atlantic this has proven quite a challenge. The fish species that appear on artificial reefs are apt to be a peculiar lot, with a guest list sometimes influenced more by timing and happenstance than the desires of human reefbuilders.[86,90,104–108] "Randall's Reef" is an early experiment in a cinder block artificial reef built by famed coral reef fish biologist Jack Randall. It sits in a sea grass bed about 10 m of water off Yawzie Point, St. John, U.S. Virgin Islands. From soon after its construction in the 1960s, one of the most prevalent fishes on Randall's Reef has been, to the bemused astonishment of ichthyologists, a colony of small groupers called mutton hamlets, *Alphestes afer*[109] (personal observation 1975). Apart from an opportunistic streak, this inconspicuous fish is otherwise restricted to seagrass beds and fleshy algal pavements. The exact placement of reef units with respect to habitat patchiness and each other can strongly influence fish recruitment and retention patterns.[110] Students of artificial reefs are also embarrassingly familiar with what happens to subsequent fish recruitment when an experimental module is occupied by a territorial damselfish: nothing (e.g., reference 106). On the other hand, algal gardening damselfishes can indirectly influence subsequent fish recruitment via the way that they modify the benthic community (e.g., reference 111). The recruitment of one species of coral reef fish can have strong positive but indirect effects on the success of others, sometimes by very simple mechanisms.[112] Such effects may be essential; simply providing a naturalistic physical habitat is not enough. In Florida, fish assemblages that accumulated on reef surfaces restored at great cost were similar to those of nearby natural hard grounds (eroded, low-relief limestone platforms) but not at all like those of nearby, relatively intact coral reefs.[113] Evidently, adjacency to other habitats, priority effects, epifaunal community structure, the ontogeny of recruitment, and structural relief are all important factors shaping a resident fish community.

Reef restorations and human-made reefs often do attract many fishes, however, so it would be advantageous if we could be more picky about who is on the guest list. The possibility of planting restoration frameworks with corals that attract clouds of zooplanktivores has already been mentioned. Fish attraction devices (FADs) can bring in a great variety of species, at least to visit, but the resulting assemblage is dominated by predatory pelagics whose cascading effects on a developing reef community could wind up diminishing the more desirable but smaller zooplanktivorous and herbivorous species.[75] Eventually, however, a reasonably natural coral reef and fish assemblage will just about have settled in when somebody brings up the next logical idea on a slippery slope: management for fish extraction. It is inevitable that restored reefs, like their natural predecessors, will be looked at in terms of their potential for fisheries and aquaculture. Restoration engineers should think about how to optimize the ratio of fish recruitment to fish attraction during the course of a project's lifetime and through the seasons of the year. Attention to the size and spacing of restoration frameworks can help to maximize the ability of the structures to recruit, and not just attract, new residents.[114]

Once an appropriate fish assemblage has been attracted, it will not necessarily remain. Transient disturbance can have a profound impact on fish communities on artificial and restored reef habitats, just as on natural ones, though with possibly greater consequences. For example, a surprisingly large proportion of the initial reef fish community can persist for at least the first few weeks or months following a catastrophic bleaching event. However, species strongly associated with live corals will not be among them.[115,116] Species composition is at first strongly influenced by chance recruitment, but once a mature assemblage is established, it remains for several years as these first arrivals mature and live out their lives.[117] As this "storage effect" is exhausted, however, the fish community is likely to undergo a profound shift, subject to the stochastic availability of replacement larvae, most likely of a different suite of species.[118] On the scale of an entire reef system this hardly matters, but it could have important consequences for the future of any one patch or a small restoration project. Coral gardeners must simply get used to the idea of natural variability, as have their terrestrial counterparts. Whether it is a border of marigolds or clouds of green chromis that you wish to maintain in a particular spot, it may be necessary for you to "plant" them manually every few years.

The maintenance of high densities of large reef fishes for recreational or food fishing is another matter entirely. Managers might consider attaching FADs to enhance fishing on managed reefs during seasons when large reef-associating pelagics are in the area but then detach and remove them just before and during pulses in larval settlement. The point here is that as for any other effort to harmonize human demands with the ability of natural systems to provide for them, balance and reason must rule. Coral reef restoration and heavy fishing pressure are not a good match. Intensively managed coral gardens within totally protected marine reserves may, however, contribute well to spillover for fisheries outside the reserve.[119,120]

7.3 CONCLUSION

The artificial facilitation of reef community development through coral reef gardening has promise as one component of a coral reef conservation strategy, beneath a larger umbrella of education, enactment, and enforcement of environmental regulations. Despite some notable successes, and notwithstanding the development of innovative methods, there remains an astonishing dearth of theoretical synthesis or rigorous experimental study on which to base new coral reef restoration efforts. In response to the title of this chapter, if you build it, "they" *will* come. However, "they" are as likely as not to be undesirables that inhibit the settlement of coral reef species and retard the transformation of the artificial structure into a living coral reef. This is particularly true for the tropical Atlantic, where reef restoration could do the greatest good if only it worked more often and more easily.

Amid an undeniable measure of success, the practice of coral reef gardening exhibits two striking failures. First, there has been a failure to integrate promising technologies into a composite, state-of-the-art approach. Second, there has been a failure to bring good science to bear, particularly in the areas of nutrient cycling and species interactions. In some circumstances, such as the restoration of Indo-Pacific reefs damaged by blast-fishing, recovery times can be shortened greatly with very simple methods that can be carried out on a village level. This is only because nature is able to bear the burden. The natural pieces of the system, the ones that Aldo Leopold so wisely suggested we save in the course of our tinkering, remain in rich abundance and diversity, free for the asking. On Atlantic reefs overrun by *Lobophora*, the natural capital was originally more modest and has since been sucked dry. Here the prospects of success are much gloomier. On the flip side of this dismal picture lies an enormous potential for experiments in reef restoration to illuminate and apply the basic processes that drive natural coral reef regeneration. Such insights would enhance coral reef conservation on many levels, from improved first aid to more effective and far-reaching conservation legislation. The greatest hope in this entire affair lies in the almost pathetic determination and relentlessness with which people insist upon trying to heal damaged coral reefs by hand. The small scale of these efforts may seem Sysyphean, but the social forces behind them bespeak an enormous potential for good works.

An era of extensive and persistent clinical trials will be needed for the development of a therapeutic protocol for coral reef restoration fitted to each large marine ecosystem. We have begun, but let us insist that each trial be intelligent, systematic, rigorous, well-documented, and built upon the trials before it. By combining human perseverance with determination to conduct every restoration project as a good scientific experiment, wonderful things can be achieved. Unless and until this change is brought about, coral reef restoration is doomed to dwell in an atmosphere of pathos, an odd and intensely human footnote to the great Holocene epoch of coral reef growth that we ourselves have brought crashing down about us.

ACKNOWLEDGMENTS

The author is grateful to many individuals who have tolerated his rants while coming to grips with the topic of this chapter, a topic he grew to dislike. Special thanks are due to John Ogden, friend and valued colleague of many years, for reminding me tirelessly that Leopold knew it all and laid it out simply for later workers to obscure. Helen Fox (Figure 7.3), Michael Moore and Mark Erdmann (Figure 7.1), and James Cervino (Figure 7.2) kindly granted permission for use of their photographs. Heather Marlow, Jesse Schwartz, Burton Shank, and Elizabeth Soule were heavily involved in the experiment that produced the data for Figure 7.4. Deep thanks also to the National Geographic Society and my field partners Tim Laman and Zafer Kizilkaya for stimulating assignments that made the extensive field observations for this paper possible in Indonesia, Micronesia, and Fiji. Jean-Francois Bertrand, Bob Hecky, John Ogden, and James Cervino were kind enough to review the manuscript prior to submission, and I thank the anonymous reviewers as well.

REFERENCES

1. Wilkenson, C.R. 1993. Coral reefs of the world are facing widespread devastation: can we prevent this through sustainable management practice? *Proc. 7th Int. Coral Reef Symp.* Guam 1:11–21.
2. Hughes, T.P. 1994. Catastrophes, phase shifts, and large-scale degradation of a Caribbean coral reef. *Science* 265:1547–1551.
3. Jackson, J.B.C. 1997. Reefs since Columbus. *Coral Reefs* 16:S23–S32.
4. Jackson, J. et al. 2001. Historical overfishing and the recent collapse of coastal ecosystems. *Science* 293:629–637.
5. Harvell, C.D., K. Kim, J.M. Burkholder, R.R. Colwell, P.R. Epstein, D.J. Grimes, E.E. Hoffman, E.K. Lipp, A.D.M.E. Osterhaus, R.M. Overstreet, J.W. Porter, G.W. Smith, and G.R. Vasta. 1999. Emerging marine diseases — climate links and anthropogenic factors. *Science* 285:1505–1510.
6. Coleman, F. and S. Williams. 2002. Overexploiting marine ecosystem engineers: potential consequences for biodiversity. *TREE* 17:40–44.
7. Guzman, H. M., C. Guevara, and A. Castillo. 2003. Natural disturbances and mining of Panamanian coral reefs by indigenous people. *Cons. Biol.* 17:1396–1401.
8. Kaufman, L.S. 1986. Why the ark is sinking. In Kaufman, L.S. and K.E. Mallory (eds.). *The Last Extinction.* Cambridge, MA: MIT Press.
9. Aronson, R.B., I.G. Macintyre, W.B. Precht, T.J.T. Murdoch, and C.M. Wapnick. 2002. The expanding scale of species turnover events on coral reefs in Belize. *Ecol. Monogr.* 72:233–249.

10. Scheffer, M., S. Carpenter, J.A. Foley, C. Folke, and B. Walker. 2001. Catastrophic shifts in ecosystems. *Nature* 413:591–596.
11. Williams, I.D. and N.V.C. Polunin. 2000. Differences between protected and unprotected Caribbean reefs in attributes preferred by dive tourists. *Environ. Conserv.* 27:382–391.
12. Spurgeon, J.P.G. and U. Lindahl. 2000. Economics of coral reef restoration. Pp. 125–136 In Cesar, H. (ed.) *Collected Essays on the Economics of Coral Reefs*. Kalmar, Sweden: CORDIO.
13. Soong, K. and T. Chen. 2003. Coral transplantation: regeneration and growth of acropora fragments in a nursery. *Restoration Ecol.* 11:62–71.
14. Bolker, B.M., C.M. St.Mary, C.W. Osenberg, R.J. Schmitt, and S.J. Holbrook. 2002. Management at a different scale: marine ornamentals and local processes. *Bull. Mar. Sci.* 70: 733–748.
15. Swearer, S.E., J.E. Caselle, D. Lea, and R.R. Warner. 1999. Larval retention and recruitment in an island population of a coral-reef fish. *Nature* 402:799–802.
16. Jones, G.P., M.J. Milicich, M.J. Emslie, and C. Lunow. 1999. Self-recruitment in a coral reef fish population. *Nature* 402:802–804.
17. van Treeck, P. and H. Schuhmacher. 1999. Mass diving tourism – a new dimension calls for new management approaches. *Mar. Pollut. Bull.* 37:499–504.
18. Epstein, N., R.P.M. Bak, and B. Rinkevich. 2001. Strategies for gardening denuded coral areas: the applicability of using different types of coral material for reef restoration. *Restoration Ecol.* 9:432–442.
19. Epstein, N., R.P.M. Bak, and B. Rinkevich. 2003. Applying forest restoration principles to coral reef rehabilitation. *Aquat. Conserv. Marine Freshwater Ecosyst.* 13:387–395.
20. Lirman, D. and M.W. Miller. 2003. Modeling and monitoring tools to assess recovery and convergence rates between restored and undisturbed coral reef habitats. *Restoration Ecol.* 11: 448–456.
21. Jaap, W.C. 2000. Coral reef restoration. *Ecol. Eng.* 15:345–364.
22. Fox, H.E. 2001. Enhancing reef recovery in Komodo National Park, Indonesia: a proposal for coral reef rehabilitation at ecologically significant scales. University of California, Berkeley and The Nature Conservancy Indonesia Coastal and Marine Program. 12 pp.
23. Hilbertz, W.H., D. Fletcher, and C. Krausse. 1977. Mineral accretion technology: applications for architecture and aquaculture. *Indust. Forum* 8:75–84.
24. Hilbertz, W. 1992. Solar-generated building material from seawater as a sink for carbon. *Ambio* 21:126–129.
25. Goreau, T.J. and W. Hilbertz. 1996. Reef restoration using sea-water electrolysis in Jamaica. *Proc. 8th Int. Coral Reef Symp.*, Panamá.
26. Holden, C. (ed.) 2004. Reef therapy. *Science* 305:1398.
27. Shashar, N. and N. Stambler. 1992. Endolithic algae within corals — life in an extreme environment. *J. Exp. Mar. Biol. Ecol.* 163:277–286.
28. Priess, K., T. Le Campion-Alsumard, S. Golubic, F. Gadel, and B.A. Thomassin, 2000. Fungi in corals: black bands and density-banding of *Porites lutea* and *P. lobata* skeleton. *Mar. Biol.* 136:19–27.
29. Bentis, C.J., L. Kaufman, and S. Golubic. 2000. Endolithic fungi in reef-building corals (Order: Scleractinia) are common, cosmopolitan, and potentially pathogenic. *Biol. Bull.* 198:254–260.
30. Svane, I. and J.K. Petersen. 2001. On the problems of epibioses, fouling, and artificial reefs, a review. *Mar. Ecol.* 22:169–188.
31. Bongiorne, L., S. Shafir, D. Angel, and B. Rinkevich. 2003. Survival, growth and gonad development of two hermatypic corals subjected to *in situ* fish-farm nutrient enrichment. *Mar. Ecol. Prog. Ser.* 253:137–144.
32. Morse, D.E., and A.N.C. Morse. 1991. Enzymatic characterization of the morphogen recognized by *Agaricia humilis* (Scleractinian coral) larvae. *Biol. Bull.* 181:104–122.
33. Morse, D.E., A.N.C. Morse, P.T. Raimondi, and N. Hooker. 1994. Morphogen-based chemical flypaper for *Agaricia humilis* coral larvae. *Biol. Bull.* 186(2):172–181.
34. Morse, A.N.C., K. Iwao, M. Baba, K. Shimoike, T. Hayashibara, and M. Omori. 1996. An ancient chemosensory mechanism brings new life to coral reefs. *Biol. Bull.* 191:149–154.
35. McClanahan, T.R., N. Polunin, and T. Done. 2002b. Ecological states and the resilience of coral reefs. *Cons. Ecol.* 6(2):18 (online).
36. River, G.F. and P.J. Edmunds. 2001. Mechanisms of interaction between macroalgae and scleractinians on a coral reef in Jamaica. *J. Exp. Mar. Biol. Ecol.* 261:159–172.

37. McClanahan, T.R., V. Hendrick, M.J. Rodrigues, and N.V.C. Polunin. 1999. Varying responses of herbivorous and invertebrate-feeding fishes to macroalgal reduction on a coral reef. *Coral Reefs* 18:195–203.
38. McClanahan, T.R., J.N. Uku, and H. Machano. 2002. Effect of macroalgal reduction on coral-reef fish in the Watamu Marine National Park, Kenya. *Mar. Freshwat. Res.* 53:223–231.
39. Nugues, M.M., G.W. Smith, R.J. Hooidonk, M.I. Seabra, and R.P.M. Bak. 2004. Algal contact as a trigger for coral disease. *Ecol. Lett.*, 7:919–923.
40. Sussman, M., Y. Loya, M. Fine, and E. Rosenberg. 2003. The marine fireworm *Hermodice carunculata* is a winter reservoir and spring–summer vector for the coral-bleaching pathogen *Vibrio shiloi. Environ. Microbiol.* 5:250–255.
41. Hay, M.E. 1997. The ecology and evolution of seaweed–herbivore interactions on coral reefs. *Coral Reefs* 16:S67–S76.
42. Lirman, D. 2001. Competition between macroalgae and corals: effects of herbivore exclusion and increased algal biomass on coral survivorship and growth. *Coral Reefs* 19(4):392–399.
43. Hay, M.E., J.D. Parker, D.E. Burkepile, C.C. Caudill, A.E. Wilson, Z.P. Hallinan, and A. D. Chequer. 2004. Mutualisms and aquatic community structure: the enemy of my enemy is my friend. *Annu. Rev. Ecol. Evol. Systemat.* 35: 179–197.
44. Cheroske, A.G., S.L. Williams, and R.C. Carpenter. 2000. Effects of physical and biological disturbances on algal turfs in Kaneohe Bay, Hawaii. *J. Exp. Mar. Biol. Ecol.* 248:1–34.
45. Smith, J., C. Smith, and C. Hunter. 2001. An experimental analysis of the effects of herbivory and nutrient enrichment on benthic community dynamics on a Hawaiian reef. *Coral Reefs* 19: 332–342.
46. Fairfull, S.J.L. and V.J. Harriott. 1999. Succession, space, and coral recruitment in a subtropical fouling community. *Mar. Freshwat. Res.* 50:35–242.
47. Wulff, J. 1997. Parrotfish predation on cryptic sponges of Caribbean coral reefs. *Mar. Biol.* 129:41–52.
48. Maida, M., P.W. Sammarco, and J.C. Coll. 2001. Effects of soft corals on scleractinian coral recruitment. II: allelopathy, spat survivorship and reef community structure. *Mar. Ecol.* 22:397.
49. Bailey-Brock, J.H. 1989. Fouling community development on an artificial reef in Hawaiian waters. *Bull. Mar. Sci.* 44:580–591.
50. Fitzhardinge, R.C. and J.H. Bailey-Brock. 1989. Colonization of artificial reef materials by corals and other sessile organisms. *Bull. Mar. Sci.* 44:567–579.
51. Sutherland, J.P. 1974. Multiple stable points in natural communities. *Am. Nat.* 108:859–873.
52. Buss, L.W. and J.B.C. Jackson. 1979. Competitive networks: nontransitive competitive relationships in cryptic coral reef environments. *Am. Nat.* 113:223–234.
53. LaPointe, B. 1997. Nutrient thresholds for bottom-up control of macroalgal blooms on coral reefs in Jamaica and South Florida. *Limnol. Oceanogr.* 42:1119–1131.
54. Williams, S.L. and R.C. Carpenter. 1997. Grazing effects on nitrogen fixation in coral reef algal turfs. *Mar. Biol.* 130:223–231.
55. McCook, L.J. 1999. Macroalgae, nutrients and phase shifts on coral reefs: scientific issues and management consequences for the Great Barrier Reef. *Coral Reefs* 18:357–367.
56. Miller, M.W., M.E. Hay, S.L. Miller, D. Malone, E.E. Sotka and A.M. Szmant. 1999. Effects of herbivores versus nutrients on reef algae: a new method for manipulating nutrients on coral reefs. *Limnol. Oceanogr.* 44:1847–1861.
57. Bellivaux, S.A. and V.J. Paul. 2002. Effects of herbivory and nutrients on the early colonization of crustose coralline and fleshy algae. *Mar. Ecol. Prog. Ser.* 232:105–114.
58. Rasheed, M., M.I. Badran, C. Richter, and M. Huettel. 2002. Effect of reef framework and bottom sediment on nutrient enrichment in a coral reef of the Gulf of Aqaba, Red Sea. *Mar. Ecol. Progr. Ser.* 239:277–285.
59. Shashar, N., S. Kinane, P.L. Jokiel, and M.R. Patterson. 1996. Hydromechanical boundary layers over a coral reef. *J. Exp. Mar. Biol. Ecol.* 199:17–28.
60. Fabricius, K.E., A. Genin, and Y. Benayahu. 1995. Flow-dependent herbivory and growth in zooxanthellae-free soft corals. *Limnol. Oceanogr.* 40:1290–1301.
61. Davis, W.P. and R.S. Birdsong. 1973. Coral reef fishes which forage in the water column. *Hegolander wiss. Werunters* 24:292–306.

62. Bray, R.N., A.C. Miller, and G.G. Geesey. 1981. The fish connection: a trophic link between planktonic and rocky reef communities? *Science* 214:204–205.
63. Fishelson, L. 2003. Coral and fish biocoenosis: ecological cells gradually maturing in complexity, species composition, and energy turnover. *Environ. Biol. Fishes* 68:391–405.
64. Pratchett, M.S., N. Gust, G. Goby, and S.O. Klanten. 2001. Consumption of coral propagules represents a significant trophic link between corals and reef fish. *Coral Reefs* 20:13–17.
65. Meyer, J.L., E.T. Schultz, and G.S. Helfman. 1983. Fish schools: an asset to corals. *Science* 220:1047–1049.
66. Meyer, J.L. and E.T. Schultz. 1985. Migrating haemulid fishes as a source of nutrients and matter on coral reefs. *Limnol. Oceanogr.* 30:146–156.
67. Lesser, M.P., C.H. Mazel, M.Y. Gorbunov, and P.G. Falkowski. 2004. Discovery of symbiotic nitrogen-fixing cyanobacteria in corals. *Science* 305:997–1000.
68. Rohwer, F., M. Breithart, J. Jara, F. Azam, and N. Knowlton. 2001. Diversity of bacteria associated with the Caribbean coral *Montastrea franksi. Coral Reefs* 20:85–91.
69. Rohwer, F., V. Seguritan, F. Azam, and N. Knowlton. 2002. Diversity and distribution of coral-associated bacteria. *Mar. Ecol. Prog. Ser.* 243:1–10.
70. Hecky, R.E., R.E.H. Smith, D.R. Barton, S.J. Guildford, W.D. Taylor, M.N. Charlton, and T. Howell. 2004. The nearshore phosphate shunt: a consequence of ecosystem engineering by dresseinids in the Laurentian Great Lakes. *Can. J. Fish. Aquat. Sci.* 61:1285–1293.
71. Ellison, A.M., E.J. Farnsworth, and R.R. Twilley. 1996. Facultative mutualism between red mangroves and root-fouling sponges in a Belizean mangal. *Ecology* 77:2431–2444.
72. Grall, J. and L. Chauvaud. 2002. Marine eutrophication and benthos: the need for new approaches and concepts. *Global Change Biol.* 8:813–830.
73. Prins, T.C., A.C. Small, and R.F. Dame. 1997. A review of the feedbacks between bivalve grazing and ecosystem processes. *Aquatic Ecol.* 31(4):349–359.
74. Dame, R., D. Bushek, D. Allen, A. Lewitus, D. Edwards, E. Koepfler, and L. Gregory. 2002. Ecosystem response to bivalve density reduction: management implications. *Aquatic Ecol.* 36:51–65.
75. Beets, J. 1989. Experimental evaluation of fish recruitment to combinations of fish aggregating devices and benthic artificial reefs. *Bull. Mar. Sci.* 44:973–983.
76. Miller, M.W. and M. Hay. 1998. Effects of fish predation and seaweed competition on the survival and growth of corals. *Oecologia* 113:231–238.
77. Lindahl, U. 2003. Coral reef rehabilitation through transplantation of staghorn corals: effects of artificial stabilization and mechanical damages. *Coral Reefs* 22:217–223.
78. Chapman, C.A., L.J. Chapman, L.S. Kaufman, and A.E. Zanne. 1999. Potential causes of arrested succession in Kibale National Park: growth and mortality of seedlings. *Afr. J. Ecol.* 37:81–92.
79. Rinkovich, B. and S. Shafir. 1998. *Ex situ* culture of colonial marine ornamental invertebrates: concepts for domestication. *Aquarium Sci. Conserv.* 2:237–250.
80. Richmond, R.H. 1999. Coral cultivation and its application to reef restoration, environmental assessment, monitoring and the aquarium trade. *Proc. Int. Conf. Scientific Aspects of Coral Reef Assessment, Monitoring, and Restoration* April 14–16, 1999, Ft. Lauderdale, FL: 161–162.
81. Delbeek, C. 2001. Coral farming: Past, present and future trends. *Aquarium Sci. Conserv.* 3:171–181.
82. Borneman, E.H. and J. Lowrie. 2001. Advances in captive husbandry and propagation: an easily utilized reef replenishment means from the private sector? *Bull. Mar. Sci.* 69:897–913.
83. Raymundo, L.J. 2001. Mediation of growth by conspecific neighbors and the effect of site in transplanted fragments of the coral *Porites attenuata* Nemenzo in the central Philippines. *Coral Reefs* 20:263–272.
84. Raymundo, L.J. and A.P. Maypa. 1999. Using cultured coral to rehabilitate a degraded reef in the central Philippines. *Proc. Int. Conf. Scientific Aspects of Coral Reef Assessment, Monitoring, and Restoration* April 14–16, 1999, Ft. Lauderdale, FL: 160.
85. Gomez, E.D., A.C. Alcala, and L.C. Alcala, 1982. Growth of some corals in an artificial reef off Dumaguete, Central Visayas, Philippines. *Kalikasan, Philipp. J. Biol.* 11(1):148–157.
86. Alcala, A.C., L.C. Alcala, E.D. Gomez, M.E. Cowan, and H.T. Yap. 1982. Growth of certain corals, molluscs and fish in artificial reefs in the Philippines, p. 215–220. In E.D. Gomez, C.E. Birkeland, R.W. Buddemeier, R.E. Johannes, J.A. Marsh, Jr., and R.T. Tsuda (eds.) *Proceedings of the 4th International Coral Reef Symposium.* MarineScience Center, University of the Philippines, Manila, Philippines 2.

87. Clark, S. 2001. Evaluation of succession and coral recruitment in the Maldives. pp. 169–175. In *Coral Reef Degradation in the Indian Ocean: Status Reports and Project Presentations 2000.*
88. Clark, S. and A.J. Edwards. 1994. Use of artificial reef structures to rehabilitate reef flats degraded by coral mining in the Maldives. *Bull. Mar. Sci.* 55:724–744.
89. Clark, S. and A.J. Edwards. 1999. An evaluation of artificial reef structures as tools for marine habitat rehabilitation in the Maldives. *Aquatic Conserv. Mar. Freshwat. Ecosyst.* 9:5–21.
90. Abelson, A. and Y. Shlesinger. 2002. Comparison of the development of coral and fish communities on rock-aggregated artificial reefs in Eilat, Red Sea. *ICES J. Mar. Sci.* 59:S122–S126.
91. Becker, L.C. and Mueller, E., 1999. The culture, transplantation, and storage of *Montastraea faveolata*, *Acropora cervicornis*, and *A. palmata*: what we have learned so far. *Proc. Int. Conf. Sci. Aspects Coral Reef Assessment, Monitoring, and Restoration Bull. Mar. Sci.* 69: 881–896 April 14–16, 1999, Ft. Lauderdale, FL: 53.
92. Edwards A., J.S. Clark, H. Zahir, A. Rajasuriya, A. Naseer, and J. Rubens. 2001. Coral bleaching and mortality on artificial and natural reefs in Maldives in 1998, sea surface temperature anomalies and initial recovery. *Mar. Pollu. Bull.* 42:7–15.
93. Woodley, J. et al. 1981. Hurricane Allen: initial impact on the coral reef communities of Discovery Bay, Jamaica. *Science* 214:749–755.
94. Kaufman, L.S. 1983. Effects of Hurricane Allen on reef fish assemblages near Discovery Bay, Jamaica. *Coral Reefs* 2:1–5.
95. Knowlton, N., J.C. Lang, and B.D. Keller. 1989. Fates of Staghorn Coral Isolates on Hurricane-Damaged Reefs in Jamaica: the Role of Predators. *Proc. 6th Int. Coral Reef Symp.* Australia, 1988 2:83–88.
96. McIlwain, J.L. and G.P. Jones. 1997. Prey selection by an obligate coral-feeding wrasse and its response to small-scale disturbance. *MEPS* 155:189–198.
97. Bruckner, A.W. and R.J. Bruckner. 2001. Condition of restored *Acropora palmata* fragments off Mona Island, Puerto Rico, 2 years after the Fortuna Reefer ship grounding. *Coral Reefs* 20:235–243.
98. Miller, M.W. 2001. Corallivorous snail removal: evaluation of impact on *Acropora palmata. Coral Reefs* 19:293–295.
99. Gochfield, D.J. and G. S. Aeby. 1997. Control of populations of the coral-feeding nudibranch *Phestilla sibogae* by fish and crustacean predators. *Mar. Biol.* 130:63–69.
100. Idrisi, N., T.R. Capo, and J.E. Serafy. 2003. Postmetamorphic growth and metabolism of long-spined black sea urchin (*Diadema antillarum*) reared in the laboratory. *Mar. Freshwat. Behav. Physiol.* 36:87–95.
101. Goreau, T.J., J.M. Cervino, and R. Pollina. 2004. Increased zooxanthellae numbers and mitotic index in electrically stimulated corals. *Symbiosis* 37:107–120.
102. van Treeck, P. and H. Schuhmacher. 1997. Initial survival of coral nubbins transplanted by a new coral transplantation technology — options for reef rehabilitation. *Mar. Ecol. Progr. Ser.* 150:287–292.
103. de Loma, T.L. and M.L. Harmelin-Vivien. 2002. Summer fluxes of organic carbon and nitrogen through a damselfish resident, *Stegastes nigricans* (Lacepede, 1803) on a reef flat at Reunion (Indian Ocean). *Mar. Fres. Res.* 53:169–174.
104. Polovina, J. 1989. Artificial reefs: nothing more than benthic fish aggregators. *CALCOFI Report* 30:37–39.
105. Bortone, S.A., R.K. Turpin, R.C. Cody, C.M. Bundrick, and R.L. Hill. 1997. Factors associated with artificial-reef fish assemblages. *Gulf Mexico Sci.* 1997:17–34.
106. Rooker, J.R., Q.R. Dokken, C.V. Pattengill, and G.J. Holt. 1997. Fish assemblages on artificial and natural reefs in the Flower Garden Banks National Marine Sanctuary, USA. *Coral Reefs* 16:83–92.
107. Rilov, G. and Y. Benayahu. 2000. Fish assemblage on natural versus vertical artificial reefs: the rehabilitation perspective. *Mar. Biol.* 136:931–942.
108. Powers, S.P., J. H. Grabowski, C. H. Peterson, and W. J. Lindberg. 2003. Estimating enhancement of fish production by offshore artificial reefs: uncertainty exhibited by divergent scenarios. *Mar. Ecol. Progr. Ser.* 264:265–277.
109. Ogden, J.C. and J.P. Ebersole. 1981. Scale and community structure of coral reef fishes: a long-term study of a large artificial reef. *Mar. Ecol. Progr. Ser.* 4:97–103.
110. Nanimi, A. and M. Nishihira. 2002. The structures and dynamics of fish communities in an okinawan coral reef: effects of coral-based habitat structures at sites with rocky and sandy sea bottoms. *Environ. Biol. Fishes* 63:353–372.

111. Fereira, E.L., A.G. Concalves, R. Coutinho, and C. Peret. 1998. Herbivory by the damselfish *Stegastes fuscus* (Cuvier, 1830) in a tropical rocky shore: effects on the benthic community. *J. Exp. Mar. Biol. Ecol.* 229:241–264.
112. Webster, M.S. and G.R. Almany. 2002. Positive indirect effects in a coral reef fish community. *Ecol. Lett.* 5:549.
113. Ebersole, J.P. 2001. Recovery of fish assemblages from ship groundings on coral reefs in the Florida Keys National Marine Sanctuary. *Bull. Mar. Sci.* 69:655–671.
114. Wilson, J., C.W. Osenberg, C.M. St. Mary, C.A. Watson, and W.J. Lindberg. 2001. Artificial reefs, the attraction-production issue, and density dependence in marine ornamental fishes. *Aquat. Sci. Cons.* 3:95–105.
115. Booth, D.J. and G.A. Baretta. 2002. Changes in a fish assemblage after a coral bleaching event. *Mar. Ecol. Prog. Ser.* 245:205–212.
116. Sano, M. 2004. Short-term effects of a mass coral bleaching event on a reef fish assemblage at Iriomote Island, Japan. *Fisheries Sci.* 70(1):41–46.
117. Mapstone, B.D. and A.J. Fowler. 1988. Recruitment and the structure of assemblages of fish on coral reefs. *Trends Ecol. Evol.* 3:72–77.
118. Warner, R.R. and P.L. Chesson. 1985. Coexistence mediated by recruitment fluctuations: a field guide to the storage effect. *Am. Nat.* 125:769–787
119. McClanahan, T.R., and S. Mangi. 2000. Spillover of exploitable fishes from a marine park and its effect on the adjacent fishery. *Ecol. Appl.* 10:1792–1805.
120. Pitcher, T.J. and W. Seaman, Jr. 2000. Petrarch's principle: how protected human-made reefs can help the reconstruction of fisheries and marine ecosystems. *Fish Fisheries* 1:73–81.

8 Legal Protections for Coral Reefs*

Mary Gray Davidson

CONTENTS

8.1 INTRODUCTION

Government authorities are becoming increasingly aware of the importance of coral reef ecosystems and of the rapid pace at which coral reefs are dying. Given the large array of local, state, and national initiatives attempting to protect and preserve these ecosystems, this chapter focuses on the primary U.S. laws and international legal instruments that do or may protect coral reefs. This focus in no way diminishes the importance of locally based initiatives, which, in the end, may be the only sustainable approach to ecological problems (Salm et al. 2001, Birkeland 1997). However, given the limitations of one chapter, I have chosen to consider programs with the widest possible application.

---

* This chapter originally appeared as an article in *The Harvard Environmental Law Review*, Vol. 26, No. 2 (2002).

8.2 U.S. LAW

Until fairly recently, many in the modern world believed that our oceans could provide limitless resources and were impervious to human activity. Only in the second half of the twentieth century did the United States begin to pay significant attention to the consequences of human activity on ocean life. First, Congress enacted a series of laws that indirectly benefited coral reefs, including the Fish and Wildlife Coordination Act of 1934,* the Fish and Wildlife Act of 1956,** the National Environmental Policy Act of 1969,*** and the Magnuson Fishery Conservation and Management Act of 1976.**** Other federal laws that provided some protection to coral reef resources include the Coastal Zone Management Act of 1972,***** the Clean Water Act,****** the Sikes Act,******* the Endangered Species Act, and the Lacey Act.******** Given the interconnected nature of all life in the ocean, these initial efforts provided tangential protection for coral reefs, but none was specifically directed at coral reefs.

8.2.1 The National Marine Sanctuaries Act

Beginning in the 1970s with the devastation caused by massive oil spills in the oceans, Congress responded with new initiatives to protect the marine environment. The most important was Title III of the Marine Protection, Research, and Sanctuaries Act (MPRSA) of 1972. The law created protected preserves that were in some ways similar to the land-based national parks system created nearly a century earlier (NOAA 2002a). Marine sanctuary designations were among the first U.S. attempts to take an ecosystem approach to protecting our ocean resources. The legislation coordinates the work of federal agencies with overlapping jurisdiction in the sanctuary areas (Birkeland 1997), providing more integrated protection to a limited number of ocean habitats.

The original MPRSA established a system of Marine Protected Areas (MPAs) and was designed to prevent the "unregulated dumping of material into ocean waters" that endangers "human health, welfare, or amenities or the marine environment, ecological systems, or economic potentialities." Title III of the MPRSA charged the Secretary of Commerce, who oversees the National Oceanic and Atmospheric Administration (NOAA), to identify, designate, and manage marine sites based on their "conservation, recreational, ecological, or esthetic values" within the U.S. ocean territories and the Great Lakes. After designating a marine sanctuary, the MPRSA authorized the Secretary to "issue necessary and reasonable regulations to control any activities permitted within the designated marine sanctuary." This was the first effort to preserve marine ecosystems as a whole, and the primary concern was the deleterious effect of actively dumping waste into the ocean.

In its reauthorization of Title III of MPRSA in 1984, Congress greatly expanded the purpose and process of designating such protected areas. The amended MPRSA provides for a balancing-of-needs inquiry before a sanctuary is added to the program. Congress instructed the Secretary of

---

* The Fish and Wildlife Coordination Act of 1934 recognizes the importance of wildlife resources and authorizes the Secretary of the Interior to provide assistance to and cooperate with federal, state, and other authorities to protect wildlife.

** The Fish and Wildlife Act of 1956 establishes, among other things, the position of Assistant Secretary of the Interior for Fish and Wildlife and a Fisheries Loan Fund.

*** The National Environmental Policy Act of 1969 declared a national policy to encourage productive and enjoyable harmony between man and his environment. NEPA requires preparation of an environmental impact statement for proposed legislation and other major federal actions significantly affecting the environment.

**** The Magnuson Fishery Conservation and Management Act of 1976 recognizes the importance of fisheries to the U.S. economy and the dangers of overfishing.

***** The Coastal Zone Management Act of 1972 recognizes the negative impact on marine resources by coastal development activities.

****** The Clean Water Act provides some protection to coral reefs by regulating discharges into U.S. waters.

******* The Sikes Act regulates Department of Defense activities affecting natural resources.

******** The Lacey Act makes it unlawful to import, export, sell, acquire, or purchase fish, wildlife, or plants taken, possessed, transported, or sold: 1) in violation of U.S. or Indian law, or 2) in interstate or foreign commerce involving any fish, wildlife, or plants taken possessed or sold in violation of state or foreign law.

Commerce to look at areas of special national significance due to their "resource or human-use values" and to consider factors such as biological productivity, ecosystem structure, and threatened species present in the area. The reauthorization also instructed the Secretary to consider an area's "historical, cultural, archaeological, or paleontological significance."

On the other side of the equation, the Secretary is to consider the negative impacts produced by "management restrictions on income-generating activities such as living and nonliving resources development; and the socioeconomic effects of sanctuary designation." Marine sanctuary status provides some protection but does not eliminate all commercial activity within sanctuary boundaries. For this reason, some argue that marine sanctuaries are more similar to national forests, where commercial logging is permitted, than national parks (Ranchod 2001).

The 1988 reauthorization of MPRSA enlarged the scope of the statute still further and allowed the sanctuaries program to collect and use funds obtained from resource damage claims. Under the amended statute, "any vessel used to destroy, cause the loss of, or injure any sanctuary resource shall be liable *in rem* to the United States for response costs and damages resulting from such destruction, loss, or injury." Thus, when vessels cause destruction through oil spills, groundings, or other actions that damage marine sanctuary resources, repairs can be made from recovered settlements. This is important because coral reefs tend to occur in shallow waters where they are more vulnerable to human activity and damage from ships.

The National Marine Sanctuaries Act (NMSA) does provide affirmative defenses for acts of God, war, third-party acts, or negligible damage. The courts, however, have interpreted these defenses very narrowly. For example, a federal district court in Florida granted summary judgment to the government in the case of the *M/V Miss Beholden.* The court found that the ship intentionally ran aground the Western Sambo Reef in the Florida Keys National Marine Sanctuary during a storm in 1993, damaging or destroying 1025 m^2 of live coral and 133 m^2 of established reef framework. The defendant ship owners were not allowed to use any of the affirmative defenses since bad weather had been forecast for the area 2 days before the accident. In 1996, the U.S. Court of Appeals for the Eleventh Circuit interpreted the NMSA as a strict liability statute and affirmed the lower court's damages award to the sanctuary when the ship *Jacquelyn L* ran aground on the same reef in the Florida Keys.

Thirteen marine sanctuaries have been established in the United States during the first 30 years of the program. The national marine sanctuaries system covers 18,000 square miles in the Atlantic and Pacific Oceans (S. Rep. No. 106-353, 2000). So far only five of the sanctuaries are home to coral reefs, including the Flower Garden Banks National Marine Sanctuary in the Gulf of Mexico and the Florida Keys National Marine Sanctuary (NOAA 2002b). Flower Garden Banks, located 110 miles off the coasts of Texas and Louisiana, harbors the northernmost coral reefs in the United States and covers 41.7 square nautical miles, containing 350 acres of reef crest (NOAA 2003).

The Florida sanctuary runs alongside the Florida Keys and extends approximately 220 miles southwest from the southern tip of the Florida peninsula. The sanctuary is home to a complex ecosystem including seagrass meadows, mangrove islands, and living coral reefs of "extensive conservation, recreational, commercial, ecological, historical, research, educational, and aesthetic values" (FKNMS 2004). The Florida sanctuary received additional protection in 2002 when the International Maritime Organization, a specialized agency of the United Nations, approved NOAA's proposal to designate the area as a Particularly Sensitive Sea Area (PSSA) (NOAA 2002c). The waters around the Florida Keys are some of the most heavily trafficked shipping areas in the world, with 40% of the world's commerce passing through the Florida Straits each year, and the Florida PSSA is one of only five such areas in the world (NOAA 2002c). It regulates ships larger than 50 m to internationally accepted and enforceable rules designed to address the harmful effects of anchorings, groundings, collisions, and discharges of harmful substances (NOAA 2002c). NOAA hopes that "PSSA status will help educate the international shipping community about the sensitivity of coral reef resources to international shipping activities and increase compliance with domestic measures already in place to protect the area" (NOAA 2001).

There may eventually be another U.S. reef included in the marine sanctuaries program. On December 4, 2000, President Clinton issued Executive Order 13,178 to establish the Northwestern Hawaiian Islands Coral Reef Ecosystem Reserve ("Reserve") (Exec. Order No. 13,178, 2000). The order recognizes that the United States holds 3% of the world's coral reefs and that 70% of the U.S. total is located in Hawaii. The order establishes an 84-million-acre reserve to protect Hawaii's reefs. It would be the second-largest MPA on Earth, exceeded only by the Great Barrier Reef in Australia (Breen 2001).

The final order establishing the Hawaiian Reserve caps the current level of commercial and recreational fishing at the amount taken in 2000, except in specific areas of the Reserve where fishing is prohibited (Exec. Order 13,178, 2000; Exec. Order No. 13,196, 2001). It prohibits all other commercial activity such as drilling, oil and mineral exploration, anchoring of boats, discharging any material into the water, or collecting items from the Reserve. President Clinton issued the executive order using his authority under a variety of laws including the NMSA, the Endangered Species Act of 1973 (ESA), and the National Historic Preservation Act. The executive order directs the Secretary of Commerce to initiate the process to designate the Reserve as a national marine sanctuary under the National Marine Sanctuaries Program Authorization Act of 1988.

The National Marine Sanctuary Program (NMSP) worked with the Reserve staff to develop the Final Reserve Operations Program (NOAA 2004b). NSMP and the Reserve have also begun the process to designate the Reserve as this country's 14th National Marine Sanctuary under the NMSA.

The national marine sanctuaries program is the best federal effort to date to protect coral reefs. Still, it alone is insufficient to ensure the preservation of the marine environment. The program would be more successful if more coordination existed with local, state, and federal authorities to reduce the amount of land-based pollution entering the sanctuaries and degrading the reefs, particularly the near-shore reefs off the coast of Florida.

8.2.2 The Antiquities Act

President Clinton took another avenue of executive power to protect coral reefs using the Antiquities Act of 1906 (16 U.S.C. §§ 431-33). Shortly before leaving office, Clinton employed the Act to establish the Virgin Islands Coral Reef National Monument (Proclamation No. 7399, 2001) and expand the Buck Island Reef National Monument in the U.S. Virgin Islands (Proclamation No. 7392, 2001). Together, the two designations set aside 30,843 marine acres as monuments. The Virgin Islands monument protects a fragile Caribbean tropical ecosystem and recognizes the interdependence of the fishery habitats, the "mangroves, sea grass beds, coral reefs, octocoral hardbottom, sand communities, shallow mud and fine sediment habitat, and algal plains" (Ranchod 2001). The expanded Buck Island Reef National Monument now encompasses "additional coral reefs ... barrier reefs, sea grass beds, and sand communities, as well as algal plains, shelf edge, and other supporting habitats not included within the initial boundary" (Proclamation No. 7392, 2001).

Clinton's use of the Antiquities Act represented a departure from the typical national monument designation (Ranchod 2001). Traditionally, national monuments were selected to preserve "curiosities ... that stand out from the landscape by virtue of their extraordinary beauty, or unusual geographic or historical value" (Graham 2000). Clinton's novel use of the Antiquities Act created national monuments that "revolve around large ecosystems that are distinct and of significance" (Ranchod 2001).* The Antiquities Act directs the president to limit the parcels or public lands set aside as monuments to "the smallest area compatible with the proper care and management of the objects to be protected" (16 U.S.C. § 431). While Clinton's expansion of the Antiquities Act from individual "curiosities" to entire ecosystems is novel, it corresponds with the growing knowledge

* It is important to note that previous presidents also used the Antiquities Act to protect public lands. Several were later redesignated as national parks by Congress, including the Grand Canyon, Zion, Bryce Canyon, and Joshua Tree.

that an individual species does not exist independent of its surroundings; rather, an ecosystem is a community in which all parts are interdependent.

Designation as a national monument under the Antiquities Act may provide greater and quicker protection for coral reefs than designation as a marine sanctuary currently provides. The Antiquities Act does not require the level of intragovernmental consultation, public participation, and congressional oversight that the NMSA requires.* Unlike the NMSA, the Antiquities Act does not require the president to consider conflicting uses of the area. However, due to the unilateral nature of the executive action under the Antiquities Act, the underwater monuments could be in greater danger than marine sanctuaries of being reversed or eviscerated by subsequent presidents or congressional action. This danger includes inadequate funding to carry out the intent of the designating executive order (Ranchod 2001). Since the passage of the Antiquities Act, 14 of 17 presidents have used it to establish 123 national monuments (Ranchod 2001). Congress has only abolished seven of those monuments, and five others have been reduced in size (Larvie 2001), which may indicate Congress's reluctance to override the executive in this area.

8.2.3 Marine Protected Areas

MPAs refer to an existing patchwork of local, state, and national efforts to protect corals. These efforts preserve, to varying degrees, certain areas of the nation's waters, including some areas with coral reefs. In the United States, MPA is an umbrella term that includes "national marine sanctuaries, fisheries management zones, national seashores, national parks, national monuments, critical habitats, national wildlife refuges, national estuarine research reserves, state conservation areas, state reserves, and many others" (NOAA 2004c).

Recognizing that the seas have generally been treated as "commons" available to everyone, whether within a country's boundaries or on the high seas, MPAs establish specific boundaries and specify the permitted and nonpermitted uses within them (Salm et al. 2001). An MPA may be established for a variety of reasons, such as maintaining fisheries through "no-take" zones, high species diversity, critical habitat for particular species, special cultural values (historic, religious, or recreational), or tourist attractions (Salm et al. 2001). Some MPAs restrict or forbid human activity within the protected area, while others simply manage an area to enhance ocean use (Salm et al. 2001).

In May 2000, President Clinton signed Executive Order 13,158 to strengthen and expand the nation's system of MPAs (Exec. Order No. 13,158, 2000). The executive order places primary responsibility for developing a national system of MPAs in the hands of the Department of the Interior (DOI) and the Department of Commerce (DOC). NOAA calls the creation of a comprehensive system of MPAs "perhaps the most important, and most challenging, ocean management effort of the 21st century" (NOAA 2004d), and one that "has never been attempted by our nation" (NOAA 2004e).

The administering departments have developed two parallel tracks to carry out the executive order. The first is an evaluation of the existing MPAs, including recommendations for improving them, and recommendations for creating new MPAs. The other is a science-based track that will develop tools and management strategies to support a national MPA network (NOAA 2004e).

One of the first tasks under the executive order is to publish and maintain a list of MPAs existing in the United States (Exec. Order 13,158, 2000). Because of the varied definitions of MPA, the executive order specifically defines MPA as "any area of the marine environment that has been reserved by federal, state, territorial, tribal, or local laws or regulations to provide lasting protection for part or all of the natural and cultural resources therein." The order further

* However, legislation introduced in Congress in 2001 would amend the Antiquities Act to require the president to allow for public participation before designating future monuments involving more than 50,000 acres and to consult with the Governor and members of Congress from the State or territory where the designated monument is located. See National Monument Fairness Act of 2001, H.R. 2114, 107th Cong.

defines "marine environment" to mean "those areas of coastal and ocean waters, the Great Lakes and their connecting waters, and submerged lands thereunder, over which the United States exercises jurisdiction, consistent with international law." The MPA Center expects the final list to identify between 1500 and 2000 MPAs when the inventory is completed (NOAA 2005). After the departments compile the list of existing MPAs, new candidates for protection can be added.

One of the major challenges in designing a comprehensive national system of MPAs will be "coordinating management efforts across areas of complex, multiple jurisdictions" (NOAA 2004d). The executive order addresses this challenge by directing the implementing agencies to create the following: a Web site to facilitate information sharing; an MPA Federal Advisory Committee to provide expert advice on and recommendations for the national system of MPAs; and a National MPA Center, whose mission is to develop a "framework for a national system of MPAs, and to provide federal, state, territorial, tribal, and local governments with the information, technologies, and strategies to support the system" (Exec. Order No. 13,158, 2000). Funding the initiative will be a further challenge, especially because the executive order does not provide for funding. President Bush's Secretary of Commerce announced in 2001 that the administration intended to retain and proceed with Executive Order 13,158 (NOAA 2004e).

The DOC has been carrying out its mandate under Executive Order 13,158. The DOC has created a Web site (NOAA 2005a), established the National MPA Center (NOAA 2004g), organized an MPA Advisory Committee with 30 members (NOAA 2004g), and convened several conferences (NOAA 2004h).

An important feature of the executive order is the requirement that federal agencies identify those actions that will "affect the natural or cultural resources that are protected by an MPA" (Exec. Order No. 13,158, 2000). The order states that in taking such actions, the agency "shall avoid harm to the natural and cultural resources that are protected by an MPA," although the agency is only required to avoid such harm "to the extent permitted by law and to the maximum extent practicable." The order further requires each federal agency affected by the order to prepare, and make public, a description of the actions taken by that agency in the previous year to implement the order.

The executive order itself does not create any right or benefit "enforceable in law or equity by a party against the United States, its agencies, its officers, or any person." However, DOI and DOC already possess some enforcement authority over MPAs. DOI has jurisdiction over "1.8 million of the nation's 4.2 million acres of coral reefs" (Craig 2000) with the authority to promulgate regulations for those designated as national parks, including fines and jail sentences for violations of the law (Craig 2000). Also, as discussed earlier, the NMSA gives the Secretary of Commerce considerable enforcement authority.

Given that Congress has not yet enacted a comprehensive, coordinated, long-term national policy to protect the nation's coral reefs (Craig 2000), the MPA executive order is an important new tool in managing ocean resources that could eventually prove beneficial to coral reefs. Its success, however, will depend on Congress's long-term willingness to fund the mandate.

The United States joins a number of other countries in experimenting with MPAs as a way of protecting important ecosystems. One of the best examples of an MPA is the Great Barrier Reef of Australia. It is cited as a "model of integrated and multiple-use management, allowing sustainable utilization of the reef by a wide range of users with numerous and often conflicting needs" (Bryant et al. 1998). Another promising example is the Bonaire Marine Park in the Caribbean, a self-funded park "supported entirely from tourist revenues (which also bring in half of that country's total gross domestic product)" (Bryant et al. 1998). In the Philippines, the Apo Island Reserve "has allowed [fish] stocks to recover sufficiently so that local fishermen operating in the surrounding areas are reporting major increases in fish yields" (Bryant et al. 1998). The United States could draw on the best practices from these successful MPAs to enhance its own fledgling system in protecting its marine resources.

8.2.4 U.S. Coral Reef Task Force

The increased awareness of the importance of coral reefs to the ocean system has spawned other federal efforts designed specifically to protect them. In 1998, the year of the mass coral bleaching event, President Clinton issued Executive Order 13,089, entitled "Coral Reef Protection." (Exec. Order No. 13,089, 1998). The order affirmatively requires all federal agencies to identify actions that may affect U.S. coral reefs and to ensure, subject to certain exceptions, that their actions will not degrade those ecosystems.

Executive Order 13,089 also created the U.S. Coral Reef Task Force (CRTF). Chaired by the Secretaries of the Interior and Commerce, the CRTF has the following responsibilities: to coordinate efforts to map and monitor all U.S. coral reefs; to research the causes of, and solutions for, coral reef degradation; to reduce and mitigate coral reef degradation from pollution, overfishing, and other causes; and to implement strategies to promote conservation and sustainable use of coral reefs internationally (Exec. Order No. 13,089, 1998).

In March 2000, the CRTF released a National Action Plan calling for 20% of all U.S. coral reefs to be designated as no-take ecological reserves by 2010 (CRTF 2000). A no-take zone is a particular type of MPA that bans all consumptive uses, including fishing and mineral extraction (Sanchirico 2000). The National Action Plan also calls for all U.S. coral reefs to be mapped by 2007; only a small percentage of U.S. reefs have been adequately mapped (Sanchirico 2000). In 2002, the CRTF published a National Coral Reef Action Strategy to assess what it had accomplished and priorities for reducing the adverse impacts of human activities (CRTF 2002).

During one of its first meetings, the CRTF voted to take complaints from members of the public who believe a federal agency has violated the executive order (*Orlando Sentinel* 1999). The first complaint came in late 1999 from the government of Puerto Rico (Commonwealth of Puerto Rico v. Rumsfeld, 2001), which accused the U.S. Navy of destroying its coral reef during bombing exercises in Vieques (*L.A. Times* 1999). The CRTF National Action Plan also states that the Department of Defense (DOD) is actively working to implement Executive Order 13,089 "to the maximum extent feasible consistent with mission requirements" (CRTF 2000). That commitment, though, gives DOD great latitude in determining what is feasible. Also, Congress recently freed the DOD from some restrictions on its activities imposed by laws that provide some protection to coral reefs and creatures that depend on reefs for their survival. The National Defense Authorization Act for Fiscal Year 2004 (NDAA) amends, among other things, the Endangered Species Act and the Marine Mammal Protection Act (MMPA) to limit their applicability to activities of the DOD. NDAA amends MMPA to exempt for up to 2 years any action "necessary for national defense" and exempts military bases from some of the Endangered Species Act's habitat-protection requirements.

The CRTF's National Action Plan was followed that same year by the Coral Reef Conservation Act of 2000 (16 U.S.C. §§ 6401–6409), the first legislation ever specifically targeted at coral reef issues. The Act incorporated by reference the provisions of Executive Order 13,089. It continues the CRTF and the U.S. Coral Reef Initiative, an existing partnership between governmental and commercial interests whose purpose is to design "management, education, monitoring, research, and restoration efforts to conserve coral reef ecosystems…." A primary objective of the Coral Reef Conservation Act is to provide matching grants, subject to the availability of funds, to coral reef conservation projects.

Because the CRTF comprises representatives from 11 agencies, it represents a more coordinated effort than in the past. Assuming the CRTF continues to receive adequate funding, it may provide much-needed leadership in responding to a growing environmental crisis both in the United States and internationally.

8.2.5 Endangered Species Act

The marine sanctuary approach to coral reef preservation attempts to conserve reefs as a whole. Regulations in the Florida Keys National Marine Sanctuary, for example, forbid removing, injuring, and even possessing any coral or live rock (15 C.F.R. § 922.163). They also forbid collecting many species of fish, anchoring on live coral, and discharging waste anywhere in the sanctuary. Another approach to conserving reefs is to protect specific coral species under the Endangered Species Act (ESA) (16 U.S.C. §§ 1531-1544). While ESA covers numerous marine creatures such as sea turtles and many species of reef and other fish, no corals have been added to the Federal Lists of Endangered and Threatened Wildlife and Plants (ESA lists).

In 1999 the National Marine Fisheries Service (NMFS) began to consider adding two species of coral found in the Caribbean to the ESA lists — elkhorn coral and staghorn coral (64 Fed. Reg. 2629, 1999). According to NMFS, nearly 96% of corals of these two species have disappeared during the last two decades due to hurricane damage, coral diseases, increased predation, boat groundings, sedimentation, and other factors (NMFS 2004 and 2004a). While scientists and activists debate whether a species-by-species approach can be effective when an entire ecosystem is under attack, many are supportive of any legal effort that enhances reef protection.

If these coral species are added to the ESA lists, they would receive protection throughout their habitat range, which includes areas currently outside the designated sanctuaries in the United States. ESA forbids importing, taking, or even possessing species on the ESA lists (16 U.S.C. §1538(a)(1)(A)-(D)). Since most of the public do not see corals and are not even aware that corals are living creatures, adding corals to ESA could also serve as notice to the public that corals in general are disappearing.

ESA's ban on the import of listed species into the United States is another important legal protection (16 U.S.C. § 1538(a)(1)(A)). Reefs in the Philippines are being decimated by activities such as harvesting for export (Broad and Cavanagh 1993). Currently the United States is the main importer of stony corals from the Philippines as curios, even though legislation such as the MPRSA bans collection on our own reefs (16 U.S.C. §§ 1436-37, 15 C.F.R. § 922.122). More than half of the Philippines' exports of ornamental coral and exotic reef fish are sent to the United States (Broad and Cavanagh 1993). If ESA listed more coral species and banned their import, then fewer corals would likely be harvested in other countries because some of the U.S. market would dry up, at least among those who wish to comply with U.S. law.

Another argument for including species of coral on the ESA lists is that while naming individual species, ESA actually provides protection for the species' entire ecosystem. As ESA states, "the purposes of this [Act] are to provide a means whereby the ecosystems upon which endangered species and threatened species depend may be conserved." (16 U.S.C. 1531(b)). ESA forbids harming species included on the ESA lists (16 U.S.C. §§ 1532(19), 1538(a)(1)(B)-(D)). Federal regulations implementing ESA define the word "harm" in the definition of "take" to mean "an act which actually kills or injures wildlife. Such acts may include significant habitat modification or degradation where it actually kills or injures wildlife by significantly impairing essential behavioral patterns, including breeding, feeding or sheltering" (50 C.F.R. § 17.3 (2000)). The U.S. Supreme Court upheld this definition in *Babbitt v. Sweet Home Chapter of Communities for a Great Oregon.* However, Justice Scalia, in a dissent joined by Chief Justice Rehnquist and Justice Thomas, argued that the destruction of habitat does not "harm" an endangered species within the meaning of ESA. This dissent coincides with growing complaints that ESA was mostly intended to preserve large animals, not tiny creatures and their habitat (Petersen 1999).

In later amendments to ESA, Congress required that critical habitat be designated at the same time a species is listed (16 U.S.C. § 1533(b)(6)(C)). These provisions could provide enhanced protection for entire reefs where staghorn and/or elkhorn corals are located. Reefs are "extremely susceptible to sewage and industrial wastes, oil spills, siltation, and water stagnation brought about by dredging and filling, thermal pollution, and flooding with low salinity or silt-laden water resulting from poor land management" (Gold 1988). If the regulating agencies could prove the necessary nexus between these harms and destruction of the endangered corals' habitat, then ESA could become a powerful tool in combating reef degradation, just as it has been in rescuing individual species threatened with extinction.

8.2.6 Conclusion

The federal provisions described above have overlapping purposes but leave many gaps in the protection of coral reefs. Table 8.1 summarizes the major domestic initiatives aimed at preserving marine resources, including coral reefs.

The federal coral reef initiatives have been piecemeal until just recently. Executive Order 13,089 directly placed an affirmative duty on federal agencies not to harm coral reefs. Equally important was its directive to the CRTF to determine the extent of the United States' coral reefs and to map them. With that information in hand, and the necessary executive and congressional will, bodies such as the CRTF can devise and implement strategies to prevent further degradation of our coral reefs and make recommendations for more comprehensive legislation to preserve our reefs for the long term.

When they receive adequate funding to enforce them, the federal protections for coral reefs are useful. But, as stated earlier, we need greater control over human activities away from the reefs that contribute to reef degradation. This includes not only land-based sources of pollution and sedimentation but also human behavior contributing to global climate change (Wilkinson 2000).

The federal efforts to protect coral reefs are not without criticism, both from those who would like to exploit the resources in the sanctuaries and from conservationists. The Eleventh Circuit Court of Appeals discussed the controversial designation of the Florida Keys National Marine Sanctuary and opposition to it by Florida's governor in the 1996 case *United States v. M/V Jacquelyn L.* As one ardent supporter of the Florida reefs complains, "Generally, I think that local, peer-supported, community-based initiatives are honored whereas mandates from afar tend to be unenforceable and unenforced in most instances unless there is a heavy enforcement hand onsite" (Quirolo, personal communication, 2001). This critic believes that areas of the Florida Keys sanctuary were actually more protected prior to the sanctuary designation because the federal law does not provide adequate enforcement measures.

The United States has made great strides in recent decades in recognizing the importance of coral reefs and attempting to provide legal protections for some of them, but these protections obviously are not enough. Sanctuary status has not prevented the precipitous decline in the Florida reef system. Five hundred ship groundings a year occur in the Florida Keys National Marine Sanctuary (Spalding et al. 2001). Agricultural runoff from the mainland and sewage dumping from 22,000 septic tanks, 5000 cesspools, and 139 marinas all contribute to pollution and eutrophication in the sanctuary (Spalding et al. 2001). The sanctuary program needs to be coordinated with state and local efforts to eliminate these sources of pollution. As of 2001, only 1% of the total sanctuary area in Florida had been designated as no-take marine reserves (Spalding et al. 2001). Those areas show signs of recovery (Spalding et al. 2001), and the no-take designations should be increased. These measures and a commitment to continue funding the efforts of the sanctuary programs, the CRTF and the MPA initiative, are essential to preserving the other U.S. coral reefs that have not yet borne the sustained assault that the Florida reef has.

The United States is slowly recognizing that coral reefs are a precious resource that requires legal protection. The Clinton administration made several attempts to address coral reefs directly. It remains up to present and future administrations to see that these laws are utilized for the maximum protection of reefs.

8.3 INTERNATIONAL PROTECTIONS FOR CORAL REEFS

A variety of international legal instruments either directly or indirectly provide protection for coral reefs. Though these measures offer promise for enhanced protection of reefs, the level of protection depends on the ratification and enforcement of these instruments.

The United Nations Convention on the Law of the Sea (UNCLOS) remains the guiding document for ocean issues, but many other specialized conventions potentially afford greater protections for coral reefs. While this chapter addresses national and international laws and conventions that can

TABLE 8.1
Summary of Domestic (U.S.) Provisions Affecting Coral Reef Conservation

Provision	Date	Scope of Provision	Terms of Provision	Effect on Corals
Marine Protection Research and Sanctuaries Act (MPRSA)	1972	Selected marine areas	Regulates activities, mainly dumping, to protect ocean habitat	Secondary benefits to reefs
Endangered Species Act	1973	Designated plant and animal species	Forbids taking or possessing of designated species	No coral reef species listed yet, but reef habitat protected through designation of marine species that share the same habitat
National Marine Sanctuaries Act (formerly Title III of MPRSA)	1992	Selected marine areas	Secretary of Commerce to designate protected sites within ocean territories and regulate activities within sites	Coral reefs located in five of the designated sanctuaries
Executive Order 13,089	1998	All U.S. coral reefs	Requires all federal agencies to ensure that their actions do not degrade reef ecosystems; creates the CRTF to research and implement strategies to map and protect coral reefs	No enforceability against noncompliant agencies
Executive Orders 13,178 and 15,158	2000	Marine reserve in northwestern Hawaiian Islands	Caps fishing at year 2000 levels and prohibits other commercial activities	Reserve is a coral reef ecosystem
Executive Order 13,158	2000	Marine Protected Areas (MPAs)	Creates advisory committee to coordinate strengthening and expanding a comprehensive system of MPAs	Benefits to coral reefs within MPAs
Coral Reef Conservation Act of 2000	2000	All U.S. coral reefs	Incorporates Executive Order 13,089 and provides matching funds for reef conservation projects.	Same as Executive Order 13,089 and provides additional resources for reef conservation
Antiquities Act	1906	Areas of extraordinary geographical, historical, aesthetic value	Authorizes president to designate sites of historical or scientific interest that are situated on federal public lands as national monuments; each site to be limited to the smallest area compatible with the proper care and management of the objects to be protected	Used by presidential proclamation in 2001 to create and expand two national monuments
Proclamation 7399: Virgin Island Coral Reef National Monument	2001	12,000 marine acres in the U.S. Virgin Islands	Same as above	Monument includes coral reefs
Proclamation 7392: The Buck Island Reef National Monument	2001	18,000 marine acres in the U.S. Virgin Islands	Same as above	Monument includes coral reefs; proclamation expands original monument thus protecting more coral reefs

protect reefs, it should be noted that traditional systems of control like customary tenure, where communities have ownership of reefs and their resources, frequently produce highly effective forms of reef management (Spalding et al. 2001).

8.3.1 United Nations Convention on the Law of the Sea

UNCLOS is the primary convention regarding the use of the ocean and its resources. UNCLOS grants every state "the right to establish the breadth of its territorial sea up to a limit not exceeding 12 nautical miles, measured from baselines determined in accordance with this Convention." The Convention states that "waters on the landward side of the baseline of the territorial sea form part of the internal waters of the State." Moreover, Articles 56 and 57 of the Convention give coastal states sovereign rights in an "exclusive economic zone" out to 200 miles. Because most reef formations are limited to waters of less than 50 m depth (Gold 1988), they tend to occur in near-shore waters. This places the majority of coral reefs within countries' internal waters and exclusive jurisdiction.

Reefs are specifically mentioned in Article 6 of UNCLOS, which states that "in the case of islands situated on atolls or of islands having fringing reefs, the baseline for measuring the breadth of the territorial sea is the seaward low-water line of the reef." Thus it might appear that UNCLOS does not provide much protection for coral reefs because they are within a state's internal waters. However, UNCLOS was a landmark treaty in the development of international environmental law because it contains many conservation-oriented provisions (Hafetz 1999). Specifically, it requires states to protect and maintain their marine species, even within internal waters.

The preamble to UNCLOS states that among the primary objectives of the 1982 convention is the "study, protection, and preservation of the marine environment." UNCLOS provides "the first comprehensive statement of international law on the issue … [and] a movement toward regulation based upon a more holistic conception of the ocean as a resource that is exhaustible and finite, and ocean usage as a resource management question" (McConnell and Gold 1991). Even within the exclusive economic zones of coastal states, UNCLOS states that "the coastal State … shall ensure through proper conservation and management measures that the maintenance of the living resources in the exclusive economic zone is not endangered by over-exploitation."

UNCLOS contains many positive obligations that affect marine resources in national waters. Part XII of the convention sets forth many of the international legal requirements pertaining to the marine environment, including a system for enforcing those requirements. Article 192 sets forth the general obligation "to protect and preserve the marine environment." Article 193 recognizes the "sovereign right [of States] to exploit their natural resources" but this is subject to the "duty to protect and preserve the marine environment." Some of the specific requirements include taking measures necessary to "prevent, reduce, and control pollution of the marine environment," and to ensure that activities "are so conducted as not to cause damage by pollution to other States and their environments." States must consider all sources of pollution to the marine environment, including the following: harmful or noxious substances from land-based sources, the atmosphere, or dumping; pollution from vessels; and contamination from other installations used to explore the seabed and subsoil.

The duties expressed in Articles 192 to 194 are binding on states–parties to the Convention. Because 157 states have signed UNCLOS and 145 have ratified it (United Nations 2004), many commentators believe that the provisions are also statements of customary international law, which would make them binding on all nations, including those countries that are not parties to the convention (Hafetz 1999, Iudicello and Lytle 1994). Therefore, even though some countries, including the United States, have not ratified UNCLOS, they may be bound by many of its principles. Interestingly, in 2004, the United States Commission on Ocean Policy, in the first major federal assessment of the oceans in a generation, called on the Congress and the Bush administration to end the United States' 22-year refusal to officially join UNCLOS (Barringer 2004). The White House responded that the President "has put the treaty on the top priority list for ratification" (Barringer 2004).

Prior to UNCLOS, little international regulation of the marine environment or its conservation existed. UNCLOS's provisions for the protection and preservation of the marine environment reflected the growing awareness of what was happening to our oceans. Unfortunately, many nations have not ratified the Convention, in part because of its controversial deep seabed provisions. Therefore, a major issue today is whether the Convention reflects customary international law so that it is binding on all nations and not just those that are parties to the convention.

8.3.2 Agenda 21

Ten years after the drafting of UNCLOS, more than 178 governments adopted Agenda 21, the final document of the United Nations Conference on Environment and Development ("UNCED") held in Brazil in 1992 (United Nations Department for Economic and Social Affairs 2002). Agenda 21 reaffirmed many of the goals of UNCLOS but also recognized that "despite national, subregional, regional, and global efforts, current approaches to the management of marine and coastal resources have not always proved capable of achieving sustainable development, and coastal resources and the coastal environment are being rapidly degraded and eroded in many parts of the world" (UNCED 1992). Chapter 17 of Agenda 21 gives the protection of coral reefs high priority and calls for an integrated, international approach for their protection and use.

To implement Chapter 17 and other international conventions, the International Coral Reef Initiative (ICRI) was created at the Small Island Developing States conference in 1994 (ICRI 2004). Through ICRI, over 80 developing countries with coral reefs "sit in equal partnership with major donor countries and development banks, international environmental and development agencies, scientific associations, the private sector, and NGOs to decide on the best strategies to conserve the world's coral reef resources" (ICRI 2004a). ICRI has developed "action plans" for all regions of the world and is now working with national governments and organizations to implement those plans (ICRI 2002). Like the CRTF, ICRI is still relatively new, and it remains to be seen whether either body becomes an important force in the fight to preserve coral reefs.

Chapter 15 of Agenda 21, titled "Conservation of Biological Diversity," calls for immediate action in protecting the diversity of plant and animal resources. Chapter 15 states:

> Despite mounting efforts over the past 20 years, the loss of the world's biological diversity, mainly from habitat destruction, over-harvesting, pollution and the inappropriate introduction of foreign plants and animals, has continued.... Urgent and decisive action is needed to conserve and maintain genes, species and ecosystems, with a view to the sustainable management and use of biological resources.

Chapter 15 is especially significant for coral reefs because of their high biodiversity.

Agenda 21 represents a major development in ocean stewardship. Where previous international agreements looked at protecting specific ocean resources such as marine mammals and fish, Agenda 21 recognizes the need for overall sustainable ocean development (Craig 2002).

8.3.3 Convention on Biological Diversity

UNCED also produced the Convention on Biological Diversity (CBD) (CBD 1992). The preamble to the convention asserts that "the conservation of biological diversity is a common concern of humankind." As the primary international agreement on biodiversity issues, the CBD's three objectives are the "conservation of biological diversity, sustainable use of its components, and a fair and equitable sharing of the benefits of genetic resources" (U.S. Department of State 1999).

The CBD does not name specific ecosystems but provides for the identification and monitoring of two distinct categories: (1) ecosystems and habitats; and (2) species and communities (CBD 1992). Among the factors to consider in identifying ecosystems and habitats for protection are those "containing high diversity, large numbers of endemic or threatened species, or wilderness; required by migratory species; of social, economic, cultural or scientific importance; or, which are representative, unique, or associated with key evolutionary or other biological processes" (CBD 1992). Species and communities covered by the CBD include those that are "threatened …; of medicinal, agricultural, or other economic value; or social, scientific, or cultural importance; or importance for research into the conservation and sustainable use of biological diversity, such as indicator species." The purpose behind these broad criteria is to ensure that the CBD encompasses all possible areas of biodiversity.

The specific terms of the Convention obligate parties to comply with a variety of provisions in addition to identifying and monitoring the components of biological diversity. These provisions involve establishing protected areas, integrating conservation and sustainable use of biological resources into national decision-making, educating the public, and facilitating access to genetic resources by other states (CBD 1992). The CBD contains no enforcement mechanism, and, as the Convention's Secretariat explains, "to a large extent, compliance will depend on informed self-interest and peer pressure from other countries and from public opinion" (Secretariat of the CBD 2004). Countries that ratify the Convention must submit regular reports on what they have done to implement its provisions (CBD 1992). This report goes to the Conference of the Parties — the governing body that oversees implementation of the CBD (CBD 1992).

As of May 2004, there were 188 parties to the Convention (Secretariat of the CBD 2004a). The United States is among the countries that have signed the convention but not ratified it. The Clinton administration urged the Senate to ratify CBD, in part because "biological diversity … represents the "raw material" for the world's agricultural and pharmaceutical industries. Organisms yet to be discovered or studied could hold the key to a future cure for some terrible disease, or their genetic material may be useful in improving crop[s] … [and] help feed the world's expanding population" (U.S. Department of State 1999).

The United States has been concerned about the CBD's impact on intellectual property rights, technology transfer, and finance provisions. While the Senate Foreign Relations Committee favorably reported the Convention to the full Senate in 1994, the Senate curtailed further consideration of the treaty due to concerns about the CBD's effect on land use and agriculture in the United States (U.S. Dept. of State 1999, 140 Cong. Rec. S13791, 1994).

The CBD is a framework treaty and has been described as containing "primarily aspirational provisions, with matters of substance left to future development by its own Conference of the Parties" (Guruswamy and Hendricks 1997). These objectives are connected through a principle known as "common but differentiated responsibility." This principle holds that "developed countries acknowledge the responsibility that they bear in the international pursuit of sustainable development in view of the pressures their societies place on the global environment and of the technologies and financial resources they command" (Rio Declaration 1992). In other words, countries such as the United States that use a disproportionate share of the world's resources have a special responsibility to find a balance between resource use and preservation. As one commentator pointed out, the most biologically diverse areas left on the planet are mostly in poorer countries of the developing world, and they understandably feel "possessive of those resources" (Guruswamy and Hendricks 1997). Since the North has already consumed much of its own biodiversity, "the South would embrace sustainable development only if the North would assume the costs, and only through projects that would not compromise a growing sense of sovereignty over natural resources" (Guruswamy and Hendricks 1997). Thus, a balanced approach to conserving biodiversity must take account of how various levels of development affect a state's management of its natural resources.

The CBD was created at a critical point because "many biologists believe we are in the midst of one of the great extinction spasms of geological history" (Wilson 2000). This would be the seventh mass extinction event, the last one occurring 65 million years ago when the dinosaurs disappeared (Wilson 2000). "This time, however, human activity, not nature, is the culprit"

(Wilson 2000). The cause of this crisis, some believe, is human population increase coupled with the destruction of natural habitats, the invasion of alien species, and pollution (Wilson 2000). The CBD recognizes that in this era of mass extinctions, it is in our self-interest to preserve as much biodiversity as possible because "the loss of biodiversity threatens our food supplies, opportunities for recreation and tourism, and sources of wood, medicines, and energy. It also interferes with essential ecological functions" (Secretariat of the CBD 2000). Moreover, scientists have only identified about 1.75 million of the estimated 13 million species that exist on earth, and the CBD is an attempt to preserve the unknown as well as the known (Secretariat of the CBD 2000).

Coral reefs, one of the most diverse ecosystems on earth, also face mass extinction. This is not because of the usual "vagaries of weather and climate" (Wilson 1992) such as hurricanes and other storms. Those natural events can actually benefit ecosystems because they "prevent a few dominant species from pushing 'inferior' ones out" (Davidson 2001). As the distinguished biologist Edward O. Wilson noted, "in normal circumstances, the reefs recover from natural destruction within a few decades. But now these natural stresses are being augmented by human activity, and the coral banks are being steadily degraded with less chance for regeneration" (Wilson 1992).

For example, in looking at the reef off Florida's Key Largo, Wilson discovered that 30% has been damaged since 1970, with the chief destructors being pollution, oil spills, "accidental grounding of freighters, dredging, mining for coral rock, and harvesting of the more attractive species for decoration and amateur collections" (Wilson 1992). These sustained assaults on the health of a reef are different in kind from the brief, intermittent disturbances that occur naturally, and the prospect for reef survival in such areas is not good. Countries that have ratified the CBD are required to develop national biodiversity strategies and action plans (CBD 1992), an integral part of which should be the prevention of further destruction of a nation's coral reefs.

Even for countries that have not ratified the CBD, debt-for-nature swaps are one promising scheme for preserving biodiversity and promoting North–South relations (Wilson 1992). In these swaps, governments, and even private organizations such as The Nature Conservancy and the World Wildlife Fund, purchase a portion of a country's commercial debt. In exchange, that country must designate territory as free from development or use the additional funds for environmental education or for the improvement of land management (Wilson 1992). In 2003, a bipartisan group of U.S. Congress members referred a bill to the House Committee on International Relations that would allow certain developing countries to honor their debts to the United States by starting coral reef conservation programs instead of exploiting their natural resources to pay off the debt (Coral Reef and Marine Conservation Act 2003). One marine biologist believes this is a workable solution. "There is no reason that these arrangements cannot work for the marine realm as they have for tropical forests. Conserving the diversity of life in the sea calls for creative solutions that appeal to individual and national needs, as it does on land" (Norse 1993). Even though the United States has not ratified the CBD, some members of Congress are applying its principles to preserve coral reefs. The CBD recognizes the importance of balancing a nation's development needs with preserving its biodiversity, thus providing a framework for conserving essential biological resources for future generations.

8.3.4 Convention on International Trade in Endangered Species

While the United States is not yet a party to the CBD, it was the first state to ratify the Convention on International Trade in Endangered Species of Wild Flora and Fauna (CITES) in 1973 (CITES 1973). CITES specifically addresses the problem of international trade in endangered species. One hundred sixty-six nations have signed CITES (CITES Secretariat 2004), which provides varying degrees of protection to approximately 33,000 plant and animal species (CITES Secretariat 2004a). CITES entered into force in 1975, and the CITES Secretariat says that "not one species protected by CITES has become extinct as a result of trade since the Convention entered into force." (CITES Secretariat 2004b).

There are approximately 230 species of coral listed by their common names on the CITES Species Database (UNEP-WCMC 2004). Among them are Helioporacea (blue corals), Tubiporidae (organ pipe corals), Antipatharia (black corals), Scleractinia (stony corals), Milleporidae (fire corals), and Stylasteridae (lace corals) (CITES Secretariat 2003).

CITES is significant for coral reefs in part because "coral reef organisms are subjected to an increasing international trade.... Live and dead marine organisms are used for multiple purposes such as aquaria, swimming pools, decoration, souvenirs, jewelry, and precious stones." (TRAFFIC 2004). The largest exporting nations of corals are Indonesia, Fiji, Vietnam, and the Solomon Islands (TRAFFIC 2004). Ironically, while the United States prohibits or strictly limits coral extraction in its own waters, the United States imports 70 to 80% of the live coral traded internationally, 95% of the live rock, and 50% of the dead (NOAA 2004j, Coral Reef Alliance 2002).

Under CITES, member countries agree to ban commercial international trade in an agreed list of endangered species and to regulate and monitor trade in others that might become endangered (CITES Secretariat 2004c). CITES protects those species listed in the three appendices to the Convention (CITES 1973). Any party to CITES may propose amendments to Appendices I and II and to Appendix III if the named species is within that party's jurisdiction. CITES forbids the trade in endangered species on the list in Appendix I except in extraordinary circumstances. The exporting and importing states must each certify that specific criteria have been met to ensure that the species is not further endangered.

CITES authorizes the trade in species listed in Appendices II and III, subject to a permit system that allows states to monitor and even limit exports, if necessary (CITES 1973). In 1985, member nations of CITES listed all stony or reef-building corals on Appendix II as a response to the effect of the coral trade on reef ecosystems (U.S. Fish and Wildlife Service 2004). Now, black corals, blue corals, and organ pipe, stony, fire, and lace corals are all listed in Appendix II of CITES (CITES Secretariat 2003) and require a permit from the country of origin in order to be traded on the international market (CITES 1973).

Enforcement of the convention is not always successful. In some cases, "coral collected in countries where collection is illegal (such as the Philippines) is often exported and sold under the pretext of having been collected legally in a different country" (Coral Reef Alliance 2002). Moreover, CITES does not list many other reef species, such as "puffer fish, seahorses, starfish, sea urchins, sea fans, sponges." (Coral Reef Alliance 2002). These reef dwellers are an integral part of the coral reef ecosystem, and the collection of them for souvenirs and private aquariums can be just as detrimental to the reefs as the collection of corals themselves.

Another enforcement problem is the difficulty identifying the corals that are listed in the CITES appendices. For example, a CITES monitoring organization found that "the trade in corals and other marine organisms is increasing, and there have been many instances where CITES-listed corals have been shipped without the necessary permits, or with incorrect permits, often resulting in sizeable confiscations" (TRAFFIC 2004a). Part of the problem has been traders claiming that hard corals are "living rock," which makes them exempt from the CITES permit requirements.

Since only specialists could differentiate between living rock and marine organisms such as corals, the CITES governing body adopted a resolution in April 2000 to include live rock in its definition of coral rock, thereby making the live rock subject to the Convention (CITES Conference of the Parties 2000). Live rock is "pieces of coral rock to which are attached live specimens of invertebrate species and coralline algae not included in the CITES Appendices and which are transported moist, but not in water, in crates" (CITES Conference of the Parties 2000). The Convention does not apply to rock that does not contain any corals or in which the corals are fossilized. CITES is an evolving instrument, and clarifications such as the above should be helpful for those officers in the field trying to enforce the Convention.

Overall, CITES is useful for regulating the trade in individual coral species, but it does not protect the entire ecosystem. Nonetheless, with effective enforcement and by raising public awareness about the need to purchase only properly documented coral species, CITES is an effective tool to fight the destruction of coral reefs.

8.3.5 United Nations Convention Concerning the Protection of the World Cultural and Natural Heritage

The United Nations Convention Concerning the Protection of the World Cultural and Natural Heritage provides another means for protecting coral reefs (World Heritage Convention 1972). The Convention notes that the world's cultural and natural heritage is "increasingly threatened with destruction" and that the "deterioration or disappearance of any item of the cultural or natural heritage constitutes a harmful impoverishment of the heritage of all the nations of the world" (World Heritage Convention 1972). The Convention defines "natural heritage" as physical and biological formations of "outstanding universal value from the aesthetic or scientific point of view." Under the Convention, an Intergovernmental Committee for the Protection of the Cultural and Natural Heritage of Outstanding Universal Value maintains a "World Heritage List" of property forming part of the cultural and natural heritage, with the consent of the state concerned. The Convention makes available technical and financial assistance. This may include assistance in getting a site included on the World Heritage List, providing experts and others to help with the preservation of a listed site, or training staff and specialists in the identification and conservation of cultural and natural heritage (UNESCO 2000).

As of 2004, there were 154 natural properties on the World Heritage List (UNESCO 2004a). Eleven of those sites contain coral reefs (UNEP-WCMC 2004a). Three are in Australia, including the Great Barrier Reef, and two are in Indonesia. Belize, Mexico, the Philippines, the United States, the United Kingdom, and Seychelles each have one coral reef site according to UNEP. However, the site listed in the United States, the Everglades National Park in Florida, only tangentially touches the Florida reef system. In reality, UNESCO has not designated any coral reefs in the United States as World Heritage Sites. As a World Heritage Site, the Everglades Park has benefited from increased federal funding and the purchase of additional land to expand the park area (UNESCO 2002). Nevertheless, the fate of Everglades National Park remains uncertain, and the "biologic death" of the park is still possible (UNESCO 1997). The World Heritage Committee responded to this crisis in 1993 and entered the Everglades on the List of World Heritage in Danger.

It is clear that World Heritage Site designation will not always protect a site. World Heritage sites may be lost through the impacts of development, as seen in the Florida Everglades, and they are sometimes purposefully destroyed, as we saw when the Taliban destroyed two giant Buddha statues in Afghanistan in 2001 (Rosenberg 2001). But for countries that do want to protect their cultural and natural heritage, World Heritage Site designation does provide a level of recognition, and even assistance, that can make the difference in saving a country's heritage for future generations.

8.3.6 Conclusion

Much like the patchwork quality of U.S. provisions, international treaties and conventions have provided some, though not comprehensive protection, for marine ecosystems. These major international initiatives are summarized in Table 8.2.

Meaningful international protection for oceans has only occurred in the last two decades. Most of the international agreements take an ecosystem approach, which is important for the long-term viability of coral reefs. UNCLOS provides the most general protection for coral reefs through its requirement to preserve and protect marine environments. Agenda 21, adopted 10 years later, built on UNCLOS and specifically identified coral reefs as an area of high priority and led to the creation of ICRI, an international task force devoted to coral reef preservation. The World Heritage Convention has, to date, named ten coral reefs as World Heritage Sites, leading to more domestic legal protection and sometimes financial and technical assistance from UNESCO. The CBD provides a

TABLE 8.2
Summary of International Provisions Affecting Coral Reef Conservation

Provision	Date	Scope of Provision	Terms of Provision	Effect on Coral Reefs
United Nations Convention Concerning the Protection of the World Cultural and Natural Heritage	1972	Natural and manmade properties of outstanding cultural, aesthetic, and scientific value	Provides technical and financial assistance for preservation of unique properties	Ten World Heritage Sites contain coral reefs
Convention on International Trade in Endangered Species (CITES)	1973	Species listed as endangered	Bans trade in species listed in Appendix I and regulates, through permits, trade in species listed in Appendices II and III	230 coral species listed in Appendices II and III
United Nations Convention on the Law of the Sea (UNCLOS)	1982	Ocean and its resources	General preservation of marine environment, conservation of marine species, and pollution reduction	General protection of reefs of states that have entered the convention; some consider UNCLOS customary international law
Agenda 21	1992	Defines general rights and obligations between states and the environment	Calls for international cooperation and action to protect the environment	Gives the protection of coral reefs high priority; led to creation of the International Coral Reef Initiative (ICRI)
Convention on Biological Diversity	1992	Ecosystems and habitats, species and communities	Framework treaty to conserve biological diversity through monitoring, habitat preservation, establishment of protected areas, sustainable use of resources, and sharing of genetic resources	Affects coral reefs because of their high biodiversity

framework for conserving coral reefs because of their high biological diversity. In addition to the ecosystem approaches in the Conventions named above, CITES provides another level of protection for coral reefs by regulating the trade in various species of coral. Overall, the international provisions will prove valuable so long as there is the international will to abide by them.

8.4 RECOMMENDATIONS FOR IMPROVING THE LEGAL PROTECTION OF CORAL REEFS

Despite the legal regimes noted in this chapter, coral reefs are not adequately protected and they are rapidly disappearing. The following are recommendations for ensuring the long-term viability of the remaining reefs.

8.4.1 Establish No-Take Zones

An emerging practice in ocean management is to establish no-take zones that prohibit harvesting of marine resources. Efforts to control fisheries in the United States and elsewhere have traditionally involved regional management councils setting "restrictions on vessel size and power, total allowable catches, types of gear, time and area closures, and size and sex of the catch" (Sanchirico 2000). Currently, less than 1% of the continental shelf is set aside in no-take zones (Bohnsack interview 1998). Some scientists believe that setting aside as much as 20% of the continental shelf as no-take zones is necessary to reestablish certain depleted fisheries (Bohnsack interview 1998). Regeneration of fish populations will occur by allowing fish to mature, breed, and produce more eggs. The goal of the no-take zones is "to make sure enough of those fish grow large and breed to maintain the population. When these no-take zones are enforced, and the breeding grounds are given a rest, scientists see real benefits" (Bohnsack interview 1998).

Researchers are beginning to witness the success of no-take zones. In the Florida Keys National Marine Sanctuary, for example, managers set aside 1% of the sanctuary in 1997 as a no-take zone (Bohnsack interview 1998). After the first full year of protection, the sanctuary no-take zones showed significantly more and larger lobsters and the greatest numbers of certain economically important reef fishes in the sanctuary (FKNMS 1999). Scientists are also monitoring the response of corals within the no-take zones, but given their slow growth rate, their response to the changes is expected to take much longer. In addition to improving the health of the ecosystems within the no-take zones, marine scientists see spillover benefits outside the protected area, due to the complex biological links, particularly fish migrations, between protected and unprotected areas (Sanchirico 2000). The results are preliminary, but expanding the number of no-take zones and monitoring the results should be done quickly to see if the zones are as beneficial as anticipated.

8.4.2 Modify Fishing Practices

In lieu of banning all fishing in no-take zones, nations with coral reefs may also limit fishing methods to reduce the damage to fish stocks. For example, the governor of American Samoa issued an executive order banning scuba divers from fishing in an effort to curb the serious problem of overfishing of reefs (Exec. Order No. 002-2001). Prior to the ban, studies showed that reef fish stocks off the main Island of Tutuila had dropped to a dangerously low level since the introduction of scuba fishing in 1994 (Cornish 2001). Prior to 1994, 1 to 3 tons of parrotfish from reefs were caught annually; from 1994 to 1998, 25 to 33 tons of parrotfish were taken, with 33 tons representing one-fifth of the total biomass of parrotfish from the reefs fringing the island (Cornish 2001).

In American Samoa, local communities believed the harm was caused principally by outsiders using scuba gear to fish their reefs, and "this ecosystem approach to management was recognized by the communities as a valuable tool to aid recovery of depleted reef fish stocks" (Cornish 2001). Banning just one known, harmful method of fishing allows other, more sustainable fishing practices to continue and does not entirely cut off the livelihood of those who depend on the reef. At the same time, it permits the fisheries to replenish themselves.

8.4.3 Add Reef Species to CITES

CITES is another avenue for protecting creatures that live and depend on the reef. Currently under CITES, of all the species of types other than fish and turtles that live in coral reefs, only hard coral and giant clams are listed in Appendix II (CITES 1973, UNEP-WCMC 2002). Because parties to CITES are obligated to document and report on the quantity of trade in each species, the international community has a good idea of the magnitude of the legal international trade in those species. However, no marine ornamental fish or invertebrates typically found on reefs are covered by CITES. Therefore, any estimates of the extent of international trade in those species are simply guesses. Because the overharvesting of any one species in the reef ecosystem could upset the system's balance, we need better data on the harvesting of all reef creatures in order to assess the harm accurately. Also, since the United States is currently considering adding elkhorn coral and staghorn coral to the ESA lists, the CITES Conference of Parties should likewise consider affording those species the enhanced protection of Appendix I.

In 2000, UNEP established the WCMC to gather information on the sustainable use of the world's living resources (UNEP-WCMC 2004b). The WCMC could be a valuable resource in determining the extent of the trade in ornamental fishes and other invertebrates from coral reefs. Nearly all marine ornamentals pass through a relatively small number of wholesalers (UNEP-WCMC 2002), and the records of those businesses would be an excellent source of material on the marine ornamentals trade. Currently the reporting is voluntary (UNEP-WCMC 2002a), but an obligatory reporting regime would be invaluable.

8.4.4 Increase World Heritage Site Designations

Given the prediction that as much as 60% of the world's reefs will be gone in 30 years (Wilkinson 2000), UNESCO should expand the protection to reefs offered by the World Heritage Site designation. The World Heritage Committee could add the most endangered reefs to the List of World Heritage in Danger under Article 11.

Article 11, paragraph four of the World Heritage Convention provides that property facing "threat of disappearance caused by accelerated deterioration, large-scale public or private projects or rapid urban or tourist development projects; destruction caused by changes in the use or ownership of the land; [or] major alterations due to unknown causes" may be included on the World Heritage List. The World Resources Institute estimates that 25% of all reefs in the world are at high risk of disappearance, with the reefs in Southeast Asia — "a global hot spot of coral and fish diversity" — most endangered, primarily from "coastal development, overfishing, and destructive fishing practices" (Bryant et al. 1998). Those reefs at high risk in Southeast Asia and elsewhere should be included on the World Heritage list. Under Articles 19 and 22, a party state containing a designated reef is eligible to request international assistance in the form of technical cooperation, loans, and even grants. These funds may be used in a variety of ways, ranging from training staff to providing experts, and even supplying equipment.

UNEP has already identified a number of coral reefs that it would like to see added to the World Heritage List, including reefs found in the Red Sea, the Indian Ocean, Indonesia, the Bahamas, and Fiji (UNEP-WCMC 2004a). It should add to its list by incorporating the results of the international collaborative study titled "Reefs at Risk," which has classified the major reefs of the world in terms of their biodiversity and level of threat from human activity (Bryant et al. 1998). UNESCO's World Heritage Program already highlights the threats to a variety of other ecosystems (UNESCO 2004). Given that coral reefs are the second most diverse ecosystem on the planet, UNESCO should give them equal priority.

8.4.5 Advance U.S. Practices

Domestically, the United States should immediately take action to foster the sustainable harvest of items imported from coral reefs. Although the United States forbids or strictly limits the extraction of hard corals in its own waters, it remains the single largest importer of coral and live rock (CRTF 2000). The United States could, in conjunction with exporting nations, help develop a certification program to reduce the amount of illegally obtained reef products imported into the country. For example, the United States could require certification that reef fish were not caught using cyanide or dynamite or that live coral and other organisms were not damaged in the harvesting process. Global efforts are already underway to tap into the "green" movement and ensure consumers that the ornamental fish they purchase did not involve harmful practices (Marine Aquarium Council 2001).

President Clinton's designations of two coral reefs as national monuments should remain in place but with more restrictions on their use, such as establishing no-take zones in portions of the monuments. Future presidents should also consider using the Antiquities Act to protect endangered coral reefs in U.S. waters. An executive order using the Antiquities Act will provide quicker protection for reefs. While such executive action may be reversed by subsequent presidents, even short-term protection could give a dying reef time to regrow and contribute to the longer time needed for the sanctuary designation process.

Congress has recently shown its willingness to protect coral reefs, but it needs to continue this trend with adequate implementation and funding. Congress's passage of the Coral Reef Conservation Act of 2000 was a good first step into an area that had received little congressional attention until the 1990s. But the Act is a limited grant-making program that provides no additional protection for the reefs. Congress needs to allocate the funding necessary to carry out the recommendations of the CRTF and its National Action Plan, including a budget for coral reef enforcement. The United States should also implement the CRTF's recommendation of designating 20% of all U.S. coral reefs as no-take ecological reserves by 2010, at least for the limited time necessary to establish their efficacy. Without widespread support and lobbying by environmental groups, though, the actual percentage of no-take reserves will likely fall short of this figure.

Several Clinton-era proposals and executive orders should be continued during the current and subsequent administrations. The goal of Executive Order 13,089, to create a coordinated system of coral reef MPAs, is important and should be realized as soon as possible. Because of the interest in coral reefs generally, Congress's initial funding of the MPA project could be supplemented or replaced to protect and sustain our coral reefs by "user fees, subscriptions, support societies, volunteer organizations, etc." (Craig 2002).

Finally, the United States, which became a global role model when it established the world's first national park on land in 1872 by creating Yellowstone National Park (National Park Service 2004), could continue its role as a preservation leader by approving the Hawaiian Marine Sanctuary designated by President Clinton in December 2000.

8.5 CONCLUSION

Coral reefs are a rare habitat but one that millions of humans depend on. Legal protection for coral reefs began fairly recently and consists of piecemeal laws and conventions that serve either directly or indirectly to protect only certain coral reefs. By all scientific accounts, coral reefs are at a crisis point, and their preservation requires more coordinated measures to protect these treasures both nationally and internationally.

The best hope for coral reefs so far seems to be in establishing more MPAs (Miller 2001). Unfortunately, examples of MPAs are few. It is estimated that only 3% of the world's coral reefs are within MPAs, and at least 40 countries have no legal protections for their reefs (Bryant et al. 1998). The MPAs that do protect reefs tend to be very small, many of them only a square kilometer in size. Only a few very large sites such as the Great Barrier Reef, the Florida Keys National Marine Sanctuary, and the Ras Mohammed Park Complex in Egypt are truly substantial MPAs. However, governments must take care not to simply create "paper parks" where "legislation is not enforced, resources are lacking for protecting these areas, or management plans are poorly conceived" (Bryant et al. 1998).

Edward O. Wilson has called for land-based reserves to be expanded from their current 4.3% of the earth's land surface to 10% to prevent these fragments from becoming "shrunken habitat islands, whose faunas and floras will continue to dwindle until a new, often lower equilibrium is reached" (Wilson 1992). The same argument applies to the oceans, and in particular, to coral reefs, those "rainforests of the sea" upon whose diversity so much life, human and otherwise, depends.

REFERENCES

140 CONG. REC. S13791 (daily ed. Sept. 30, 1994).
Antiquities Act of 1906, 34 Stat. 225 (1906) (codified as amended at 16 U.S.C. § 431 et seq. (2004)).
Babbitt v. Sweet Home Chapter of Communities for a Great Oregon, 515 U.S. 687 (1995) (Scalia, J., dissenting).

Barringer, F., Federal Study Urges New Strategy for Safeguarding the Seas, *New York Times,* Apr. 20, 2004.
Birkeland, C., *Life and Death of Coral Reefs*, Chapman & Hall, New York, 1997, 297, 383.
Bohnsack, J., Biologist, National Marine Fisheries Service, interview in Miami, FL (Jan. 23, 1998).
Breen, T., Coral Reefs: Enviros Concerned by White House Review of Hawaiian Reserve, *Greenwire*, July 24, 2001, at LEXIS, News Library, Wire Service Stories File.
Broad, R. and Cavanagh, J., *Plundering Paradise: The Struggle for the Environment in the Philippines*, University of California Press, Berkeley, 1993, 38, 37.
Bryant. D. et al., *Reefs at Risk: A Map-Based Indicator of Threats to the World's Coral Reefs*, World Resources Institute, Washington, 1998, 44, 20-21.
CITES, Mar. 3, 1973, 27 U.S.T. 1087, 1089, 993 U.N.T.S. 242, 244.
CITES Conference of the Parties, Trade in Hard Corals, CITES Res. 11.10, 11th Sess. (2000).
CITES Secretariat, Appendices I and II (valid as of Oct. 16, 2003), at http://www.cites.org/eng/append/1+11_0700.shtml.
CITES Secretariat, Appendices I, II and III (valid from Oct. 16, 2003), at http://www.cites.org/eng/append/appendices.shtml.
CITES Secretariat, Member Countries (May 30, 2004), at http://www.cites.org/eng/parties/index.shtml.
CITES Secretariat, The CITES Species (May 30, 2004a), at http://www.cites.org/eng/disc/species.shtml.
CITES Secretariat, What is CITES? (May 30, 2004b), at http://www.cites.org/eng/disc/what.shtml.
CITES Secretariat, How CITES Works (May 30, 2004c), at http://www.cites.org/eng/disc/how.shtml.
Clean Water Act, Pub. L. No. 92-500, 86 Stat. 896 (1972) (codified as amended at 33 U.S.C. §§ 1251-1387 (2004)).
Coastal Zone Management Act of 1972, Pub. L. No. 89-454, 86 Stat. 1280 (1972) (codified as amended at 16 U.S.C. §§ 1451-1465 (2004).
Commonwealth of Puerto Rico v. Rumsfeld, 180 F.Supp.2d 144 (D.D.C. 2001), *judgment vacated and dismissed as moot*, 2003 WL 21384576 (D.C. Cir. 2003).
Convention on Biological Diversity (CBD), June 5, 1992, 1760 U.N.T.S. (entered into force Dec. 29, 1993).
Coral Reef Alliance, Trade in Coral Reef Species (Apr. 28, 2002), at http://www.coralreefalliance.org/aboutcoralreefs/trade.html.
Coral Reef Conservation Act of 2000, Pub. L. No. 106-652, tit. II, § 202, 114 Stat. 2800 (codified at 16 U.S.C. § 6401-6409 (2000)).
Coral Reef Task Force (CRTF), The National Action Plan to Conserve Coral Reefs, 20, 30, 2000.
Coral Reef Task Force, National Coral Reef Action Strategy, 2002.
Cornish, A., Chief Biologist, Department of Marine and Wildlife Service, American Samoa andy_cornish@yahoo.com, to coral-list-daily@coral.aoml.noaa.gov (posting to listserv Apr. 8, 2001).
Craig, R.K., The Coral Reef Task Force: Protecting the Environment through Executive Order, 30 *Envtl. L. Rep.* (Envtl. L. Inst.) 10,343, 10,357, 2000.
Craig, R.K., Sustaining the Unknown Seas: Changes in U.S. Ocean Policy and Regulation Since Rio '92, 32 *Envtl. L. Rep.* (Envtl. L. Inst.) 10,190, 10,191, 2002.
CRTF National Coral Reef Action Strategy (June 2002), at http://coris.noaa.gov/activities/actionstrategy/action_reef_final.pdf.
Davidson, O.G., *Fire in the Turtle House*, Public Affairs, New York, 2001, 178.
Endangered Species Act, Pub. L. No. 93-205, 87 Stat. 884 (1973) (codified as amended at 16 U.S.C. §§ 1531-1544 (2004)).
Endangered and Threatened Species; Request for Information on Candidate Species List Under the Endangered Species Act, 64 Fed. Reg. 2629, 2629-30 (Jan. 15, 1999).
Endangered and Threatened Wildlife and Plants, 50 C.F.R. § 17 et seq. (2004).
Exec. Order No. 002-2001, Apr. 6, 2001 (executive order by Tauese P.F. Sunia, Governor of American Samoa).
Exec. Order No. 13,089, 63 Fed. Reg. 32,701 (June 16, 1998).
Exec. Order No. 13,158, 65 Fed. Reg. 34,909, 34,909 (May 26, 2000).
Exec. Order No. 13,178, 65 Fed. Reg. 76,903 (Dec. 4, 2000).
Exec. Order No. 13,196, 66 Fed. Reg. 7395 (Jan. 18, 2001).
Fish and Wildlife Act of 1956, Act of Aug. 8, 1956, ch. 1036, § 2, Pub. L. No. 8401924, 70 Stat. 1119 (codified as amended at 16 U.S.C. §§ 742a-j (2004)).
Fish and Wildlife Coordination Act of 1934, Act of Mar. 10, 1934, ch. 55, Pub. L. No. 85-624, 48 Stat. 401 (codified as amended at 16 U.S.C. §§ 661-667d (2004)).
Florida Keys National Marine Sanctuary (FKNMS), Press Release, First-Year Results Show Sanctuary No-Take Zones Beginning to Change Fish and Lobster Populations, Mar. 4, 1999.

Florida Keys National Marine Sanctuary Regulations: Final Rule, 15 C.F.R. § 922 (2000).

Florida Keys National Marine Sanctuary (FKNMS), Visitor Information (April 28, 2004), at http://floridakeys.nos.noaa.gov/visitor_information/welcome.html.

Gold, E. (ed), *A New Law of the Sea for the Caribbean: An Examination of Marine Law and Policy Issues in the Lesser Antilles*, Springer, Berlin, 1988, 126, 128.

Graham, J., U.S. Land Protection Plans are Monumental, *Chicago Tribune*, Feb. 18, 2000, 3.

Guruswamy, L.D. and Hendricks, B.R., *International Environmental Law in a Nutshell*, West, 1997, 91–93.

H.R. 1721, 108th Cong. (2003), Coral Reef and Marine Conservation Act (amending the Foreign Assistance Act of 1961 to provide for debt relief to developing countries who take action to protect critical coral reef habitats).

Hafetz, J., Fostering Protection of the Marine Environment and Economic Development: Article 121(3) of the Third Law of the Sea Convention, 15 Am. U. Int'l L. Rev. 583, 596, 597, 1999.

ICRI, ICRI Achievements 1995-1998 (Apr. 28, 2002), at http://www.icriforum.org/router.cfm?Show=secretariat/sec_home.html&Item=1.

ICRI, International Coral Reef Initiative. What is ICRI? (May 30, 2004), at http://www.icriforum.org/ router.cfm?Show=secretariat/sec_home.html&Item=1.

ICRI leaflet, International Coral Reef Initiative (May 30, 2004a), at http://www.icriforum.org/router.cfm?Show=secretariat/sec_home.html&Item=1.

Iudicello, S. and Lytle, M., Marine Biodiversity and International Law: Instruments and Institutions That Can Be Used to Conserve Marine Biological Diversity, 8 *Tul. Envtl. L. J.* 123, 132, 1994.

L.A. Times, Navy Accused of Blowing up Reefs, A31, Nov. 4, 1999.

Lacey Act, 16 U.S.C. §§ 1371–3378.

Larvie, V., Protection of Cultural Resources: the Antiquities Act and National Monuments, Paper Presented at American Law Inst.–American Bar Association Federal Lands Law Conference (Oct. 18-19, 2001), LEXIS, Secondary Legal Library, Combined ALIABA Course of Study Materials File.

Magnuson Fishery Conservation and Management Act of 1976, Pub. L. No. 94-265, 90 Stat. 331 (codified as amended in scattered sections of 16 U.S.C. (2000)).

Marine Aquarium Council, MAC Certification (2001), at http://www.aquariumcouncil.org/subpage.asp?Section=13.

Marine Protection Research, and Sanctuaries Act of 1972, Pub. L. No. 92-532, titl. III, 86 Stat. 1052 (codified as amended at 33 U.S.C. §§ 1401-1421, 1441-1445, 2801-2805 (1994 & Supp. V 1999), and at 16 U.S.C. §§ 1431-1445, 1447 (2000)).

McConnell, M. and Gold, E., The Modern Law of the Sea: Framework for the Protection and Preservation of the Marine Environment? 23 *Case W. Res. J. Int'l L.* 83, 84–85, 1991.

Miller, S., Director, National Undersea Research Center, University of North Carolina at Wilmington, smiller@gate.net, to coral-list-daily@coral.aoml.noaa.gov (posting to listserv Oct. 4, 2001).

National Defense Authorization Act of 2004, Pub. L. No. 108-136, 117 Stat. 1392 (2003).

National Environmental Policy Act of 1969, Pub. L. No. 91-190, 83 Stat. 852 (1970) (codified as amended at 42 U.S.C. §§ 4321-4370f (1994 & Supp. V 1999).

National Marine Fisheries Services (NMFS), Endangered and Threatened Species; Request for Information on Candidate Species List Under the Endangered Species Act, 64 Fed. Reg. 2629, 2629-30 (Jan. 15, 1999).

National Marine Fisheries Services (NMFS), Species of Concern and Candidate Species: Staghorn Coral, (April 13, 2004), at http://www.nmfs.noaa.gov/prot_res/species/inverts/staghorncoral.html.

National Marine Fisheries Service (NMFS), Species of Concern and Candidate Species: Elkhorn Coral, (April 13, 2004a), at http://www.nmfs.noaa.gov/prot_res/species/inverts/elkhorncoral.html.

National Park Service, Yellowstone National Park (May 30, 2004), at http://www.nps.gov/yell/.

National Park System, 16 U.S.C. § 16 1a-1 (2005).

NOAA, Press Release, NOAA Receives Initial Green Light to Extend International Protection to the Florida Keys (July 10, 2001), at http://www.sanctuaries.nos.noaa.gov/news/pressreleases/pressreleases 7_10_ 01.html.

NOAA, Marine Protected Areas: Status of NOAA Activities Under Executive Order 13158 (2001a), at http://mpa.gov/information_tools/pdf/mpa_factsheet_oct_2001.pdf.

NOAA, Welcome to the National Program (Aug. 6, 2002a), at http://www.sanctuaries.nos.noaa.gov/natprogram/natprogram.html.

NOAA, National Sanctuaries Programs: History (Sept. 25, 2002b), at http://www.sanctuaries.nos.noaa.gov/natprogram/nphistory/nphistory.html.

NOAA Press Release, Florida Keys Coral Reefs First in U.S. to Receive International Protection (Nov. 13, 2002c), at http://sanctuaries.nos.noaa.gov/news/pressrelease11_13_02.html.

NOAA, NMS —Marine Sanctuaries —Flower Gardens (Feb. 12, 2003), at http://www.sanctuaries.nos.noaa.gov/oms/omsflower/omsflower.html.

NOAA, State of the Sanctuary Report 2003/2004, (March 5, 2004a), at http://sanctuaries.noaa.gov/library/national/sots04.pdf.

NOAA, Northwestern Hawaiian Islands Coral Reef Ecosystem Reserve (May 30, 2004b), at http://www.hawaii-reef.noaa.gov.

NOAA, What Is a Marine Protected Area (Feb. 12, 2004c), at http://mpa.gov/information_tools/archives/ what_is_mpa.html#varying.

NOAA, Marine Protected Areas Information and Tools—Archives (Feb. 12, 2004d), at http://mpa.gov/information_tools/archives/challenges.html.

NOAA, Marine Protected Areas—Information and Tools—Archives, (Feb. 12, 2004e), at http://mpa.gov/information_tools/archives/initiative.html.

NOAA, Inventory of Sites—Status of the Inventory, (April 22, 2004f), at http://mpa.gov/inventory/status.html.

NOAA, MPA Advisory Committee—About the FAC (Feb. 12, 2004g), at http://www.mpa.gov/fac/ about_fac.html.

NOAA, Marine Protected Areas: Information and Tools—Conferences and Workshops (May 30, 2004h), at http://www2.mpa.gov/mpa/mpaservices/information_tools/conf_and_workshops.lasso.

NOAA Fisheries, Office of Protected Resources, International Trade in Coral Reef Species (last visited May 30, 2004i), at http://www.nmfs.noaa.gov/prot_res/PR/tradeincorals.html.

NOAA, Marine Protected Areas of the United States, (Jan. 14, 2005), at http://mpa.gov.

Norse, E.A. (ed), *Global Marine Biological Diversity*, Island Press, Covelo, 1993, 217.

Orlando Sentinel, 1st Task Force Case: Vieques Coral Reef, Nov. 4, 1999, A12.

Petersen, S., Congress and Charismatic Megafauna: A Legislative History of the Endangered Species Act, 29 *Envtl. L.* 463, 1999.

Proclamation No. 7392, 66 Fed. Reg. 7335 (Jan. 22, 2001).

Proclamation No. 7399, 66 Fed. Reg. 7364 (Jan. 22, 2001).

Quirolo, D., Executive Director, Reef Relief, e-mail correspondence to author (Mar. 21, 2001).

Ranchod, S., The Clinton National Monuments: Protecting Ecosystems with the Antiquities Act, 25 *Harv. Envtl. L. Rev.* 535, 552–55, 569, 581, 585–87, 2001.

Rio Declaration on Environment and Development, Principle 7, June 14, 1992, 31 I.L.M. 874, 877.

Rosenberg, T., Destroying history's treasures, *New York Times*, Mar. 15, 2001, A24.

S. Rep. No. 106-353, 2000.

Salm, R.V., Clark, J.R., and Siirili, E., *Marine and Coastal Protected Areas: A Guide For Planners and Managers,* IUCN, Washington, 2001, 13–15, 65.

Sanchirico, J., Marine Protected Areas as Fishery Policy: A Discussion of Potential Costs and Benefits (unpublished discussion paper, Resources for the Future), 1–3, 5–6, 11, 2000.

Secretariat of the CBD, Sustaining Life on Earth: How the Convention on Biological Diversity Promotes Nature and Human Well-Being 4, 5, 2000.

Secretariat of the CBD, International Action (May 4, 2004), http://www.biodiv.org/doc/publications/guide.asp?id=action-int.

Secretariat of the CBD, Parties to the Convention on Biological Diversity/Cartagena Protocol on Biosafety (2004a), at http://www.biodiv.org/world/parties.asp.

Sikes Act, Pub. L. No. 86-797 (1960) (codified as amended at 16 U.S.C. § 670(a)-(o) (2004)).

Spalding, M.D., Ravilious, C. and Green, E.P., *World Atlas of Coral Reefs*, University of California Press, Berkeley, 2001, 97–98, 66.

TRAFFIC USA, Legal Determination of Coral and Marine Organism Identification in the Netherlands (May 30, 2004), at http://www.traffic.org/making-CITES-work/mcw_nl-coral.html.

TRAFFIC USA, Making Cites Work (May 30, 2004a), at http://www.traffic.org/making-CITES-work/mcw_nl-coral.html.

UNEP-WCMC, Global Marine Aquarium Database, Background (Aug. 7, 2002), at http://www.unep-wcmc.org/marine/GMAD/background.html.

UNEP-WCMC, Global Marine Aquarium Database, Description of GMAD (Jan. 14, 2002a), at http://www.unep-wcmc.org/marine/GMAD/description.html.

UNEP-WCMC, CITES-listed Species Database: Fauna (May 30, 2004), at http://www.cites.org/eng/resources/species.html.

UNEP-WCMC, A Global Overview of Wetland and Marine Protected Areas on the World Heritage List tbl 8 (May 30, 2004a), at http://www.unep-wcmc.org/wh/reviews/wetlands/t8.htm.

UNEP-WCMC, What UNEP-WCMC Does (Oct. 18, 2004b) at http://www.unep-wcmc.org/reception/does.htm.

UNESCO, Everglades National Park/World Heritage Site, Two Anniversaries, *The World Heritage Newsletter* (The World Heritage Center, UNESCO), Oct. 1997.

UNESCO, The World Heritage Fund (Mar. 3, 2000), at http://www.unesco.org/whc/ab_fund.htm.

UNESCO, Everglades National Park (July 17, 2002), at http://www.unesco.org/whc/sites/76.htm.

UNESCO, Our Past, Our Future (May 30, 2004), at http://www.unesco.org/whc/nwhc/pages/sites/main.htm.

UNESCO, The World Heritage List (May 30, 2004a), at http://www.unesco.org/whc/heritage.htm.

United Nations Conference on Environment and Development (UNCED), Agenda 21, ch. 17.4, at 131 (Vol. II), U.N. Doc. A/CONF.151/26 (1992).

United Nations Convention on the Law of the Sea (UNCLOS), Dec. 10, 1982, 1833 U.N.T.S. 397 (entered into force Nov. 16, 1994).

United Nations Department of Economic and Social Affairs, Division for Sustainable Development, Agenda 21 (Apr. 28, 2002), at http://www.un.org/esa/sustdev/agenda21.htm.

United Nations, Oceans and Law of the Sea, Chronological List of Ratifications, accessions and successions to the Convention (Jan. 16, 2004), at http://www.un.org/Depts/los/reference_files/status2002.pdf.

U.S. Department of State, Biodiversity and the Convention on Biodiversity (Jan. 5, 1999), at http://www.state.gov/www/global/oes/fs-biodiversity_990105.html.

U.S. Fish and Wildlife Service, Coral Trade (May 30, 2004), at http://news.fws.gov/issues/coral.html.

United States v. M/V Jacquelyn L, 100 F.3d 1520 (11th Cir. 1996).

Wilkinson, C. (ed), Status of Coral Reefs of the World: 2000 Report of the Global Coral Reef Monitoring Network 7, 2000.

Wilson, E.O., *The Diversity of Life*, W.W. Norton & Co., New York, 1992, 270, 337.

Wilson, E.O., Doomed to Early Demise, *UNESCO Courier*, 22, 2000.

World Heritage Convention, Nov. 16, 1972, 27 U.S.T. 37, 1037 U.N.T.S. 151.

9 Streamlined Injury Assessment and Restoration Planning in the U.S. National Marine Sanctuaries

Lisa C. Symons, Alice Stratton, and William Goodwin

CONTENTS

9.1 INTRODUCTION

Physical impacts to coral reefs occur from pollution, weather, and impacts of human uses and activities such as fishing, marine transportation, and recreation. The National Oceanic and Atmospheric Administration (NOAA), as the natural resource trustee, is responsible for responding to and, where appropriate, restoring coral injuries in the National Marine Sanctuary System in order to protect and preserve coral reefs. This chapter describes typical impacts from mechanical injuries, primarily resulting from marine transportation and recreational activities, and outlines NOAA's efforts to streamline assessment and restoration activities.

This chapter will not address pollution, weather, and global warming impacts to corals and coral ecosystems. While significant injury can be caused by extreme climatic (i.e., hurricanes, El Niño) or biological (i.e., bleaching and disease) events, these are natural processes that are not appropriate for restoration. Also, spills of oil or other hazardous substances and environmental alterations, such as sedimentation and nutrient enrichment, are not covered by this document. Injuries from spills cause surficial impacts and are addressed through the Oil Pollution Act; environmental injuries result from larger alterations in the region and cannot be directly addressed through restoration measures.

Mechanical (i.e., physical) injuries to coral reefs caused by direct human intervention include the most visible and common injuries, such as impacts from vessel groundings, anchors and lines, aquaria harvest activities, and fishing gear, but also direct impacts resulting from snorkeling and diving activities, anthropogenic turbidity, and acoustic activities. These injuries occur for a number of reasons. Sometimes the circumstances make their occurrence unavoidable, but more often than not these incidents are attributable to human error. NOAA has dealt with vessel groundings due to navigational errors resulting from having no one at the helm, misdirected autopilots, using incorrect (or no) nautical charts, inexperience, and choosing "short cuts." Bad weather, lack of familiarity with local waters, and inebriation are also often factors. Vessels of all types and sizes can end up where they cause significant environmental harm. Anchoring can also cause damage from the anchor, anchor chain, and cables undercutting, toppling, and crushing coral colonies. Damage of this type is generally due to lack of awareness of how destructive the practice is and to deliberately not adhering to charted "no anchoring" prohibitions.

Even though protected as a national marine sanctuary, the Florida Keys are severely impacted by both human activities and large-scale ecosystem changes in the Atlantic and Caribbean. Exacerbating this stress by not restoring physical injuries when possible is irresponsible trusteeship, particularly given the specific legal mandate within the National Marine Sanctuaries Act (NMSA) (16 U.S.C.§ 1441, et seq.). Over the last 20 years, NOAA has developed and refined injury assessment and restoration planning practices for both coral and seagrass habitats. These are constantly refined to address new situations, technologies, and challenges as they arise; this chapter summarizes the current status of those efforts for coral. NOAA's recent efforts to standardize injury assessment methods and to develop a programmatic environmental impact statement (PEIS) that evaluates and delineates the restoration techniques for varying types of coral injuries are briefly described here.

It is hoped that the PEIS, once completed, will facilitate timely restoration planning and implementation. Numerous small- to medium-size groundings occur on a yearly basis in the Florida

Keys, and in general, restoration times have decreased significantly as NOAA works with cooperative responsible parties to immediately implement salvage, emergency triage, and restoration work. Prompt restoration, including the ability to implement emergency restoration, can save significantly more coral tissue and decrease the loss of ecological services. The recent anchoring injury from the MSC *Diego* in the Tortugas (±1200 m^2 corals scarred, dislodged, toppled, and fractured) is a good example. Reattachment of over 1000 toppled corals was accomplished in a matter of months rather than years due to the cooperation of the responsible party. Because of NOAA's desire to respond to injuries more quickly and efficiently, the agency has implemented some smaller restorations prior to settlement with the responsible party. This requires subsequent billing of the responsible party for the costs of the restoration, as well as the response, injury assessment, compensatory restoration, and monitoring costs. This increase in efficiency and efficacy is critical, especially when an ecosystem is under significant ecological pressure. While NOAA would like to have the fiscal flexibility to do this for all cases, the funds are simply not available for that to occur.

9.2 SCOPE AND SCALE OF THE ISSUE

In 1972, Congress passed the Marine Sanctuary Resource Protection Act, later reauthorized as the NMSA. (16 U.S.C.§ 1441, et seq.). This Act directs NOAA to identify, designate, and manage marine protected areas of special national significance. In 2005, there are 13 national marine sanctuaries and one coral reef ecosystem reserve currently in the sanctuary designation process. These include several sites in which the tropical coral reef ecosystems are managed and protected: Florida Keys National Marine Sanctuary (FKNMS), Flower Garden Banks National Marine Sanctuary (FGBNMS), Fagatele Bay National Marine Sanctuary (FBNMS), and the Northwestern Hawaiian Islands Coral Reef Ecosystem Reserve (NWHI CRER).* Of the sites, FKNMS is the most active for vessel groundings, with over 600 reported groundings a year. About 40% of these occur in coral reef substrate or hardground habitats that support coral development. These groundings range in size and type from gouges, small fractures, and abrasions to large-scale crushing injuries.

After working through a number of both large- and small-scale restorations and the associated legal environmental planning processes, NOAA has explored options to streamline the review of these requirements for each case. Toward that end, in 2000, NOAA undertook an effort to develop a pair of PEISs in an effort to expedite coral as well as seagrass restoration planning in the national marine sanctuaries. The purpose of a PEIS for coral reef restoration is to use the regulatory planning process under the National Environmental Policy Act (NEPA) and the NMSA to evaluate ahead of time the range of coral restoration alternatives that can be used in the FKNMS and the FGBNMS. Then, when an incident occurs, restoration plans specific to the incident can be written much more quickly by incorporating the PEIS by reference. It also allows the agency to evaluate new technologies in a more thoughtful and reasoned manner, encompassing current scientific and engineering principles, rather than under pressure to save live coral tissue. While this streamlined planning process requires a significant investment of time and resources upfront. it is expected to decrease restoration planning time and expense considerably over the long run.

Without a standard set of assessment methods and protocols, restoration methods, and effective contracting and implementation mechanisms, precious time and live coral tissue are lost after an incident occurs. In addition, the scientific validity of an injury assessment is time sensitive and must be done as soon after the injury as possible. An injury assessment done 12 to 14 months after an incident is not descriptive of the original injury and may not be as defensible in litigation. Likewise, restoration plans developed 12 to 14 months after the assessment may not adequately address the existing injuries.

The NMSA is very clear in its expectations of NOAA as a responsible resource trustee. Under Section 312 of the NMSA, NOAA is specifically directed to undertake restoration of injured

* This chapter does not address impacts to more temperate deepwater corals.

sanctuary resources. Section 312 directs NOAA to seek damages from the responsible parties that injured sanctuary resources and use the recovered funds to recoup emergency response, damage assessment, restoration, and monitoring costs.* Restoration planning undertaken under the §312 authority and funded from those Natural Resource Damage Assessment (NRDA) funds must be done in accordance with a range of legal requirements, including the NEPA, the NMSA, and any relevant state statutes if restoration activities are to take place in state waters. This often makes restoration planning and engineering on a large scale difficult to accomplish quickly. Environmental planning processes, developed to ensure appropriate consideration of all options, take time and often leave resource trustees with great plans but no remaining live coral tissue to restore and/or an injury that is significantly larger than the one originally designed to be addressed. Once a restoration plan is developed, federal procurement and permitting requirements can also cause additional delays.

The goal of an NRDA is to assess the extent of the injury, recover response and damage assessment costs, and implement primary and compensatory restoration to make the environment and public whole as a result of the injury. "Primary restoration" refers to restoration activities at the actual injury site. The goal of restoration activities is to return injured coral communities to preinjury conditions. For coral reef communities, "baseline" refers to the level of ecological services (type, quality, and coverage of coral) existing prior to the incident. Baseline conditions are typically measured via field assessment techniques in the undisturbed reef communities bordering the grounding site.[1] In many circumstances, without primary restoration, the injured reef communities are subject to redisturbance by storms that could slow recovery and/or expand the size of the injury. "Compensatory restoration" refers to a restoration project, typically offsite, that compensates the public for the lost interim ecological services as a result of the time it takes for the original "primary" injury (with or without primary restoration) to return to baseline conditions. Funds collected for small compensatory restoration projects may be pooled together for the implementation of a larger compensatory restoration project. Funds are also collected to cover emergency response and monitoring costs. The monitoring costs generally will cover both the primary and compensatory restoration actions.

9.3 NOAA'S CORAL ASSESSMENT AND RESTORATION EXPERIENCE

The first element of assessment is response. Having standardized response objectives is critical. This section describes response in terms of three factors: minimizing impact, gathering information, and the continuum between response and injury assessment.

Very little time is available during an incident to consider a broad range of response options and procedures. If appropriate decisions are not made during the removal of a vessel and during the collection of data both prior to and after removal, significant additional injuries can occur and information critical to the development of a legal case are lost. For instance, salvage efforts will often incur the least additional damage if the vessel can be removed along the entry path. That is not always possible, so specific efforts have to be made to find the next-least-injurious route. In addition, when the vessel is on the reef, measurements need to be taken to understand the original footprint and outline of the injury so it can later be determined how the footprint may have changed during the vessel removal process. This is especially important if a vessel remains *in situ* for several days, continuing to impact resources. These activities are critical for implementing an effective response, as well as for working up an accurate injury assessment.

If injury assessments are not done immediately during and after a grounding incident, the information gained may not be not accurate or reflective of the injury at hand but may be reflective of other confounding elements. These could include additional injury incidents in the same place,

* Under the NMSA, the damages are defined to include: (1) the costs to restore, replace, or acquire the equivalent of the resources injured; (2) the value of the lost use of the resources pending restoration; (3) the costs of assessing the injuries to those resources; (4) the cost of monitoring the restoration; (5) the costs of enforcement. 16 U.S.C. § 1443 (a).

disturbance of the site by weather events, algal colonization of injury surfaces, or recruitment of other benthic organisms not representative of the baseline conditions. Of course, sound judgment and common sense must be exercised by the assessment personnel so as not to jeopardize their safety or the safety of other response team members. Heavy seas, thunderstorms, fuel spills, and other hazardous situations or conditions may delay the assessment process in deference to safety concerns. However, assessment activities should commence as soon as it is safely possible to carry them out.

Without the ability to reliably return to an injury site, it is difficult to prove injury. Locational accuracy is also critical for the accurate description of the areal extent of an injury and to differentiate one injury from another and distinguish the injured area from the unimpacted reef. It also facilitates comprehensive restoration planning and restoration monitoring. This is particularly important in high-traffic areas ("hot spots") that have sustained multiple injuries. Differential geographic position system, or dGPS, is the current state of the art in defining location and should be used by both law enforcement and injury assessment personnel.

Personnel must be appropriately trained in the recognition of corals and other benthic species endemic to the injury area. In addition, they must be trained to "read" an injury once the vessel is no longer on site. It is important to understand how the injuries occurred, as that can determine what type of injuries are found and where they may be found.

An injury assessment should be conducted in a manner that will support litigation and the necessary standards of proof. Accordingly, assessment methods must be generally accepted and replicable; they must provide the necessary information to develop an accurate injury assessment and to develop appropriate restoration plans. While it may be necessary to update an assessment if a long period has lapsed between the injury and the restoration, the initial assessment should be done as soon as possible following an injury. Chapter 8 discusses legal authorities around the world.

Careful review of more complex restoration plans is especially critical. In some instances, methods and materials have not performed as expected. In the *Contship Houston* grounding, the responsible party quickly undertook the restoration. That restoration specifically facilitated the use of some new techniques, some of which have not held up well over time. Funds set aside for monitoring have been used twice to fund midcourse corrections, and NOAA is considering whether additional corrective action is necessary or appropriate. The length of time necessary to implement some restorations left injuries, at some sites, vulnerable to additional impacts, from new groundings and from weather. Hurricane Georges increased the depth and areal extent of the *R/V Columbus Iselin* site by nearly 100%,[2] and NOAA estimates that 75% of the area restored at the *M/V Wellwood* site was injured postgrounding.[3] Restoration delays can be caused by something as simple as a graduated settlement structure, which may not convey enough funds to cover restoration costs until several years after settlement. The federal procurement process is also difficult to synchronize with a limited field season. Weather delays significantly impact costs and can only be estimated at the time of settlement. User communities can be suspicious and even resentful when popular dive areas have restricted access during construction. Some have been known to take malicious action toward contractors during restoration activities. Taking the time to reach out to this community is critical, as these users generally become strong advocates once educated.

9.4 ASSESSMENT AND RESTORATION IMPEDIMENTS

Throughout coral management agencies and authorities, a number of common impediments prevent appropriate and prompt ecosystem restoration. These include untrained personnel, lack of necessary equipment, and the lack of standardized techniques.

Many smaller resource management agencies do not have personnel specifically dedicated to damage assessment or restoration activities. Thus when incidents occur, untrained personnel are required to manage the situation without necessarily having experience or understanding of the processes and legal constraints involved, or having appropriate training to use required mapping and assessment tools. NOAA has largely addressed this in the FKNMS, where seven personnel are specifically dedicated to damage assessment and restoration. Other personnel may be pressed into service as caseloads require, but someone with appropriate training and experience always leads the effort. Staffing does remain an issue in some other NMS sites, where resource injuries are not as frequent, and even within the FKNMS it can be difficult to ensure that all personnel remain trained on new or updated equipment or safety requirements (e.g., new dGPS units, mapping [GIS] software or digital photo/video gear). This includes not only injury assessment work, but also certification as an NOAA Working Diver and in boat handling practices. Given that law enforcement personnel are often the first on scene and play a critical role, they too must be trained to undertake the initial documentation of an injury.

Personnel should be prepared to provide advice on the most appropriate salvage methods or document an injury in a legally defensible manner. Some examples of appropriate salvage include avoiding the use of propulsion systems where propeller wash can scour the bottom, use of ground tackle to provide maneuvering and pull, using ingress paths where possible, choosing appropriate tow paths, even the building of access platforms in limited circumstances.[4]

Lack of appropriate equipment, specifically vessel availability, underwater photo/video cameras, and current electronic positioning equipment (dGPS) for mapping and relocating injury areas, can also be problematic. In the FKNMS, access to boats is generally not a problem, and all equipment has recently been upgraded, but in times when multiple incidents occur, other commitments intervene, or vessels are down for maintenance, boat scheduling conflicts may delay assessing injuries in a timely fashion. This can be a much more difficult issue when dealing with deepwater or remote sites that may require multiple days offshore to complete the necessary assessment dives. Often the amount of equipment, staffing, and time needed for such incidents can be daunting.

Coral injury assessment and restoration is a relatively new field. As with any new field, there are numerous innovations and changes in technology that can improve one's ability to do injury assessments and restorations. Unless all staff are equally trained, inconsistencies in assessment techniques and methodologies may develop; this is not necessarily a problem until and unless it compromises an NRDA case.

Often restoration plans use an approach or technique that has worked in another site; this is particularly true for injuries that have a significant structural component. In some instances adapting a technique to site-specific circumstances can require extensive coordination with marine engineers to determine whether the restoration design and materials will perform as expected. Very few individuals have extensive training in coral restoration design and implementation if it involves anything more than "simple" reattachment. The types of materials and techniques that are used in coral environments are quite specialized. Larger projects may require the use of heavy equipment such as barges, cranes, cement trucks, specialized pumps, surface or mixed gas dive support services and other specialized techniques. The stringent requirements for operations in a marine protected area such as a national marine sanctuary are not typical for most marine contractors, who may be needed to implement larger restoration projects. Permitting agencies and reviewers are generally unfamiliar with these types of activities as well. Education of these individuals takes time that can result in loss of live coral tissue. Implementation of poor designs may lead to additional costs down the road as midcourse corrections and additional repairs are required.

If an injury is left unrestored for extended periods, particularly over winter or storm seasons, there is an extremely high likelihood additional injury will result. This secondary injury is generally of two types: resuspension of loose materials within a structural injury and the subsequent scouring of adjacent unimpacted reef in a "sandblasting" effect (thereby exacerbating the impacts on surrounding healthy tissue), and tissue death from smothering or shading by debris. These changes to the injury site often necessitate alterations to a restoration plan, from just increasing the amounts of necessary materials to changing entire design elements. These types of changes may also in turn trigger additional permit reviews, necessitate additional or different contractors, and almost always increase costs.

With the exceptions of photographic documentation (both aerial and subsurface) and visual inspection, activities associated with injury assessment and particularly restoration in the Sanctuaries will require permits. Timely permitting is imperative to facilitate timely assessment and restoration. Some of the activities associated with the assessment and emergency response phases are covered under "blanket" sanctuary management permits, such as the deployment of measuring tapes and transect lines and placing quadrats and handling corals for reattachment; however, these generally only cover activities for NOAA personnel. Larger-scale, much more intrusive and manipulative activities, such as installation of permanent marking monuments, deployment of manmade reef replacement structures, or filling of trenches and scars with quarried limestone material, require a sanctuary permit on a case-by-case basis. Also, certain activities, such as removal of large amounts of loose coral rubble with a barge and crane (i.e., dredging), may fall under the regulatory authority of another government agency, such as the Army Corps of Engineers (ACE). Time delays in this part of the process can have direct negative impacts on the success of the restoration. NOAA has worked closely with the ACE to have coral and seagrass restoration activities covered in a nationwide blanket permit. Conferring with state agencies for National Historic Preservation Act and Coastal Zone Management Act consistency rulings also takes time. This type of permit coordination or "streamlining" significantly improves timeliness in restoration implementation.

Delays in implementation lead to increased costs due to basic inflation, increased injury area, and higher staff costs for assessment and redesign time. Large sites are more likely to have significant changes to the injury footprint over time. Settlements must include sufficient contingencies to address injury increases, inflation, and increased construction costs as well as monitoring. Settlements may also not provide adequate funding for long-term monitoring, a critical component in determination of restoration effectiveness. Graduated settlements can be an option to fund initial expenses while paying the resource management agency back over a prescribed number of years. In some instances, the responsible party will not have any insurance or sufficient assets to pay for costs. Unless a resource management agency has an alternate funding source available, such injuries often go unrestored.

9.5 INJURY TYPES AND INJURY ASSESSMENTS

Accurate quantification and documentation of vessel-inflicted injuries are critical to the success of any legal claims or actions taken by the trustee agency against a responsible party, as well as providing a solid foundation for primary restoration and monitoring planning. Although the size, shape, and location of reef injuries vary considerably, the basic elements of an assessment are essentially the same for all incidents. The primary objectives of any assessment are to:

- Pinpoint the geographic location of the site (georeferencing)
- Accurately document the size and geometry of the injury
- Describe the nature of the injury (e.g., dislodging and overturning of coral colonies as opposed to crushing and fracturing types of impacts)
- Identify and quantify the corals and other benthos impacted (which includes description of the reef habitat type)
- Quantitatively describe the adjacent unimpacted reef substrate (i.e., determination of what was lost to the incident through examination of the intact portion of the same reef and extrapolating that information to the injury)
- Qualitatively document the injury (photographically and/or with videography) before camouflaging by biofouling marine organisms occurs

Finding submerged cultural or historical resources at an injury site adds some additional assessment, recovery. and curation issues beyond the scope of many injury assessment personnel and should be handled where possible by underwater archeologists.

Vessel damage to reefs results primarily from contact between the vessel's hull and the reef substrate, but impacts from running gear and propellers can produce significant injury as well. Injuries typical of a smaller vessel are scraping and gouging of reef surfaces (both living and nonliving), overturning and displacement of loose or weakly cemented reef components (e.g., churning or plowing of coral rubble on the reef flat), breakage and scattering of outer portions of nonliving reef framework, breakage and fracturing of corals, and the dislodging and toppling of entire coral colonies. The condition of the vessel can also play a role in the ultimate extent of injury, as a hull which is already weak or structurally compromised in any way can, through wave action and pounding on the hard reef substrate, begin to breach, and in some cases, completely disintegrate. In this event, the contents of the vessel spill out onto the surface of the reef, where components of the vessel's rigging and running gear, as well as hull fragments, often join them. This usually compounds the extent of injury to the reef, as this debris can migrate away from the vessel's final resting site and cover, shade, and abrade corals that might otherwise have been out of harm's way (thus further substantiating the urgency for removing a grounded vessel from the reef as quickly as possible).

The methodologies and equipment used to collect and analyze assessment data may vary from case to case and in some instances must be tailored more or less to the characteristics of a specific incident. It is not within the scope of this discussion to attempt to describe all of the possible injury scenarios and assessment techniques that may be encountered by those involved in the management of coral reef resources. A broad summary of injury types and potential restoration options can be found in Table 9.1. Experience, logic, practicality, and sound judgment must guide the assessors in their choices of assessment tools and methods appropriate to the situation. However, the observation and analysis of hundreds of vessel contact injuries to coral reefs occurring within the Florida Keys and Flower Gardens National Marine Sanctuaries over a period of two decades have led to classifying these impacts into three broadly defined groups, based on vessel size. In general, three basic categories of reef injuries by vessels exist:

- Groundings by small- to medium-sized vessels (less than or equal to 75' [23 m] in length)
- Groundings by large vessels (those greater than 75' [23 m] in length)
- Large vessel (freighters and other ships) anchor damage

While not always the case, when vessels in the small-to-medium category run aground, the damage generated tends to be much less extensive than that inflicted by a commercial freighter or other type of ship. Vessels in the ≤ 23 m category are relatively shallow-draft vessels that usually draw less than a couple of meters of water (in many cases a meter or less), although some sailboats may have keels that further deepen their drafts to beyond 2 m. It follows that reef groundings involving boats in this size class occur on shallow patch reefs and the shallowest zones of bank and fringing reefs, i.e., the reef crest, the reef flat (back reef rubble zone), and the shallowest portions of the fore reef. The configuration and condition of this size-class vessel can play a major role in the amount of resulting injury from a grounding incident. Given similar conditions and circumstances, a 10-m fiberglass-hulled, shallow-draft sport fishing vessel is likely to cause less injury than a similar-sized sailboat with a heavier composition hull and a keel that draws 2 m.

The proximity of many coral reefs to major shipping lanes places them in jeopardy of impacts from large vessels. In fact, such collisions have been a chronic situation for centuries on a number of the world's reefs, historically spawning local industry to deliberately wreck vessels for their valuable cargo. In spite of technological advances in modern charting and navigational electronics, large commercial vessels can, and on occasion, do, stray from a safe course and end up hard aground on reefs, almost invariably resulting in catastrophic injury to coral resources. Individual

TABLE 9.1
Injury Categories vs. Restoration Options

				Primary Restoration Options						
	Injury Category	**Characteristics**	**Examples and Alternative Terms**	**Rubble Removal and/or Stabilization**	**Framework Cementation**	**Framework Infill**	**Relief Reconstruction**	**Coral Reattachment**	**Biological Enhancement[a]**	**No Action**
Reef Framework	Fracturing	Cracks through surface framework, loose reef rock blocks	Cracking Splitting	X	X		X			X
	Displacement	Displacement of reef rock by hull, anchor, or propwash, includes back reef rubble	Gouging Grooving Hull scars Blowout Craters	X		X	X			X
Coral Tissue and Reef Framework	Toppling	Detached, but otherwise intact colonies	Tumbling	X				X		X
	Destruction	Colony obliterated	Scraping Lesions	X	X		X		X	X
Coral Tissue and Cover	Abrasion	Soft tissue damage only, no or minimal skeletal damage, bottom paint residue	Gashes						X	X
	Gouging	Soft tissue and colony skeleton damaged	Fracturing Cracking Splitting				X		X	X
	Fragmentation	Colony split, or breakage and dismemberment of branches	Crushing					X		X
	Destruction	Colony obliterated		X	X		X		X	X

[a] Including but not limited to coral transplantation, larval seeding, chemical attractants, and removal of bottom paint or other toxic substances.

injuries from small vessels appear superficial in comparison to the magnitude of a container ship or oil tanker grounding.

Vessels of this size range are of sufficient draft, mass, and momentum to destroy thousands of years' worth of accumulated reef structure in a matter of minutes. Due to their typically deep draft (sometimes 10 m or more), the contact and resulting damage from ships will more likely occur in deeper water, thus affecting reef structures unreachable by small- to medium-size class vessels. Massive, centuries-old coral colonies can be broken off their bases and scattered in disarray. Other similar corals will often be thoroughly fractured and shattered, looking as if an explosive device had been detonated from within. Large fragments of reef framework may be cleaved from the sidewalls of coral spur formations and toppled into adjacent grooves and sand channels. The heat of friction created as the vessel grinds its way to a halt can leave large smears of bottom paint "branded" onto flattened reef surfaces. The final resting area beneath the ship's hull (the "footprint") is often expressed as large (up to hundreds of square meters) flattened areas where all of the reef's living and nonliving components have been ground and pulverized into a flat, planar surface (sometimes referred to in assessment jargon as "the parking lot"). Deep stress fractures from the impact can cause weakening of the reef framework to a depth of a meter or more. The gravel, cobble, and boulder-sized rubble generated by the "bulldozing" effect of the ship's hull can be plowed up into berms a meter or more deep around the inbound grounding track and the final resting area. Alternatively, this material may spill down the flanks of coral spurs to form talus piles that bury the flanks of the spurs. In both cases, living corals and other sessile benthic organisms will be smothered and perish, further increasing the collateral damage from the incident.

It is also common for the vessel captain to attempt to "power-off" of the reef, using full forward and/or reverse throttle in a fruitless effort to free the ship. The enormous propeller wash generated by their powerful propellers has sufficient force to excavate craters in the reef substrate that can be several meters deep and tens of meters in diameter. The ejecta blown from these craters (or "blowholes") can settle as mounds or berms of coral rubble, which if left in place can smother living coral or be resuspended and "sandblast" adjacent unimpacted reef areas during major storm events. If the vessel is allowed to stay in contact with the reef for a matter of days or weeks, wave energy acting on the hull can cause the vessel to rock to and fro, creating a "mortar and pestle" action that can grind deeper and deeper into the already-compromised reef structure, removing even more material that has taken millennia to develop. Additionally, the ship's hull will shade surviving corals in the areas immediately adjacent to the vessel's final resting location and in the topographic lows beneath the hull. If this situation continues for more than a few days, these corals can become stressed and often will bleach (expel their symbiotic zooxanthellae) completely.

Another factor that can significantly increase the amount of indirect, collateral reef damage is the use of heavy, steel-braided towing cables by salvors attempting to extricate the grounded ship from the reef. The draping of these cables across the unimpacted reef as they are made fast to the grounded vessel, and the subsequent tensioning of the same cables as the salvage vessel(s) begin pulling on the ship, can result in grievous injury to live corals in the forms of breakage, abrasion, and dislodging of entire colonies. In some instances, the damage from nonbuoyant lines and cables has rivaled or even surpassed the amount of injury from the grounded vessel itself. Needless to say, a grounding of this immensity is one of the most serious emergencies a coral reef resource manager or assessment biologist will ever face.

The third category of reef injury is that of large vessel anchor damage. The most significant feature of ship anchor impacts is the depth of water in which they usually occur. Large commercial vessels will often deploy their anchors on submarine ridges and rises near active seaports as they await clearance from the port authorities to make entry into the harbor. This allows the vessel operators to drop less anchor chain, saving time and effort when they weigh anchor and head for port. Unfortunately, these topographic highpoints are sometimes deeper-water (20 to 30 m+) coral reef structures, which can be devastated by these activities. The massive anchors and chains that these ships employ weigh thousands of pounds, and when dropped on or dragged across a reef can easily make deep abrasions in the surfaces of living corals; dislodge, overturn, and scatter entire coral colonies; and plow up tens to hundreds of square meters of coral reef substrate as easily as a tractor tills topsoil. Because of the depth at which this category of injury typically occurs, large-vessel anchoring damage assessments necessitate the use of SCUBA by the investigators. Decompression limits dictate that the dives by assessment personnel be kept considerably shorter than those for shallower water incidents. Consequently, multiple dives spread over several days are usually required to collect all of the necessary measurements and adequately document the full extent of the anchor-inflicted damage.

Information regarding the extent and severity of the damage that must be collected is the same for all three categories described above. However, there are differences in the techniques; tools and scale of effort may differ. Therefore, a discussion of the three categories of injuries that highlights their similarity, while simultaneously contrasting the differences between them, is best presented in the framework of a "generic" vessel impact, wherein the step-by-step assessment procedure is outlined and the aspects specific to the three size categories are noted. This assessment protocol outline (located in Appendix A of this chapter) assumes a medium- to large-sized vessel grounding, and differences in procedure or technique for smaller vessel groundings and large vessel anchoring injuries are noted appropriately.

9.5.1 Restoration Alternatives and Implementation

The ultimate goal of the response and injury assessment process is to implement an effective coral reef restoration. NOAA has worked over a number of years to develop a series of techniques that are matched to the types of injuries, proven effective, and both ecologically viable and aesthetically pleasing. Experience in dealing with delays on a number of larger groundings has prompted NOAA to develop some standardized tools, such as a PEIS that summarizes what is known about the specific coral ecosystems, categorizes the injury types, and also describes and evaluates the varying restoration options. The PEIS analysis can be used for both primary and compensatory restoration planning. Table 9.2 references the criteria NOAA uses for evaluating which restoration options are most appropriate for a given injury site.

The goal of an NRDA is to assess the extent of the injury, recover response and damage assessment costs, and implement primary and compensatory restoration to make the environment and public whole as a result of the injury. "Primary restoration" refers to restoration activities at the actual injury site. The goal of restoration activities is to return injured coral communities to preinjury conditions. For coral reef communities, "baseline" refers to the level of ecological services (type, quality, and coverage of coral) existing prior to the incident. Baseline conditions are typically measured via field assessment techniques in the undisturbed reef communities bordering the grounding site.[1] In many circumstances, without primary restoration, the injured reef communities are subject to redisturbance by storms that could slow recovery and/or expand the size of the injury. "Compensatory restoration" refers to a restoration project, typically offsite, that compensates the public for the lost interim ecological services as a result of the time it takes for the original "primary" injury to return to baseline conditions. Funds collected for small compensatory restoration projects may be pooled together for the implementation of a larger compensatory restoration project.

TABLE 9.2
Criteria for Evaluating Coral Restoration Options

Criteria	Definition
Technical feasibility	Likelihood that a given restoration action will work at the site and the technology and management skills exist to implement the restoration action
Recovery time	Measures that accelerate or sustain the long-term natural processes important to recovery of the affected resources and/or services injured or lost in the incident
Additional injury	Likelihood that the requirements, materials, or implementation of a restoration action minimizes the potential for additional injury
Aesthetic acceptability	Restoration alternative that creates substrates and topography that most closely resemble the surrounding habitat and minimize visual degradation
Site-specific context	Environmental conditions at the site including but not limited to location, extent, and severity of the injury; hydrological characteristics; seagrass species composition; and other social and resource management concerns

NOAA has been and continues to evaluate reef restoration alternatives from throughout the Caribbean and other reef areas, and is working to refine those and develop new techniques for the specific habitats and management requirements of the NMSP.[5–9] This has been an evolutionary process, as NOAA personnel strive to implement restoration techniques that provide the greatest overall environmental and socioeconomic benefit in the most cost-effective manner. This includes the use of both new and modified physical (structural) restoration techniques as well as innovative biological techniques. Traditional biological methods have involved only the transplantation of coral fragments; as the understanding of reef ecology continues to expand, the options for biological restoration have expanded as well.

The goal of the restoration activities at an injury site is to recreate, to the extent practicable, suitable habitat that enhances recolonization on the injured reef without attempting to completely compensate for the loss of the thousands of years of growth that created the physical structure of the reef.

9.5.2 Programmatic Environmental Impact Statement (PEIS) Alternatives

The goal of the coral PEIS is to describe a range of coral restoration techniques that may be used for both primary and compensatory restoration projects that may be implemented in the sanctuaries and to analyze the environmental and socioeconomic impacts of each technique. This document is meant to be inclusive of current technology, although NOAA is constrained to implementing proven techniques.

The coral restoration projects described in the PEIS will be implemented within the FGBNMS and the FKNMS with funds collected through NRDA settlements for injuries to corals in those sites. The NMSA establishes a hierarchy of how funds received under Section 312 of the NMSA are to be spent, prioritizing the restoration of the injured site, restoration of similar injuries within the same sanctuary, or other management purposes. As a matter of practice, NOAA generally restores the injured site or a similar one within the same sanctuary, rather than diverting funds for general management responsibilities.

The PEIS is intended to:

- Provide a comprehensive review of restoration techniques and potential impacts
- Provide a preapproved NEPA analysis applicable to future injuries
- Provide a framework for comparative analysis of emerging technologies, when appropriate

Therefore, the PEIS speeds restoration plan development, NEPA review, and permitting and thus is expected to allow faster implementation of restoration projects and save more coral. Restoration techniques considered in the PEIS are described below.

The Council on Environmental Quality regulations prescribe inclusion of the no-action alternative. It serves as a benchmark against which the preferred action can be evaluated. The no-action alternative would leave the reef in its current condition, allowing natural recovery processes to occur. The no-action alternative could have two general outcomes: natural recovery on a longer time scale or further deterioration of the reef system.

A no-action alternative relies on natural settlement and growth of coral species, natural sediment filling of blowholes and propeller scars, and natural redistribution and consolidation of loose rubble. A no-action alternative can increase the risk of secondary injury to nearby coral communities from the unstable conditions created by fractured framework and loose rubble from crushed corals and reef framework at the grounding site. Progressive deterioration of coral injuries from storm and hurricane force wave energy has been documented to expand coral injuries in such cases.[5,6] The no-action alternative is most often used for grounding cases when NOAA determines that an injury site is likely to recover in a short period of time with a low likelihood of injury expansion, or where other social, environmental, or logistical considerations dictate that no action is the best course.

Determination of a no-action alternative for primary restoration does not necessarily preclude the assessment of compensatory restoration requirements.

9.5.3 Physical Restoration Alternatives

It has long been known that in order for corals to successfully recruit and grow in a disturbed area, the substrate on which the corals settle must be conducive to their establishment.[10] The presence of coral rubble and other reef debris, particularly coral sand, in the injury area is not conducive to coral settlement because water movement resuspends the sand, resulting in coral colonies being buried and killed. As for rubble, if storm- or current-generated water movement is forceful enough, coral fragments can serve as projectiles to be smashed into the nascent coral colonies, fracturing or dislodging them.[11] Some authors have termed this a downcurrent a "domino effect," i.e., where one colony is shattered, its fragments can cause a cascade of damage to nearby colonies before coming to rest on the benthos. Anthropogenic disturbances that produce great quantities of coral rubble and fine sediment material include large-vessel groundings and anchoring.

Research has revealed that rubble created by mechanical injury to a reef (usually in the form of hurricane damage) is the major component that goes into making up future reef framework development,[12,13] especially for *Acropora* and other species that reproduce through fragmentation. Following mechanical injury and rubble generation, the rubble layer is often quickly colonized by crustose coralline algae (e.g., *Neogoniolithon* sp., *Tenarea* sp., *Fosliella* sp., *Lithophylluma* sp., *Porolithon* sp., and *Hydrolithon* sp.) as well as noncolonial encrusters such as foraminifera, bryozoa, and serpulids.

However, stabilization of the rubble (as well as some fraction of the sediment) does not occur until these organisms have had a chance to cement the reef debris together,[12,13] a process that can take many years. Therefore, unstable substratum will sometimes persist for long periods of time, and thus coral reestablishment will likewise be delayed.[14] Also, in cases of anthropogenic rubble creation, the amount of rubble created is often far larger than might be created naturally and subsequently requires much longer time frames for stabilization.

In general, managers have advocated clearing the site of rubble and doing it as soon as possible after the injury.[15–17] This prevents collateral injury and also allows preservation of large pieces of live coral. Small quantities can be removed by divers using baskets, lift bags, etc., while larger amounts require using a barge with suction, guided by underwater divers. Divers can also cache live corals nearby until a full restoration can be undertaken, thereby reducing tissue loss.

When larger areas of framework are injured, the rubble and sediment may be too large for removal and need to be stabilized in place. To stabilize and fill in small (less than 1 m^2) areas, limestone boulders can be placed in the injured area and then stabilized with a tremie pour of concrete around them. The boulders can be designed and stacked so as to recreate the look of and replace the relief destroyed. Plastic composite rebar (which is lighter, easier to use, and more durable than steel) can be placed in the concrete for improved attachment between boulder and concrete layers. The result looks like naturally bare coral heads, which enhances the opportunities for benthic recolonization and provides holes or openings as habitat for cryptic organisms.

For large areas, preformed modules ("modules") can be designed to fill depressions and create relief. Modules are typically created from a combination of materials, including concrete, limestone boulders or reef rubble, and steel or composite reinforcing rods. Modules are similar to boulders but larger and heavier, and they can be designed in a variety of shapes and sizes, depending on the nature of the restoration needs. Once in place, tremie is poured around them for stabilization. In addition to creating relief, because of their greater weight, modules can also be valuable for stabilizing underlying framework injury.

Several alternatives prepared by filling tubes with concrete are collectively grouped as "concrete pillows." These methods are used to protect the perimeter of injured areas from further deterioration. The pillows, mattresses, or tubes are constructed out of reinforced material, such as Kevlar, to add external support and filled with nonseparable marine concrete. The filling operations are fairly complex and would typically require concrete preparation onsite and require a downhill gradient. Due to the complexity of installation, pillows are not likely to be a cost-effective option for small-scale restoration projects of the type typically required in sanctuary sites. This technology also entails the use of unnatural materials (Kevlar), which is generally not preferred in sanctuary sites. Injuries for which this would be a reasonable alternative include a large blowhole or trench, especially in a situation where additional restoration may not be implemented for an extended time and the perimeter needs temporary stabilization.

Gabions are prefabricated steel or tenser grid cages containing loose rubble that are used to stabilize the surface of disturbed areas and can be useful for bringing a hole back up to surrounding grade. The standard design for gabions consists of a very sturdy plastic webbed mattress that is filled with gravel, or sand if an internal filter fabric is used. Due to their weight, it is uncertain whether gabions would crush or crack the underlying reef structure. Gabions are generally considered a temporary measure or interim relief until a more permanent structure can be constructed. The technique entails the use of unnatural materials and has the potential for inflicting additional framework injury. Injuries for which this would be a reasonable alternative include the presence of a large blowhole or trench that requires filling, or large rubble zones.

Revetment mats consist of concrete blocks, usually 1 ft^2, interconnected by flexible polypropylene, Kevlar, or similar cables. These mats are usually assembled on land and then installed in place from a construction platform using a crane and spreader bar. They are relatively flexible structures that conform to the shape of the natural contour. As with gabions, it is uncertain whether revetment mats would crush or crack the underlying reef structure. Revetment mats were used in the restoration of the *Contship Houston* site in 1997.[18] These mats later required some relocation and rearticulation after Hurricane Georges caused the leading edge of one mat to be lifted and broken apart; in future uses, the mats will be more securely fastened to the bottom. Injuries for which this would be a reasonable alternative include sites with a large rubble zone in a low energy environment where rubble needs to be stabilized. These mats are not aesthetically similar to surrounding area, an important consideration in areas where aesthetics, as well as habitat value, are of concern.

9.6 BIOLOGICAL RESTORATION

Even though physical restoration has traditionally been the focus of coral reef restoration, the length of time required for a coral community to develop can mean that even after the physical environment has been restored, the injured site will take many years to resemble a coral ecosystem. Therefore, NOAA is very interested in the use of techniques to enhance the rate at which the coral community will redevelop in an injured area.

Transplantation of coral is an effective technique to facilitate the redevelopment of coral communities on injured or degraded reefs.[19–21] In addition to onsite coral fragments, there are several other potential sources for donor corals, including coral pieces collected from small-scale groundings and held until restoration is implemented; corals taken from artificial substrates such as seawalls, piers, or pilings; corals at risk from human disturbance (i.e., dredging); and selected coral colonies that have been held for culturing, research, or rescue purposes and are now available for return to the reef. The colonies used for transplantation are not taken from donors on surrounding reefs and thus do not increase injury to reef habitat. This ensures that restoration is not accomplished at the expense of surrounding habitats.

All fragments are attached using either underwater epoxy (for fragments <10 cm) or Portland Type II cement, and can be attached directly to any solid substrate. In addition, branching corals may also be reattached to appropriate substrate using wire and/or cable ties or by wedging fragments into small crevices and voids. Transplantation would be primarily of small coral colonies or fragments. Every effort is made to transplant the diversity of species found at the original site, but the diversity and number of each species transplanted ultimately depend on the donors available at the time of transplantation.

In addition to the scleractinian corals, gorgonian coral colonies can also be transplanted. The transplanted colonies may be entire dislodged colonies, or clippings may be taken from local mature colonies without endangering the donor colony. The branching nature of gorgonians means that clippings can readily be taken from local donors without causing additional habitat injury. Gorgonian transplantation is accomplished using epoxy to reattach an entire colony to solid substrate, or by inserting the axis of a clipping into a hole and affixing it in place using epoxy.[22]

Although several new techniques currently being developed and tested should expand the range of biological alternatives available for restoration, NOAA is limited in its use of experimental techniques and can only consider for use techniques that have a proven track record of successful implementation. Descriptions of several of these new techniques are provided below; NOAA is actively following the testing and evaluation of these techniques and will analyze them further once they are proven successful.

One relatively new idea for structural restoration of coral reefs is the *in situ* formation of semiartificial substrate by electrolysis.[23] The calcium and magnesium ions present in seawater can be deposited on a template by electrochemical deposition. In studies by van Treeck and Schuhmacher,[23–25] the template utilized was nongalvanized steel mesh (e.g., chicken wire). A current is passed through the template, which acts as a cathode where brucite [$Mg(OH)_2$] and aragonite ($CaCO_3$) are "electroplated." The accreted material consists mainly of calcium carbonate and is chemically similar to reef limestone. These investigators also combined elements of biological restoration with deposition by inserting coral nubbins into the mesh. Although spontaneous settlement by benthic organisms can take place on the precipitated substrate without the nubbin transplantation, use of the nubbins is a recovery-hastening method. This technique has not been used in the field but may be appropriate for small-scale projects, depending on the coral species impacted.

There are several new technologies under development that focus on increasing the settlement of coral larvae; NOAA is particularly interested in the potential use of larval flypaper and settlement tents. The larval flypaper technique[26,27] requires the placement of chemical stimuli (metamorphic inducers) on the reef surface to attract coral larvae. The use of the chemical stimuli increases the likelihood that coral larvae will settle onto the selected reef location. This has been tested successfully in Pacific areas but has not been used with Caribbean corals. The settlement tent technique uses larvae collected during spawning events and held in a lab until they are competent to settle. The larvae are then introduced into a fine-mesh net enclosure deployed over the injured area and held for another few days until they have settled onto the substrate. Constraining the larvae in a tent for a few days over the reef provides greater likelihood of their settling on the selected reef substrate than if they were free floating and subject to current and tidal influences. These two techniques take different approaches to capitalizing on the natural abundance of coral larvae throughout the Keys, with the intent of increasing the density of settlers at a specific site. Chapter 12 describes this technique in greater detail.

In some situations, the entire ecosystem is in decline and although direct restoration of an injury would be appropriate, it would not be as effective as predicted due to the declining community structure. In these cases, action to address the community structure would be more beneficial than injury-specific restoration. At least one method being developed to reverse this decline is the reintroduction of the long-spined sea urchin, *Diadema antillarum*. In the mid-1980s, *D. antillarum*, a keystone predator throughout the Caribbean, suffered a basin-wide mass mortality;[28,29] it has yet to recover in most locations. *Diadema* is an herbivore; it played an essential role in regulating algal communities on coral reefs. Since the die-off, many reefs have suffered from algal overgrowth that is believed to be a direct result of loss of urchin populations. The reintroduction of *Diadema* has the potential to reduce algal overgrowth, increase settlement of juvenile corals, and increase overall percent cover of stony corals, thus increasing the overall health of coral habitats. Two efforts are underway to investigate the potential for *Diadema* restoration:

translocation of wild juveniles and transplantation of cultured juveniles. Initial work translocating wild juveniles has shown that translocation of even small numbers can show dramatic benthic changes in 1 year.[30] Rearing cultured *Diadema* has shown some promise, but transplanting cultured juveniles to the field has had limited success (see Chapter 12) and efforts to refine these techniques continue.

9.7 CONCLUSIONS

NOAA uses a variety of tools developed over many years of fieldwork and casework to implement small- to large-scale coral restorations in an effective, efficient, and timely manner. Our efforts to standardize assessment techniques have enabled collection of consistent and legally defensible data and provided a more accurate basis for restoration. Recent efforts to upgrade field equipment and maintain appropriate staff training have contributed to that effort. Several ongoing activities, including streamlining permitting processes and publishing the PEIS, will continue our efforts to facilitate coral restoration. It is hoped that this short description will be of interest and utility to other coral resource managers and personnel in their efforts to assess coral injuries and to implement restorations.

APPENDIX A: INJURY ASSESSMENT PROTOCOLS

This appendix assumes a medium- to large-sized vessel grounding, and differences in procedure or technique for smaller vessel groundings and large-vessel anchoring injuries are noted appropriately.

9.A.1 Injury Location: Vessel Still Aground

The location of the injury should be precisely documented for evidentiary purposes and to facilitate injury assessment and restoration planning. When the vessel is still aground, the following steps should be taken.

9.A.1.1 Georeferencing and Site Marking

Acquire dGPS coordinates of the grounding site, including the location of the endpoints of the path the vessel traveled, the positions of the ship's bow and stern, and enough data points to sufficiently locate/describe the major features of the injury for comparison to like coordinates taken after the removal of the vessel. This will help determine whether any additional injury occurs during the salvage operation. Search at the location of the last set of coordinates (the grounding site) to ensure that all damage has been located. Use all sets of coordinates and a navigational chart to plot out the path traveled before the vessel ran aground to identify possible areas of additional resource injury. Examination of the areas adjacent to the vessel while it is still on the reef can yield critical information for the removal of the vessel, as well as for subsequent damage assessment.

In addition to getting dGPS fixes of the various reference points, marking of the site with visual reference aids should also be performed. For small-vessel situations, simply marking key features with weighted buoys may be sufficient, whereas marking the area of a freighter grounding site requires a much more robust marking system because the marking devices will need to be in place over the duration of what will undoubtedly be a somewhat lengthy assessment process. In this situation, large, sturdy buoys affixed to the periphery of the grounding site with heavy lines running to permanently installed eyebolts or some other heavy-duty anchoring system are in order. These markers may also be useful when interpreting aerial photographs of the site. This kind of sturdy, highly visible marking system is a must in ship anchoring investigations, as they occur in deeper water, thereby making relocation of the site difficult. It also aids the returning assessment personnel in minimizing the amount of time looking for features and thus maximizing precious bottom time.

9.A.1.2 Vessel's Direction of Travel and Compass Bearings

Use the information and a navigational chart to identify possible areas of resource injury. Not applicable to anchor injury incidents.

9.A.1.3 Description of Grounding Incident

Response officials or trained law enforcement personnel should prepare an in-depth description of the grounding incident. Use the information and a navigational chart to identify possible areas of resource injury. This may or may not be of value in anchoring incidents.

9.A.1.4 Plotting on a Navigational Chart

Use a combination of dGPS coordinates, direction of travel, compass bearings, and a description of the incident to plot onto a navigational chart and extrapolate for the entrance and exit paths. A portable (laptop) computer with navigational and/or GIS software installed can greatly expedite this step. From the chart identify possible areas of resource injury.

9.A.1.5 Overflights

Helicopters/fixed wing aircraft can be used to fly over a grounding location if good weather (good visibility, relatively calm sea state) conditions exist; they can assist in locating an injury and entrance and exit scars.

Photo document the site. If true vertical photography is not possible with available aircraft, high-quality oblique photos and even video can be useful for evidentiary documentation (particularly if the vessel is still hard aground) and provide useful information for subsequent assessment activities.

Injuries from small-to-medium size class vessels are often too small-scale to appear very prominently in aerial photography, although an overflight may assist in the location of additional injuries, especially in cases where damage inflicted was heavy.

In ship anchoring situations, overflights may or may not provide any useful information, as the depth of water will usually preclude sufficient light penetration for adequate visibility.

9.A.1.6 Samples/Photographs and Video

Samples and photographs or video should be taken of relevant debris (pieces of propellers and other running gear, spilled or jettisoned vessel contents, bottom paint smear samples, etc.) and injuries to reef substrate. Acquisition of this documentation at this time is particularly important as vessel debris can be lost over time due to storm events, and reef injuries will become obscured by epibiotic overgrowth. All appropriate chain of custody protocols should be followed.

9.A.1.6.1 Paint Chips/Scrapings/Pieces of Vessel from Injury Sites

- Assign exhibit number of sample before collection.
- Take photos of sample before collection.
- Use underwater writing tablet to display the exhibit number for that sample in photos.
- Take close-up photo from distance to give a reference of the location of the sample being collected.
- If possible use the same tool (i.e., dive knife) to collect paint samples from features too large to collect.
- Document which tool was used to collect each sample. Store samples in sea water and refrigerate.

9.A.1.6.2 Document Compass Bearings of Paint Striations or Grounding Scars

Take photos of each compass bearing obtained. This information can be used to help determine exit/entrance direction.

9.A.1.6.3 Evidence Should Be Numbered in the Field

Carry a copy of the evidence log in the field (allows for consistent numbering).

9.A.2 Removal of Vessel

Coral reef resource managers and/or response personnel must be prepared to address considerations for removing vessels specific to coral reef groundings, such as removing vessels at high tide; using incoming tract for outbound removal (unless there is a shorter, less injurious exit path to suitably deep water); not allowing the grounded vessel to "power-off" (using forward and/or reverse throttle to attempt extrication); obtaining as much towing and salvage assistance as necessary; conducting operations so that damage is minimized (including retrieval of large-vessel anchors that will cause the least amount of additional injury); etc.

Other possible concerns/considerations are:

- Determination of direction and whether likely to refloat at high tide
- Whether sea state or impending foul weather will result in more serious resource injury if the vessel is left aground until the optimal tide is reached
- Whether refloating/removal is likely to cause spills or breakup of vessel
- Whether necessary to lighter/remove fuel/cargo
- Ensuring that salvage lines do not cause collateral damage

9.A.3 Injury Location: Vessel No Longer Aground

When the formerly grounded vessel has been removed or has extricated itself, the following information and techniques may be used to establish the footprint of a grounding. If the assessment personnel were not present while the vessel was aground, it is important that a member or members of the initial response team accompany them to the site to aid in locating it as well as to point out primary injury features. They may also be able to provide valuable information about the direction of travel of the vessel prior to grounding, the final resting area and orientation of the grounded vessel, the direction of removal of the vessel, etc.

9.A.3.1 dGPS Coordinates, Including the Location of the Grounding Site or the Path the Vessel Traveled

Conduct a thorough search at the location of the last set of coordinates (the grounding site) to determine if all injury has been accounted for within the presumed terminus of the grounding track. This is also a good opportunity to make a preliminary swim-through of the grounding track to determine the boundaries of the injury, and it helps plan for the establishment of an assessment baseline (or baselines) through the most reasonable (usually the longest or most linear) axis of the grounding path. If GPS coordinates are not available, use the information listed below.

9.A.3.2 Vessel's Direction of Travel and Compass Bearings

Use the information and a navigational chart to identify possible areas of resource injury. In the case of large-vessel anchoring situations, it must be kept in mind that although the vessel was presumably not underway while at anchor, she may have dragged anchor during its deployment, or the ship may have swung while at anchor, causing a "windshield wiper" effect, that is, a broad arc of damage as opposed to a simple linear one. Topic 9.A.3.3 below may help assessors make these kinds of determinations, but diver inspection is the surest method to locate the totality of the injury.

9.A.3.3 Description of Grounding Incident

A sketch of the grounding track and the footprint of the vessel's hull should be made while the injury is "fresh" to assist assessment personnel with the determination of the boundaries of the area of impact if a complete assessment is to be conducted at a future date. Also, a general description of the type(s) of injuries within the sketched area should be included to further aid any future determinations of extent of injury.

9.A.3.4 Plotting on a Navigational Chart

Use a combination of dGPS coordinates, direction of travel, compass bearings, and a description of the incident to plot onto a navigational chart; extrapolate for entrance and exit paths. From the chart identify possible areas of resource injury. Electronic charting software or GIS installed on a laptop computer can greatly expedite this procedure.

This may or may not be of particular assistance in the assessment of large-vessel anchor damage, although it is possible that plotting the position of the anchor site on a detailed bottom-chart may suggest nearby topographic highs that could have been impacted if the anchor dragged or the ship swung while at anchor.

9.A.3.5 Search Patterns by Boat

Using towlines attached to the stern of the assessment vessel, pull snorkelers behind the boat at a slow, steady rate of speed; most efficient if pulling two people. While pulling snorkelers, boat driver should run parallel transects through search area. Have floats and weights ready to drop onto injury sites when located by snorkelers. Lobster/crab trap floats or similar buoys tethered to large-sized lead dive weights with sturdy line, such as parachute cord, are suitable for temporary markers. To avoid duplication of effort, buoys of a different color may be placed to delineate previously inspected transect lines.

This technique may or may not be of value in the case of deeper-water anchoring injuries. If the assessment team has access to them, electrically powered underwater scooters can be of great assistance in covering large areas at depth in delineating the total extent of damage.

9.A.3.6 Overflight

Helicopters/fixed wing aircraft can be used to fly over the grounding location. If the weather is good and sea conditions right, they can assist in locating large-scale injury features, as well as entrance and exit scars. True vertical, low-altitude, high-resolution aerial photographs of the site can be utilized to analyze, quantify, and interpret the injuries, especially if they are of the extensive type most often caused by freighters and other large commercial vessels. These photos can be digitized and imported into computer-aided drafting (CAD) software packages that allow the injuries to be outlined and measured, or better still, fed into a GIS system that will not only allow for making area measurements, but will also georectify the image and plot it onto a nautical chart, a benthic map of the area, or a preexisting photographic basemap, or layer it with any combination of these. Aerials need to be georeferenced to be useful in calculating injury size. dGPS-referenced aerial photo targets (at least a meter across for most digital interpretation applications, painted white or some other highly reflective color) can be affixed to the bottom in shallow water at key points within the injury site. Alternatively, large, brightly colored surface buoys can be tight-line rigged to the bottom with anchoring devices within the grounding site and their locations fixed by dGPS. However, the types of targets that are fixed to the bottom provide more precise referencing, as tidal movement, wind and waves, or currents do not influence them. Whichever aerial photo target methodology is used, it is generally agreed upon by those who work in the field of geospatial analysis that the more targets the better, but a minimum of four reasonably spaced targets is necessary to properly georectify a true vertical remote image.

Aerial photography is generally of little aid to the assessment of ship anchoring injuries, due to the depth of water in which they typically occur. Likewise, aerial photography may or may not be appropriate to document smaller vessel groundings, as the injuries are usually not of a scale that can be resolved even in low-altitude photographs. Aerial photography is relatively expensive and therefore should only be utilized in situations appropriate for its use. However, there is no question that in the case of large-vessel groundings producing extensive footprints and scars, high-quality aerial photos may be a cost-effective alternative to hours and hours of field measurements by assessment personnel, as well as serving as dramatic qualitative documentation of a catastrophic event.

9.A.4 Injury Assessment

9.A.4.1 Types of Possible Injury

- *Barge tow cable markings*: Characteristic injury from cable strikes will show as striations in coral and on hard surfaces. Cables can impart extensive damage to stands of branching corals. The cable will also injure or remove other invertebrates such as soft corals or sponges.
- *Scraping from bottom of vessel*: This injury can be, depending upon the configuration of the hull, characterized by dislodged and/or overturned coral colonies and reef rock; "flat topped" areas where all living invertebrates and framework have been removed to a certain depth leaving a flat surface; or broken and shattered ("exploded") coral colonies. Larger vessels will create extensive scarified areas of reef substrate, or "footprints."
- *Scars from propellers or keel of vessel*: Fractured, crushed, or trenched substrate where vessel's props, running gear, and keel contacts substrate, generating very characteristic injuries. The fractures that penetrate the nonliving coral framework can sometimes be less obvious than the broken and crushed reef substrate.
- *"Blowhole" (excavation crater)*: Created by vessel propeller wash as vessel operator attempted to "power-off" the reef. In some cases may represent the most serious component of reef framework injury from a vessel-grounding incident.
- *Sediment pile or berm*: Created from displaced material (ejecta) generated by a vessel that excavates a "blowhole." These features can also be generated by vessel's hull as it plows up and throws to side loose reef substrate material.
- *Displaced features*: Whole coral colonies or sections of coral heads that have been dislodged and displaced, or large pieces of reef framework that have been fragmented and removed from their original location.
- *Anchor injuries*: Can include scrapes, strikes, fracturing, or toppling of individual heads. May also be characterized by broad, arcing areas of injury caused by the chain as the vessel swings on anchor ("windshield wiper effect").

9.A.4.2 Establish Baseline

A master transect line should be established centrally through the length of the injury site (usually the longest axis, starting at the beginning of the inbound path and terminating at the distal end of the final resting place). This baseline can serve as a referencing tool for mapping the site; making measurements of area of injury; location and documentation of important injury features outside of the main body of injury; and relocation of key features during subsequent assessment field visits; as well as aiding future restoration and monitoring efforts. In some cases, such as injuries involving very long inbound paths, a number of very large injury features, or a "confused" site characterized by segmented grounding tracks that change direction and/or scattered areas of intermittent damage, multiple baseline transects may be required to adequately describe the site.

Even if aerial photography and digital analysis will be used to measure large injury features, a master transect line is indispensable for detailed field mapping and groundtruthing for remote analysis. Of course, a baseline is an absolute must if aerial photography is not an option and manual field mapping and measuring is the only alternative.

For small-to-medium vessel groundings, where it is known that the assessment will only require a day or two of fieldwork, then a temporary baseline may be adequate. This can be as simple as a fiberglass meter tape stretched tight and made fast to stakes (sharpened stainless steel rods work well) driven into the reef substrate at the beginning and ending points of the transect.

Assessments of large-vessel and ship's anchor injuries often require multiple site visits to complete, and the baseline may need to stay in place for days, weeks, or even months. Additionally, there is a high likelihood that restoration and monitoring efforts will be instituted in the future, and the original master transect is an invaluable tool for reorienting to the site. Therefore, the transect endpoints and incremental points between them need to be permanently established by installing stainless steel or fiberglass stakes in the reef substrate, using a predrilled hole and quick-setting cement or specially formulated underwater epoxy. Once the master stakes have been established, fiberglass meter tapes can be deployed between them as baselines, or sturdy cordage marked off in highly visible increments can be used and left in place over longer periods of time. A surface buoy tethered to the stakes will help in relocation on future visits.

In all cases, the baseline endpoints should be georeferenced by taking dGPS coordinates from the surface as vertically as possible above them and a compass heading for each baseline noted.

9.A.4.3 Site Characterization

Site characterization includes mapping and describing the injuries. Once the master transect line(s) have been established, these tasks can be accomplished:

Mapping/referencing of injuries: If the injury or injuries are broad and extensive, as is typical of larger-vessel grounding and anchor damage, and aerial photography is not available, then the method known as "fishbone" mapping can be utilized. In this technique, one member of the assessment team follows the baseline carrying another meter tape and at an appropriate increment (usually 1 m, unless the injury is very large and a broader increment is acceptable), the end of the meter tape is made fast to the baseline and a second member of the team uses the tape to make a perpendicular measurement out to the edge of the injury. Repeat this procedure for the other side of the baseline, recording the measurements and corresponding incremental positions along the baseline on waterproof data sheets. From this data, a base map of the injury site can be generated.

This technique can be used for mapping larger injury features of small- and medium-sized vessel groundings, such as inbound tracks, final resting area footprints, and propwash excavation craters (blowholes). In many instances, smaller-vessel damage is somewhat scattered, disjointed, and intermittent. In this case, individual injuries can be referenced back to the baseline with a distance and compass-bearing notation. This is a useful technique for mapping features associated with large-scale incidents that lie outside the main body of damage.

Numbering injury features: Many people, including the investigating officer, biologist and, possibly, restoration contractors, will need to reference the injured sites and significant injury features. By enumerating these immediately as they are located, all references to the sites will correspond.

Use metal stakes or spikes hammered into the bottom with numbered heavy plastic or stainless steel tags attached, or glue numbered tags directly to prepared surfaces of reef substrate.

Nature of injuries: Using the terminology and definitions outlined in Section 9.A.4.1, describe the injuries found within the site(s). Any given site may display a variety of damage types.

Determine species impacted: Identify injured corals and other organisms to species, if possible.

Determine percent cover of species and species diversity: Use accepted methods of determining percent cover, species diversity, and relative abundance of live corals and other benthic reef species. These include visual inspection estimates of randomly placed meter square quadrats (Braun–Blanquet or similar methodology), digital analysis of photo quadrats, or use of video transects point-counting computer software.

In many cases, especially when larger vessels are involved, the coral reef substrate within the injury site may be in such disarray (or in some instances, completely obliterated) that any determination of preexisting species cover, diversity, and relative abundance is impossible. In these instances, this information can be gathered from unimpacted reef substrate immediately adjacent to the injury site, as it is reasonable to assume that this closely approximates the pregrounding condition of the area within the injury site.

Obtain an estimate of rugosity/three-dimensional relief: Use standard, accepted method to determine rugosity (index of substrate complexity) within the injury site and in the adjacent unimpacted reef as a measure of loss of three-dimensional relief. A commonly used, widely accepted technique for rugosity determination is the chain-and-tape method. This is accomplished by deploying a length of surveyor's tape (usually 10 m) across the reef substrate of interest, then laying a light chain (brass works well) along the length of the tape, paying out as much chain as required to conform to the substrate profile over the same distance of the meter tape. The measurement of the length of chain used is divided by the straight-line distance. The resulting number is an index of "wrinkledness" (rugosity) of the substrate. The closer this number approaches unity, the flatter the surface. Average indices within injury sites can be compared to those acquired from immediately adjacent uninjured areas to describe the change in substrate complexity caused by the grounding or anchoring incident.

9.A.4.4 Quantification of Injuries

Measurement and quantification of injuries can be accomplished by a variety of methods, and the particular techniques used are determined by the scale of the injuries, as well as the tools and technology available to the assessment personnel. Some of the most commonly employed techniques are as follows:

Small quadrats and meter tapes: Measuring and quantifying minor injuries can be made quickly and efficiently with the use of a surveyor's (meter) tape, a meter stick, and a quadrat (a square constructed of polyvinyl chloride (PVC) pipe, 0.25 m^2 in area, makes a convenient field model). The tape and rule can be employed to make length and width measurements, and the small quadrat can be utilized to visually estimate areas of minor injury.

Meter-square quadrat: A square PVC pipe frame, 1 m on each side, is used as a visual aid for quantifying area of larger injuries, as well as arriving at cover/diversity estimates, as described in Section 9.A.4.3. The percent cover figure from the adjacent uninjured reef can be multiplied by the square meter area of injury measurements to calculate the amount of living coral (and other benthic species) lost to the incident.

Photo documentation: Scaled, vertical photographs of injury bounded by a quadrat can also be used to measure injuries, although it can be somewhat difficult to get a truly vertical, undistorted photograph under typical field conditions.

Fishbone: The fishbone method of injury mapping, as described in Section 9.A.4.3, can also be used to arrive at area measurement. The polygon generated from the measurements collected can be plotted onto a sheet of scaled graph paper and the area within the polygon calculated. Similarly, this data can be plotted with CAD software (such as Canvas™ by Deneba) and the area within the polygon calculated by the program's Object Specifications function. The polygon created can then be used to produce a diagram of the injury site, with significant features, scale information, north arrow, and other relevant information included.

Aerial photos: If injury features are of sufficient scale, ground-truthed, true vertical aerial photos of known and verifiable scale can be digitized and imported into CAD programs that allow the user to orient the photo to a compass direction (usually due north) and trace the outlines of the features of interest. The square meter areas described by the polygons thus created can be calculated almost instantaneously by the program's Object Specs function.

Digitized, georeferenced aerial photos can also be imported into GIS programs where the injury area can be oriented, measured, and layered with other geospatial information for that locality.

9.A.4.5 Qualitative Documentation

This primarily involves photo and/or video documentation of significant features of the injury. Photo/video documentation of the injury with an underwater still camera or video camcorder in a waterproof housing provides a permanent record of the "look" of the site that can be referred back to at any future phase of the assessment, legal case development, or restoration and monitoring planning. High quality 35-mm or digital images can be used to review the injury if questions regarding species identification arise or if details about the nature of the injuries are lacking from field notes. Perhaps most importantly, photographs and video acquired while the site is "fresh" provide a dramatic, graphic representation of the injury for anyone involved in the review of the case that could not observe the damage firsthand ("a picture is worth a thousand words").

Video transects: Following the baseline from the beginning (the zero mark), make slow, steady, continuous video (digital format if at all possible) of the injury, stopping to shoot extra footage of particularly important features. If videography of key features takes the camera away from the baseline, return to the same position along the transect when the digression was made, noting the meter mark upon return (get close-up of meter increment) and proceed with videography along the baseline. The field of view of the housed camcorder may not encompass the breadth of injury, and it may be necessary to slowly, steadily pan or "sweep" from one side of the injury to the other as the transect line is followed, thereby documenting the entire scene. When the procedure is followed carefully and methodically, an invaluable visual archive of the entire injury site is created.

Qualitative footage of the unimpacted reef adjacent to the injury should also be acquired as a visual reference of what the site would have presumably looked like before the incident.

Newer generations of digital video camcorders have the ability to make high-resolution digital still images in addition to the regular video mode. This can be a very useful feature when high-quality still images of key features are needed, or, when used in conjunction with an attached framing rod and visual scale, images for quantitative interpretation can be acquired. Alternatively, reasonably good-quality still images for illustrative purposes can be "lifted" from video with the frame-grabbing feature included in most video editing software for personal computers.

Software packages, such as Ravenview™, enable vertical video transects to be digitally rectified for moderate variances in distance from subject and angle of view, and then "stitched" or collaged together to form a digital image map of the site. Other available software allows vertical video transects to be used in statistical pointcount interpretation.

Still images: High-quality digital or 35-mm images of the site are desirable in that they can be utilized as an adjunct to species identification or for use as illustrations included in the production of injury assessment documents. With the proper leveling and framing apparatus and inclusion of a visual scale device, still images can be used to make measurements of injury or sizes of coral colonies, as well as use in making photo quadrats that can be used to estimate percent cover and relative abundance calculations within the injury and the adjacent unimpacted reef.

When making video, use a color-correcting filter in water deeper than 3 to 5 m, and when taking close-ups with a 35-mm still camera, a strobe should be used to provide necessary color and detail. When taking close-ups of either injured or unimpacted organisms, a second photo should be taken at distance to show the location of the organisms in reference to other areas. An item of known length should be used in photos to show scale. Photos should be developed and labeled as soon as possible;

the same goes for downloading/filing digital images and archiving ("duping") video. Negatives and slides, as well as digital images and video, should be treated as "crime-scene" evidence.

9.A.4.6 Field Note Management

Field notes can be made on underwater slates or on waterproof paper. Prepared data sheets can be photocopied onto waterproof photocopy paper, such as NeverTear® sheets by Xerox™, which can be filled underwater with an ordinary pencil.

In addition to data specific to the injuries, information to be included with every day's site visit includes:

- Name of site and latitude/longitude reference
- Date
- Time of day
- Tide state
- Average depth of water over site
- Sea state
- Names of assessment personnel (or others assisting)

As is the case with physical samples and photo/video documentation, data sheets become legal documents and must be treated the same as other pieces of evidence.

9.A.4.7 Create Photo Mosaic of Injured Area

In good weather, when time is limited and other technology is not available, the mosaic can be completed by one diver/snorkeler swimming on the surface holding another diver at a consistent arm's length below the surface. The diver below the surface takes continuous overlapping photos. The diver on the surface guides the photographer through the injured area. This technique allows for consistent depth for photos.

9.A.4.8 Daily Field Notes

Upon completion of each day's field work, the biologist should create sketches of each new injured area. On each sketch the biologist should include:

- The location and direction each photo was taken
- Photo numbers that correspond to the numbers on each labeled photo
- The direction of each video transect
- The location where evidence was collected and corresponding evidence number (obtained from the investigating officer)
- A drawing of the injured site with the location identification of injured organisms

The storage of original copies of sketches, drawings, and field notes should be done in a manner compatible with evidence handling practice.

REFERENCES

1. Hudson, J.H. and Goodwin, W.B., Assessment of vessel grounding injury to coral reef and seagrass habitats in the Florida Keys National Marine Sanctuary, Florida: protocol and methods, *Bull. Mar. Sci.*, 69, 509, 2001.
2. National Oceanic and Atmospheric Administration, Environmental assessment, *R/V Columbus Iselin* grounding site restoration, Florida Keys National Marine Sanctuary, Monroe County, FL, prepared by TetraTech, Fairfax, VA, 1999.

3. National Oceanic and Atmospheric Administration, Environmental assessment: *M/V Wellwood* grounding site restoration, Florida Keys National Marine Sanctuary, Monroe County, FL, 2002.
4. Michel, J. and Helton, D., Environmental considerations during wreck removal and scuttling, Meeting of the American Salvage Association, 2003.
5. National Oceanic and Atmospheric Administration, Environmental assessment for the structural restoration of the *M/V Alec Owen Maitland* grounding site, Key Largo National Marine Sanctuary, Florida, prepared by Industrial Economics, Inc., Cambridge, MA, 1995.
6. National Oceanic and Atmospheric Administration, Habitat equivalency analysis: an overview, http://www.darp.noaa.gov/library/pdf/heaoverv.pdf. Damage Assessment Center, National Ocean Service, NOAA. Silver Spring, MD, 1995 (revised 2000).
7. National Oceanic and Atmospheric Administration, Environmental assessment for the structural restoration of the *M/V Elpis* grounding site, Key Largo National Marine Sanctuary, Florida, prepared by Industrial Economics, Inc., Cambridge, MA, 1995.
8. National Oceanic and Atmospheric Administration, *R/V Columbus Iselin* grounding lost use damages Looe Key Reef, Looe Key National Marine Sanctuary, prepared by Norman Meade, NOAA Damage Assessment Center, Silver Spring, MD, 1996.
9. National Oceanic and Atmospheric Administration, Florida Keys National Marine Sanctuary, Final management plan/environmental impact statement, Volume II, NOAA. Silver Spring, MD, 1996.
10. Crisp, D. and Ryland, J., Influence of filming and surface texture on the settlement of marine organisms, *Nature*, 185, 119, 1960.
11. Endean, R., Destruction and recovery of coral reef communities, in *Biology and Geology of Coral Reefs. Volume III: Biology 2,* Jones, O. and Endean, R. Eds., Academic Press, New York, 1976.
12. Perry, C., Reef framework preservation in four contrasting modern reef environments, Discovery Bay, Jamaica, *J. Coastal Rec.,* 15, 796, 1999.
13. Blanchon, P., Jones, B., and Kalbfleisch, W., Anatomy of a fringing reef around Grand Cayman: storm rubble, not coral framework, *J. Sed. Res.* (*A: Sedimentary Petroleum Processes*), 67, 1, 1997.
14. Riegl, B. and Luke, K., Ecological parameters of dynamited reefs in the northern Red Sea and their relevance to reef rehabilitation, *Mar. Poll. Bull.,* 37, 488, 1998.
15. Gittings, S., et al, The recovery process in a mechanically damaged coral reef community: recruitment and growth, in *Proc. 6th Int. Coral Reef Symp.,* Choat, J.H., et al., Eds., Townsville, Australia, 1998, 225.
16. Zobrist, E., Coral reef restoration and protection from vessel groundings, *Gulf Res. Reports,* 10, 85, 1999.
17. Hudson, J.H. and Diaz, R., Damage survey and restoration of *M/V Wellwood* grounding site, Molasses Reef, Key Largo National Marine Sanctuary, Florida, in *Proc. 6th Int. Coral Reef Symp.,* Choat, J.H., et al., Eds., Townsville, Australia, 1998, 231.
18. ECM/Hudson Maritime Services LLC, *Contship Houston* restoration project: second reef monitoring program event, including new baseline survey, Report to the National Marine Sanctuaries Division, Office of Coastal Resource Management, NOAA, Wilton, CT, 1999.
19. Gleason, D.F., Brazeau, D.A., and Munfus, D., Can self-fertilizing coral species be used to enhance restoration of Caribbean reefs? *Bull. Mar. Sci.,* 69, 933, 2001.
20. Jaap, W. C., Coral reef restoration, *Ecol. Eng.,* 15, 345, 2000.
21. Jaap, W. C. and Morelock, J., Two-year monitoring report: Soto's Reef restoration project, George Town, Grand Cayman Island, British West Indies, Lithophyte Research and Marine Research Associates, 1998.
22. Stratton, A, An innovative method to restore coral reefs using gorgonian clippings for transplantation. M.S. Thesis, University of Maryland, College Park, 2001.
23. Van Treeck, P. and Schuhmacher, H., Initial survival of coral nubbins transplanted by a new coral transplantation technology — options for reef rehabilitation, *Mar. Ecol. Prog. Ser.*, 150, 287, 1997.
24. Van Treeck, P. and Schuhmacher, H., Mass diving tourism — a new dimension calls for new management approaches, *Mar. Poll. Bull.*, 37, 499, 1998.
25. Van Treeck, P. and Schuhmacher, H., Artificial reefs created by electrolysis and coral transplantation, *Estuarine Coastal Shelf Sci.,* 49, 75, 1999.
26. Morse, D. et al., Morphogen-based chemical flypaper for *Agaricia humilis* coral larvae, *Biol. Bull.,*186, 172, 1994.

27. Morse, A. and Morse, D., Flypapers for coral and other planktonic larvae, *BioScience*, 46, 254, 1996.
28. Bak, R., Recruitment patterns and mass mortalities in the sea urchin *Diadema antillarum*, in *Proc. of the 5th Int. Coral Reef Cong.*, Harmelin-Vivien, M., et al. (Eds.), Antenne Museum-EPHE, Moorea (French Polynesia), 1985, 267.
29. Lessios, H.A., *Diadema antillarum* 10 years after mass mortality: still rare, despite help from a competitor, *Proc. R. Soc. Lond. B,* 259: 331, 1995.
30. Nedimyer, K. and Moe, M.A. Jr., Techniques development for the reestablishment of the long-spined sea urchin, *Diadema antillarum*, on two small patch reefs in the upper Florida Keys, Report to NOAA, 2003.

10 Aesthetic Components of Ecological Restoration

Jessica Tallman

CONTENTS

10.1 INTRODUCTION

The beauty of coral reefs has appealed to the human psyche for centuries. In the late 1800s, Alfred Russel Wallace described a coral reef as "one of the most astonishing and beautiful sights I have ever beheld. The bottom was absolutely hidden by a continuous series of corals, sponges, actinae [sea anemones] and other marine productions, of magnificent dimensions, varied forms, and brilliant colours … . It was a sight to gaze at for hours, and no description can do justice to its surpassing beauty and interest."[1] Beholding a relatively healthy coral reef, such as the one described by Wallace, or the one pictured in Figure 10.1, can be like swimming amid a rainbow. Slices of spectrum undulate in an enticing dance due to whirlpools of fish spiraling in parades of color and flashing their iridescent scales against the sunlight. Coral growth forms the basis of this ecosystem, having developed for thousands of years into magnificent pinnacles, caverns, towers, and peaks. Amid these castles one feels the only way to explore is to slip into a school of vibrant fish and glide alongside their finned rhythm. Multitudes of people visit coral reefs to experience immersion in this world of abundant and color-infused marine life.

Due to their beauty and biological bounty, coral reef ecosystems provide services to millions of people. Businesses based near reefs create millions of jobs and contribute billions of dollars in tourism-dependent revenue annually to the world's coastal regions. Unfortunately, the rapidly declining health of the world's coral reefs is largely due to human use. Some diving destinations have far exceeded their ecologically bearable capacity.[2] Significant effects from tourism are especially found on major reef-building hard corals.[3] Yet marine-related tourism is the cornerstone of many national economies, and the loss of this revenue is too great to end reef-related activities in order to preserve coral reefs.[4] The reefs of southeast Florida, alone, are deemed to have an asset value of $7.6 billion.[5] Sustaining this industry that generates billions of dollars is dependent on maintaining existing reefs and rehabilitating the lost splendor of this fragile ecosystem.

Every restoration project has a unique set of goals, and while it is hoped that biological success is the top priority of all undertakings, aesthetics should also be of importance. Incorporating beauty

FIGURE 10.1 A natural coral reef off of Discovery Bay, Jamaica (Joshua Idjadi, with permission).

and viewer interest in project plans is profitable for tourism-dependent operations and also raises awareness of conservation issues. Conversely, aesthetics increases recreational activities that can be damaging to restoration efforts and therefore should not be incorporated into all restoration projects. Where appropriate, however, increasing aesthetic beauty of the world beneath the waves enhances the experience of reefs for millions of people.

10.2 ARTIFICIAL REEFS

Artificial reefs are major tools of reef restoration efforts. Artificial reefs have been in use since the 1700s, when Japanese fishermen shaped bamboo into structures attractive to fish. In 1830, artificial reefs were used in the United States for the first time off of South Carolina. These reefs of pine boughs were also meant to attract fish. Fish aggregative devices (FADs) are still very common uses of artificial reefs; however, in the present day, artificial reefs may be used for many other reasons as well. For example, some artificial reefs are a means of generating tourism as they are created to heighten interest and beauty for snorkellers and scuba divers. Today, "materials of opportunity" such as oil rigs, sunken ships, vehicles, and railway tracks are dumped into the ocean, or their presence there is exploited. With a greater environmental effort in mind, purpose-designed reefs sculpted out of concrete, limestone rocks, metal, or bamboo are more favored for building ecologically sound reefs on which marine life and human visitor appeal can thrive.

10.2.1 MATERIALS OF OPPORTUNITY

Materials of opportunity are commonly sunk to form artificial, or "found," reefs. Many scientists believe artificial reefs made of surplus materials are ocean pollutants destroying marine habitat.[6,7] Some found reef material contains substances that may be toxic to biota such as heavy metals in cables, paints, and alloys.[8] Additionally, artificial reefs are often longlasting, if not permanent intrusions into the benthic habitat.[9] Permanent structures with such repercussions have little, if any, ecological value. Even if reefs made from materials of opportunity did not pose an ecological threat, they do little to restore coral cover. They are also often inappropriate for providing fish recruits with necessary protection found in a complex habitat due to large open spaces and lack of small recesses within the structure. Without fish, algae will likely grow to cover the structure, thus smothering coral recruits.

FIGURE 10.2 A poor substitute for a coral reef (©2004 www.ecoreefs.com, with permission).

As seen in Figure 10.2, reefs made of materials of opportunity can be unsightly. People dive and snorkel in tropical waters to see splendor, not someone's old Chevy parked where there could be a flourishing reef. Many dive operations, hotels, and other tourism-dependent businesses are opting for more ecologically sound artificial reefs. In a study on visitor experience and perceived conditions for tourists of the Great Barrier Reef, natural conditions were overwhelmingly the most important influence on enjoyment.[10] The ecological need to mimic natural conditions coincides with aesthetic desires of reef visitors. The tourism value of found reefs falls far short of the alternatives, purpose-designed artificial reefs or increasing efforts to preserve naturally healthy reefs. Found reefs contribute very little ecologically and more often than not, have negative impacts on the visual beauty of the ocean floor. Improving upon these surplus material reefs with purpose-designed reefs will help to achieve not only ecological but aesthetic goals.

10.2.2 Purpose-Designed Reefs

There is a considerable demand for artificial reefs to be aesthetically pleasing from the moment they are first deployed. It can, however, take decades for an artificial reef to look like and take on some of the functions of a natural reef. Researchers, however, have discovered methods to hasten the time it takes for a dive site to be aesthetically intriguing. Today, aquascapists design snorkel trails and transplant benthic organisms to artificial or damaged reefs, and underwater sculptors create topographically diverse artificial reefs. The present technology in ocean engineering and construction allows almost any restoration design to be implemented.[11] With scientific basis, reefs can be constructed that not only meet scientific objectives but also create a beautiful environment for visitors to experience. Knowledge of successful and unsuccessful restoration methods allows for a multidisciplinary approach and incorporation of goals that may have been overlooked when reef restoration was a science in its beginning. Socioeconomic aims can now take on greater importance in restoration plans by creating beautiful reefs not only attracting marine life but also human visitors.

Harold Hudson (aka "Reef Doctor") of the Florida Keys National Marine Sanctuary has been researching restoration modules for years. His research has resulted in a terrific example of an artificial reef built with biological goals as a top priority while being mindful of the aesthetic beauty of a natural reef formation. Along with the FKNMS, Hudson has created modules of preformed concrete casts. Figure 10.3 shows an example of the units made of small limestone boulders, rebar, concrete, and sand. The conglomerates form a medium able to take on a number of shapes. In the case of the NOAA *Wellwood* restoration project, the casts mimic the spur-and-groove formation of the natural reef at the grounding site, thus mindful of the necessity to create maximum habitat for marine life.

Hudson has also worked to develop benthic organism transplanting techniques.[12] Nearly every client funding a coral reef restoration effort where tourism is involved insists on aesthetic components and natural aesthetics takes time. Transplants can hasten the process and are utilized

FIGURE 10.3 A material primarily of concrete and limestone boulders allows designers to create modules that imitate natural coral reef topography (FKNMS/NOAA, with permission).

in a number of restoration efforts using purpose-designed artificial reefs and damaged natural reefs. A major goal of transplanting is to reduce the time it takes for an artificial reef to take on some of the functions of a natural reef. Transplanting also coincides with aesthetic goals as the reef mimics a natural reef and thus speeds the attraction to humans. Divers, like the one pictured in Figure 10.4, employ quick-setting cement to adhere hard and soft corals to reef substrate. Bolting corals has also been relatively successful but unsightly.[13] To further emulate the natural reef, the finished concrete surfaces of some modules have smaller rocks pressed into them or rocks secured with mortar. Aesthetics are heightened with these methods by reducing concrete exposure. Increasing the amount of exposed limestone surfaces, coral, sponges, and algae not only catalyzes biological recovery, it lessens the visually stressing sight of human damage.

Another group utilizing the transplanting technique is the Reef Ball Foundation, Inc. Reef Balls™ are concrete artificial reefs that now number over half a million throughout the world. Aquascapists of the Reef Ball Foundation Coral Team employ their creative and scientific backgrounds to cover reefs with coral, algae, and sponge transplants. Reef Balls are designed to form the base of a reef system as any coral structure would. Transplanted organisms are selected with specific traits for various ecological reasons such as compatibility, resistance to sedimentation, and fast growth rates. Aesthetic selection of specific corals helps to create a design attractive to divers and snorkellers. Colorizing a reef in designed schemes is carried out in much the same way a landscaper would sculpt grounds. Thus, interesting topography and coloration patterns can be somewhat predetermined by transplanting, at least for the time being before successionary processes progress.

FIGURE 10.4 Transplanting live coral (FKNMS/NOAA, with permission).

FIGURE 10.5 A standard Reef Ball (www.reefball.com, with permission).

The morphologies of Reef Balls are also designed to function as the natural bases of coral fossils and rock outcroppings. Marine-friendly concrete is poured into fiberglass molds containing buoys. In the standard design, a buoy is surrounded by various sizes of inflatable balls that, once removed, result in openings. As with natural reefs, Reef Balls vary in shape, hole density, and texture. This flexible technique allows for custom designs to be easily created. Figure 10.5 and Figure 10.6 are examples of the diversity of forms into which Reef Balls can be molded. There are 10 different sizes of Reef Balls and 20 different patterns. More specific options are also available because patterns can be mixed and matched into endless combinations. As reef visitors desire variation in topography, reef restoration efforts can increase aesthetics by including a range of Reef Ball™ morphologies.

The Reef Ball Development Group, Ltd. has created a number of purely aesthetic components that can be incorporated into their projects. Reef Balls can be fashioned into blowholes, and lights can be added for a nighttime effect. A project where visual impact was of top priority was the creation of an underwater sculpture park at Club Cozumel Carib. The Reef Ball Development Group, Ltd. taught cement artisans eco-safe construction methods to use in creating underwater art. The structures were immediately attractive to human visitors and will develop into reefs that will hopefully grow to mimic some of the functions of a natural reef. Due to the combination of creativity and aptitude for ecological restoration, Reef Balls are incredibly popular with marine tourism operations throughout the world.

FIGURE 10.6 A layer cake Reef Ball (www.reefball.com, with permission).

FIGURE 10.7 Installation of EcoReef modules (©2004 www.ecoreefs.com, with permission).

From the time they are first deployed, one of the most morphologically convincing purpose-designed reef types is the EcoReef® module designed by president Michael Moore. Clients are drawn to the branching coral morphology of EcoReefs. Modules look instantly organic, as seen in the picture of an EcoReef team installing the EcoReef modules in Figure 10.7. EcoReefs, Inc., has the ability to create many coral forms. Efforts to date have been using the branching coral module, seen in Figure 10.8, while other morphologies continue to be researched. Diving operations, hotels, and other tourism-dependent operations are finding this design to be an aesthetically favorable solution.

The branching coral morphology is designed with both short- and long-term aesthetic goals in mind. It has been seen through time, branching corals are among the first corals to recolonize a disturbed area.[14] These pioneer species provide a large surface area for coral recruits to settle and protection for fish and invertebrates. EcoReefs catalyze the reef successionary process and break down, as would any branching coral, when other coral species take over. Thus, the common EcoReef module

FIGURE 10.8 An EcoReef module ready for deployment (©2004 www.ecoreefs.com, with permission).

is an appropriate shape, considering a natural reef's successionary process. The reefs are not meant to last hundreds of years as with other artificial reef modules. As a naturally beautiful reef takes over, the EcoReef performs its duty as a substrate, functioning well both ecologically and aesthetically.

An important function of the EcoReef modules is the ability to create a large ecological footprint in a variety of areas. Many artificial reefs are limited in their impact on the ecosystem because of their small size. Eco Reefs can also be placed on slopes, a rare option for artificial reefs. Eco Reefs not only create something immediately attractive, they do so over a large area no matter what the existing topography.

A smaller footprint with excellent potential for aesthetic grandeur is the mineral accretion method first proposed by Dr. Wolf Hilbertz in 1976.[15] Hilbertz developed the method of *in situ* formation of semiartificial substrate by electrolysis. When a conductive material, such as steel mesh, is connected to a DC power supply, magnesium and calcium minerals precipitate upon the material. This accreted material consists mainly of calcium carbonate that has chemical properties similar to reef limestone. The matrix can be morphed in a multitude of shapes. As with most restoration projects, the structure can be integrated into the natural topography of nearby reefs. Reef organisms benefit from a variety of morphologies from tunnels creating whirlpools that increase food availability to nooks and overhangs offering protection. As with limestone boulders, Reef Balls and EcoReefs, coral fragments can be rescued from damaged reef areas and placed on the conductive material to help speed the coral cover of the reefs. Mineral accretion is very compatible with the desire of some reef projects to be aesthetically interesting.

Dr. Peter van Treeck and Dr. Harold Schuhmacher of the University of Essen are among a number of researchers currently exploring the possibilities of mineral accretion. Van Treeck and Schuhmacher believe that many underwater coastal activities do not require coral reefs.[16] Any three-dimensional structure may be sufficiently attractive to the dive tourists upon whom many economies with reef-lined coasts are dependent. The University of Essen's Department of Hydrobiology has developed a concept for the installation of attractive recreation areas for divers as a way of lessening the impact of divers on natural destinations. Save COral REefs (SCORE) modules are metal structures with platforms. The researchers aim to place these in unobjectionable areas such as sand flats. Van Treeck and Schuhmacher hope these underwater parks will meet the aesthetic desires of divers while lessening biological degradation of natural reefs.

10.2.3 Underwater Artists

A number of professional artists are celebrating the world's marine environment. Ocean-inspired sculptures form reefs for the purpose of creating art while attempting to mimic some of the ecological functions of a natural reef. Some provide an interesting structure, helping to lessen the amount of human traffic to natural reefs, while others are specifically designed to increase habitat for marine organisms. All are creative approaches, enticing visitors to explore the underwater environment.

In 1992, artist Ann Lorraine Labriola produced Stargazer, the world's largest underwater sculptured reef. Stargazer is 200 ft long, 70 ft wide and 10 ft high. It is located 5 miles off the coast of Key West, Florida, between Rock Key and Sand Key. Ten steel pieces each weighing as much as 6 tons and as long as 168 ft are welded together to form Stargazer. The structure is anchored to the ocean floor 22 ft below the ocean surface. When seen from the air, Stargazer points southwest toward the Sand Key Lighthouse. Labriola's piece is a symbol of movement, honoring early mariners who navigated by the stars. Stargazer was immediately attractive to visitors after its installation with its holes cut in the pattern of a star constellation. The sculpture has since attracted more permanent life with a diverse accumulation of marine organisms encompassing the structure. Stargazer has grown to be a much-loved reef system among divers and snorkellers.

An angel is watching over the reefs off of Nassau, Bahamas. The Angel of Harmony, the wings of which are pictured in Figure 10.9, was created by Dave Caudill. Caudill is a sculptor from

FIGURE 10.9 Dave Caudill's Angel of Harmony (www.caudillart.com, with permission).

Louisville, Kentucky who also forms land sculptures, many with an open-weave construction of stainless steel rods. Both above and below the water's surface, Caudill aims to integrate natural beauty with artistic form. He hopes his Angel of Harmony will eventually become one with the seabed and viewers will have to ask Is this a man-made or natural form? A few years after installation, the Angel of Harmony has taken on a full spectrum of colors, encrusted in corals, sponges, and algae. Marine organisms are especially attracted to the hollow steel cylinders that form the wings of the sculpture. Sea creatures dart in and out of these pipes covered with marine growth. People can also be found circling the angel. Located in 14 ft of water, the Angel of Harmony is easily accessed by people who enjoy the sculpture, many playfully mimicking the form by striking an angel pose. The Angel of Harmony is an enticing attraction to the coast of Nassau, augmenting the beauty of nearby natural coral patch reefs.

Bronze sculptures are another means of artistically enhancing the seafloor. They provide little biological habitat yet serve an ecological purpose. Not only do they encourage divers and snorkellers to visit marine areas, thus raising awareness of marine issues, they also create an exciting attraction for divers while they improve their buoyancy skills. Inexperienced divers can have a ruinous impact on natural reefs. *Oceanic Voyagers* created by Dale Evers and *Mermaid* by Simon Morris are some of the most loved underwater sculpture installations in the Cayman Islands. These works compel visitors to seek the beauty of the underwater world that has inspired their artists to create these pieces. Bronze sculptures are a valuable and beautiful union between art and ocean.

10.3 RECOMMENDATIONS

Many believe the design of a reef should be attractive to visitors from the very first days of a restoration effort without depending on coral growth or fish population to create an appealing environment.[17] Designers of reefs take into account a variety of considerations to heighten the interest of a dive site. Naturalness, nonrepetitive topography, and diversity and abundance of species are high on the list of visitor wishes. Familiar species, such as *Acropora* spp. corals, are also key ingredients in the mix of an ecosystem visitors will perceive as healthy and attractive.

It is hard to improve upon the beauty of a natural reef like the one in Figure 10.10. A natural appearance is a goal for many restoration efforts. Using the percent of live coral cover as a measure of marine environmental quality, a study in Roatan, Honduras looked at the correlation of reef attributes with scuba diving behavior.[18] The study found that marine environmental quality is a significant predictor of dive site visitation. Thus, reefs that mimic the beauty of nature are a likely success with eco tourists. Many restoration efforts include ecological goals of creating topographies similar to those of reefs before damage occurred. This is in line with aesthetic desires of divers and snorkellers, who seek naturalness in their eco tourism destinations. Limestone modules have

FIGURE 10.10 A topographically diverse reef off of Discovery Bay, Jamaica (Joshua Idjadi, with permission).

proven to be excellent means to achieve similar topography to that of natural reefs. EcoReefs, as seen in Figure 10.11, mimic the branching form of corals that divers consider a classic shape and associate with high-quality snorkeling and diving sites.[6] Molding concrete and transplanting benthic organisms also aid in restoring natural appearance.

Multidimensional and nonrepetitive habitats are a great attraction of reefs and can be achieved by variation of reef structures. Modules such as EcoReefs and Reef Balls are available in such an assortment of forms, they can create a multitude of shapes across the ocean floor. Using mineral

FIGURE 10.11 EcoReefs mimic branching forms of natural coral (©2004 www.ecoreefs.com, with permission).

FIGURE 10.12 An unsightly artificial reef (©2004 www.ecoreefs.com, with permission).

accretion methods, an entire underwater village could be built. For tourist divers, the more interesting the structure, the greater likelihood of the success as a dive site.[8] Both natural-looking and purely aesthetic reefs that are topographically diverse will pique the interest of visitors. With today's techniques, a designer's creativity is the only constraint on possible morphologies.

High coloration is a guaranteed aesthetic success. Transplanting can quickly infuse a reef with color. Creativity can abound in patterns of colorful biota. As coloration can also serve as visual cues, transplanting specific colors can help to steer divers away from more fragile areas and preserve the beauty of natural reefs. Transplanted biota become a living palette for marine restoration "artists" to beautify underwater restoration projects.

Technology has no doubt advanced to a state where reef restoration projects can combine aesthetics and scientific techniques. Ecological restoration performed by scientists in conjunction with artists can create a reef that satisfies the aesthetic objectives so many tourism-dependent operations require. With so much of conservation driven by tourism dollars, aesthetics are an objective that should be considered in many restoration efforts. No longer do sunken materials of opportunity need to be considered in many restoration projects. These junk reefs tend to permanently structure a reef in an unnatural and unsightly way such as the tire reef in Figure 10.12. Purpose-designed artificial reefs are far more attractive and much more compatible with ecologically sound reef rehabilitation projects. Purpose-designed artificial reefs and mitigation to damaged natural reefs are intelligent approaches to ecological restoration that can create visual excitement for divers. While natural beauty is by far the greatest attraction under the surface of our water world, technology is allowing us to replace some of the brilliance that has vanished beneath the ocean's surface.

ACKNOWLEDGMENTS

I would like to thank Todd Barber, Michael Moore, Dave Caudill, Nadav Shashar, Sue Wells, and Ann Lorraine Labriola for their time spent talking with me. I am also grateful to Joshua Idjadi, and The Florida Keys National Marine Sanctuary for their help in acquiring images.

REFERENCES

1. Wallace, A.R., quoted in Davidson, O.G., *The Enchanted Braid: Coming to Terms with Nature on the Coral Reef.* John Wiley & Sons, New York, 1998, chap. 1.
2. Raphael, A.B., and Inglis, G.A., Impacts of recreational scuba diving at sites with different reef topographies. *Biol. Cons.,* 82, 329, 1997.
3. Tratalos, J.A., and Austin, T.J., Impacts of recreational SCUBA diving on coral communities of the Caribbean island of Grand Cayman. *Biol. Cons.,* 102 (1), 67, 2001.
4. Maglio, C., The effects of environment on artificial reefs in Sarasota Bay. B.A. thesis, Division of Biology, New College, Sarasota, 2001.
5. Johns, G.M. et al., Socio-economic study of reef resources in Southeast Florida and the Florida Keys, Broward County Department of Planning and Environmental Protection, Fort Lauderdale, FL, 2001.
6. Moore, M., personal communication, 2004.
7. Weisburd, S., Artificial reefs. *Sc. News,* 59, Jul. 26, 1986.
8. Wells, S. et al., Toolkit for MPA managers in the Western Indian Ocean. IUCN — The World Conservation Union, 2004.
9. Buckly, R., Artificial reefs should only be built by fishery managers and researchers. *Bull. Mar. Sci.,* 44, 1054, 1989.
10. Schafer, C.S. et al., Visitor experiences and perceived conditions on day trips to the Great Barrier Reef. Technical Report, CRC Reef Res. Cent., 21, 1998.
11. Precht, W.F., Call for abstracts: coral reef restoration and remediation mini symposium 4–20, 10th ICRS. Coral-List. Oct 2003.
12. Hudson, H.J. et al., Building a coral reef in Southeast Florida: combining technology and aesthetics. *Bull. Mar. Sci.,* 44, 1067, 1989.
13. Jaap, W.C., Coral reef restoration. *Ecol. Eng.,* 15, 345, 2000.
14. Precht, W.F., personal communication, 2003.
15. Hilbertz, W.H., Electrodeposition of minerals in sea water: experiments and applications. *IEEE J. Oceanic Eng.*, OE-4(3), 94, 1979.
16. Van Treeck, P., and Schuhmacher, H., Artificial reefs created by electrolysis and coral transplantation — an approach ensuring the compatibility of environmental protection and diving tourism. *Est. Coast. Shelf Sci.,* 49,75, 1999.
17. Shashar, N., Guidelines for deployment of artificial reefs in the Gulf of Eilat. The Ministry of the Environment, Israel, 2004.
18. Pendelton, L.H., Environmental quality and recreation demand in a Caribbean coral reef. *Coast. Mgmt.,* 22, 399, 1994.

11 International Trends in Injury Assessment and Restoration

Greg E. Challenger

CONTENTS

11.1 INTRODUCTION

Trends in coral reef injury assessment and restoration are broad and can encompass many varied goals and objectives. Injury to coral reefs can occur by many means, natural and anthropogenic, and from direct insult and cumulative nonpoint sources. Physical insults to coral reefs include natural events such as hurricanes and typhoons, and injury from anchors, divers, boat hulls, destructive fishing techniques, dredging, coral mining, and installation of underwater pipelines and cables. Restoration projects for physical damage vary based on the degree of injury, environmental factors that influence the potential success of the project, and the means available to those undertaking the project. This chapter will focus primarily on injuries resulting from direct anthropogenic physical insults such as vessel groundings, since much of the current understanding of restoration stems from attempts to restore damages from large vessel groundings.

Understanding coral reef injury magnitude, duration, recovery, and regeneration in the context of the incident setting and how active restoration techniques may enhance recovery are critical to help guide future restoration projects. Trends in injury assessment and restoration techniques are discussed in this context with general examples of lessons learned from recent incidents where relevant. In addition to the biological and technical aspects driving assessment and restoration programs, potential political aspects may interact in positive and negative ways affecting the assessment and restoration planning process. Examination and resolution of these issues will be important to ensure appropriate, efficient, and practicable restoration in future incidents throughout the world.

11.2 VESSEL GROUNDINGS

Vessels have run aground ever since people began using the sea for transportation and commerce. The Florida State University Index of Shipwrecks lists over 1000 wrecks in nearshore waters of the U.S. State of Florida alone between 1828 and 1911.[1] This does not include vessels that grounded and were removed from their strand. Casualties have become far less frequent among large vessels in the past century with improved vessel construction, navigation, and charting and the prevalence of navigational aids. However, increases in vessel traffic, uncontrollable forces of nature, and occasional equipment malfunction and error mean that large vessel groundings still occur. For example, the United States Coral Reef Task Force[2] lists 13 major groundings in the Florida Keys since 1984, nine major groundings in the Hawaiian Islands National Wildlife Refuge since 1970, and three major groundings in Midway Atoll and Rose Atoll National Wildlife Refuge. In Australia, the Great Barrier Reef Marine Park Authority reports five major groundings in the Park since 1995.[3] In many regions, vessel groundings are not widely reported. Additionally, the traffic of small recreational and commercial vessels has increased in many locations. The Florida Department of Environmental Protection estimates that approximately 500 small-vessel groundings in the Florida Keys are reported annually.[2] The number of unreported groundings could be substantially higher. The impetus to avoid and minimize damage to the environment from vessel groundings has historically relied upon the financial interest of owners for their vessels and cargo. Shipowners could ill-afford vessel casualties on a frequent basis. While technological advancements in navigation have protected from losses, accidents and casualties that cause substantial injury to the marine environment still occur, resulting in the need for protection against this risk by vessel owners and coastal nations.

The need to protect the resources injured by vessel groundings has been realized more recently due to a growing awareness and understanding of the interconnected functions and values of natural resources. As a result of the increasing legal protection afforded coral reefs from grounding injuries in many regions, industry awareness has been elevated among operators, insurers, and cargo and vessel owners, which has served to promote increased attention to risk management.

11.3 INJURY

When a large vessel grounds or becomes stranded on a coral reef, it may result in a variety of injuries dictated by water depth, vessel draft, hull weight, impact speed, and the biological composition and health of the affected habitats. Trends in injury assessment and restoration of injury are dictated by our evolving understanding of injury and the potential for regeneration or recovery. Injuries from vessels can occur from four general types of actions: improper anchoring, running aground, action of the vessel while aground, and salvage of the vessel from the strand (Figure 11.1). The initial grounding may or may not have been unavoidable, but injury as a result of the actions of the vessel master and salvage vessels can be minimized or prevented. Types of possible injuries to the coral reef from anchor and vessel contact include abrasion and tissue damage, fragmentation and overturning of colonies, creation of rubble and sediment, direct loss and burial of live reef organisms, fracturing of underlying reef structure, flattening of diverse habitat, and the possible discharge of cargo or the presence of toxic antifouling paint. Indirect effects of tissue damage can lead to potential further loss from susceptibility to disease. If the vessel master attempts to remove the vessel from the strand unsuccessfully, or if the vessel is lively in the surf, propeller damage and vessel movement can exacerbate damage and enlarge the injured area. The action of the propeller, if engaged, can create a blowhole in the relatively soft limestone reef structure, scattering rubble and sediment to other nearby areas. Salvage vessels can cause damage to bottom habitats with nonbuoyant towlines and from sediment disturbance and burial of nearby corals (Figure 11.2). The alteration of habitat can vary widely in each case, leading to varying degrees of injury.

The duration of the injury will also vary depending on the severity and type of injury. Under optimal conditions in some locations, extensive coral recovery has been reported within 5 years following storm damage, but in others recovery has been estimated to require 40 to 70 years.[4] Rogers and Garrison[5] report that the effects of damage from a cruise ship anchor in the U.S. Virgin Islands remain after 10 years as a possible result of the planar aspect of the scar. Aronson and Swanson[6] found that unrestored grounding sites in the Florida Keys more closely resembled hard

FIGURE 11.1 Sediment disturbance from the action of salvage vessels. Photo by Gary Mauseth.

bottom communities than nearby spur and groove coral reefs after 10 years. Lasting impacts at two grounding sites in the Red Sea have been reported after 20 years due to the low availability of stable substratum.[7] Rubble fields created by destructive fishing practices over approximately 15 to 20 years in shallow waters of Indonesia showed no signs of recovery.[8]

The absence of stable substratum has also been reported to delay recovery and cause ancillary damage in several cases.[9–11] In addition to stable substratum, recovery of coral reefs following some types of disturbance events is often related to the survival of fragments.[12] Coral reef recovery estimates of 5 to 10 years when many fragments survive and 20 to 50 years in areas with few surviving fragments have been reported.[13]

Recovery times from vessel groundings are also affected by pregrounding disturbance conditions and may be rapid in areas dominated by small colonies accustomed to frequent natural disturbance and protracted in areas with long-term stability and large and diverse coral colonies. The type of restoration program may vary based on the frequency of the disturbance regime. Biological conditions in the affected area can also be a determinant of injury duration. Pioneer species of reef algae that often colonize sites following groundings have been reported to both delay and accelerate recovery, depending on the type of algae. A long-term shift from coral to algal habitats without recovery of the coral community has been reported at vessel-damaged sites on the Great Barrier Reef.[14] Dominance of macroalgae and algal mats has been suggested to be inhibitory

FIGURE 11.2 Tow line scarring inflicted while removing a vessel. Photo by Greg Challenger.

to coral recruitment.[5,15] The presence of other types of algae may accelerate coral recovery. Crustose coralline algae have been documented to contain chemical inducers that facilitate coral larval settlement and metamorphosis.[16] Miller and Barimo[17] found that differences in postrestoration coral assemblages are associated with benthic algal assemblages, which may mediate differential coral recruitment success. The change in the bottom topography and substrate composition can also result in changes in organism indices that can affect the coral and algal community, such as fish and invertebrate assemblages.[18] In turn, the loss of herbivorous fish and invertebrates can result in prohibitive algae growth and delayed recruitment.[19–22] This can be confounded by the potential reduction in recruitment as a result of the scarcity or loss of adult colonies that serve as a local source for sexual recruitment.[23] In addition to the magnitude of the injury, preexisting disturbance regimes, baseline biological factors and secondary effects from changes in community structure, and potential lingering contamination from unknown discharges or antifouling bottom paint may also be prohibitive to recovery. Sediments containing antifouling paint from a vessel grounding on the Great Barrier Reef in 2000 were found to significantly inhibit larval settlement and metamorphosis in laboratory studies.[24]

Injuries to coral reefs can also result in economic effects. Coral reefs provide numerous goods and services. While some are traded in the marketplace, many are not, making it difficult to estimate the economic value of a reef and even more difficult to quantify the value of a loss of a localized area of the reef. While there are techniques for estimating economic losses, a number of them are controversial and have limitations. Nonetheless, a project that emphasizes environmental restoration should also protect against potential future economic effects.

11.4 LEGISLATION

Natural resource trustees charged with protecting the public interest in the resource have become increasingly interested in developing legislation to provide guidelines for conducting coral reef injury assessment and restoration planning as a means to recover damages. When discussing international trends, it is useful to examine existing legislation in the United States that may influence development of vessel grounding assessment and restoration programs in other regions. In the United States, protection is afforded coral reefs under a broad variety of regulation. Executive Order 13089, signed by the President in 1998, directs federal agencies to utilize programs and authorities to protect and enhance coral reefs.[25] Numerous federal programs are available. Section 404 of the Clean Water Act provides for a permit program to regulate dredged and fill material in coastal waters. The Marine Protection, Research, and Sanctuaries Act (Sections 102 and 103) provides criteria to avoid harmful effects of ocean disposal programs. Section 10 of the Rivers and Harbors Act provides a means for agency and public interaction and comment regarding potential nearshore project impacts. Federally approved state coastal management plans are reviewed for consistency with federally conducted or supported projects through the Coastal Zone Management Act. Under the Oil Pollution Act of 1990 (OPA), the U.S. may recover removal costs and damages for discharge of oil or for substantial threat of discharge of oil. If injuries to corals or seagrasses occur during the removal of the threat, guidelines for seeking compensation for injuries are provided. Congress passed the Omnibus Parks and Public Lands Management Act in 1990, amending the Park System Resource Protection Act of 1990 (Park Act) to authorize the National Park Service to seek compensation from third parties for resource damage in any park.

Many recent coral grounding cases in the United States have followed the Natural Resource Damage Assessment Guidelines under OPA, or when in a National Park, the Park System Resource Protection Act. The National Oceanic and Atmospheric Administration (NOAA) administers the damage assessment and restoration process under OPA, and the Department of Interior, National Park Service under the Park Act. Both formal processes provide guidelines for the assessment while allowing for the use of a variety of techniques. Both equate damages with the reasonable cost of assessment and restoration. Restoration is defined as *primary* for actions that seek to return the affected area to its original condition and *compensatory* for actions that provide services equal to those lost while the reef recovers.

Internationally, the 1969 Convention on Civil Liability for Oil Pollution Damage as amended by the 1992 Protocol, and the 1971 Fund Convention allow affected parties from signatory countries in participating nations to seek reimbursement for environmental damage caused by oil spills and are limited to costs incurred for reasonable measures to reinstate the contaminated environment. They may allow expenses incurred for preventive measures to be recovered, such as those for removing the vessel, even when no spill of oil occurs, provided there was grave and imminent threat of pollution damage. The Conventions are also limited to vessels of certain gross tonnage. Although funds for vessel removal are sometimes available, few laws in nations that are not signatories of the Conventions provide a mechanism expressly for compensation for injuries to corals from vessel groundings. In nonsignatory nations, injuries are often treated as a third-party claim against the vessel for reinstatement of the environment. This has resulted in cases in which considerable uncertainty exists and no clear guidelines to reach settlement are available.

The U.S. program has been successful in its ability to recover damages from vessel groundings. Between 1990 and 2000, NOAA recovered more than 10 million USD (U.S. dollars) from shipowners to restore damage to coral reefs,[2] more than many regions have available to allocate to manage and protect coral reefs. These settlements resulted from a relatively small number of cases that addressed restoration of localized injured areas. There are over 17,000 km^2 of coral reefs in U.S. waters.[26] The 2001 annual budget request to restore coral reefs provided 16 million USD for NOAA and 10 million USD for the Department of Interior to implement the recommendations of the U.S. Coral Reef Task Force included in the Action Plan for Coral Reef Conservation.[2] The total funds provided by NOAA in their Coral Reef Conservation Grant Program from 2000 to 2003 for coral reef research, management, and conservation was approximately 10.7 million USD distributed among 83 grants. The funds recovered from vessel groundings in the U.S. are substantial when compared to budgetary constraints of many programs in the U.S. and abroad. The potential accomplishments that can be achieved by research and conservation programs with similar funding are attractive to many regions. Legislation development is underway in many U.S. territories, the Republic of Palau, Mexico, and other nations. While there is an understandable desire to receive adequate compensation to reinstate injuries to the environment, questions remain about the practicality and efficacy of some extensive and elaborate restoration programs and their potential application to other areas. In addition, many vessels operating in remote regions are fishing vessels or small cargo carriers that may not have adequate means to ensure elaborate programs consistent with some grounding case histories. New legislation in many regions should recognize technically sound yet practical site-specific approaches to assessment and restoration and consider alternative strategies when on-site restoration may not be successful.

11.5 ASSESSMENT

Understanding the nature of the injury to the coral reef and its recovery potential is essential to the development of successful restoration strategies. Injury assessment should be focused on developing effective projects with the highest present-day values or a reasonable chance of success in the short term. If the area is subject to ongoing cumulative processes of degradation, an elaborate restoration scheme may have little value and alternative approaches may be considered. Vessels that run aground often do so near where they operate, in the entrances to ports or in shipping channels that do not experience high tourist visitation or commercial fishing use, or are often degraded from a variety of preexisting anthropogenic disturbances. Corals that are chronically degraded may not be effectively restored at any cost. In locations that do not experience high tourist use, stabilization of large colonies and substrate has been considered sufficient.[11] At sites with high tourist use, aesthetics may play a large role in restoration costs.

Many guidelines for assessment of coral reef conditions exist worldwide, each with a specific objective. For instance, the Atlantic and Gulf Rapid Reef Assessment Protocol (AGRRA) is a means for providing basic information regarding coral reef species, size, distribution, and cover, as well as information regarding algae and fish assemblages, and is intended to facilitate long-term monitoring of regional changes.[27] Specific strategies to assess injuries from coral reef groundings are not well documented. Hudson and Goodwin[28] provide a protocol for measuring injuries from vessel grounding in the United States Florida Keys National Marine Sanctuary. This protocol focuses on determining the size of the injured area using permanent markers and transects and other established means of documenting biota lost. More recent techniques using highly accurate acoustic methods of measuring the area of the injury polygon have been employed in cases where the size of the injured area is in dispute.[29] Neither of these methods considers the other factors that affect restoration potential, recovery, and regeneration, recruitment-limiting factors and confounding natural or anthropogenic disturbance regimes. Historically, professional judgment has played a substantial role in determining appropriate restoration tasks.

In addition to a need to assess relevant biological factors affecting regeneration potential, there are sometimes political hurdles that interfere with assessment goals. Local authorities are commonly aware of reported settlements in other cases on a cost-per-unit-area basis. Settlement amounts from high-profile cases have been widely reported.[30] Despite the wide range of restoration costs stemming from site-specific differences, the awareness of the high potential monetary value in some cases has been a hindrance to achieving reinstatement of the resource in many regions. In many recent vessel-grounding cases, assessment of the coral injury has been focused largely on the size of the injured area, with less attention to the degree of injury and the characteristics of the site that dictate potential restoration success. The size of the area is sometimes applied to the reported cost per unit area of high profile cases to develop a damage claim. In other instances, cost estimates were developed using the known laundry list of available restoration options in the literature with little consideration of their potential effectiveness at the site in question. There are also dramatic site-specific differences in labor costs and logistics between regions and grounding sites. Those responsible are reluctant to undertake programs to reinstate the environment without meaningful substantiation for the program effort.

Given the urgency for rapid restoration in cases where secondary injury and delayed recovery could occur, the importance of the initial approach to the injury assessment is critical. Many regions do not actively involve the participation of representatives of the vessel. Recent grounding assessments in Nevis and St. Kitts, the Republic of Palau, and the Federated States of Micronesia were conducted independently and resulted in considerable disagreement. A cooperative approach to assessment and restoration among authorities and those responsible can avoid misunderstanding and lengthy debate over multiple assessment conclusions. These problems may ultimately result in litigation and delayed or nonexistent restoration efforts. Cooperative approaches are recommended in the United States and have more recently been used in the Cayman Islands and Mexico. While the cooperative approach is more commonly employed during the assessment phase, it is often neglected during the restoration-planning phase. Following the grounding of a cargo vessel in Mexico in 1999, detailed assessments involving a wide variety of data collection were employed to assess injuries. In this case, the responsible party and the authorities worked cooperatively to conduct studies of the damage. Assessment programs examined important parameters, such as reference conditions and recovery potential. Although the assessment results yielded little disagreement, restoration proposals were completed independently and resulted in disagreement and debate. Although settlement was reached, it is uncertain whether restoration was subsequently conducted.

Assessment should be cooperative and have a reasonable expectation of achieving a sound conclusion. Specifically, assessment should help determine the magnitude and duration of injury and provide information useful to help determine the potential success of various restoration techniques. Most programs of assessment lag behind emerging issues from restoration studies in the literature. Recommendations for assessment programs may be developed and revised after reviewing the lessons learned from restoration studies.

11.6 RESTORATION

Many different strategies have been employed in coral restoration cases throughout the world and have been discussed in detail throughout this text. Restoration strategies for localized injuries have included removing unconsolidated sediment and rubble, adding structure, transplanting and translocating corals to increase coral cover, introducing herbivores to enhance coral settlement, enhancing coral settlement by using larval chemoreception mechanisms, and using electrolysis to induce calcium carbonate mineralization. Many of these techniques have been very successful. Choosing restoration strategies that will most likely be practicable and effective can be challenging. Each vessel grounding case is different due to the complex and variable nature of the setting and the injury. Recent strategies for restoration are briefly reviewed below with implications for the setting in which they may or may not be useful and trends in how they are used.

Scientists are generally in agreement with the importance of rapid salvage of live corals or other emergency restoration actions immediately following an incident to prevent ancillary damage and promote rapid recovery. Emergency restoration actions have included removal, attachment, or relocation of live coral fragments to positions where they are more likely to survive and relocation or removal of unstable substratum. Potential problems arise between the authorities and the responsible party since fast action to save surviving fragments requires vessel representatives to trust the proposed emergency triage program of local authorities as appropriate and effective. In some cases there has been demonstrated success,[31] but often documentation is limited and often does not include an assessment of the overall changes in ecological services as a result of the program. When emergency restoration can reduce the recovery time of the community, it will correspondingly reduce the injury and ultimate damages and has been widely adopted in many regions.

Artificial material has been used to stabilize rubble and provide new framework in many cases. Materials such as concrete structures, artificial reef modules, and reticulating concrete blankets have been effective at stabilization.[32] When structure is added to increase habitat complexity and/or stabilize substratum, larval settlement may be more often associated with natural coral rubble or limestone when compared to concrete and other materials.[17,33] Decisions between artificial and natural materials may be balanced between ecological goals and the potential need for substantial engineered stabilization, and a wide variety of products continue to be used worldwide.

Rubble created following vessel groundings has been reported to cause ancillary injury and delay coral recruitment and recovery. Delayed recovery has been documented on rubble fields created by destructive fishing methods.[8] Qualitative reports of enlargement of vessel scars from a lack of stabilization have also been reported.[9,11] However, the size of unconsolidated debris may be an important consideration when determining appropriate emergency restoration actions. In studies in Indonesia, larger rubble from unconsolidated areas has been piled on top of smaller unconsolidated rubble and found to enhance coral settlement better than other techniques to stabilize rubble such as netting or cement slabs.[34] Jaap and Hudson[10] reported that rubble in depths greater than 8 m remained stable following the passage of a moderate hurricane. Most large vessels draft over 8 m, and not all groundings may result in rubble in shallow locations where it will cause additional damage, even in storms. Reconfiguring available on-site material into forms that provide structure and coral settlement surfaces may represent a low-cost and potentially more effective method of stabilization and regeneration (Figure 11.3). In many cases, local authorities assume all rubble should be removed or stabilized by artificial means simply because it has been reported in other cases, yet these actions may be unnecessary. Further research on rubble size and depth requirements and optimal oceanic conditions will improve the decision-making process regarding the presence of coral rubble as ancillary injury or potential regeneration tool. There is evidence to suggest it may be both under different circumstances.

FIGURE 11.3 Large fractured coral colonies that may remain stable and provide future habitat value and recruitment surfaces. Photo by Greg Challenger.

Transplantation of broken or dislodged fragments of live coral, cultured corals from aquaria, or "gardened" corals grown in off-site locations has had widely varying degrees of success depending on size, species, location, and technique. Survivorship as high as 87 to 100% has been reported in varying depths among hermaphroditic brooding corals with high self-fertilization rates.[35] Self-brooding corals may produce larger numbers of offspring in isolation, which may reduce the number of transplants required for restoration projects.[36] However, hermaphroditic brooding corals are not typically the reef framework builders in the Caribbean. Nonetheless, coral colonization may be enhanced by the presence of other nearby species,[37,38] although the relationship is not well understood. An average 69% survival of a transplanted coral was recently reported for transplanted red sea corals, however, an estimated *in situ* nursery period of 8 years was recommended for small fragments,[39] which is likely not practicable for many projects. Guzman[40] found 80% survival of transplants of the Pacific coral *Pocillapora* spp. in Costa Rica from nearby donor reefs. Additionally, natural fragmentation from the transplants themselves caused an estimated 41 to 115% increase in new colonies. Survival rates of acroporids of generally over 80% in offshore and nearshore waters of the Florida Keys have also been reported, with survival dramatically affected by a hurricane in some locations,[11] stressing the dangers of funding programs that may be lost by frequent natural events. Morse[41] reported losing 100% of salvaged transplanted fragments of *Acropora palmata* in a recent study.

Transplantation can also be expensive and labor intensive and may exceed available funds in many regions.[42] Several cost-effective means have been examined. Unattached fragments were scattered in 5 to 6 m of water along the coast of Curacao, Netherlands Antilles. Survival of all fragments ranged from 20 to 49% after 4 months, which is lower than in many studies using fragment attachment techniques.[43] However, the study included transplantation onto sediment and loose rubble, which is known to inhibit survival. Scattering unattached fragments of *Acropora* spp. on unstable rubble was found to be effective in locations where recruitment is limited[13] and may preclude the need for stabilization of unconsolidated rubble as a universally accepted requirement. In unstable areas, growth and survival are related to sediment cover and fragment movement. In a study of 276 unattached fragments of *Acropora* spp., 84 were swept away by a hurricane, and many of those survived.[13] Harriott and Fisk[44] found little difference in survival of fragments attached, carefully placed, and randomly scattered in a Great Barrier Reef study. This method may be considered as a cost-effective tool in locations with appropriate environmental conditions.

Other studies have found reduced survival and growth of transplanted fragments[45] and increased mortality and reduced reproductive potential of donor colonies.[46] As a result, translocation from donor reefs has not recently been considered as a vessel grounding restoration option in most regions since it may only serve to redistribute the loss of coral over a larger area.[11] Culling at-risk broken corals has been used in some case studies and is often the preferred option in injury cases when transplantation is considered. Using available broken fragments injured by the vessel grounding is preferred whenever possible, stressing the importance of emergency restoration.

Consideration of the current and future benefits of the increased success of transplants versus natural recruitment has been suggested.[30] If transplantation does not improve the long-term recovery substantially, the discounted future benefit of the up-front cost may indicate that effort may be better focused on programs with more immediate benefits such as education, management, enforcement, protection from ongoing anchor damage, and other projects. In areas that are not recruitment limited, transplantation is sometimes not considered sufficiently effective to warrant the effort.[34,47] The number of larvae in the water column is a likely determining factor of benthic community structure and is dependent on a nearby local source. Recruitment rates in the central and western Pacific have been reported to be sufficiently high to preclude the need for coral transplantation following injury.[47]

Some research indicates that remediation efforts such as debris removal and coral transplantation at grounding sites have little effect on fish abundance and richness. Fish assemblages at remediated grounding sites with rubble removal and coral transplantation versus unremediated grounding sites in the Florida Keys were found to be similar.[18] Differences in fish assemblages may play a role in recovery rates since the action of herbivorous invertebrates and fishes on reducing algal cover may contribute to coral settlement and growth. Numerous studies have suggested the importance of herbivory on coral settlement.[19–22] Ebersole[18] found structural complexity explained differences in fish assemblages, indicating adding structural complexity alone may be sufficient at returning herbivorous fish and invertebrates to pregrounding conditions and positively influencing coral settlement.

A growing body of literature suggests that local conditions should shape restoration techniques.[39] Spieler et al.[48] suggest relatively little is understood about the interaction of the artificial substrate with the ecology of the setting and remains a "best guess" endeavor. Given the expense of some cases, it has been suggested that sequestering funds from grounding damages for further research may ultimately have greater benefit.

Several new techniques are also emerging that may have benefit but have not yet been attempted for vessel grounding sites. Moe[49] reports that reintroduction of herbivorous grazers such as *Diadema antillarum*, the long-spined sea urchin, in locations in the Florida Keys can enhance coral recruitment. A Caribbean-wide die-off of the urchin in 1983 resulted in encroachment and competition of fleshy algae. The study involved the translocation of at-risk juveniles from shallow habitats to deeper reef areas. The translocation sites with higher *D. antillarum* density exhibited a 1-year increase in coral cover of nearly 6%, a 59% increase over baseline conditions. Coral cover decreased on control reefs by 24.5% during the study period. The study reefs also showed substantially higher cover of crustose corraline algae, also known to enhance larval settlement and metamorphosis. Studies of other herbivores that have been historically more abundant may lead to other new techniques. As mentioned, Ebersole[18] reported that herbivorous fish assemblages are returned to a damaged area by adding structure alone. Electrolytic precipitation of limestone has also recently been used to accelerate coral growth.[50] Metallic structures with transplanted fragments are attached to a power source to create enhanced mineralization by electrolysis. One potentially major drawback is that a land-based power source is required for electrolysis, which may make it impractical for many vessel grounding cases. Another potential technique is the use of the larval chemoreception to enhance settlement and metamorphosis of coral larvae. Morphogens in crustose coralline red algae have been identified that promote coral larval settlement.[51] While no large-scale projects have been attempted, it is believed that large surface areas covered by monomolecular layers of the inducer can be constructed.[52]

Whatever techniques are available, the long-term benefit per unit cost should be predictable to increase the chance of success. The costs of vessel grounding restoration actions vary enormously. Since coral reef recovery may take a long time with or without restoration, the discounting of future values of reef restoration in cost-benefit analysis yields lower present-day values of reef restoration.[30] By improving techniques and reducing initial costs, and by better understanding the site-specific success potential of alternatives, restoration projects will result in higher present-day benefits.

11.7 DISCUSSION

While our understanding of coral reef ecology and appropriate restoration actions for injured reef sites has improved substantially in recent decades, vessel grounding injury studies and restoration planning do not always match current scientific understanding. Given the successful ability of the U.S. program at recovering damages for groundings, numerous other regions are drafting or considering legislation. Awareness of cases with high settlements has sometimes resulted in conflicting approaches to determining restoration and compensation by diverting focus from the environmental goal. In some instances, the awareness of high-profile cases has prompted a desire to seek compensation by estimating restoration costs consistent with settlements in other cases without the intent to conduct restoration. In addition, past settlement amounts are not always equal to the cost of restoration and may have been the result of political, criminal, or competing financial concerns.

When an observable and measurable ecological loss from a vessel grounding is present, the goal of restoration should be to equate the restoration scheme with the loss, develop an approach with the highest benefit, and to undertake the restoration. These goals are more effectively accomplished with a cooperative process of assessment and restoration planning. Although a cooperative model for assessment of injuries is more frequently used in some regions, restoration planning often remains a unilateral exercise among local authorities. Those responsible are better able to understand the decision-making process and the need for restoration when all parties are involved in the planning process. As discussed herein, site restoration potential and strategies that are likely to be successful are not the same in every case. Variability of the baseline condition, ecological injury, logistics, costs of labor and materials, and human use with each incident and each region must be considered when determining the assessment strategy and considering restoration alternatives.

When grounding injuries are equated with practical and beneficial projects for reinstatement of the injury, the process will likely be met with the least resistance from those responsible. When restoration at the injured site is feasible, practicality should be considered by evaluating the frequency of disturbance, presence or absence of biological factors that enhance or hinder recruitment, the potential for ancillary injury, and projections of recovery and long-term benefit. Management programs and other alternative restoration strategies may also provide valid benefits to the resource and comprise reinstatement. Currently, consideration of how many alternative strategies can be scaled to the injury is lacking. There are typically no legal mechanisms in many regions that direct authorities how to scale the injury or allocate settlement monies. The U.S. system specifies that the responsible party provide equivalent resources to the injury and define damages as the cost of providing equivalent resources. However, this approach and common mechanisms of scaling the injury to a restoration or compensation project such as Habitat Equivalency Analysis sometimes limit the ability to select alternative restoration options. Alternatives to primary restoration of the injury should be available if primary restoration is deemed cost prohibitive or offers limited effectiveness. There is a need for innovative ways to determine the relative value of programs such as education, navigation aids, reef management, and salvage protocols. Data regarding vessel grounding frequency and damage may be useful in placing a value on aids to navigation that are aimed at preventing future groundings. New legislation in other regions may consider a means of providing for expedited assessment and alternatives to traditional restoration.

REFERENCES

1. Florida State University. 1988. Ship Wreck Index 1828-1911. Compiled by Tom Hambright, Monroe County Public Library, Key West, FL. Admiralty Final Record Book for U.S. District Court for the Southern District of Florida, volumes 2 thru 19, Microfilm Copy Number 1360, rolls 2 thru 19.
2. United States Coral Reef Task Force. 2000. Working Groups; Coastal uses working group status report, Fishing pressures, coastal development and shoreline modification, vessel traffic. Located at: http://208.139.192.240/workgroup.cfm.
3. Chadwick, H.V. and J. Storrie. 2001. Shipping issues within the great barrier reef. Shipping in the Asia-Pacific Arena — Conference Papers, International Symposium on National Shipping Industry Conference; 8–9 March 2001.
4. Dollar, S.J. and G.W. Tribble. 1993. Recurrent storm disturbance and recovery: A long-term study of coral communities in Hawaii. *Coral Reefs* 12:223–233.
5. Rogers, C.S. and V.H. Garrison. 2001. Ten years after the crime: Lasting effects of damage from a cruise ship anchor on a coral reef in St. John, U.S. Virgin Islands. *Bull. Mar. Sci.* 69: 793–803.
6. Aronson R.B. and D.W. Swanson. 1997. Disturbance and recovery from ship groundings in the Florida Keys National Marine Sanctuary. Dauphin Island Sea Lab Tech Rep 97-002, National Undersea Research Center — University of North Carolina, Wilmington.
7. Riegl, B. 2001. Degradation of reef structure, coral and fish communities in the Red Sea by ship groundings and dynamite fisheries. *Bull. Mar. Sci.* 69: 595–611.
8. Fox, H.E., J.S. Pet, R. Dahuri, and R.L. Caldwell. 2003. Recovery in rubble fields: long-term impacts of blast fishing. *Mar. Poll. Bull.* 46: 1024–1031.
9. Zobrist, E.C. 1998. Coral reef restoration and protection from vessel groundings. Gulf Estuarine Research Society Spring Meeting 1998, Galveston, TX (U.S.A.), 26–28 Mar 1998, Gulf-Research-Reports, 1999, 10, 85.
10. Japp, W.C. and J.H. Hudson. 2001. Coral reef restoration following anthropogenic disturbances. NCRI Special Session Chaired by Walter C. Japp and J. Harold Hudson. *Bull. Mar. Sci.* 69: 333.
11. Becker, L.C. and E. Mueller. 2001. The culture, transplantation and storage of *Montastraea faveolata, Acropora cervicornis* and *Acropora palmata*: what we have learned so far. *Bull. Mar. Sci.* 69: 881–896.
12. Edmunds, P.J. and J.D. Witman. 1991. Effect of Hurricane Hugo on the primary framework of a reef along the south shore of St. John, U.S. Virgin Islands. *Mar. Ecol. Prog. Ser.* 78: 201–204.
13. Bowden-Kerby, A. 2001. Low-tech coral reef restoration methods modeled after natural fragmentation processes. *Bull. Mar. Sci.* 69:915–931.
14. Hatcher, B.G. 1984. A maritime accident provides evidence for alternate stable states in benthic communities on coral reefs. *Coral Reefs* 3: 199–204.
15. Bechtel, J.D. 2002. The recovery of *Diadema antillarum* in Discovery Bay, Jamaica: impacts and implications for reef management. Ph.D. Dissertation, Boston University, Boston, MA. DAI-B 62/12, p. 5491, June 2002.
16. Morris, D.E., N. Hooker, A.N.C. Morse, and R. Jensen. 1988. Control of larval metamorphosis and recruitment in sympatric agariciid corals. *J. Exp. Mar. Biol. Ecol.* 116: 193–217.
17. Miller, M.W. and J. Barimo. 2001. Assessment of juvenile coral populations at two reef restoration sites in the Florida Keys National Marine Sanctuary: Indicators of success? *Bull. Mar. Sci.* 69: 395–405.
18. Ebersole, J.P. 2001. Recovery of fish assemblages from ship groundings on coral reefs in the Florida Keys National Marine Sanctuary. *Bull. Mar. Sci.* 69: 655–671.
19. Liddell, W.D. and S.L. Ohlhurst. 1986. Changes in community composition following the mass mortality of *Diadema* at Jamaica. *J. Exp. Mar. Biol. Ecol.* 95: 272–278.
20. Carpenter, R.C. 1990. Mass mortality of *Diadema antillarum.* I. Long term effects on sea urchin population dynamics and coral reef algal communities. *Mar. Biol.* 104: 67–77.
21. Miller, M.W. and M.E. Hay. 1996. Coral/seaweed/grazer/nutrient interactions on temperate reefs. *Ecol. Monogr.* 66: 323–344.
22. Hixon, M. 1997. Effect of reef fishes on corals and algae. Pp. 230–246 in Birkeland, C. (ed.) *Life and Death of Coral Reefs.* Chapman and Hall, New York.
23. Hughes, T.P., D. Ayre, and J.H. Connell. 1992. The evolutionary ecology of corals. *Trends Ecol. Evol.* 7: 292–295.

24. Negri, A.P., L.D. Smith, N.S. Webster, and A.J. Heyward. 2002. Understanding ship-grounding impacts on a coral reef: potential effects of antifoulant paint contamination on coral recruitment. *Mar. Poll. Bull.* 44: 111–117.
25. U.S. Environmental Protection Agency. 1994. Memorandum to the Field: http://www.epa.gov/owow/wetlands/guidance/coral.html. Executive Order 13089: Coral reef protection. Located at: http://208.139.192.240/execorder.cfm.
26. Yozell, S.J. 2001. Text of the keynote presentation to the international conference on scientific aspects of coral reef assessment, monitoring, and restoration. *Bull. Mar. Sci.* 69(2): 295–303.
27. Ginsburg, R.N. 2000. *Atlantic and Gulf Rapid Reef Assessment.* MGG-RSMAS, University of Miami. Miami, FL.
28. Hudson, J.H, and B.W. Goodwin. 2001. Assessment of vessel grounding injury to coral reef and seagrass habitats in the Florida Keys National Marine Sanctuary, Florida: protocols and methods. *Bull. Mar. Sci.* 69: 509–516.
29. Challenger, G.E. 2004. Natural Resource Injury Assessment for Injuries to Coral Reefs from Large Vessel Groundings: Determining Appropriate Restoration. 2nd National Conference on Coastal and Estuarine Habitat Restoration, Seattle, WA.
30. Spurgeon, J.P.G. 2001. Improving the economic effectiveness of coral reef restoration. *Bull. Mar. Sci.* 69: 1031–1045.
31. Bruckner, A.W. and R.J. Bruckner. 2001. Condition of restored *Acropora palmata* fragments off Mona Island, Puerto Rico, 2 years after the *Fortuna Reefer* ship grounding. *Coral Reefs* 20: 235–243.
32. Jaap, W.C. 2000. Coral reef restoration. *Ecol. Eng.* 15: 345–364.
33. Reyes, M.Z. and H.T. Yap. 2001. Effect of artificial substratum material and resident adults on coral settlement patterns at Danjugan Island, Philippines. *Bull. Mar. Sci.* 69: 559–566.
34. Fox, H.E., J.S. Pet, R. Dahuri, and R.L. Caldwell. 2001. Coral reef restoration after blast fishing in Indonesia. *Proc 9th Int. Coral Reef Symposium,* Fort Lauderdale, FL.
35. Gleason, D.F., D.A. Brazeau, and D. Munfus. 2001. Can self-fertilizing coral species be used to enhance restoration of Caribbean reefs? *Bull. Mar. Sci.* 69: 933–943.
36. Rinkevich, B. 1995. Restoration strategies for coral reefs damaged by recreational activities: the use of sexual and asexual recruits. *Res. Ecol.* 3: 241–251.
37. Lindahl, U. 1998. Low-tech rehabilitation of degraded coral reefs through transportation of staghorn corals. *Ambio* 27: 645–650.
38. Gittings, S.R., T.J. Bright, A. Choi, and R.R. Barnett. 1988. The recovery process in a mechanically damaged coral reef community: recruitment and growth. In Choat, J.H., Barnes, D., Borowitzka, M.A., et al. (eds.), Proceedings of the Sixth International Coral Reef Symposium, 8–12 August 1988, Townsville, Australia.
39. Rinkevich, B. 2000. Steps towards the evaluation of coral reef restoration by using small branch fragments. *Mar. Biol.* 136: 807–812.
40. Guzman, H.M. 1991. Restoration of coral reefs in pacific Costa Rica. *Conserv. Biol.* 5: 189–195.
41. Morse, A.N.C. 2000. Opportunities for biotechnology for coral and reef restoration. *Opportunities for Environmental Applications of Marine Biotechnology, Proceedings of the October 5–6, 1999, Workshop,* pp. 74–84.
42. Hatcher, B.G., R.E. Johannes, and A.I. Robertson. 1989. Review of research relevant to the conservation of shallow tropical marine ecosystems. *Oceanogr. Mar. Biol. Annu. Rev.* 27: 337–414.
43. Nagelkerken, S., S. Bouma, S. van den Akker, and R.P.M. Bak. 2000. Growth and survival of unattached *Madracis mirabilis* fragments transplanted to different reef sites, and the implication for reef rehabilitation. *Bull. Mar. Sci.* 66: 497–505.
44. Harriott, V.J. and D.A. Fisk. 1988. Coral transplantation as a reef management option. *Proc. 6th Int'l. Coral Reef Symp.,* Fort Lauderdale, FL. 2: 375–379.
45. Custodio, H.M. and H.T. Yap. 1997. Skeletal extension rates of *Porites cylindrical* and *Porites (Synaraea) rus* after transplantation to two depths. *Coral Reefs* 16: 267–268.
46. Epstein, N., R.P.M. Bak, and B. Rinkevich. 2001. Strategies for gardening denuded coral reef areas: the applicability of using different types of coral material for reef restoration. *Restoration Ecol.* 9: 432–442.
47. Kojis, B.L. and N.J. Quinn. 2001. The importance of regional differences in hard coral recruitment rates for determining the need for coral restoration. *Bull. Mar. Sci.* 69: 967–974.

48. Spieler, R.E., D.S. Gilliam, and R.L. Sherman. 2001. Artificial substrate and coral reef restoration: What do we need to know to know what we need? *Mar. Sci.* 69: 1013–1030.
49. Moe, M.A. Jr. 2003. Coral reef restoration: returning the caretakers to the reef. *SeaScope,* 20(4).
50. Hilbertz, W. and T. Goreau. 2001. Pemuteran coral reef restoration project progress report: May 29, 2001. Located at the Global Coral Reef Alliance (GCRA) website; Restoration Papers link: http://www.globalcoral.org/index.html
51. Morse, D.E., A.N.C. Morse, P.T. Raimondi, and N. Hooker. 1994. Morphogen-based chemical flypaper for *Agaricia humilis* coral larvae. *Biol. Bull. Woods Hole* 186: 172.
52. Morse, A.N.C and D.E. Morse. 1996. Flypapers for coral and other planktonic larvae. *BioScience* 46: 254–262.

12 Lessons Learned from Experimental Key-Species Restoration

Margaret W. Miller and Alina M. Szmant

CONTENTS

12.1 INTRODUCTION

The mortality of reef-building corals, particularly in the Caribbean, has been a persistent phenomenon over the past two decades[1] despite increasing implementation of conservation measures, including the establishment of protected reef areas.[2] Increasing conservation efforts, therefore, have coincided with practically monotonic reef decline. Conservation measures alone thus seem insufficient to prevent or reverse the decline in coral cover. We suggest that restoration is a necessary next step to rebuild coral populations and reestablish reef resiliency.

As large reef-building corals have died, new ones have failed to replace them. Where coral recruitment has been noted, it is generally of the smaller brooding coral species that are minor contributors to reef framework construction. There is evidence of dramatic decrease in overall coral recruitment success in the Caribbean,[3] suggesting a generalized loss of resilience of Caribbean reef systems.[4] Several interacting factors likely contribute to this recruitment failure in reef-building corals. As live coral cover declines by the loss of adult corals, and/or adult corals are stressed and have reduced fecundity, the areal production of coral gametes decreases,[5] which can increase the likelihood of fertilization limitation in broadcast-spawning species (the so-called Allee effect). Fertilization success has also been found to be low when the spawning corals have been recently bleached.[6] Thus, a smaller supply of larvae will be available to settle onto reefs, and recruitment may be supply-side limited. There is concern, but few data to support the possibility, that exposure to xenobiotics or other aspects of poor water quality may compromise reproductive effort and effectiveness.[7] Lastly, Caribbean reefs have manifested dramatic changes in substrate quality following the mass mortality of the important grazing sea-urchin, *Diadema antillarum*, in 1982 to 1983,[8] namely, the increase of macroalgal cover and loss of substrate types associated with intense grazing (e.g., live corals, crustose coralline algae [CCA] and fine turfs). Some corals exhibit a requirement for certain types of CCA as substrate cues to induce settlement;[9–11] therefore, loss of CCA cover may further limit coral recruitment rates. Recent observations of increased coral recruitment (though mostly small brooders, not broadcast-spawning, reef-building species) in patchy areas where *Diadema* have for the most part recovered[12] indicate that this so-called "phase shift" from coral to macroalgal dominance may be reversible and suggest that *Diadema* grazing may be an important aspect of reef recovery and resilience.

In responding to acute coral loss, such as that caused by anthropogenic physical disturbances (e.g., ship groundings), most restoration efforts have focused on transplantation of coral colonies or fragments from adjacent reefs,[13] although the net ecological benefit of transplantation has been questioned.[14] More recently, response to acute disturbances has included rescue and reattachment of impacted corals.[13,15] Active propagation of coral fragments to build source populations for transplant efforts is also under study.[16] However, these strategies are particularly compromised in the Caribbean by a dearth of healthy donor populations and a species pool which is depauperate in fast-growing, branching species that lend themselves to asexual propagation and transplantation. In fact, both of the main fast-growing branching corals in the Caribbean, *Acropora cervicornis* and *A. palmata,* are in the process of being listed as threatened under the U.S. Endangered Species Act (Federal Register, 50 CFR Part 223).

Given the clear need for intervention to help speed up the restoration of coral abundance on Caribbean reefs, the practical limitations of asexual propagation and transplantation in this region, and the clear importance of the grazing urchin *D. antillarum* in maintaining substrate quality suitable for successful coral recruitment, we have chosen to pursue experimental restoration approaches that include restocking *Diadema* to restore substrate quality, followed by seeding of restored substrate with sexual propagules (planula larvae) of reef-building coral species.

12.2 SEXUAL PROPAGATION AND SEEDING OF BROADCAST-SPAWNING CORALS

Scleractinian corals display two broad categories of life history strategies (Figure 12.1). One includes opportunistic species with generally small colony size, which have internal fertilization and brood their larvae to a fairly advanced stage and release planula larvae that can settle and metamorphose shortly after release from the mother colony (Cycle II in Figure 12.1). These species, known as brooders, generally recruit quite effectively. The second broad category includes the majority of coral species that spawn their gametes into the water column. Known as broadcast-spawners, these species undergo fertilization and a week or more of larval development in the water column, subject to its dilution, currents, and predators. The potential for fertilization limitation (especially in populations with low adult density), together with advection away from reef areas, and expected high mortality for planktonic larvae, could lead to poor recruitment by these species. Broadcasting species are the corals that tend to attain large colony size and thus, are responsible for most reef accretion. In the Indo-Pacific region, broadcasters achieve high rates of recruitment, even in recent years.[17–20] By contrast, most of the major broadcasting reef-builders in the Caribbean are infrequently observed in recruitment studies.[21,22]

The main justification for our interest in developing a methodology for larval seeding of coral-poor reef substrates is based on the premise that a major bottleneck in coral life history and, therefore, a major aspect of recruitment failure of reef-building species, lies in the presettlement phase, specifically insufficient larval supply (for all the reasons described above). Therefore, if a large number of viable larvae can be introduced to a substrate in need of restoration, coral recruitment of reef-building species can be enhanced. In our approach, fertilization success and larval survival are increased by collecting and maintaining high gamete concentrations during the

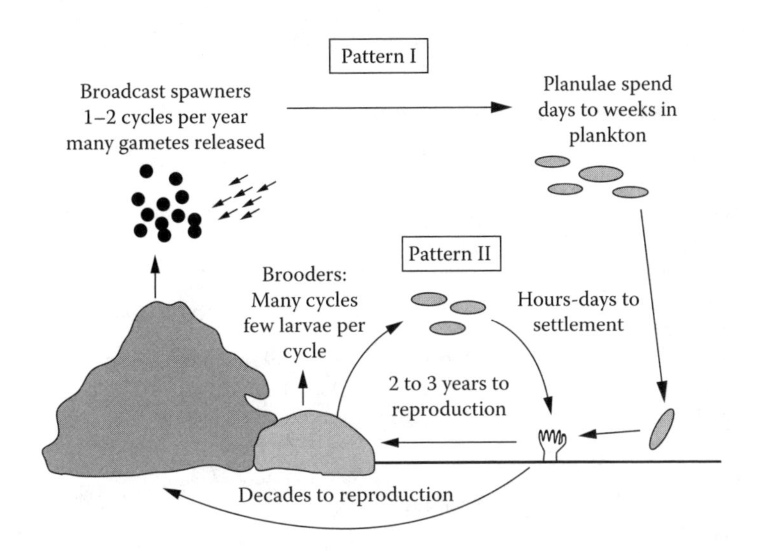

FIGURE 12.1 Two major reproductive cycles of Scleractinian corals: I. *Broadcasting* of gametes or gamete bundles into the water column where fertilization and larval development take place. This pattern is generally associated with a long-lived, large-colony-size life history strategy; II. *Brooding* of gametes, where fertilization is presumably internal, and embryos are retained for weeks to months until they develop to a fully competent planula larva stage. Our restoration efforts have focussed on Pattern I species.

fertilization phase and avoiding the high natural mortality expected in the planktonic phase. Based on two decades of work,[23–25] we can predict the spawning nights within a narrow window. Thus, we deploy teams of divers to collect broadcast-spawned gametes as they are released, fertilize them in buckets or coolers, culture them through the planktonic phase to competency under protected conditions (either in the laboratory or in field enclosures), and settle them onto experimental tiles or directly onto natural or "restored" reef substrates. Our efforts have focused on the two most important (and poorly recruiting) reef-building corals in the Caribbean, *Montastraea annularis* complex and *Acropora palmata.*

12.2.1 *Acropora palmata*

Acropora palmata, along with its congener *A. cervicornis*, has been responsible for much of the structural accretion of Caribbean reefs in the Holocene.[26] Both species have branching morphologies and high growth rates and were formerly highly abundant. However, these species have undergone such drastic decline over the past two decades that they are in process of being listed as threatened under the U.S. Endangered Species Act. Disease, bleaching, and various sources of physical disturbance have been instrumental in this decline.[27,28]

At the present time, most remaining populations of *A. palmata* are small, consisting largely of isolated colonies. Because *A. palmata* often propagates by fragmentation, some of these populations have extremely limited genotypic diversity. This species requires out-crossing for successful fertilization;[24] populations consisting of a single clone will thus not produce any larvae. These factors of rarity, isolation, and limited genotypic diversity suggest that its potential for successful sexual reproduction may be at present limited. The same considerations apply to *A. cervicornis.* We have focused our efforts in the Florida Keys on *A. palmata* because *A. cervicornis* is so rare that no adult source populations were available.

In the Florida Keys, *A. palmata* is predicted to spawn in the range of three to five nights after the full moon in August or September. However, the first challenge to working with this species is its unreliability in spawning time and synchrony (summarized in Table 12.1). In some years

TABLE 12.1
Recent Spawning Observations for *Acropora palmata* in Key Largo, Florida

Year	Site	Date and Timing
1994	Key Largo Dry Rocks	None on August 25 to 26 (5 to 6 nights AFM): watched from 2130 to 2300
1995	Ball Buoy Reef, BNP; Key Largo area; others	August 14 to 18 (4 to 8 nights AFM); 2150 to 2230; a few colonies spawned
1996	Key Largo Dry Rocks	August 3 to 4 (4 to 5 nights AFM); 2145 to 2230 September 1 to 2 (4 to 5 nights AFM); 2150 to 2230
1997	Key Largo Dry Rocks	August 20 to 21 (2 to 3 nights AFM)
1998	Horseshoe	August 11 (3 nights AFM); 2150 to 2245
1999		No observations
2000	Horseshoe, Florida and Atravesado, Puerto Rico	None on August 6 to 8 (3 to 5 nights AFM)
2001	Horseshoe	August 8 (5 nights AFM); 2233 to 2300
2002	Sand Island and French	None on August 24 to 27 (2 to 5 nights AFM)
2003	Sand Island Horseshoe Little Grecian	July 21 (9 nights AFM) 3/60 colonies on Sand Island (SI) August 17 (5 nights AFM) 8/60 colonies on SI; 50% on Horseshoe and Little Grecian; 2200 to 2230
2004	Horseshoe, Little Grecian, Key Largo Dry Rocks (KLDR)	Heavy spawn (>80% of colonies) at Horseshoe and Little Grecian (fewer at KLDR) on Aug 4 (4 nights AFM); 2200 to 2230.

Note: AFM = After full moon; BNP = Biscayne National Park.

(e.g., 2002), no spawning was observed throughout the entire predicted window. In 2003 we had observations for the predicted window for both July and August. One colony at Sand Island reef, Key Largo, released some spawn on nights 4 and 5 with three additional colonies spawning on night 9 after the full moon in July (I. Baums, personal communication). In August, the main spawn occurred on night 5 with about 50% of colonies at Horseshoe and Little Grecian reefs spawning but only eight of 60 colonies spawning at Sand Island. At least one large colony at Sand Island had ripe gametes based on histological examination earlier in July (I. Baums and S. Colley, personal communication) but was not observed to spawn, suggesting that additional asynchronous spawning occurred on nights we were not present to observe.

Besides timing, another challenge in culturing *A. palmata* larvae has been to collect adequate amounts of spawn to work with. It is currently quite rare, and colony condition is poor in many areas. Up until 1998, we had successfully collected large quantities of viable *A. palmata* spawn at Key Largo Dry Rocks, which had extensive thickets of this species. The site also has mooring buoys and is shallow and somewhat protected, making nighttime field operations feasible there under most weather conditions. The *A. palmata* stand at Key Largo Dry Rocks was decimated in 1998 by severe bleaching and Hurricane Georges. Thus, in 1999 to 2001 we shifted our collecting efforts to nearby Horseshoe Reef, where a robust and fairly dense *A. palmata* stand persisted. We were able to make large collections in 2000 and 2001, but fertilization was poor and no larvae resulted.

New information on *A. palmata* population genetics became available in 2002. The extensive *A. palmata* thicket at Horseshoe Reef is in fact monotypic, comprised of a single genetic individual.[29]

This hermaphroditic species cannot self-fertilize (Szmant[24] and later observations) so it is no surprise that our collections in 2000 and 2001 failed to produce larvae. In subsequent years, we have doubled our logistical burden to collect spawn at two different sites in order to assure outcrossing. During the summers of 2003 and 2004, we were able to raise several thousand larvae of *A. palmata*, some of which were used in laboratory experiments,[30] with the remainder seeded onto substrates that were deployed at the *Wellwood* grounding site (see Section 12.2.4 below).

12.2.2 *Montastraea* spp.

The reef coral *M. faveolata* is another primary reef-building coral in the Florida Keys and throughout the Caribbean, often occurring as enormous colonies several meters high. Many of these colonies are hundreds of years old and have large cavities that are important shelter and substrate for reef fishes and invertebrates. There are also large numbers of mid-sized colonies 1 m or so in diameter, but smaller (<10 cm) colonies of *M. faveolata* are virtually absent, suggesting that these corals have not been able to successfully recruit to Florida reefs in recent decades.[31–33] Meanwhile, this species complex has suffered a major decline in live cover over the past two decades due to bleaching stress and diseases.[34,35] Higher recruitment rates than those observed over the past decades are required to prevent local ecological extinction[3] on Florida reefs. Given the structural and ecological importance of these corals to Florida and Caribbean reefs, we have focused effort on culturing and seeding this species, as well as research to understand the ecological factors that affect their ability to achieve successful sexual recruitment. Fortunately, *M. faveolata* spawning is highly predictable, occurring either the sixth or seventh nights following the full moons of August or September.[25,36]

12.2.3 Larval Culture

At dusk on the nights of expected spawning, we deploy specialized collectors over individual colonies on the reef. Each collector is a tent-like, conical-shaped structure made of cloth or fine mesh with a float at the top to keep it upright and a plastic collector cup at the tip (Figure 12.2). The collectors are attached either by cinching them down with drawstrings around the base of the colony (for the mounding *Montastraea*) or by tying down strings attached around a hoop at the bottom of the collector (for the branching *A. palmata* colonies). When the colonies release their buoyant gamete bundles (eggs plus sperm), these rise gently into the collector cups. Divers then seal the cups with lids and retrieve them back to the boat. The spawn from multiple colonies (or for *A. palmata*, multiple reefs) is combined to achieve high rates of cross-fertilization. Prior to breakup of the bundles, the collector cups are opened carefully on the boat and the gamete bundles are decanted into a larger fertilization container in order to keep the spawn concentrated. The bundles break down in this container (ca. 40 to 60 minutes after release for *Montastraea* or more than an hour for *A. palmata*), and eggs and sperm from the different colonies are thus mixed for fertilization. The spawn is kept concentrated for ca. 1 hr, and then clean filtered seawater is used to flush out the excess sperm so it does not foul the culture.

We have attempted to culture the larvae both in the laboratory and in field chambers. The laboratory culture involves keeping the larvae at relatively low concentrations (500 to 1000 larvae per liter) in filtered seawater. One key to improve the success of these laboratory cultures is to have a high fertilization rate (high sperm concentrations of 10^6 sperm/ml for ca. 1 hr in the fertilization step) in order to minimize the amount of decaying material in the early stages of the cultures. It is beneficial to keep the culture containers in motion (e.g., on a roller or shaker table), as the buoyant larvae tend to get caught in surface slicks, especially along the walls of a container. Frequent and labor-intensive water changes are needed, especially in the early stages, as any unfertilized eggs break down and foul the water. Water changes are accomplished by gently siphoning a portion of the culture water out through a fine Nitex sieve (<120 μm for the smaller *Montastraea* larvae, 200 μm for the larger *A. palmata* larvae). Microbial infection is also a concern in the lab cultures,

(a)

(b)

FIGURE 12.2 A spawn collecting net deployed by a diver over a colony of *Acropora palmata* (a) and captured gamete bundles (b).

and antibiotic cocktails can be used. It is imperative to remove any debris from the cultures to prevent larvae from getting stuck and fungal infections from getting started. Culture containers need to be thoroughly cleaned at each water change by rinsing with very hot tap water and deionized water. Soap should only be used to clean the containers when hot water does not remove the organic buildup, and then must be followed by extensive tap water rinses.

We have also made multiple attempts to keep coral larval cultures in field chambers in order to scale up the number of larvae we can raise with reduced labor input. Our multifaceted efforts in this regard have met with very little success. We have piloted both surface ponds modeled after Heyward et al.[37] and small subsurface Nitex cylinders suspended in the water column on a mooring line (Figure 12.3). Though Heyward et al.[37] report success with the former approach in Australia's

(a)

(b)

FIGURE 12.3 (See color insert following page 240.) Two chamber types that we tried to use to culture coral larvae in the field. (a) Large floating chamber built after the design presented by Heyward et al.[37] The orange coloration is from the larval "slick" contained[44] in the chamber. (b) Cylindrical Nitex mesh chambers deployed either attached to the substrate or at the surface (as in a).

Great Barrier Reef, in three different years our attempts have yielded cultures that fared well for a couple of days but then crashed for unknown reasons. Perhaps a passing rain squall or simply inadequate water exchange via the small mesh windows of the enclosure was responsible. This prompted us to attempt keeping the field cultures in a subsurface position, but this has also met with little success, perhaps due to overstocking.

(a)

(b)

FIGURE 12.4 Small basket and net enclosures used to seed substrates with larvae of *Montastraea faveolata* (a) and *Acropora palmata* (b).

12.2.4 Seeding Efforts

The larvae gradually lose their buoyancy and complete their development to swimming planulae in 3 to 5 days.[38] When they are competent to settle, they begin swimming toward and "probing" the bottom. To date, we have made two attempts to deploy lab-cultured larvae in seeding reef substrates in the field. For *Montastraea*, we have used small Nitex chambers (a plastic basket frame with Nitex lining ~15 x 30 cm, Figure 12.4a) secured with nails and lead line to small portions of the reef restoration structure at the *Wellwood* grounding site in the Florida Keys National Marine Sanctuary. A 60-ml syringe was used to deploy ~5000 competent larvae into the chamber where they were left for 48 hr before removing the chamber. For *A. palmata*, Nitex-lined chambers with drawstrings (actually mosquito head nets, Figure 12.4b) were placed over individual branches of dead *A. palmata* skeletons, and the larvae were enclosed for 48 hr to allow an opportunity for settlement.

Newly settled coral larvae are very small (0.4 to 1 mm diameter), have little coloration, and often settle in small cervices (Figure 12.5). They can be difficult to see even with a light microscope. Neither *Montastraea* nor *Acropora* larvae have the zooxanthellae that give them more color until about 1 to 2 weeks after settlement. Because of this, there is a fundamental uncertainty involved in assessing the success of these field seeding efforts. In order to gain some insights into settlement success and survivorship, we also settle the lab-cultured larvae onto experimental plates, which can be placed in the field under different conditions. We have used various preconditioned substrates for larval settlement (limestone plates, clay tiles, and natural reef rubble). After settlement, spat are mapped under a microscope, characterized, placed in the field under various conditions and monitored by periodically retrieving them and examining them under a microscope. With this approach we were able to determine survivorship rates of only 1 to 2% for *Montastraea* over the first 3 months. For *A. palmata*, 16% of 129 spat survived after 2 months and 3% after 9 months.[30] The largest of these was ca. 5 mm in diameter and had 16 polyps (Figure 12.5d). These very preliminary observations give hope that with a greater effort to collect larger quantities of *A. palmata* spawn from as many locations as possible, now-dead patches of *A. palmata* skeletons could be reseeded with cultured larvae. This species has the fastest growth rate of any Caribbean coral.[39] Within 5 years, a successfully reseeded patch could have a significant live coral cover.

12.3 RESTOCKING OF *DIADEMA ANTILLARUM* IN THE FLORIDA KEYS

Although *D. antillarum* has finally begun to manifest a notable degree of recovery in many areas of the Caribbean within the past 5 years,[12,40,41] this has not been the case in the Florida Keys.[42] *Diadema* sightings on Florida Keys fore-reefs are still quite rare, though individual back reef sites (e.g., Conch back reef, Pickles back reef) fairly consistently have small *Diadema* populations (personal observation). We undertook experiments to explore the feasibility of enhancing reef substrate quality for coral settlement by transplanting (i.e., concentrating) large adult sea urchins in small-scale enclosures and by outplanting laboratory-cultured sea urchins of various sizes into various habitat types. The laboratory culture was performed by Tom Capo at the University of Miami Experimental Hatchery using broodstock collected from the Florida Keys. In all cases, sea urchin loss after release was dramatic.

In the adult transplants, we initially stocked roofless enclosures (~8 to 10 m^2) at Little Grecian reef (where we observed no *Diadema* prior to the experiment) at over 4 sea urchins m^{-2} with animals transplanted from other areas. However, this density declined over the following months to a level just over 1 m^{-2} (Figure 12.6a). Some sea urchins did escape the enclosures, but we conducted careful searches of the surrounding reef and replaced any recovered *Diadema* into the nearest enclosure. Density remained constant at this level for several months prior to our first restocking of the enclosures. In the series of restockings and declines over the 1.5 years of the experiment, the density always "relaxed" to a level around 1 m^{-2}, where it remained fairly stable. One hypothesis

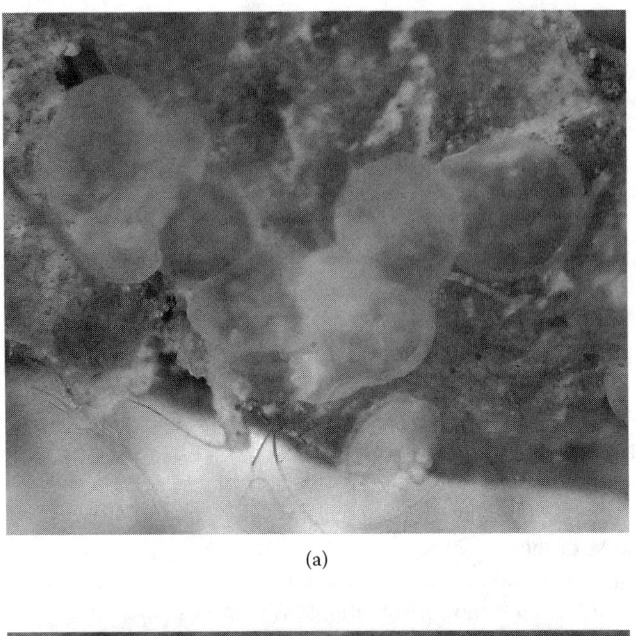

(a)

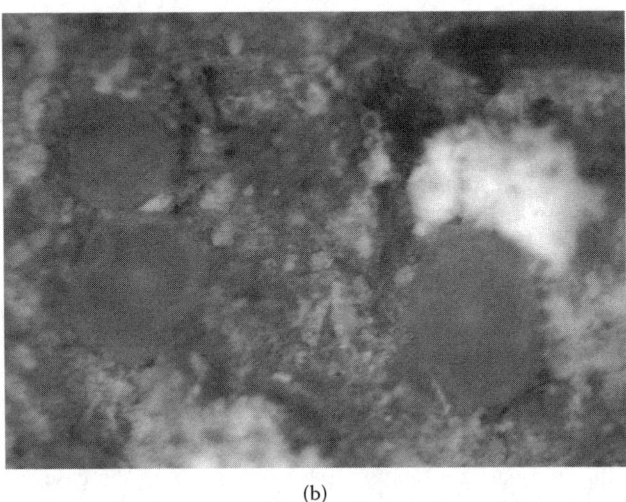

(b)

FIGURE 12.5 (See color insert following page 240.) Photomicrographs of newly settled polyps and older juvenile colonies of *Montastraea faveolata* and *Acropora palmata.* (a) Ten-day-old *M. faveolata.* (b) One-month-old *M. faveolata.* (c) Two-week-old *A. palmata.* (d) Nine-month-old *A. palmata.*

is that this density represents some sort of a carrying capacity for *Diadema* in this habitat, but what little information is available from before the 1983 die-off (mostly anecdotal) suggests that higher densities were common in the Florida Keys at that time.[43,44] Although we rarely observed predation "in the act" (despite many hours of video surveillance), we consistently found indirect evidence of predation, including spines and crushed bits of sea urchin test in the enclosures, suggesting that predation was a major cause of sea urchin loss in our enclosures.

Due to this fluctuating sea urchin density over the course of the experiment, it is difficult to assess the impact on macroalgal abundance in the enclosures. In the few months following the initial transplant, when *Diadema* densities were at their highest, there was an obvious impact on

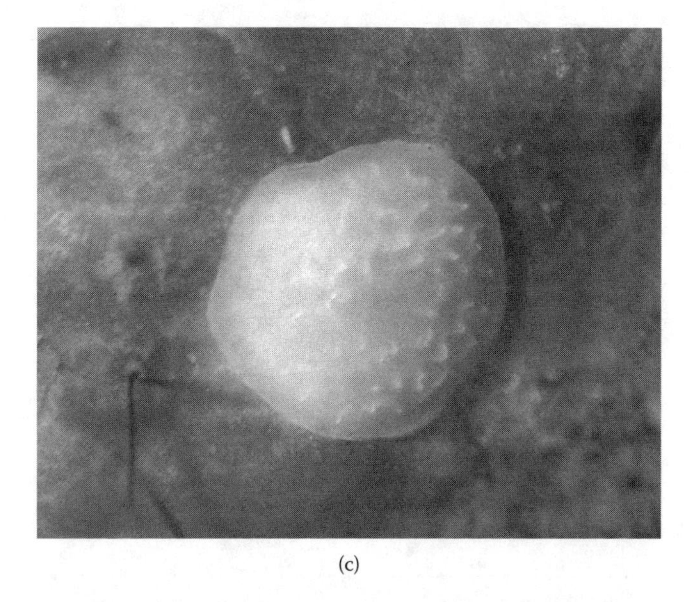

(c)

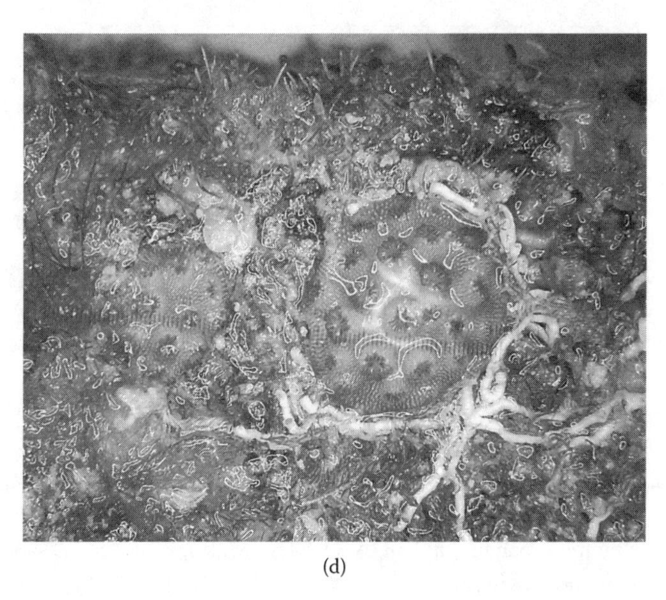

(d)

FIGURE 12.5 (Continued).

benthic communities, with the substrates appearing visibly clearer (Figure 12.7). Overall, about one third of the variation in macroalgal cover in the *Diadema* enclosures was attributable to *Diadema* density averaged over the interval from the last algal survey (Figure 12.6b). This relationship was likely affected by some undetermined degree of time lag needed for *Diadema* to crop down the macroalgae to a noticeable degree and by seasonal fluctuations in baseline macroalgal abundance.[45]

A separate experimental *Diadema* restoration project in the uppper FKNMS conducted by Moe and Nedimyer transplanted naturally recruiting juvenile sea urchins from reef crest rubble zones, where mortality from winter storm disturbance is high, to deeper patch reefs.[46] Interestingly, the overall density of urchins transplanted in this experiment declined gradually from a total potential density of 4.5 m^{-2} and 3.2 m^{-2} on each of two experimental patch reefs to about 1 m^{-2} over a

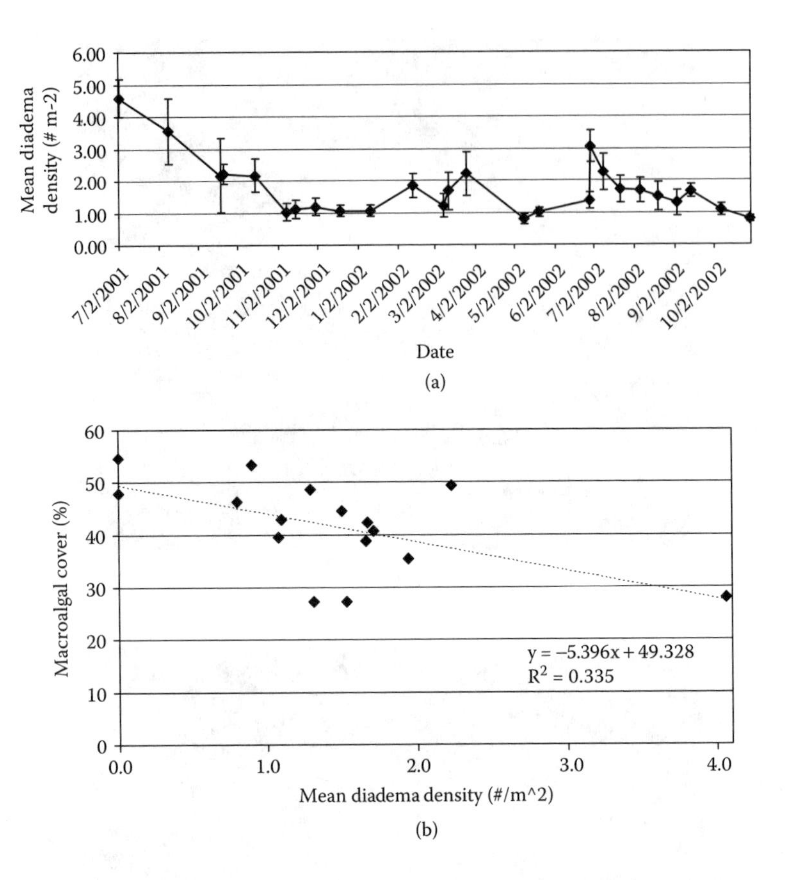

FIGURE 12.6 *Diadema* density fluctuated over time (a) as transplants were lost (largely due to predation) and we periodically undertook restocking. (b) Relationship of *Diadema* density in stocked enclosures with macroalgal cover between May 2001 and Oct 2002 (n = 4 to 6). Two macroalgal surveys (at zero *Diadema* density) were conducted prior to initial transplant of large adult *Diadema* in July 2001.

17-month duration, a similar final density as resulted in our nearby experiment in a different reef habitat. These researchers also report a decline in algal cover over the first year after transplant, though they do not report the specific urchin densities during this time frame.[46]

We also conducted three separate outplant experiments with lab-cultured *Diadema* ranging in size from 12 to 35 mm test diameter. These outplants were conducted in a variety of habitats including high relief fore-reef, back reef, patch reefs, artificial rubble mounds on sand, and seagrass margins. In one of these experiments, persistence of outplanted sea urchins was 8% over 2 months; the other two experiments had much lower persistence (e.g., 0 to 10% persistence over 5 days). Repeated observations of spines and fresh test fragments over the course of these experiments, as well as smaller outplant attempts utilizing cages, suggest that predation was a major factor in lack of *Diadema* persistence.

12.4 CONCLUSIONS

The development of restoration-scale methods for enhancement of sexual recruitment by broadcast spawning corals is still in its infancy but shows signs of potential. It will be more difficult to accomplish in the Caribbean where the species pool and overall coral abundance are depressed. In some Pacific areas, such as the Australian Great Barrier Reef, where coral cover is high and diverse and spawn easy to collect, success at seeding reef substrates has been reported.[37] However, the depressed status of Caribbean reefs makes such efforts all the more important in spite of the greater effort, and likely

(a)

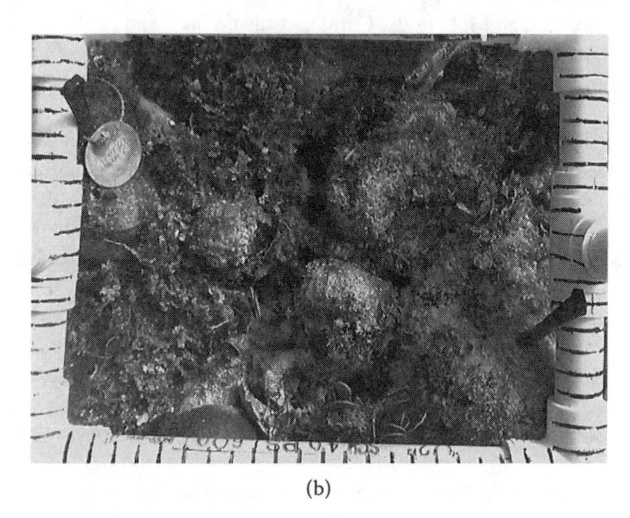

(b)

FIGURE 12.7 (See color insert following page 240.) Permanent plots within a *Diadema* corral taken at the time of *Diadema* stocking (a) Sept. 13, 2001 and 1 week later (b) Sept. 21, 2001.

also cost, that will be required. Attempts to restore *Diadema* to the Florida Keys reef, however, need further consideration. It is not clear whether the realized density of *Diadema* in transplant experiments (i.e., ~1 m^{-2}) is adequate to favorably impact reef benthic communities over more than a few-month period. The high rates of predation and the high cost at this time to raise laboratory sea urchins suggest that this avenue of restoration is not cost effective. Transplanting at-risk juveniles or concentrating adult *Diadema* into breeding colonies in nearshore areas, where predation appears to be lower, might help increase the local larval supply to a level where it is large enough to overcome the high predation pressure. It will then be up to Mother Nature to gradually increase the densities of *Diadema* on the offshore reefs where their grazing action is greatly needed.

ACKNOWLEDGMENTS

This work would not have been possible without a host of diligent assistants and colleagues. Specifically, Ernesto Weil, Liz Colon, Nicole Fogarty, and Charles Fasano all made important contributions. Field assistance was also provided by Amanda Bourque, Robert Carter, Iliana Baums,

Brad Buckley, Patrick Biber, Lou Kaufman, Loretta Lawrence, Dana Williams, Mark Vermeij, Aaron Bartholomew, Kathy Morrow, Justin Campbell, Benjamin Mason, Shauna Slingsby, Tracy Prude, and volunteers coordinated by Sherry Dawson and Brad Rosov of The Nature Conservancy. Tom Capo was responsible for culturing the *Diadema* used in the outplant experiments. Funding and logistic support were provided by the National Sea Grant Fisheries Habitat program (Grant # 155-NSGP-UNCW 1998-02), NOAA-Fisheries Coral Reef Conservation Program, and the Florida Keys National Marine Sanctuary. We are indebted to all. Views expressed including any errors of fact or judgment are solely our own.

REFERENCES

1. Wilkinson, C., *Status of Coral Reefs of the World: 2000*, Australian Institute of Marine Science, Townsville, 2000, pp. 363.
2. Rogers, C.S. and Beets, J., Degradation of marine ecosystems and decline of fishery resources in marine protected areas in the U.S. Virgin Islands, *Enviornmental Conservation* 28, 312–322, 2001.
3. Hughes, T.P. and Tanner, J.E., Recruitment failure, life histories, and long-term decline of Caribbean corals, *Ecology* 81, 2250–2263, 2000.
4. Nystrom, M., Folke, C., and Moberg, F., Coral reef disturbance and resilience in a human-dominated environment, *Trends in Ecology & Evolution* 15, 413–417, 2000.
5. Szmant, A.M. and Gassman, N.J., Caribbean reef corals. The evolution of reproductive strategies, *Oceanus* 34, 11–18, 1990.
6. Omori, M., Fukami, H., Kobinata, H., and Hatta, M., Significant drop of fertilization of *Acropora* corals in 1999: an after-effect of heavy coral bleaching? *Limnology and Oceanography* 46, 704–706, 2001.
7. Richmond, R.H., Reproduction and recruitment in corals: critical links in the persistence of reefs, in *Life and Death of Coral Reefs*, Birkeland, C. E., ed., Chapman and Hall, New York, 1997, pp. 175–197.
8. Lessios, H.A., Mass mortality of *Diadema antillarum* in the Caribbean: what have we learned? *Annual Review of Ecology and Systematics* 19, 371–393, 1988.
9. Heyward, A.J. and Negri, A.P., Natural inducers for coral larval metamorphosis, *Coral Reefs* 18, 273–279, 1999.
10. Morse, D., Morse, A., Raimondi, P., and Hooker, N., Morphogen-based chemical flypaper for *Agaricia humilis* coral larvae, *Biological Bulletin, Marine Biological Laboratory, Woods Hole* 186, 172–181, 1994.
11. Morse, A. and Morse, D., Flypapers for coral and other planktonic larvae, *Bioscience* 46, 254–262, 1996.
12. Edmunds, P.J. and Carpenter, R.C., Recovery of *Diadema antillarum* reduces macroalgal cover and increases abundance of juvenile corals on a Caribbean reef, *Proceedings of the National Academy of Sciences* 98, 5067–5071, 2001.
13. Jaap, W.C., Coral reef restoration, *Ecological Engineering* 15, 345–364, 2000.
14. Edwards, A.J. and Clark, S., Coral transplantation: a useful management tool or misguided meddling? *Marine Pollution Bulletin* 37, 474–487, 1998.
15. Bruckner, A.W. and Bruckner, R.J., Condition of restored *Acropora palmata* fragments off Mona Island, Puerto Rico, 2 years after the *Fortuna Reefer* ship grounding, *Coral Reefs* 20, 235–243, 2001.
16. Epstein, N., Bak, R.P.M., and Rinkevich, B., Applying forest restoration principles to coral reef rehabilitation, *Aquatic Conservation: Marine and Freshwater Ecosystems* 13, 387–395, 2003.
17. Harrison, P.L. and Wallace, C.C., Reproduction, dispersal and recruitment of scleractinian corals, in *Ecosystems of the World, Vol 25: Coral Reefs,* Dubinsky, Z., ed., Elsevier, New York, 1990, pp. 133–208.
18. Baird, A.H. and Hughes, T.P., Competitive dominance by tabular corals: an experimental analysis of recruitment and survival of understory assemblages, *Journal of Experimental Marine Biology and Ecology* 251, 117–132, 2000.
19. Hughes, T.P., Baird, A.H., Dinsdale, E.A., Moltschaniwskyj, N.A., Pratchett, M.S., Tanner, J.E., and Willis, B.L., Patterns of recruitment and abundance of corals along the Great Barrier Reef, *Nature* 397, 59–63, 1999.
20. Hughes, T.P., Baird, A.H., Dinsdale, E.A., Motschaniwskyj, N.A., Pratchett, M.S., Tanner, J.E., and Willis, B.L., Supply-side ecology works both ways: the link between benthic adults, fecundity, and larval recruits, *Ecology* 81, 2241–2249, 2000.

21. Kojis, B. and Quinn, N., The importance of regional differences in hard coral recruitment rates for determining the need for coral restoration, *Bulletin of Marine Science* 69, 967–974, 1999.
22. Rogers, C., Fitz, H.I., Gilnack, M., and Hardin, J., Scleractinian coral recruitment patterns at Salt River Submarine Canyon, St. Croix, U.S. Virgin Islands, *Coral Reefs* 3, 69–76, 1984.
23. Szmant, A.M., Sexual reproduction by the Caribbean reef corals *Montastrea annularis* and *M. cavernosa, Marine Ecology Progress Series* 74, 13–25, 1991.
24. Szmant, A.M., Reproductive ecology of reef corals, *Coral Reefs* 5, 43–54, 1986.
25. Szmant, A.M., Weil, E., Miller, M.W., and Colon, D.E., Hybridization within the species complex of the scleractinian coral *Montastraea annularis, Marine Biology* 129, 561–572, 1997.
26. Adey, W.H., Coral reef morphogenesis: a multidimensional model, *Science* 202, 831–837, 1978.
27. Bruckner, A.W., Proceedings of the Caribbean Acropora Workshop: Potential Application of the U.S. Endangered Species Act as a Conservation Strategy, NOAA Technical Memorandum NMFS-OPR-24, 2002.
28. Aronson, R.B. and Precht, W.F., White-band disease and the changing face of Caribbean coral reefs, *Hydrobiologia* 460 (1–3), 25–38, 2001.
29. Baums, I.B., Hughes, C.R., and Hellberg, M., Mendelian microsatellite loci for the Caribbean hard coral *Acropora palmata, Marine Ecology Progress Series* 288, 115–127, 2005.
30. Szmant, A.M. and Miller, M.W., Settlement preferences and postsettlement mortality of laboratory cultured and settled larvae of the Caribbean hermatypic corals *Montastraea faveolata* and *Acropora palmata* in the Florida Keys, U.S.A., *Proceedings 10th International Coral Reef Symposium,* in press.
31. Smith, S.R., Patterns of coral settlement, recruitment, and juvenile mortality with depth at Conch Reef, Florida, *Proceedings 8th International Coral Reef Symposium* 2, 1197–1202, 1997.
32. Miller, M.W., Weil, E., and Szmant, A.M., Coral recruitmant and juvenile mortality as structuring factors for reef benthic communities in Biscayne National Park, U.S.A., *Coral Reefs* 19, 115–123, 2000.
33. Chiappone, M. and Sullivan, K., Distribution, abundance, and species composition of juvenile scler-actinian corals in the Florida reef tract, *Bulletin of Marine Science* 58, 555–569, 1996.
34. Ginsburg, R.N., Gischler, E., and Kiene, W.E., Partial mortality of massive reef-building corals: an index of patch reef condition, Florida reef tract, *Bulletin of Marine Science* 69, 1149–1173, 2001.
35. Porter, J.W. and Meier, O.W., Quantification of loss and change in Floridian reef coral populations, *American Zoologist* 32, 625–640, 1992.
36. Van Veghel, M.L.J., Multiple species spawning on Curacao reefs, *Bulletin of Marine Science* 52, 1017–1021, 1994.
37. Heyward, A.J., Smith, L.D., Rees, M., and Field, S.N., Enhancement of coral recruitment by *in situ* mass culture of coral larvae, *Marine Ecology Progress Series* 230, 113–118, 2002.
38. Szmant, A. and Meadows, M., Developmental changes in coral larval buoyancy and vertical swimming behavior: implications for dispersal and connectivity, *Proceedings 10th Annual Coral Reef Symposium,* in press.
39. Gladfelter, E.H., Monahan, R.K., and Gladfelter, W.B., Growth rates of five reef-building corals in the northeastern Caribbean, *Bulletin of Marine Science* 28, 728–734, 1978.
40. Aronson, R.B. and Precht, W.F., Herbivory and algal dynamics on the coral reef at Discovery Bay, Jamaica, *Limnology and Oceanography* 45, 251–255, 2000.
41. Miller, R.J., Adams, A.J., Ogden, N.B., Ogden, J.C., and Ebersole, J.P., *Diadema antillarum* 17 years after mass mortality: is recovery beginning on St Croix? *Coral Reefs* 22, 181–187, 2003.
42. Chiappone, M., Swanson, D.W., and Miller, S.L., Density, spatial distribution and size structure of sea urchins in Florida Keys coral reef and hard-bottom habitats, *Marine Ecology Progress Series* 235, 117–126, 2002.
43. Bauer, J.C., Observations on geographic variations in population density of the echinoid *Diadema antillarum* within the western North Atlantic, *Bulletin of Marine Science* 30, 509–515, 1980.
44. Bohnsack, J.B., White, M.W., and Jaap, W.C., Status of selected coral resources, in *Resource Survey of Looe Key National Marine Sanctuary 1983*, Bohnsack, J.A., Cantillo, A.Y., and Bello, M.J., eds., NOAA Technical Memorandum NMFS-SEFSC-478, 2002, pp. 289–322.
45. Lirman, D. and Biber, P., Seasonal dynamics of macroalgal communities of the northern Florida reef tract, *Botanical Marina* 43, 305–314, 2000.
46. Moe, M.A., Coral reef restoration: returning the caretakers to the reef, *Sea Scope* 20, 1–4, 2003.

13 Cooperative Natural Resource Damage Assessment and Coral Reef Restoration at the Container Ship *Houston* Grounding in the Florida Keys National Marine Sanctuary

George P. Schmahl, Donald R. Deis, and Sharon K. Shutler

CONTENTS

13.1 BACKGROUND

13.1.1 THE GROUNDING

On the night of February 2, 1997, the 600-ft (183-m) container ship *Contship Houston* ran aground on the Florida Keys reef tract between Maryland and American Shoals (near 24° 30.59' N, 81° 34.28' W), resulting in serious natural resource injury in water depths ranging from 18 to 35 ft (5.5 to 11 m). The grounding site is located approximately 13 miles east southeast of Key West (Figure 13.1) and lies within the Florida Keys National Marine Sanctuary (FKNMS), a marine protected area administered by the National Oceanic and Atmospheric Administration (NOAA) and managed in cooperation with the State of Florida. The *Contship Houston*, owned by Transportacion Maritima Mexican S.A. de C.V. and registered in Liberia, was en route from New Orleans, Louisiana to Valencia, Spain, carrying over 1400 containers of general cargo (Figure 13.2). The event was reported to the U.S. Coast Guard at 5:30 A.M. on February 3, 1997. The submerged habitat in the area is characterized as a low profile "spur and groove" reef formation with associated hard-bottom habitats. The hard substrate is considered relict or "drowned" coral reef, colonized by hard and soft corals, sponges, and a variety of other benthic species.

The primary cause of the grounding was reportedly a gross navigational error resulting from setting a course a full degree latitude north of the desired heading. Inadequate pilot oversight was obviously also a factor because the vessel went hard aground within 2 miles of an operational

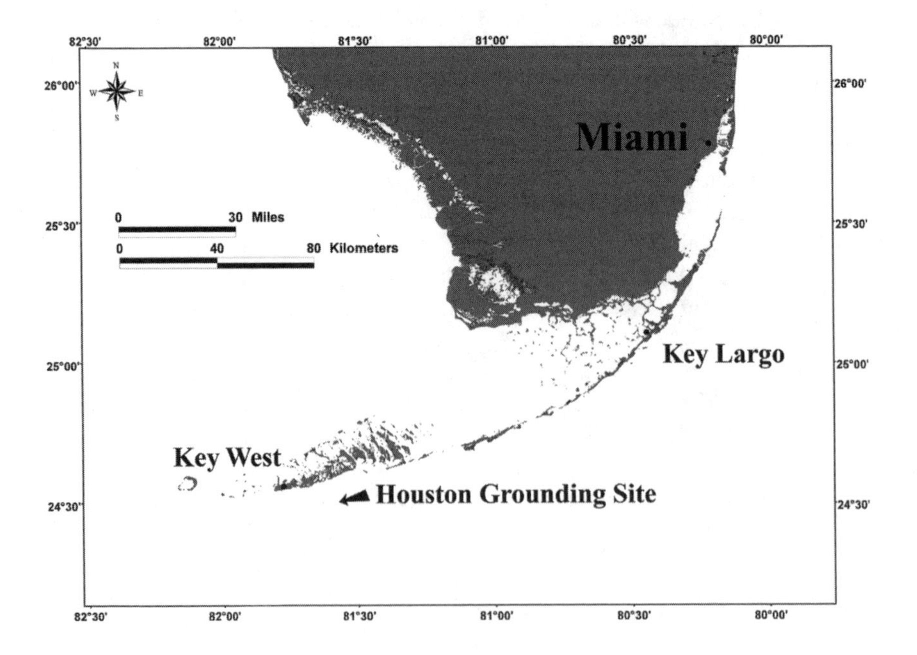

FIGURE 13.1 General location of the grounding of the *Contship Houston*.

FIGURE 13.2 The *Contship Houston* aground near Maryland Shoal, Florida Keys, February 3, 1997.

109-ft-tall lighthouse located at American Shoal. The *Houston* approached the reef tract from the southwest, traveling from the deep water of the Florida Straits onto the shallow shelf-edge reef bank at a speed of approximately 16 knots. The vessel's draft (depth below the waterline) was approximately 30 feet, and as it approached shallow water it contacted the reef intermittently at places where the bottom contour was elevated above that depth. As the water became shallower, more consistent contact with bottom features occurred until the vessel encountered water depths of less than 20 ft, at which point the *Houston* became hard aground. The entire length of the grounding tract, from initial impact to the vessel resting area, was approximately 600 m. After the grounding, the *Houston's* captain attempted to free the ship by discharging ballast water to lighten the vessel, putting the engines in reverse, and applying full engine power. This action did not free the vessel but did result in a slight pivoting of the ship's stern to the north, causing significant additional injury from the vessel propeller wash and mechanical movement. It was not until after the vessel attempted to free itself that the U.S. Coast Guard was notified of the incident.

At the time of the grounding, the *Houston* was carrying over 2500 tons (810,000 gallons) of fuel and 153 tons of lubrication oil. Due to the threat of a pollutant spill, the U.S. Coast Guard initially took control of the incident and established a unified command system where all relevant parties were represented. Trustee responsibility for this case was shared by the federal government and the State of Florida through NOAA/Florida Keys National Marine Sanctuary (FKNMS) and the Florida Department of Environmental Protection (FDEP) (collectively, the Trustees). The owner of the vessel, as the Responsible Party (RP), was represented by an agent and a representative of its insurance company. To avoid movement of the grounded vessel that could cause additional injury, the *Houston* was initially reballasted with seawater while a plan for its removal was formulated. Over the next 6 days, over 500,000 gallons of fuel was offloaded from the vessel, hydrographic surveys of the area were conducted, the hull condition was assessed, and the ship was prepared for removal from the reef. When preparations were complete and the weather conditions allowed, the ship was stabilized by tugboats and deballasted. The *Houston* was removed from the reef on February 8, 1997.

To forestall the arrest of the vessel by the government and ensure accountability by the RP, the vessel owners signed a letter of undertaking (LOU) developed by NOAA and the Department of Justice on February 7, 1997. The LOU guaranteed payment of up to $6 million to satisfy any settlement or judgment in favor of the United States for damage claims arising from the grounding. It also required the ship's owner to waive jurisdictional defenses should the Trustees have to litigate. In return for those agreements, the LOU provided that the owner could remove the vessel from the reef and operate it within the waters of the United States. Subsequent to the execution of the LOU, the ship's owners and insurers agreed to work cooperatively with the Trustees to assess the damages and restore the injured area.

13.1.2 Preliminary Assessment

Preliminary scientific investigations were performed between February 4 and 8, 1997, by biologists from the FKNMS and environmental support services contractors during the response effort. Divers provided photo documentation of the area around the ship and outlined the footprint of the injured area with subsurface buoys. These buoys marked the nearest undisturbed features (typically, large coral heads) to the injured area. The markers were used for orientation for subsequent investigations and to determine potential additional injury from ship removal. The maximum total affected area from the grounding was initially estimated to be approximately 15,200 square meters.

The initial surveys indicated that the approach of the ship, the vessel grounding, and the subsequent removal attempt resulted in several general categories of natural resource injury. In the inbound track area, injury resulted from intermittent contact with the reef surface and with coral heads and other benthic features that were in the vessel's path. In some areas, portions of the reef surface were "scraped clean" of any marine organisms, while in other areas, coral heads were knocked loose from the bottom, broken in various places or crushed into smaller pieces. In some cases, the tops of coral heads and large sponges were sheared off, while the rest of the organism remained attached to the substrate. In the area where the vessel came to rest, the bottom features were completely crushed and flattened, destroying all living organisms beneath it. This area became known as the "parking lot," and included portions of at least two low-relief coral spurs. Coral spurs are typically comprised of an accumulation of loosely consolidated material "capped" by a solid, external limestone crust. As the top crust layer is compromised, the underlying less consolidated material is exposed. This material is then subject to continual erosion from storms and could result in subsequent injury to the reef system. A third category of injury was related to the action of the vessel during the attempted removal. This resulted in the creation of large piles of reef rubble from the physical movement of the hull and from the scouring action of prop wash. Several linear tracts of reef rubble were generated, and substantial areas were excavated, some as deep as 10 to 15 ft deep in the surrounding reef zone.

13.1.3 Cooperative Natural Resource Damage Assessment

The agreement by the RP to work cooperatively with the Trustees paved the way for the establishment of a formal cooperative natural resource damage assessment (NRDA). Meetings between representatives of the RP and the Trustees (the Parties) to initiate the cooperative injury assessment began on February 7, 1997. At that and subsequent meetings, an outline of the tasks and responsibilities was developed and agreed upon by both parties. This agreement became the basis for a Memorandum of Understanding (MOU) between the Parties. The MOU provided a framework by which the RP agreed to implement injury assessment and restoration tasks under Trustee oversight. All subsequent tasks not originally identified were developed in accordance with the MOU and incorporated through appendices after approval by the Trustees.

Pursuant to the MOU, RP representatives agreed to perform the injury assessment and primary restoration (coral repair and rubble removal or stabilization) prior to the beginning of the peak of hurricane season in Florida (September 1). In order to achieve this objective, a project time line was proposed, as shown in Table 13.1.

The value of the cooperative relationship between the Trustees and the RP related to injury assessment and primary restoration cannot be overestimated. An agreement was reached early in the process that was advantageous to both Parties and resulted in more timely restoration than had

TABLE 13.1
Proposed Project Timeline

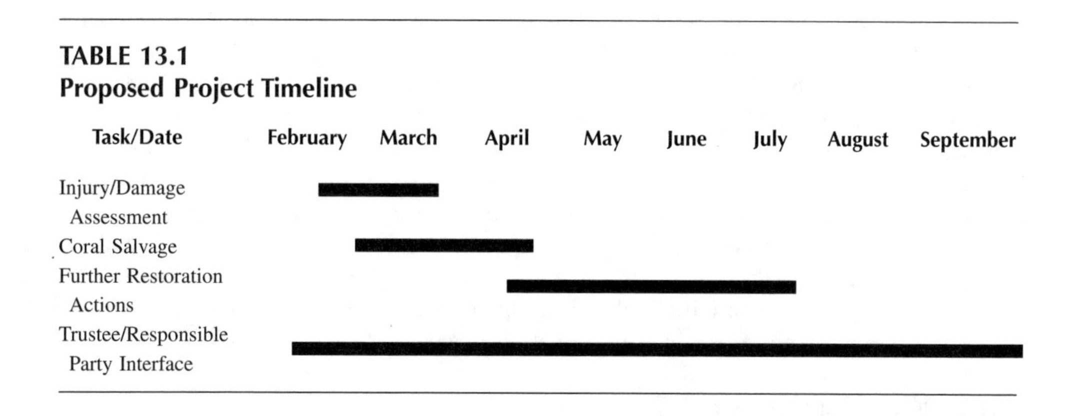

the relationship been adversarial. In several previous grounding cases, the vessel owners refused to enter into cooperative damage assessments. Thus, salvage and restoration activities were delayed while the Parties argued about the amount of injury and the appropriate restoration options. A timely injury assessment is a key to facilitating restoration activities and reducing mortality to injured coral. The cooperative nature of the *Houston* damage assessment is a success story and should serve as a model for future resource injury cases.

13.2 PRERESTORATION ACTIVITIES

13.2.1 Emergency Coral Salvage

Personnel from the Trustee agencies agreed to be responsible for carrying out emergency salvage of injured coral colonies, primarily along the inbound track of the grounding area. This involved uprighting dislodged coral heads and coral pieces in order to protect the living tissue from burial or contact with the substrate and to prevent further mortality. This allowed time for the RP to mobilize contractors to perform the injury assessment and permanent stabilization. The uprighted coral pieces were clearly tagged to indicate that they had been moved from their original positions. The Trustees also agreed to collect limited quantities of injured coral pieces, less than 6 in. in length, within the damage area for experimental rearing for future use in reef restoration. These coral pieces were taken to Mote Marine Laboratory on Summerland Key for propagation experiments. The emergency salvage operation occurred between February 24 and March 4, 1997. During that time, the Trustees also established a baseline the length of the grounding track by permanently installing stainless steel eyebolt markers at 25-m intervals that were drilled and secured into the reef substrate with marine epoxy. This baseline served as a reference for injury assessment and other activities, as well as providing attachment points for mooring buoys that could be used by vessels involved with restoration activities. During this period, an initial map was developed to aid divers in orientation to the site. It was also agreed that as the injury assessment was being conducted, coral heads and pieces that were not salvaged by the Trustees were to be uprighted and clearly tagged for salvage and restoration by the RP.

13.2.2 Injury Assessment

13.2.2.1 Site Preparation and Mapping

Immediately following the emergency coral salvage operations, contractors for the RP began preparation for a detailed injury assessment. The scale of the project necessitated extensive preliminary work before the assessment could begin. A system was developed to refine the initial site map and "ground truth" the available aerial imagery. The mapping system utilized a Differential Global Positioning System (DGPS) to geographically locate all points of interest within the site, using the central baseline as the reference. This information was incorporated into a Geographic Information System (GIS) in order to produce map products (Shaul et al. 1999).

Prior to beginning baseline mapping, large surface buoys were installed using the available mooring eyebolts established by NOAA personnel at the bow and stern positions of the grounded vessel and at the seaward end of the inbound tract. Additional buoys were added at 50-m intervals along the inbound tract through to the initial bow location for visual alignment. All mooring eyebolts were numbered sequentially and identified using underwater markers for reference. The baseline mapping effort was performed along transects spaced at intervals, determined by underwater visibility, perpendicular to the baseline over the entire grounding site. A scientific diver was towed slowly over the injured area to initially document the nature and extent of injury and to establish injury classification categories for future detailed assessment.

13.2.2.2 Injury Classification

A matrix of resource injury categories was used to classify the grounding impacts observed within individual locations in each area (Table 13.2). This matrix was the result of collaboration between the Trustees and representatives of the RP and was based on preliminary field surveys by both parties. The matrix provided a description of the types of injury that might be encountered during the detailed assessment and the information necessary to be collected for each injury category. The information generated from this matrix was converted into a computer database for collection of data during the detailed injury assessment. A scientific diver used the matrix categories to assign an injury classification to each observed incident type within the grounding area.

13.2.2.3 Detailed Area Assessments

An innovative tool that had been tested previously on smaller sites was used at the grounding site for detailed mapping and injury assessment. This tool, known as the Integrated Video Mapping

TABLE 13.2
Injury Classification Matrix

Non-Injured	Injured	
Coral Reef Habitat (Spurs) Sediment Deposits (Grooves)	1 Minor Rubble/Sediment Deposits	2 Minor Rubble/Sediment Deposits
	1. Minor scars and hull paint on reef platform	
	2. Major scars and hull paint on reef platform	
	3. Areas excavated by propeller wash	
	4. Surface areas are void of sediment and/or attached organisms	
	5. Dislodged or damaged octocorals and other attached organisms	
Coral Size Categories (meter squared)	3	4
1: <0.25	1. Corals dislodged from the reef platform	Dislodged or damaged sponges
2: 0.25–0.5	2. Corals with minor scrapes and hull paint	
3: 0.5–0.75	3. Corals crushed, fractured or fragmented (still intact)	
4: 0.75–1.0	4. Corals crushed, fractured, or fragmented (in pieces)	
5: >1.0		

(a)

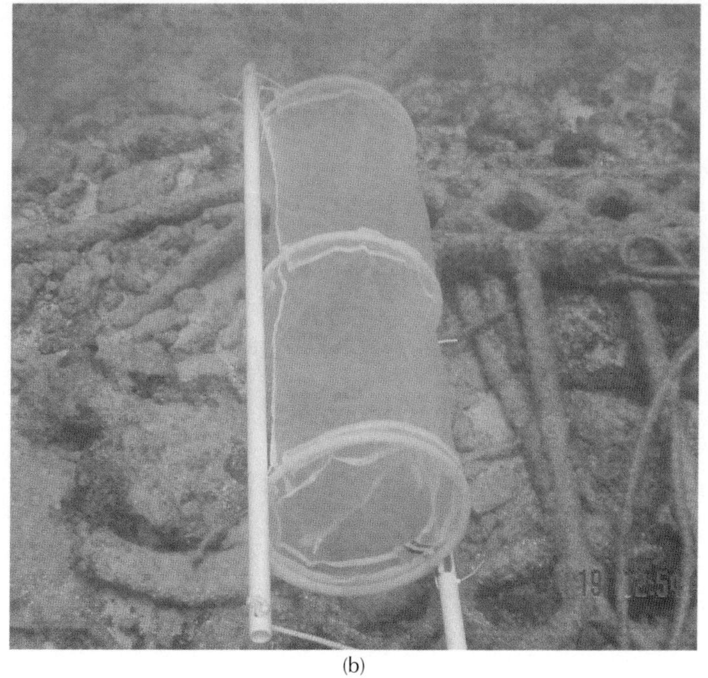

(b)

COLOR FIGURE 12.3

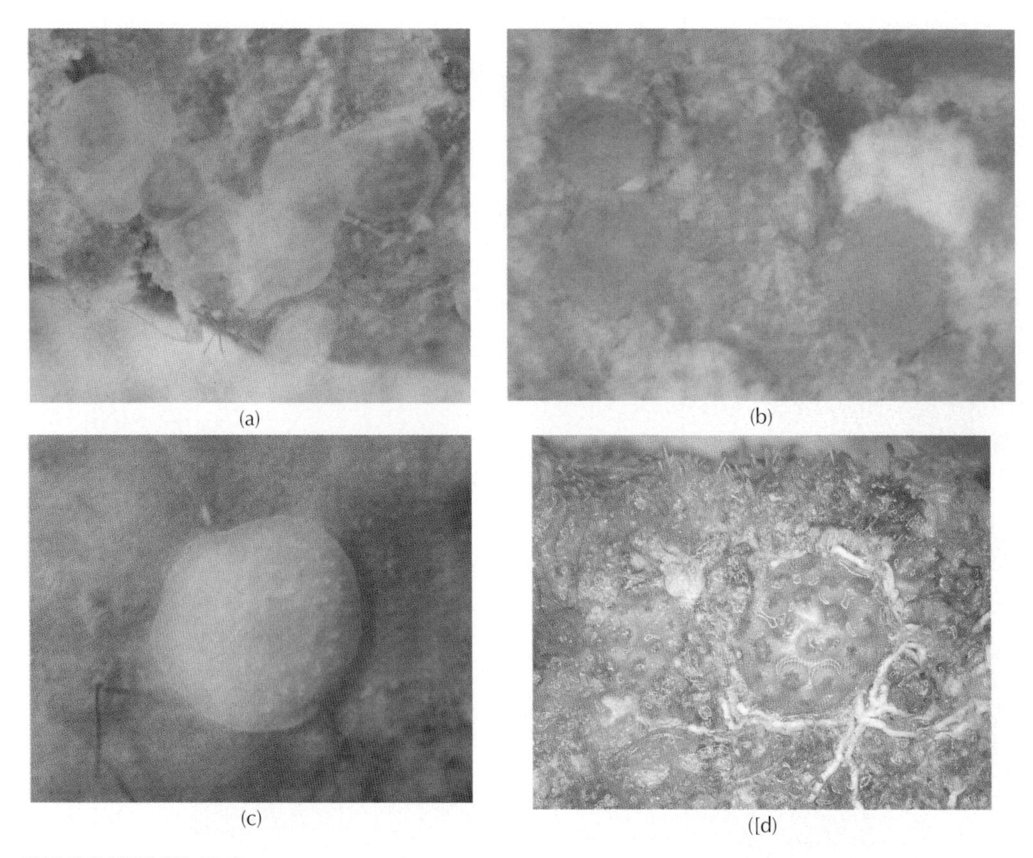

COLOR FIGURE 12.5

COLOR FIGURE 12.7

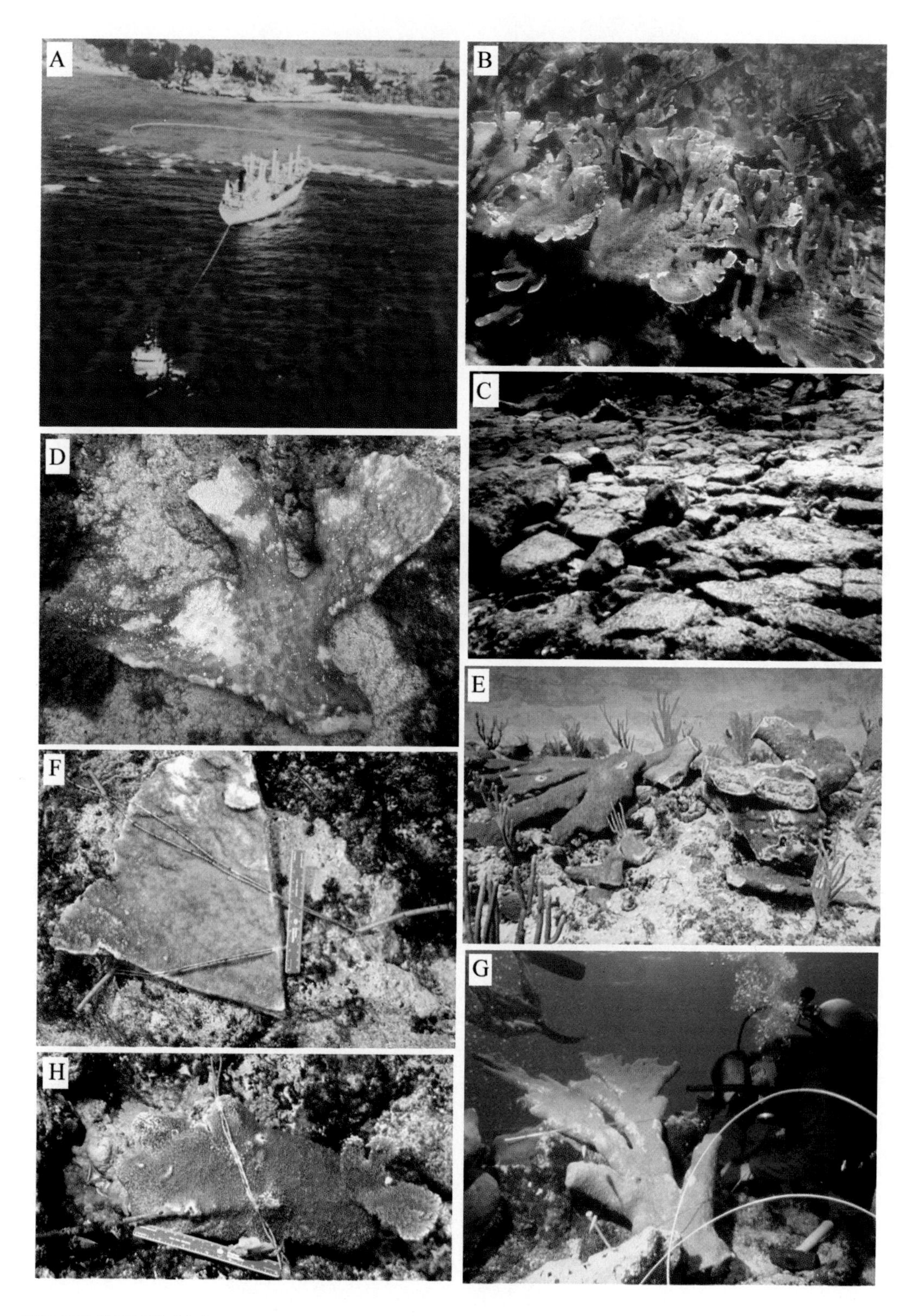

COLOR FIGURE 14.2

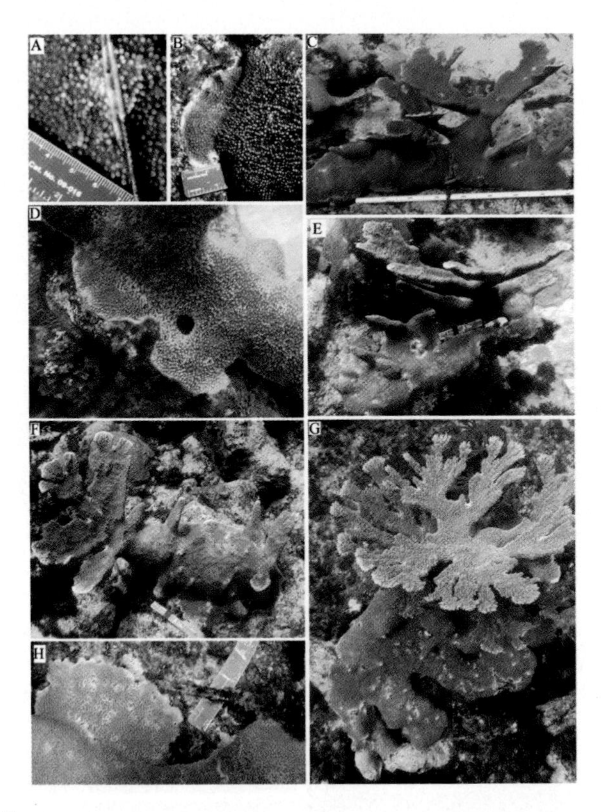

COLOR FIGURE 14.5

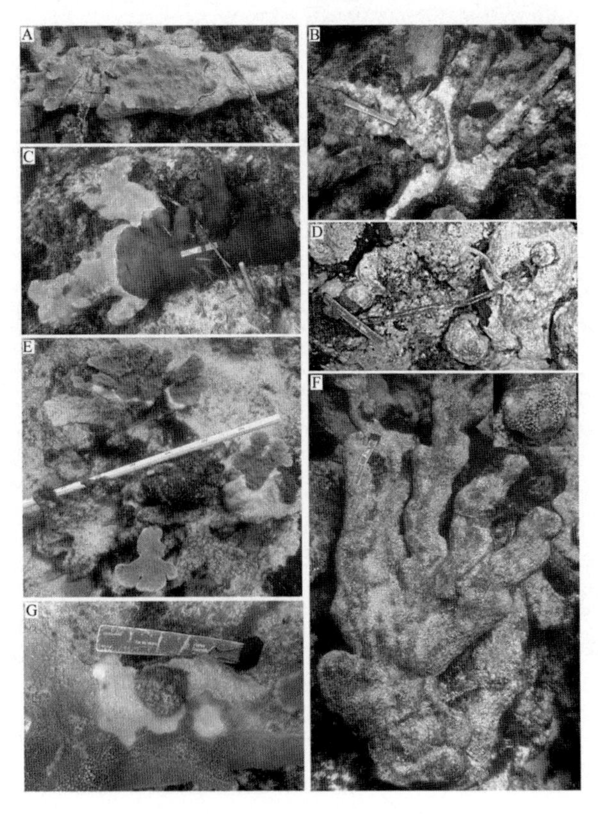

COLOR FIGURE 14.8

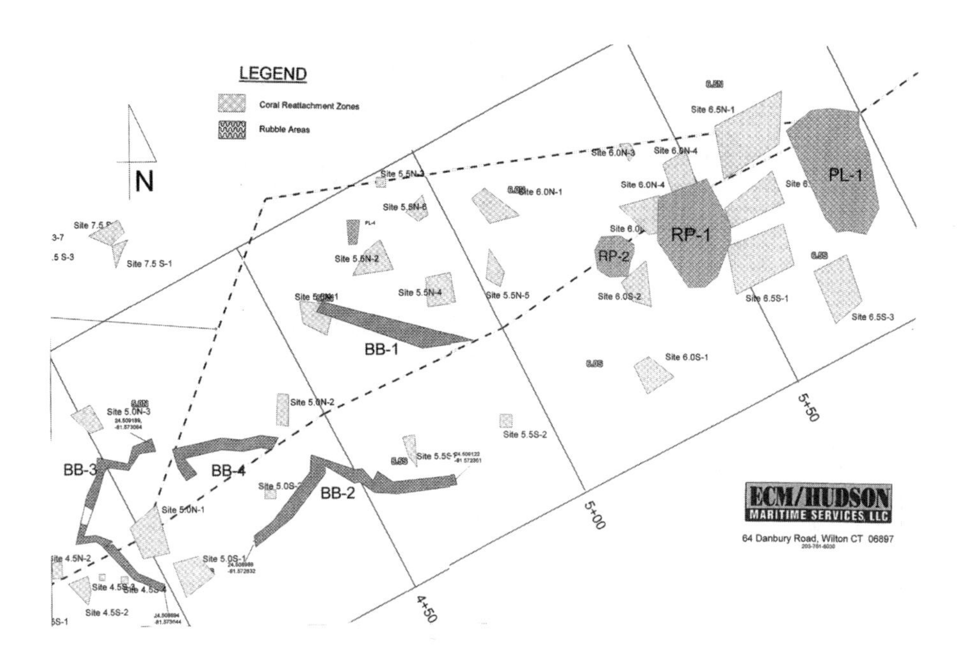

FIGURE 13.3 Portion of a detailed injury assessment map, showing the general location of the primary vessel grounding area, major vessel impact categories, and coral re-attachment sites. The bow was located to the upper right of the figure in the vicinity of the area marked PL-1. The stern was located near the linear rubble berms BB-3 and BB-4, which were created by propeller wash.

System (IVMS), incorporates the use of video data, DGPS, and GIS technology to provide detailed mapping and characterization of injury sites (Shaul et al. 1999). In this system, a scientific diver is deployed from a support vessel connected by an umbilical cable that provides surface-supplied breathing air and "hardwired" communication equipment. A floating DGPS receiver antenna is towed directly above the diver to provide accurate location information. As injury areas are encountered, each site is marked, measured, and photographed by the diver using a helmet-mounted video camera, as the diver describes the site using audio communication. All of this information is integrated as it is directly recorded on videotape and captured into a computer database aboard the support vessel. Using the IVMS, scientific divers identified, measured, and videotaped detailed locations (e.g., particular coral heads, scarified areas, rubble piles) based upon the injury classification categories. Analysis and postprocessing of the data generated an accurate map of the impacts at the grounding site (Figure 13.3). The site map was organized into 50-m by 50-m grids, originating from the initial point of impact and ending at the final bow location. The grids extended north and south of a baseline that ran the length of the grounding tract.

As the data were analyzed, a general description of the injured habitat was generated and the injury related to the ship's activities during and after grounding could be reconstructed. The primary injury zones were categorized as follows:

1. Stern grounding area — This was the largest zone and included the area directly impacted by the propellers as the ship's crew attempted to back the ship off of the reef, the area affected by the vessel prop wash (including linear rubble berms designated BB-1 through BB-4) and associated injuries (Figure 13.4 and Figure 13.5). This area encompassed approximately 6725 m^2. Injury was not uniform within this area and ranged from complete destruction to intermittent impacts.

FIGURE 13.4 Accumulated coral rubble, comprised primarily of dead staghorn coral (*Acropora cervicornis*) skeletons generated from propeller action in the stern grounding area.

2. Bow grounding area — This zone was comprised of the forwardmost reef spur (designated PL-1 for "parking lot") that was crushed when the ship grounded. Additional injury was caused when the ship pivoted during attempts to "power off" the reef. This area comprised approximately 548 m^2.
3. Coral spur area — Two spurs, measuring approximately 390 m^2, were reduced to piles of rubble (designated RP-1 and RP-2) when the ship pivoted from south to north as the ship's crew attempted to maneuver to get off of the reef (Figure 13.6).
4. Inbound path — This zone included all of the scars generated on coral spurs and hardbottom along the inbound path of the ship. These intermittent areas were measured during the injury assessment and cumulatively totaled approximately 569 m^2.

13.2.2.4 Injury Summary

The injury from the grounding of the *Contship Houston* is summarized in Table 13.3. The entire affected injured area totaled 8,326 square meters (2.06 acres). This figure was later adjusted to account for scattered zones within the area that had escaped injury, and for hard bottom density (percent living cover) within the grounding area. The final adjusted injured area was 5,206 square meters (1.29 acres), comprising one of the largest coral reef injuries when compared to other recent grounding incidents within the Florida Keys. Only the injury from the impact of the *Mavro Vetranic*, which ran aground in the Dry Tortugas in 1989, was larger. The injured area associated with recent major ship grounding incidents are summarized in Table 13.4.

FIGURE 13.5 Section of coral reef framework exposed due to propeller action in the stern grounding area.

FIGURE 13.6 Low-relief coral spur reduced to rubble from the grounding and movement of the *Contship Houston*. Vessel hull is visible in the photo.

TABLE 13.3
Summary of Injured Areas

Injury Type	Area (m^2)
Minor scar areas	
Initial grounding	569
Vessel salvage	94
Major scar areas	
Resting and pivot areas	938
Stern/propwash	6725
Total affected area	8326
Adjusted injured area	5206

TABLE 13.4
Major Vessel Groundings in the Florida Keys

Vessel	Date	Injury Area (m^2)
Mavro Vetranic	1989	5475
Contship Houston	1997	5206
Wellwood	1984	4379
Elpis	1989	2605
Igloo Moon	1996	2360
Miss Beholden	1993	1026
Maitland	1989	681
Columbus Iselin	1994	345

13.3 RESTORATION

13.3.1 Restoration Plan

A restoration plan was formulated by the RP in consultation with the Trustees. It consisted of three main components:

1. Reattachment and repair of coral heads and broken coral pieces along the inbound path and on the perimeter of the vessel resting area
2. Stabilization and/or removal of rubble generated by movement of the ship and propeller wash
3. Stabilization of the substrate in the vessel resting area, along with re-creation of lost habitat relief

The first component of the restoration plan consisted of reattachment of dislodged coral heads and coral pieces and the repair, when possible, of broken and crushed coral colonies. Considering the size of the affected area and the number of coral colonies involved, this effort proved to be one of the largest coral reattachment projects ever conducted. Time was of the essence to prevent coral mortality from wind and wave action. Both marine epoxy and underwater cement were considered for use in coral reattachment. Underwater cement was agreed upon as the preferred option due to the number and large size of many of the coral heads. It was important that all reattached corals be documented precisely so that project success could be analyzed in the future. The coral reattachment process was initiated immediately after the injury assessment was completed, and prior to the formulation and approval of the other components of the restoration plan.

The second component of the restoration plan related to the fate of the rubble piles or "berms" generated by the prop wash and physical movement of the ship. The original plan called for the removal of the rubble from the site, with subsequent disposal at a preapproved offshore area that was being considered for development of an artificial reef. Rubble removal would require the use of marine barges and cranes. Permits were requested and received from the Army Corps of Engineers and the FDEP, and a project scope of work was submitted by the RP to the Trustees in June 1997. The proposal for rubble removal was approved by the Trustees on July 22. However, subsequent to this approval, the RP submitted an alternate proposal to stabilize the rubble on site rather than remove it. This proposal was generated by the marine contractor to the RP and was based in part on cost factors relative to rubble removal. The RP proposed to use a "pourable" marine epoxy to be "drizzled" over the rubble piles in a crosshatched pattern. Analysis by marine engineers under contract to the RP indicated that this method would stabilize the berms and enhance the natural cementation process of these areas. The RP argued that this would keep all the reef material on site where it could serve as habitat for small fish and other organisms and reduce collateral injury to surrounding areas. A demonstration project was carried out on July 30 on one of the rubble berms. After some discussion with the Trustees, the marine epoxy method was approved to stabilize the rubble on site instead of removing and disposing of the rubble in a deep-water location.

The stabilization of the crushed coral spurs underneath the ship's bow in the final vessel resting area comprised the third component of the restoration plan. Associated with this was the reestablishment of some three-dimensional substrate complexity to replace lost structural habitat caused by the grounding. The first option, rubble removal, was discarded. As mentioned previously, the external limestone crust had been crushed in this area, revealing loosely consolidated rubble. Complete removal of this rubble was probably not possible due to the unknown extent of this stratum and raised the spectre of "digging to China," a prospect neither the RP nor the Trustees considered appropriate. The second option proposed use of precast cement "caps" similar to those used in the restoration of the *Columbus Iselin* grounding at Looe Key reef. The caps would be supplemented by the construction of low relief concrete habitat modules. Primarily due to cost considerations, the contractors for the RP proposed an alternate solution to stabilize these areas.

They proposed application of flexible, articulated concrete mats, similar to those used in shoreline stabilization projects, which would be placed over the destabilized spurs in the bow grounding area. The mats would be held down by their own weight and by the addition of large (1- to 1.5-m diameter) limestone boulders, obtained from south Florida rock quarries, which would be placed in rows along the interior edges of the mats. This type of stabilization had never been used in a coral reef restoration project before, and the Trustees had some concerns about the stability of the mats. However, a marine engineering analysis provided by the RP indicated that at the water depths of the restoration site, the articulated mats would be stable under a 100-year storm event. In November 1997, the Trustees approved the utilization and installation of the concrete mats.

"Pourable" marine epoxy for stabilization of the rubble berms and flexible concrete mats for coral spur stabilization were untested coral reef restoration methods. However, the Trustees were willing to consider "experimental" methods in order to further the range of possible options available for future coral reef restoration projects. Factors related to the location of this grounding site supported allowing an experimental approach to some aspects of the project. For example, the site of the grounding of the *Houston* is slightly deeper than many of the previous grounding locations and thus is less exposed to wind and wave action. In addition, the surrounding area is a low-relief spur-and-groove coral reef with associated hardbottom communities and exhibits relatively low living coral cover. While these habitats are extremely important in the Florida Keys, this area was not considered one of the "climax" coral reef communities exemplified by some of the more prominent areas in the lower Keys (such as Looe Key reef). Accordingly, the area attracts few recreational SCUBA divers and has no mooring buoys. Thus, certain aspects of aesthetics were not considered as important as in other coral reef areas in the Florida Keys. Therefore, given the high costs of concrete modules and caps as well as the other considerations, the Trustees were willing to consider less expensive options provided such options met the Trustees' restoration objectives.

13.3.2 Coral Reattachment

To facilitate the reattachment of dislodged and broken coral heads along the inbound path and in the vicinity of the vessel resting area, a series of restoration sites were established. Injured corals were collected from the surrounding area and consolidated into these sites for reattachment. Each restoration site was chosen based on natural features that allowed efficient coral reattachment, such as the availability of level topography with adequate open hard substrate, and measured approximately 3 to 4 m in diameter (Figure 13.7). The restoration sites were chosen to accommodate the injured corals associated with a particular hard-bottom spur. Each repair area was clearly marked with a central monument. The monument was established utilizing a 0.6-m threaded eyebolt set 0.3 m into the hard bottom. The eyebolt was secured into the rock substrate using marine epoxy.

FIGURE 13.7 Coral heads re-attached to reef substrate utilizing underwater cement.

A tag identifying the site by a number and a subsurface buoy were attached to the eyebolt. All monuments were referenced and mapped relative to the center baseline of the site. All reattached and/or repaired coral heads or benthic organisms within a restoration site were then referenced to the central monument using a compass heading and distance.

At each repair location, coral heads and dislodged hard bottom with attached benthic organisms were reattached to the reef surface using Portland Type II cement. The process resulted in the repair and/or reattachment of at least 3220 corals. The reattached corals ranged in size from 10 cm to 7 meters in diameter. The following 17 species of corals were represented in the reattachment effort: *Montastraea annularis* (complex), *M. cavernosa, Colpophyllia natans, Meandrina meandrites, Porites astreoides, P. porites, Diploria labyrinthiformis, P.* cf. *divaricata, Agaricia agaricites, Siderastrea siderea, Dichocoenia stokesi, Stephanocoenia intersepta, Acropora cervicornis, Eusimilia fastigiata, Favia fragum, Mycetophyllia* sp., and *Diploria strigosa.* The most common species represented were *Montastraea annularis* (complex), *M. cavernosa*, and *C. natans*.

At intervals during the repair process, a scientific dive team was deployed to document the repair sites using the IVMS mapping system described previously. The injury assessment provided information and data for the pre- and postrestoration condition of individual sites. Scientific divers located, described, and recorded on videotape repair locations using the mapping system. A total of 86 restoration sites was established and documented along the grounding corridor.

13.3.3 Rubble Stabilization

As discussed previously, two types of rubble existed on the site:

1. Rubble associated with the vessel resting area and generated primarily by the physical movement of the vessel as it ran aground and pivoted at the grounding site.
2. Rubble generated and piled up from the action of the ship's propeller wash. This second type of rubble primarily consisted of accumulations of branches of previously dead staghorn coral (*A. cervicornis*) skeletons, and came to be known as the "bone berms" (Figure 13.4).

The concern for these rubble areas was the potential for remobilization of the material during storms, causing additional future damage to the reef. The original plan developed for these areas was to remove most of the material from the bottom.

As previously described, subsequent to the approval of the rubble disposal proposal, the marine contractor representing the RP proposed an alternate plan that would stabilize the rubble on site instead of removing it. This proposal included the placement of some of the larger rubble into the deep cuts in the reef framework created by the action of the ship's propeller. The "bone berms" were then proposed to be stabilized using a liquefied marine epoxy mixture to be "drizzled" in a crosshatched pattern over the piles. A test application of the epoxy technique occurred at the end of July. Upon inspection by Trustee representatives, it was recommended that additional epoxy was necessary to achieve stabilization. The rubble berms were treated with the epoxy material in August 1997. Four linear rubble piles (designated as BB-1 through BB-4), totaling approximately 650 linear feet (197 m), and ranging from 8 to 20 ft (2.5 to 6 m) wide, were treated in this fashion (Figure 13.8).

13.3.4 Substrate Stabilization/Habitat Replacement

Articulated concrete mats were used to stabilize unconsolidated substrate in the vicinity of the bow grounding area on two separate low-relief coral spurs, designated as PL-1 and RP-1. Each mat unit consisted of a network of 90 concrete blocks, similar to concrete blocks used for a variety of construction purposes, attached together in two dimensions by cables to form an articulated mat.

FIGURE 13.8 Coral rubble treated with pourable epoxy material for stabilization.

The mats were flexible enough so that they could be "draped" over the disturbed substrate and conform to the underlying topography (Figure 13.9). Each mat measured approximately 20 by 8 feet (6 by 2.5 meters). A total of 40 articulated mats were used in this project, covering an area of approximately 8100 ft^2 (752 m^2). To provide additional three-dimensional habitat, a series of large limestone boulders was placed in rows between the articulated mats. The mats and boulders were lifted to the stabilization site by a barge and crane and guided to their location by divers (Figure 13.10). Substrate stabilization commenced on November 8 and was completed by November 13, 1997.

13.3.5 Compensatory Restoration

In addition to the primary restoration activities that are designed to restore the injured resources and services to the preinjury condition, the National Marine Sanctuaries Act, the underlying statutory authority for this action, requires the RP to compensate the public for the interim losses that occur while the injured resources are returning to their baseline condition. Such projects are known as compensatory restoration. In this case, the Trustees recommended and the RP agreed to purchase a Racon navigation system to be installed on the light towers and other structures located at major reefs along the Atlantic side of the Florida Keys. Racons provide a warning system to vessels that they are in or are approaching an area of special navigational concern. Racons, also

FIGURE 13.9 Flexible concrete mats and limestone boulders placed over injured coral spurs and rubble in the bow grounding area.

FIGURE 13.10 Surface supplied diver directing the placement of flexible concrete mats and limestone boulders on injured coral spur.

called radar beacons, radar responders, or radar transponder beacons, are active transponder devices that transmit unambiguous signals that show up clearly on a shipboard marine radar display. A Racon responds to a received radar pulse from a ship by transmitting an identifiable mark back to the radar set. The displayed response on the radar display is encoded as a Morse code character beginning with a dash for identification. Racon locations and their identifying marks are also indicated on marine charts. The RP purchased ten Racon transponders, eight of which were installed on prominent navigation structures along the Florida Keys reef tract, and two of which were held for backup and replacement purposes. In addition to the purchase, the RP agreed to pay for the deployment and 10 years of maintenance of the system. The U.S. Coast Guard agreed to install and maintain the system for the service duration. The locations and range of the Racon beacon system are shown in Figure 13.11. Assuming a 15-nautical-mile range for each beacon, most of the coral reef in the Florida Keys is covered by the system. The existence of this system will compensate the public by preventing the loss of coral reef resources from a portion of potential

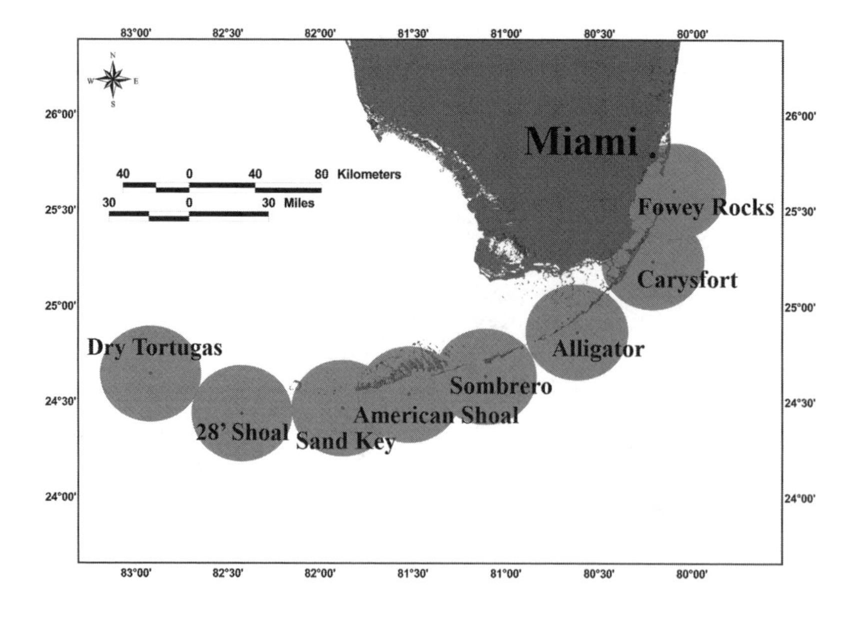

FIGURE 13.11 Locations of Racon navigation beacons installed as compensatory restoration to warn ships of sensitive coral areas in the Florida Keys. Circles indicate 15 nautical mile radius of beacon transmission.

future vessel groundings in the Florida Keys. It was estimated that if this system had been in place previously, it could have prevented 56% of the groundings that occurred along the reef system in the Florida Keys since 1984. Installation and activation of the Racon system was completed in March 1999.

13.4 MONITORING

13.4.1 Monitoring Plan

A monitoring plan was developed to assess the performance of the restoration for 10 years after completion of the project. It was designed to monitor stability and biological recovery of reattached corals, recruitment of new organisms associated with various components of the restoration, and physical stability of the rubble berms, concrete mats, and boulders. The monitoring plan also set out threshold performance values that if exceeded would require corrective action. The plan required eight monitoring events including a "baseline" survey by June 1998 (within approximately 6 months of the completion of the restoration project). Monitoring events were then scheduled after 6 months, 1 year, and 2 years subsequent to the initial survey, and at 2-year intervals thereafter. The major components of the monitoring plan are as follows:

Biological recovery — The plan required that 100 reattached corals and 50 uninjured controls be monitored, including photo-documentation, to assess colony stability and coral vitality. In addition, 30 injured barrel sponges (*Xestospongia muta*) and 15 controls would be assessed. Any detached corals observed would be reattached during the monitoring event, if still present. If a cumulative total of 20% or more of the monitored corals became detached from the substrate, all reattached corals would have to be reassessed and those detached would be stabilized.

Recruitment — Recruitment of benthic organisms was to be monitored on or within 1) 30 sections of rubble in berms not subjected to the epoxy treatment; 2) 30 sections of rubble in berms subjected to the epoxy treatment; 3) 30 concrete mat blocks; 4) 45 boulders; and 5) 30 natural substrate control areas. There were no corrective actions associated with thresholds related to this data.

Physical stability — The concrete mats, boulders, and rubble berms were to be assessed for physical stability. For the concrete mats, the two primary stabilization areas would be documented by video and assessed for movement or damage. The plan specified that any movement of concrete mats would require corrective action. For the boulders, a subset was selected for monitoring, accounting for approximately 30% of the total number. If more than 20% of these boulders moved, all boulders would be reassessed, and any boulders that moved would have to be relocated to their original positions. If excessive movement (greater than 50%) of the boulders was observed, further stabilization measures or removal would be required. For the rubble berms, a volumetric analysis was required at each monitoring event. If the rubble in the piles was reduced by 20% or greater, the berms would be restabilized or removed.

13.4.2 Hurricane Georges

The coral reef restoration effort for the *Houston* was subjected to a formidable test with the occurrence of Hurricane Georges on September 25, 1998, which passed over the lower Florida Keys in close proximity of the grounding site. Georges was classified as a strong category 1 hurricane on the Saffir–Simpson scale as it passed over the area. The meteorological station located

FIGURE 13.12 Damage to flexible concrete stabilization mats caused by Hurricane Georges.

at Sombrero Reef in the middle Florida Keys recorded a maximum sustained wind of 94.4 mi/h, with a peak gust of 105.8 mi/h. Hurricane-force winds were experienced in the lower Florida Keys for more than 10 hours. A survey was conducted by representatives of the RP shortly after the storm passed to inventory and quantify its effects. As a result of the storm, six of the 40 articulated concrete mats either moved or were damaged. Two of the mats, located on the northern edge of the easternmost restoration area (known as "PL-1"), were completely lifted up and transported approximately 25 to 30 ft (8 to 10 m) to the north. Four additional mats, two each located on the southern edges of the two bow resting areas ("PL-1" and "RP-1"), remained on the site, but were either broken apart, folded over on themselves, or crumpled in place (Figure 13.12). There was also movement of some of the boulders within all three of the areas where they had been deployed. In addition to the movement and damage to the mats, the survey found that 18 of the 100 (18%) monitored reattached corals became detached and were lost, as were 10 of the 53 (19%) control (uninjured) corals. It was also discovered that a significant amount of the rubble subjected to the pourable epoxy treatment was lost.

The monitoring program provided for actions that would be required subsequent to a catastrophic event such as Hurricane Georges. If the catastrophic event resulted in destabilization or movement of any of the concrete mats or limestone boulders, repair and/or restabilization of those areas was required. Subsequent to the damage to the restoration components inflicted by Hurricane Georges, the RP returned the concrete mats that had moved to their original positions and repaired the broken mats with new connecting material. The rubble displaced by the storm had been so widely dispersed that no restabilization was possible.

13.4.3 Monitoring Effectiveness

Three of the scheduled monitoring program events have been carried out to date: an initial (baseline) survey, with follow-up surveys at 6 months and 1 year. In addition, there was one unplanned, or catastrophic, monitoring event after the passage of Hurricane Georges. In June 1999, representatives for the RP negotiated a consent decree with the U.S. Department of Justice to resolve all claims by the United States resulting from the grounding of the *Contship Houston*. As part of the settlement agreement, the RP paid slightly over $1.4 million to cover remaining response costs by the government and to pay for future long-term monitoring of the restoration project. Plans are now underway to reassess the site in light of hurricane damage, undertake midcourse corrections, if appropriate, and resume a full monitoring program.

13.5 EVALUATION

13.5.1 Restoration Performance

13.5.1.1 Project Timeline

As previously indicated, the proposed project time line was developed to complete all primary restoration (injury assessment, coral repair, and rubble stabilization) prior to the peak of hurricane season (September 1, 1997). The actual time line for major milestones in the primary restoration effort is shown in Table 13.5. With the exception of the stabilization of the vessel resting area with the articulated concrete mats, project time lines were met. All primary restoration efforts were completed by mid-November 1997, slightly over 9 months after the vessel was removed from the reef. This is the fastest implementation of a major coral reef restoration project related to a vessel grounding in the Florida Keys to date.

13.5.1.2 Coral Reattachment

The coral reattachment effort was extremely successful. Over 3220 coral heads and coral pieces were cemented to the substrate in the area of the grounding. Even after the passage of Hurricane Georges, monitoring data indicate an extremely low failure rate of reattached corals that is not significantly different from the failure rate of naturally occurring coral heads. The health and vitality of reattached coral heads is also quite good, as evidenced from monitoring data collected approximately 2 years after project completion. At that time there were only three cases of coral mortality observed among the reattached corals. No control corals experienced mortality during the same period. Although this may indicate a slightly higher mortality rate for reattached corals, there is no significant difference between the two treatments. In addition, many of the reattached corals were severely injured prior to reattachment, so it should be expected that those corals would be under more stress and therefore more susceptible to subsequent mortality from disease, bleaching, algal competition, or other factors.

Early on, it was determined that marine epoxy was not the most effective method for coral reattachment for the extremely large-scale nature of this application. While epoxy is very good for attaching small- to medium-sized corals (Jaap 2000), it is not a cost-effective approach for the large size of the corals and the extensive area involved in the grounding of the *Contship Houston*. For epoxy to be effective, both attachment surfaces must be clean and free of appreciable amounts of loose sediment. This requires much more effort on the part of the divers involved in the reattachment. Epoxy is also

TABLE 13.5
Timeline for Restoration Actions

Action	Date
Vessel grounds	February 2, 1997
Letter of understanding	February 7, 1997
Vessel removed	February 8, 1997
Preliminary injury assessment	February 10, 1997
Emergency coral salvage	February 24 to March 4, 1997
Cooperative injury assessment	March 11 to April 1, 1997
Coral reattachment	May 12 to July 21, 1997
Rubble stabilization	July 31 to August 13, 1997
Primary restoration/stabilization	November 8 to 13, 1997

more difficult to work with in large amounts, and usually hardens faster than the time needed for positioning large coral heads. Finally, marine epoxy is much more expensive than underwater cement.

Underwater cement is not without its problems. In a few instances, "fallout" of fine-grained unconsolidated cement material was observed to cover the surrounding substrate within a meter or two of the subject coral. In addition, underwater cement is known to be highly toxic to some coral reef organisms. It can cause almost immediate tissue mortality when it comes in direct contact with many species of gorgonians, and it has been observed to completely clog up the incurrent pores of marine sponges. Collateral injury to nearby reef organisms was not quantified during the reattachment portion of this project, but the environmental marine contractors undertook special effort to ensure that cement fallout did not remain in contact with living tissue by "fanning" it away from areas in which it had accumulated. However, some minor impacts from the use of cement were observed.

13.5.1.3 Rubble Stabilization

The technique for stabilization of the linear rubble piles utilizing "pourable" marine epoxy did not perform well. During the passage of Hurricane Georges, a large percentage of the rubble treated in this fashion disappeared. The marine epoxy, for the most part, did withstand the effects of the storm, but the rubble not directly covered by the epoxy was mobilized and strewn over the surrounding area. It is possible that performance would have improved had more area of the rubble piles been covered with epoxy. Large-scale application of epoxy, however, would not have been cost effective. There is also evidence that the epoxy used in this application is at least mildly toxic to some marine life and may inhibit subsequent recruitment and recovery. Shortly after the epoxy was applied to the rubble piles, we observed that common colonizing algae would not grow directly upon the epoxy areas. Subsequent observation has confirmed less recruitment directly on epoxy than on the adjacent rubble. Material that inhibits recruitment and colonization of marine organisms should not be used in exposed areas for restoration projects. Additional analysis of the toxicity of marine epoxy to marine organisms is needed.

13.5.1.4 Substrate Stabilization

Utilization of articulated concrete mats did not perform as expected; these mats are not recommended for general use in coral reef restoration projects in the future. In spite of engineering analysis supporting their stability, some of the mats moved and broke apart under storm conditions. One factor in this failure is that the edges of the mats were not secured to the substrate appropriately. The mats should have been tied into stainless steel eyebolts that had been drilled into solid limestone and cemented with marine epoxy or other adhesive. There is evidence that during the storm event, substantial erosion occurred beneath the edges of the mats, allowing wave action to lift the edges of the mats, causing subsequent movement. Also, the mats are difficult to repair when damaged, and if the connecting cable material breaks, the individual block components can be strewn about the area, causing additional injury to the surrounding coral reef community. Finally, the mats themselves do not provide three-dimensional habitat necessary for restoration. If properly secured to the substrate, the articulated mats may serve the purpose of stabilization of the underlying substrate; however, they are aesthetically unappealing. Although aesthetics had not been clearly identified as a primary consideration in coral reef restoration projects at the time of this project, it is now obvious that it is an extremely important element of the perceived "success" of a restoration.

13.5.1.5 Recruitment

Recruitment of benthic organisms, primarily corals and sponges, has been surprisingly high at this restoration site (Figure 13.13 and Figure 13.14). Surveys conducted in 2004, 6 1/2 years after restoration, found that recruitment at the *Contship Houston* site was significantly higher than that associated with the restoration projects for the *Elpis* (restored in 1995) or the *Columbus Iselin*

FIGURE 13.13 Recruitment of coral and other benthic invertebrates on concrete mat components in 2004.

(restored in 1999). Coral recruitment at the *Houston* site included 12 genera and averaged 110 individuals per square meter (ind/m^2) on the concrete mats, and 35 ind/m^2 on the boulders, compared to an average of 23 ind/m^2 at the *Elpis* site and 9 ind/m^2 at the *Columbus Iselin* site (A. Moulding, University of Miami, personal communication). The most common corals that have recruited to the mats and boulders are *Siderastrea siderea* (comprising up to 78% of the recruits) and *Porites astreoides*. The epoxy used to stabilize the rubble berms has exhibited very low recruitment of benthic organisms (Figure 13.15). It is interesting that the concrete mats and boulders at this restoration site appear to exhibit significantly higher coral recruitment than other restoration sites studied. This aspect of the project must be analyzed, and the responsible factors should be incorporated into other project designs.

13.5.2 Cooperative Natural Resource Damage Assessment

There were obvious and significant benefits associated with the cooperative NRDA process demonstrated in the restoration of the *Contship Houston* grounding site, but there were also some drawbacks. In a cooperative NRDA process, the RP agrees to carry out the assessment and restoration program in consultation with and per the approval of the Trustees. From the outset, this

FIGURE 13.14 Section of flexible concrete mat and boulders in 2004 showing recruitment and growth of coral reef organisms.

FIGURE 13.15 Pourable epoxy treatment in 2004. Note loss of rubble material and relative lack of recruitment directly on epoxy surfaces.

type of relationship is desirable because it avoids significant time and expense related to litigation. Cooperative NRDA actions have been extensively used in oil spills and hazardous waste cases, resulting in timely settlement and restoration. In the case of the *Contship Houston*, the entire injury assessment and restoration project was completed in a little over 9 months. By contrast, coral injuries resulting from the 1994 *Columbus Iselin* grounding were not restored until 1999. In part, the delay of the *Iselin* restoration project, which was not subject to a cooperative NRDA process, resulted from protracted settlement negotiations and the lengthy contracting process required by Federal Acquisition Regulations. Clearly, the speed at which a restoration project can be implemented when the RP agrees to carry out restoration is an advantage. With federal procurement and contracting procedures eliminated, the injured resources can be restored more quickly. In the case of the *Contship Houston* restoration, the quick stabilization of the dislodged and injured corals by the RP made the difference between life and death of those corals.

Despite the many successes, the *Contship Houston* cooperative approach had some drawbacks and provides some "lessons learned" for future resource damage actions. First, the Trustees relied too heavily on the RP's technical engineering expertise with respect to stabilization of the vessel resting area and rubble berms. The RP proposed the articulated mats as a cost savings option that would have the desired stabilizing effect. As demonstrated by Hurricane Georges in 1998, the articulated mats were not the best option to withstand the wave action and turbulence created by major storm events. In addition, the poured epoxy treatments also proved to be ineffective at stabilizing rubble during such conditions.

Second, concern for cost effectiveness needs to be balanced with some realistic expectation of success. RPs will typically propose the least expensive restoration alternative. However, the least expensive option does not necessarily address whether the proposed technique is the best "value" for the money (Spurgeon and Lindahl 2000). Accordingly, the Trustees must develop clear restoration success criteria by which to evaluate the likelihood of success. When that likelihood is high, the Trustees should select the most cost-effective alternative. The RP and Trustees will both benefit in this situation: the RP has reduced costs, and the Trustee resources are restored more quickly. However, where the ability of the least expensive alternative to meet the restoration criteria is unknown, the Trustees should not feel obligated or pressured to approve it.

Third, even in a cooperative assessment and restoration process, it is critical that the Trustees retain the lead in making restoration decisions and providing the necessary oversight. When a particular restoration strategy raises doubt with the Trustees, they must be willing to undertake an independent evaluation of the biological efficacy and the physical feasibility of the strategy. This may entail hiring an independent consultant. If, after such evaluation or consultation, the RP proposal is in question, the Trustees should notify the RP accordingly. The Trustees can work with the RP to come up with an acceptable restoration strategy without divesting themselves of their decision-making authority.

Finally, it is very important to understand the true costs of any restoration project so it can be compared against other coral reef restoration efforts. Unfortunately, it is not possible to assess the cost efficiency of the *Contship Houston* project because the RP was not required to fully share this information with the Trustees. Some of the restoration methodologies utilized were purported to be less expensive options than those originally considered. While this may be true, there is no way of knowing the ultimate costs of the project without total cost information being released to the public. It is necessary in future NRDA cases that all costs associated with the project be fully reported to all parties.

13.6 RECOMMENDATIONS

Cooperative NRDA and restoration implementation should be pursued whenever possible. The benefits associated with this relationship far outweigh the negative aspects. It is especially crucial regarding the natural resource injury assessment, so that the extent of the injured resources can be quickly determined and is not subject to argument or debate by the Parties. The following recommendations are offered based on the experience with cooperative NRDA associated with the grounding of the *Contship Houston*:

First, immediately explore the opportunity of a cooperative NRDA with the RP. If the RP is interested, a Memorandum of Understanding should be executed at the outset that defines the relative roles of the Parties and makes clear that primary decision-making rests with the Trustees. The Trustees must establish clear standards and restoration success criteria that include an aesthetic component. In developing these success criteria, the Trustees must keep in mind the primary goal of restoration: to return the injured resources to as close to baseline condition as is possible. This goal must drive all components of the restoration effort. While circumstances may necessitate changes in some aspects of the restoration project due to cost considerations or other factors, the emphasis must remain on the optimal restoration of the coral or other natural resources.

Second, the fate of injured or dislodged coral colonies must be addressed immediately, regardless of other legal or administrative issues. Emergency stabilization and reattachment must occur quickly in order to prevent further coral mortality. Therefore, emergency restoration should be viewed as "Phase One" of the restoration process and should be undertaken even before the development of a full restoration plan, provided that the location of the reattached corals will not interfere with any subsequent restoration. A necessary first step in this process is the implementation of a cooperative resource injury assessment that is agreeable to both the RP and Trustees. Reattachment of coral colonies cannot take place until the complete injury is assessed because the reattachment process will alter the injured area. In the meantime, corals must be stabilized and "cached" if necessary in a safe location until the injury assessment is complete.

Third, the Trustees must ensure that a comprehensive restoration plan is developed that articulates clear success criteria. At a minimum, this should include provisions for the long-term stabilization of the injured substrate, re-creation of lost habitat values, and enhancement of natural recruitment of coral reef organisms to the area. Aesthetics should be included as one of the criteria. The goal in this regard should be that after the estimated period of recovery, the restored area should be indistinguishable from the surrounding uninjured coral reef. The Trustees should hire independent contractors if necessary to evaluate the feasibility of options that have not been used in prior coral reef restoration projects. This is particularly true where the Trustees have concerns about the efficacy of an option proposed by the RP or other party. While this may slow down some phases of the restoration, it is better to ensure that the restoration options selected will withstand the tests of time and weather.

The benefits of cooperative NRDAs cannot be overemphasized. At the *Contship Houston* grounding site, the rapid resource injury assessment and emergency reattachment of injured corals facilitated by this process ensured that coral mortality was minimized. The stabilization of coral reef substrate and rubble generated by the grounding occurred much more quickly through a cooperative NRDA than had the Trustees litigated the case or had the RP simply entered into a "cash-out" settlement agreement with the Trustees. The RP is typically able to enter into engineering and construction contracts far more expeditiously than can the Trustees. The cooperative approach can also provide an opportunity to explore some unconventional techniques of coral reef restoration. The Trustees, however, must remain vigilant about establishing clear restoration goals and agreeing to projects only if the Trustees are reasonably sure the projects will achieve those goals. In the meantime, more dedicated management and enforcement is necessary to prevent the careless operation of vessels in coral reef areas so that this type of restoration is not needed.

REFERENCES

Jaap, W.C., Coral reef restoration, *Ecological Engineering*, 15: 345, 2000.

Shaul, R., Integrated video mapping system (IVMS): a tool for coral reef injury assessment, in *Proc. Int. Conf. on Scientific Aspects of Coral Reef Assessment, Monitoring and Restoration*, National Coral Reef Institute, Nova Southeastern University, Dania, FL, 1999, 178.

Shaul, R., Waxman, J., Schmahl, G.P., and Julius, B., Using GIS to conduct injury assessment, restoration and monitoring during the *Contship Houston* grounding, in *Proc. Int. Conf. on Scientific Aspects of Coral Reef Assessment, Monitoring and Restoration*, National Coral Reef Institute, Nova Southeastern University, Dania, FL, 1999, 177.

Spurgeon, J.P.G. and Lindahl, U. Economics of coral reef restoration, in *Collected Essays on the Economics of Coral Reefs*, Cesar, H., Ed., CORDIO, Kalmar University, Kalmar, Sweden, 2000, pp. 125–136.

14 Restoration Outcomes of the *Fortuna Reefer* Grounding at Mona Island, Puerto Rico

Andrew W. Bruckner and Robin J. Bruckner

CONTENTS

14.1 INTRODUCTION

On July 24, 1997, the *M/V Fortuna Reefer* ran aground on a fringing reef located off the southeast coast of Mona Island (18°02'N; 67°51'W), 65 km west of mainland Puerto Rico (Figure 14.1). The 326-foot freighter remained grounded for 8 days within the island's largest remaining *Acropora palmata* (elkhorn coral) stand. Although the *Fortuna Reefer* was not carrying any cargo at the time of the grounding, some of the approximately 100,000 gallons of fuel oil and 33,000 gallons of marine diesel were removed to prevent a spill and lighten the vessel prior to removal. Steel cables were attached between the stern of the *Fortuna Reefer* and two tugboats to stabilize and extract the vessel. During removal, the extraction path did not follow the original collision path and the steel cables dragged across the reef surface, further expanding the area of injury.

The grounding and subsequent removal of the *Fortuna Reefer* impacted 6.8 acres of shallow fore reef habitat.[1] The reef substrate was crushed and fractured along the inbound track of the vessel, with additional damage occurring while the ship remained grounded as a result of waves and swell that caused the hull to rise and fall and shift sideways (Figure 14.2C). Total coral destruction occurred along the inbound and outbound paths of the vessel, extending from the reef crest approximately 300 m seaward (2 to 4 m depth) and up to 30 m in width. Collateral injuries from the steel cables extended beyond the perimeter of the vessel tracks, to 6 m depth. Entire

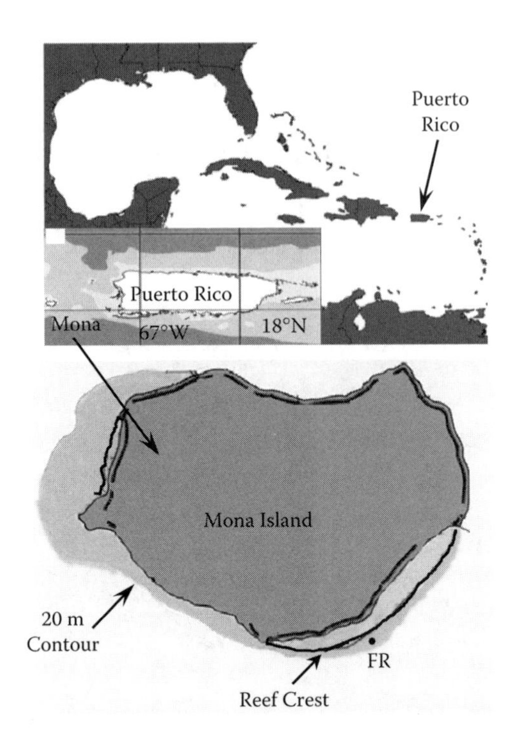

FIGURE 14.1 Location of the *Fortuna Reefer* restoration site off the southeast coast of Mona Island, Puerto Rico.

colonies of *A. palmata*, many that were several meters in diameter, were crushed or dislodged and fractured by the boat, and the cables sheared off hundreds of additional *A. palmata* branches (Figure 14.2E). Brain corals (primarily *Diploria strigosa*) and other benthic organisms were also abraded, shattered, or detached from the reef.

14.2 INITIAL RESTORATION

A team of experts from the Commonwealth of Puerto Rico and the National Oceanic and Atmospheric Administration (NOAA) Damage Assessment and Restoration Program (DARP) conducted a natural resource damage assessment of the grounding site and determined that an emergency repair of damaged corals was warranted. Coral fragments were scattered across the reef, and many had collected in sand channels. The surge continued to shift and overturn fragments, abrading their coral tissue and minimizing the likelihood of reattachment (Figure 14.2D). Without intervention, a high percentage of fragments may have died due to sand scouring or been removed from the site during periods of high wave action; securing coral fragments was predicted to accelerate recovery of the injury to the reef.[1] Under the Oil Pollution Act (OPA) of 1990, NOAA expedited a settlement with the responsible party amounting to U.S. $1.25 million for primary and compensatory restoration, including $650,000 to conduct an emergency restoration of coral resources injured as a result of the incident.[2]

The objectives of the emergency restoration were to reestablish the structural relief of the coral reef community and reduce coral mortality by reattaching loose *A. palmata* branches.[1] Between September 24, 1997, and October 14, 1997, a team of 19 marine engineers and biologists stabilized 1857 *A. palmata* coral fragments. Stabilized coral ranged from 15 cm to 3.4 m in length; all detached fragments larger than 1 m, 80% of the fragments between 0.5 and 1.0 m,

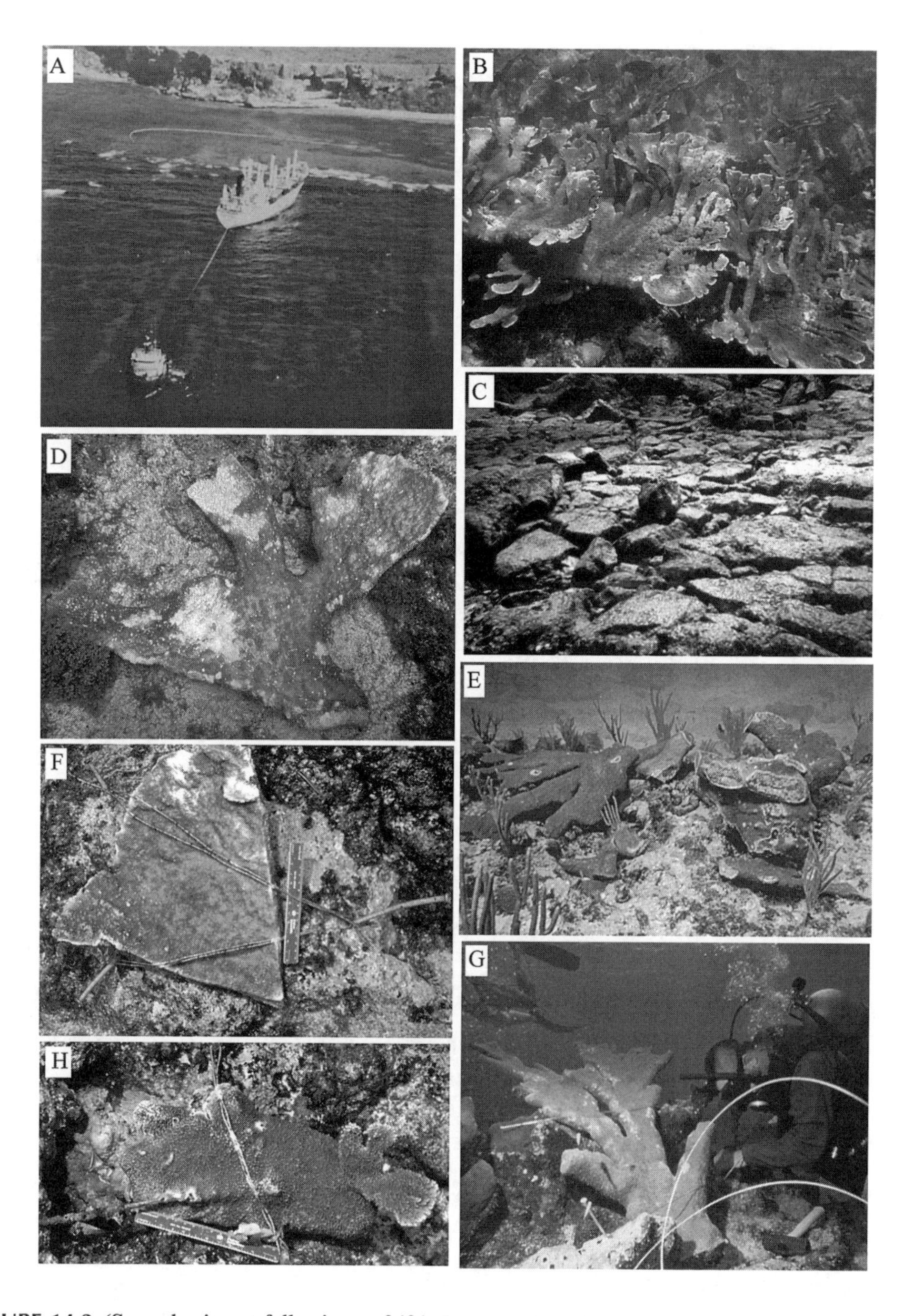

FIGURE 14.2 (See color insert following p. 240.) Impacts of the *M/V Fortuna Reefer* to shallow reefs. A. The ship grounded in the reef crest. Steel cables were attached between the vessel and tugboats to facilitate removal. Photo courtesy of the NOAA Restortation Center B. Undamaged elkhorn coral thicket adjacent to the grounding. C. The condition of the substrate after removal of the vessel. The reef substrate was crushed and fractured. D. Example of a fragment that landed in the sand and was being scoured and buried. E. *Acropora palmata* fragments sheared off by the cables. Photo courtesy of Dr. Jack Morelock. F. Example of a fragment attached to nails inserted into the reef substrate. Blue scale bar is 15 cm. G. Example of a fragment oriented upright that was being attached to a dead skeleton. Photo courtesy of Coastal Planning & Engineering, Inc. H. Example of a fragment that was reattached in 1997 and had overgrown the wire but was secured with additional wire during the midcourse correction in July 2000. Blue scale bar is 15 cm.

and 50% of the fragments up to 0.5 m in length were restored. Fragments were secured to the relict reef substrate (reef) or to dead, standing *A. palmata* skeletons (skeleton). Fragments attached to skeletons were expected to exhibit enhanced survivorship associated with a reduction in scouring, improved water circulation, increased light exposure, and possibly a reduced exposure to pathogens and benthic predators.

To secure fragments to the reef, stainless steel wire was extended across each fragment and then wrapped around stainless steel nails that were cemented into holes drilled in the reef (Figure 14.2F). Plastic cable ties were used initially to secure fragments to skeletons, with subsequent stabilization using wire that was wrapped around the fragment and skeleton.[2,3] Fragments were reattached either in an upright or downward position (with respect to their original orientation) such that the live, unbleached tissue faced upwards (Figure 14.2G). They were firmly anchored to withstand normal surge and wave action typical of shallow, exposed reef environments. Due to the high wave exposure observed during the restoration effort, cement or epoxy was not used, as the amount of time required for these materials to harden was not feasible.[1]

14.3 APPROACH TO EVALUATE FRAGMENT SURVIVORSHIP

Fragment survival and patterns of coral recovery were evaluated approximately 2 years after the grounding, and in May and August 2000, 2001, and 2003. The initial survey involved an assessment of the number, size, and condition of fragments that remained attached, and the number of fragments that were broken loose and displaced or missing. Detached fragments were estimated by:

1. Tallying groupings of nails within the reef to which fragments were no longer attached
2. Counting remnant wire on skeletons that was not associated with fragments
3. Counting detached fragments with remnants of attached wire

For each remaining fragment, measurements of the size (maximum length to nearest centimeter), orientation (up, down, or sideways with respect to their orientation prior to breakage), location of attachment (reef or skeleton), and condition (live or dead) were recorded. Dead fragments were marked with numbered aluminum tags to avoid recounting these on subsequent surveys. Live fragments were examined and evaluated for tissue growth over the wire, presence of protobranches, natural cementation (fusion) to the reef or skeleton, and growth onto the substrate or skeleton.

Estimates of remaining tissue and percent mortality were made from a planar perspective. For each fragment, a 1-m bar (divided into 1-cm increments) was oriented along the center of the long axis of the fragment to measure length, tissue survivorship, and tissue loss. Partial mortality was recorded as the percent loss of tissue from the upper surface of the reattached branches and does not include mortality to branch undersides. All fragments were presumed to have 100% of their upper surface covered with tissue when first reattached in 1997. Causes of partial or total mortality were identified as disease (white-band disease [WBD] or other syndromes); growth abnormalities (e.g., neoplasia or hyperplasia); overgrowth by boring sponges (*Cliona* spp.); predation by snails (*Coralliophila abbreviata*), polychaete worms (*Hermodice carunculata*), or parrotfish (*Sparisoma viride*); macroalgal competition; or presence of three-spot damselfish (*Stegastes planifrons*) territories. If a cause of mortality could not be determined, it was recorded as unknown.

14.3.1 Early Patterns of Fragment Survival

More than half (57%) of the fragments were alive 2 years after the restoration effort, while the remainder had died (26%) or became detached and removed from the site (17%).[4] Fragments secured to the reef had lower rates of early mortality (27%) and a higher rate of detachment due to wire breakage (24%) than fragments attached to skeletons. Surviving fragments had tissue covering a mean of 50% of the original upper branch surface, with 23% of the fragments exhibiting little (<5%) or no partial mortality. In addition, 19% of the live fragments exhibited vertical growth features in the form of small protobranches that were 2 to 5 cm in length.

14.3.2 Midcourse Correction

Interim monitoring reports from 1999 indicated a fairly good retention of wired fragments (estimated loss of 17%) even though the site had been affected by a category 5 storm (Hurricane Georges) and several severe winter storms.[4] However, broken wire was noted throughout the site, and only a small proportion of the fragments had successfully fused to the reef (10%) or to standing skeletons (7%). Between August 1999 and May 2000, another 8.3% of the fragments were detached and overturned or removed from the site. The wire continued to corrode and break, suggesting that sustained wire failure during periods of high wave action could hinder long-term recovery. A midcourse correction was conducted in July 2000 to prevent additional fragment loss. The midcourse correction primarily involved further stabilization of fragments with Monel 400, a more durable wire consisting of a copper–nickel alloy (Figure 14.2H). This wire was chosen because of its noncorrosive features, good flexibility, and excellent strength.

14.3.3 Patterns of Survival and Recovery over 6 Years

Six years after the restoration, 20.3% (377) of the restored fragments were living (Figure 14.3). Fragments had an average of 60% live tissue (Figure 14.4), although 33% had little or no mortality, and 22% showed signs of resheeting over previously denuded skeleton (Figure 14.5D). In addition, 30% (114) had solidly fused to the substrate (reef and skeletons) and exhibited new growth that was expanding outward, onto the reef or dead branches. More than half (58%) of the fragments had developed multiple branches that resembled the typical treelike morphology typical of adult colonies (Figure 14.5C). On average, fragments had four protobranches (maximum of 30) each, ranging in size from a mean of 21 to a maximum of 73 cm (Figure 14.6). Larger fragments appeared to have a greater number of protobranches ($r^2 = 0.58$, $p < 0.01$), and these were larger in size than those observed on small fragments, as estimated by regression ($r^2 = 0.60$, $p < 0.01$). Partial loss of tissue appeared to have less of an effect on the number of protobranches produced ($r^2 = 0.12$, $p < 0.0001$). While protobranch size was highly variable between fragments, protobranches on individual fragments were similar in size, suggesting that most protobranches on a single fragment formed at the same time.

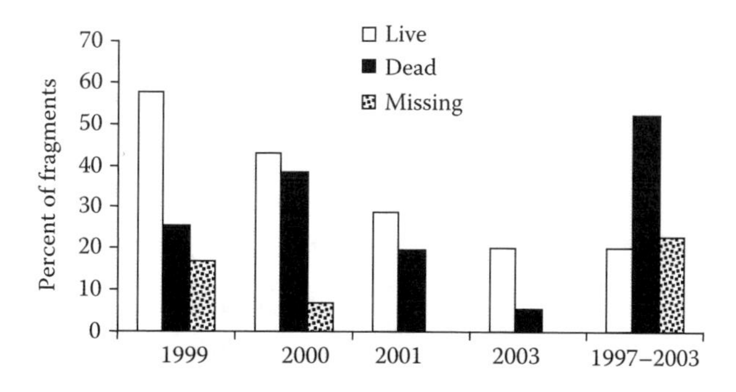

FIGURE 14.3 Proportion of fragments that were living (white bars), dead (black bars), or missing (spotted bars) during surveys in 1999, 2000, 2001, and 2003.

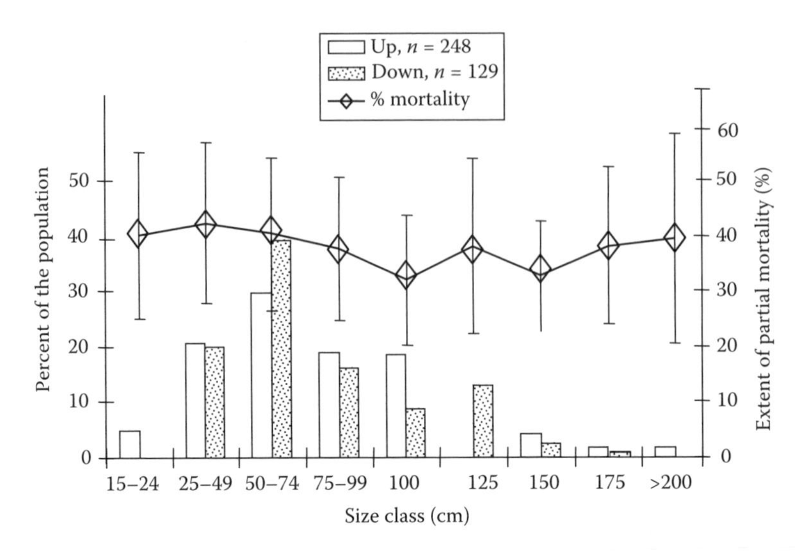

FIGURE 14.4 Size and condition of remaining live fragments in August 2003. The bars illustrate the percent of fragments in each size class oriented upright (white bars) and oriented down (spotted bars). The diamonds refer to the mean percent partial tissue mortality for each size class. Standard error is presented for partial mortality.

14.3.4 Causes of Mortality

Some of the most important factors contributing to fragment mortality were overgrowth by the brown boring sponge *Cliona* spp., predation by corallivorous gastropods, diseases, algal competition, and abrasion from the wire (Figure 14.7). Over the 6-year study, 20% of all fragments were killed by *Cliona* spp., and 5% of the remaining live fragments were affected by this sponge in August 2003 (Figure 14.8C). The sponge was most abundant on those fragments attached to dead standing colonies, advancing from the skeleton to the fragment. Many of these colonies were colonized by the sponge prior to the restoration effort, as evidenced by video documentation of the restoration. Disease was observed throughout the site at a low prevalence until 2003, when an outbreak of WBD was recorded on both restored fragments and unrestored colonies within and outside of the grounding site (Figure 14.8E). In August 2003, 4% of the living fragments and 15% of the standing colonies were affected by WBD.

The wire used in the initial restoration effort often negatively affected fragments. Tissue abrasion associated with wire and/or fragment movement during periods of surge was noted under and adjacent to the wire in 73% of the live fragments during initial surveys (Figure 14.8A). However, this was not always the case, as 22% of restored fragments had completely or partially overgrown the wire. Conversely, the wire used during the midcourse correction (July 2000) did not appear to negatively affect fragment survivorship, as tissue overgrew the wire within 30 to 45 days. Most wire breakage and fragment loss occurred prior to the midcourse correction (23%), with detachment of <0.5% of the remaining fragments occurring between August 2000 and August 2003.

14.3.5 Effects of Size, Orientation, and Attachment Site

Fragment length, attachment site, and orientation appeared to affect patterns of survival over 6 years. The highest rates of mortality were recorded between 1999 and 2001 (20 to 38% per year); these were primarily small fragments (mean = 59 cm) and fragments attached to skeletons. Additional fragments that died between August 2001 and August 2003 (5.4%) were larger than those that died in the first 3 years (t-test, $p < 0.01$), but they were still smaller than the remaining fragments

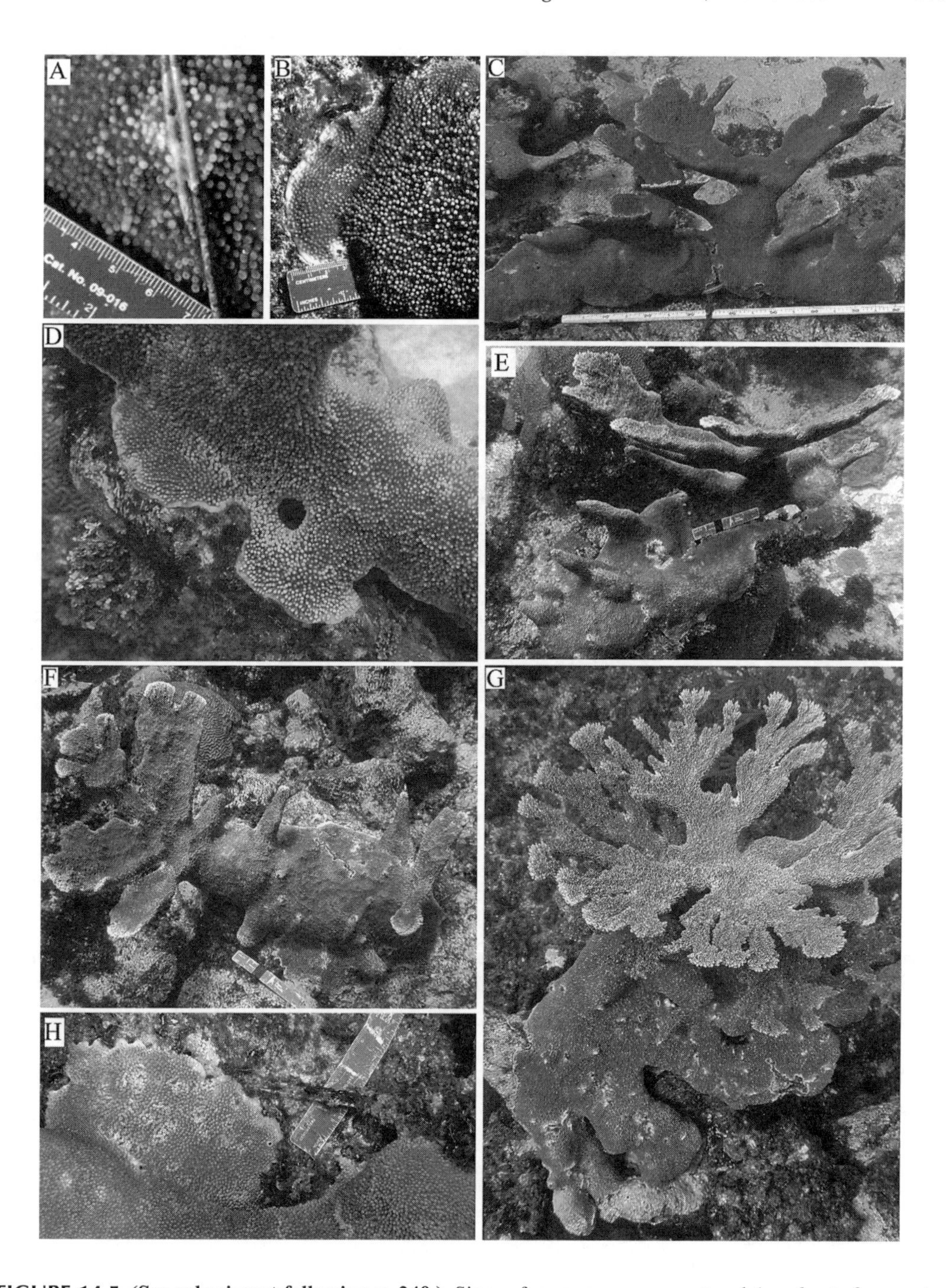

FIGURE 14.5 **(See color insert following p. 240.)** Signs of recovery among restored *A. palmata* fragments. Ruler is 15 cm. A. Growth of tissue over the wire. B. Fusion with the substrate. C. Fragment secured to the reef with numerous protobranches as of 2002. D. Growth of a fragment onto an elkhorn skeleton (2003). E. Fragment attached to a standing elkhorn skeleton (2003). The fragment has grown over the skeleton and produced multiple branches. F. Planar view of a restored fragment secured to the reef. The fragment is firmly cemented to the reef and has produced new branches (2003). G. Restored *A. palmata* fragment with a complex branching pattern (2003). H. Fragment that has begun to resheet over previously denuded skeleton and the wire.

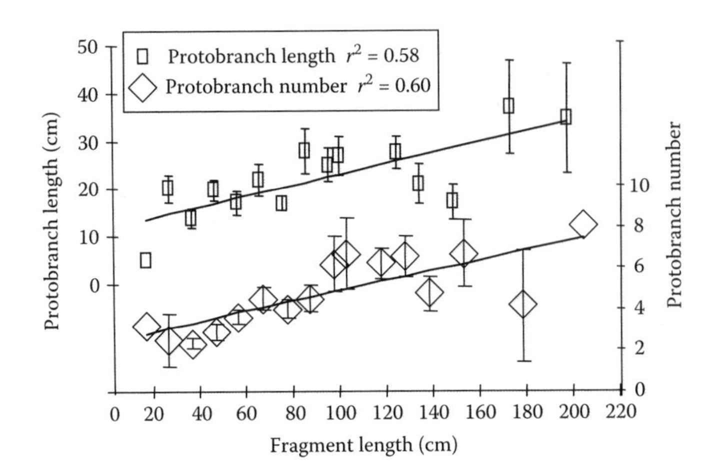

FIGURE 14.6 Relationship between fragment length and length of protobranches (rectangles) and number of protobranches (diamonds).

(Table 14.1). Fragments attached to the reef and oriented upright exhibited the highest rate of survival (33.7%), while fragments attached to skeletons and oriented down had the highest rate of mortality (85%) (Table 14.2). A significant relationship was observed between size and survival, with larger fragments surviving better than small fragments over 6 years. In contrast, there did not appear to be any interaction between attachment site and orientation (Table 14.3). Survival was also related to attachment site and orientation (Table 14.4); however, this may be due to variations among treatments, as:

1. A greater number of restored fragments were originally attached to the reef (56%).
2. Fragments attached to the reef were larger in size (75 cm) than fragments attached to skeletons (53 cm).

14.3.6 Effect of Depth

Restored fragments were reattached throughout the grounding site from about 2 to 6 m depth. Fragments attached to the reef in the shallow depths were detached more frequently than were fragments in deeper water and fragments attached to skeletons.[4] Fragments at intermediate depths (3 to 4 m) exhibited the highest survivorship, with losses in excess of 85% in areas deeper than 4 m.

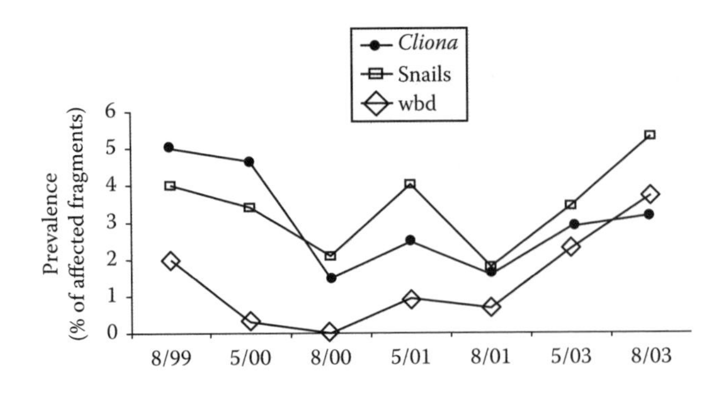

FIGURE 14.7 Prevalence of disease (open diamonds), the gastropod *Coralliophila abbreviata* (squares), and sponge (*Cliona* spp.) overgrowth (solid circles) during semiannual surveys.

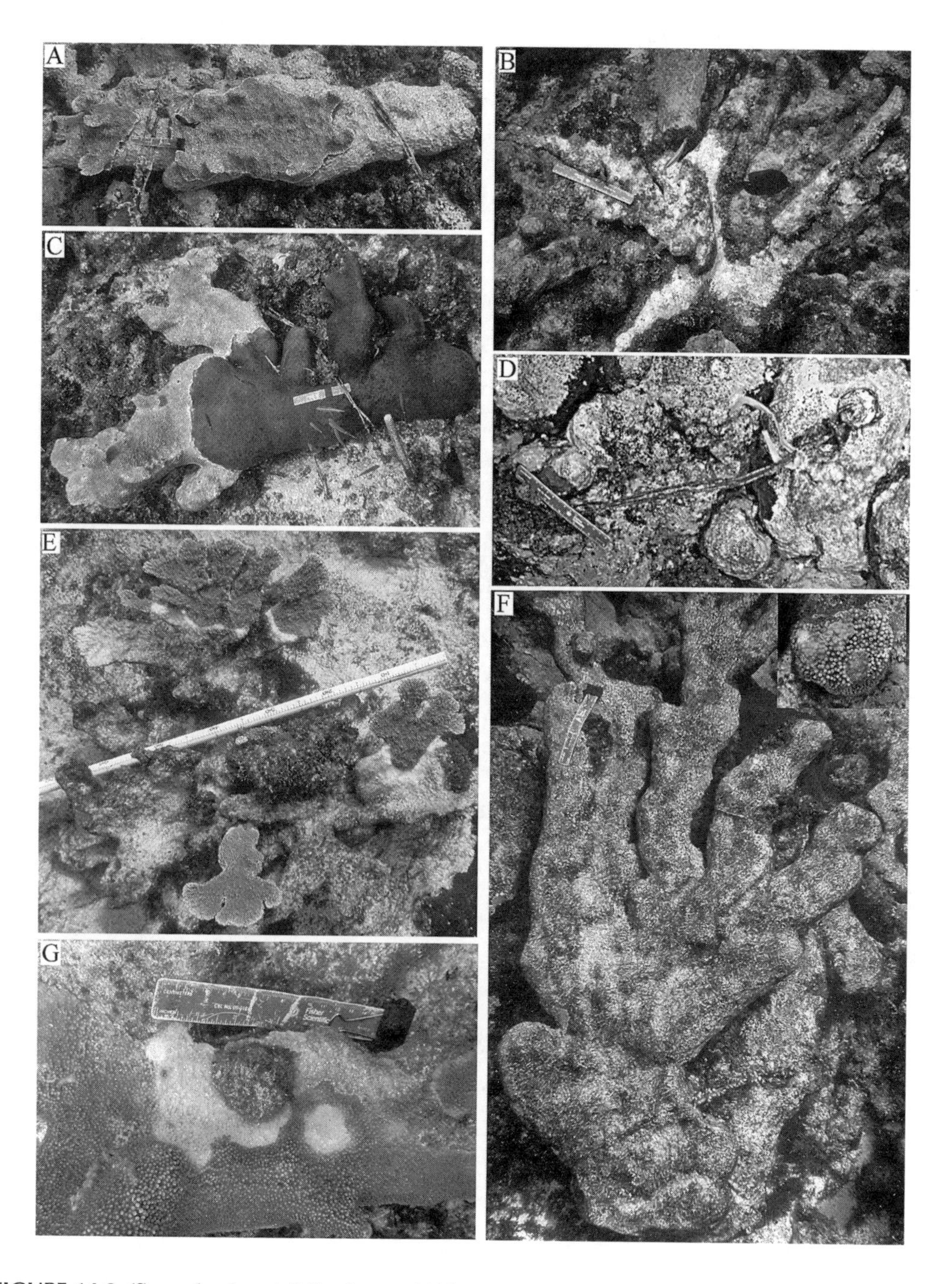

FIGURE 14.8 (See color insert following p. 240.) Factors contributing to fragment mortality. A. Partial tissue mortality associated with abrasion from the wire used to secure fragments. B. Fragment attached to a skeleton that died. C. Overgrowth of a restored fragment by *Cliona* spp. D. Missing fragment identified by the presence of nails and wire. E. Restored fragment with white-band disease. F. Large fragment that has experienced 99% partial mortality but continued to hang on. The insert in the upper right is a close-up of the remaining living tissue. G. Predation scar from *Coralliophila abbreviata* gastropods.

TABLE 14.1
Condition Fragments in August 2003

Attachment	Orientation	No. Live	Length of Live Fragments	Percent Old Mortality	Percent Recent Mortality	Percent Dead Fragments	Length of Dead Fragments
Reef	Up	134	93 (3.5)	36.4 (2.9)	1.0 (0.5)	66.3	68 (1.9)
	Down	55	80 (5.0)	39.8 (3.9)	0.8 (0.3)	74.8	68 (2.5)
Skeleton	Up	149	65 (3.2)	40.5 (2.3)	0.4 (0.1)	70.0	49 (1.4)
	Down	34	62 (4.9)	48.6 (6.0)	0.3 (0.3)	85.0	48 (1.5)

Note: The total number, mean length (cm), and degree of partial mortality are shown for live fragments attached to the reef or skeletons and oriented up or down, followed by the percent of dead fragments and their length (cm). Standard error is in parentheses.

TABLE 14.2
Extent of Growth and Partial Mortality for Live Fragments in 2003

Attachment	Orientation	No.	Length (cm)	Percent w/Protobranches	No Proto-branches	Size of Proto-branches (cm)	Percent Mortality
Reef	Up	134	93 (3.5)	66	4.9 (0.4)	24 (1.6)	37.4 (2.9)
	Down	55	80 (5.0)	49	3.8 (0.4)	19 (2.2)	40.6 (3.9)
Skeleton	Up	149	65 (3.2)	59	4.3 (0.4)	20 (1.4)	40.9 (2.3)
	Down	34	62 (4.9)	38	3.5 (0.6)	18 (0.6)	48.9 (6.0)

Note: The amount of partial mortality and extent of new growth (protobranches) are shown for the remaining live fragments that were attached to the reef or skeleton and oriented up or down with their original orientation on the source.

TABLE 14.3
Three-Factor Analysis of Variance (ANOVA) Testing for Relationships between Fragment Length and Orientation (up or down), Attachment Site (Reef or Skeleton), and Survival (Live or Dead)

Source of Variation	SS	df	MS	F-Ratio	*P*
A. Orientation	10.631	1	10.631	3.353	0.067
B. Attachment	344.444	1	344.444	108.630	<0.0001
C. Survival	196.488	1	196.488	61.968	<0.0001
A × B	6.555	1	6.555	2.07	0.151
B × C	0.897	1	0.897	0.28	0.595
A × C	0.048	1	0.048	1.6	0.2
A × B × C	6.759	1	6.759	2.132	0.145
Error	419	1323	0.983		

Note: Length (cm) data were log transformed.

TABLE 14.4
Two-Factor Analysis of Variance (ANOVA) Testing for Relationships between Survival and Attachment and Orientation

Source of Variation	SS	df	MS	F-Ratio	*P*
A. Attachment	1.320	1	1.320	6.68	0.010
B. Orientation	3.976	1	3.976	20.113	<0.001
A × B	0.316	1	0.316	1.598	0.206
Error	262.330	1327	0.198		

14.4 EFFICACY OF THE RESTORATION

Some of the most important considerations for future restorations involving *A. palmata* include selection of suitable attachment sites, choice and logistics governing materials to secure fragments to minimize negative impacts and maximize tissue survival, and strategic orientation of fragments to facilitate rapid fusion and growth. Although nearly 80% of the restored fragments died or were removed from the site, survivorship is comparable to or higher than that reported in other studies examining the fates of unrestored corals. For instance, about half of all fragments died within 4 months after a hurricane in Belize,[6] while 35 to 65% of the fragments in two locations in U.S. Virgin Islands died within 11 months.[7] In Florida, 57% of the storm-generated fragments landing on hard bottom substrates were removed within 11 months, and all remaining fragments were removed after 3 years.[5]

Rapid and near-total fragment cementation has been noted within weeks of a disturbance when fragments land on a suitable substrate (e.g., rubble appears to be better than hardground), are advantageously oriented, and have live tissue in direct contact with the attachment surface.[5] In contrast, very few fragments at the *Fortuna Reefer* site were fused after 2 years, and these were primarily those that had been positioned such that tissue at the edge of a branch was in contact with the reef or skeleton. Securing the fragments with wire retained a greater proportion of fragments than would be expected with no reattachment efforts, in light of the high wave exposure at this site combined with the low rates of fusion observed initially.

The use of wire appeared to be effective over the short term for securing coral fragments and may be better than cement or epoxy in comparable high-surge environments. Nevertheless, wire corrosion, breakage, and tissue loss associated with wire abrasion hindered the effectiveness of this technique during the first 2 years. Negative impacts associated with the wire used in the midcourse correction were minimal, possibly because the wire had a different composition and the fragments had not been recently exposed to a stressful incident with the magnitude of a major ship grounding. Other approaches used to secure fragments, including cement or epoxy, cable ties, and attachment to artificial structures such as cement rosettes may reduce partial tissue loss. However, some of these may be impractical in remote locations due to the additional weight associated with transport and availability of the preferred material (e.g., Portland type II cement). Also, proper adhesion requires additional preparation time to ensure adhesion (e.g., removal of algae, sediment, and other biotic agents), and high wave conditions may prevent the cement/epoxy from properly setting. While fragments grow over plastic cable ties more rapidly than wire, constant surge may cause continued stretching and loosening of plastic cable ties.

Fragments attached to the reef appeared to exhibit higher survivorship over 6 years than did those attached to standing skeletons. While this is the opposite of what had been anticipated, the difference may be related to the high prevalence of the boring sponge (*Cliona* spp.) on dead standing elhorn skeletons. *Cliona* that overgrew fragments was most frequently associated with dead standing colonies and less common on relict reef substrates. In addition, fragment survivorship increased with fragment size, and fragments attached to the reef were larger than those attached to skeletons. Plastic-coated wire may be the most preferable alternative, combining the durability and ease of attachment of wire with a material that does not abrade coral tissue and is readily overgrown by the coral.

Placing fragments on standing skeletons may speed up rates of ecological recovery by reducing time required to restore living topographic relief as is present with the tree-like morphology of intact *A. palmata* colonies. While fragments did resheet onto dead branches, this was limited to areas adjacent to the fragment. However, fragments must allocate energy toward regeneration and resheeting over standing skeletons in addition to new growth associated with linear extension of branches. Over the long term, increasing weight associated with continued growth of fragments, failure of fragments to completely resheet over the skeleton and continued bioerosion of tissue-denuded skeletons by sponges and other bioeroding organisms may contribute to breakage of skeletons and ultimate loss of these restored fragments.

Due to the limited observations of sexual recruitment in *A. palmata*, recovery of this species after a ship grounding or other physical disturbance (e.g., storms) is highly dependent on survivorship and growth of *A. palmata* fragments. In light of the large regional losses sustained by *A. palmata* populations, efforts to collect "fragments of opportunity" that may otherwise die (e.g., those that land in the sand) and stabilize these in optimal locations, may provide a low-impact source of fragments that could be available to rebuild populations.

14.5 CONCLUSIONS

This study provides valuable information on the benefits and drawbacks of the techniques used at the *Fortuna Reefer* site to guide future efforts and identifies some factors limiting survivorship that could be minimized to increase the viability of restored corals. The main limitation of this study is the inability, due to the lack of suitable controls, to quantify the absolute benefit of the restoration effort as compared to natural patterns of recovery if the restoration had not been undertaken. It is also difficult to conclusively demonstrate which of the two substrates was most suitable for attachment. Fragment size differed among substrate types, and fragments attached to skeletons frequently died from a biological agent (boring sponge), that could have been avoided with better fragment placement and that was less common on reef substrates. Future restoration efforts would benefit from using a science-based experimental hypothesis, if progress is to be made in developing methods to truly enhance ecological function of degraded reefs. These efforts must ensure that rigorous, quantitative monitoring data is obtained for restored fragments as well as undamaged and damaged but unrestored controls. Reattached corals must also be assessed to determine if natural recruitment processes are accelerated, and whether the restored fragments provide sufficient habitat complexity to enhance other aspects of community recovery.

ACKNOWLEDGMENTS

Funds for these surveys were provided by Earthwatch Institute, the NOAA Coral Reef Conservation Program, and the NOAA Restoration Center. Special thanks to Paco Garcia and staff of Mona Aquatics, Earthwatch Volunteers, and University of Puerto Rico graduate students. We are grateful for the field assistance provided by Ron Hill, Michelle Sharer, and Michael Nemeth. The opinions and views expressed in this document are those of the authors and do not necessarily reflect those of the National Marine Fisheries Service, the National Oceanic and Atmospheric Administration, or the U.S. Government.

REFERENCES

1. NOAA, National Oceanic and Atmospheric Administration Damage Assessment and Restoration Program (DARP) Web Site: http://www.darp.noaa.gov/southeast/fortuna/restore.html updated: 2005.
2. NOAA, *Fortuna Reefer* Settlement Agreement http://www.darp.noaa.gov/southeast/fortuna/pdf/fortcd.pdf 1997.
3. Iliff, J.W., Goodwin, W.B., Hudson, J.H., and Miller, M.W., Emergency stabilization of *Acropora palmata* with stainless steel wire and nails: impressions, lessons learned, and recommendations from Mona Island, Puerto Rico. *International Conference on Scientific Aspects of Coral Reef Assessment, Monitoring, and Restoration*, National Coral Reef Institute, Fort Lauderdale, FL, 1999, p. 110.
4. Bruckner, A.W. and Bruckner, R.J., Condition of restored *Acropora palmata* fragments off Mona Island, Puerto Rico, 2 years after the *Fortuna Reefer* ship grounding, *Coral Reefs* 20, 235–243, 2001.
5. Lirman, D., Fragmentation in the branching coral *Acropora palmata* (Lamarck): growth, survivorship, and reproduction of colonies and fragments, *Journal of Experimental Marine Biology and Ecology* 251, 41–57, 2000.
6. Highsmith, R.C., Riggs, A.C., and D'Antonio, C.M., Survival of hurricane-generated coral fragments and a disturbance model of reef calcification/growth rates, *Oecologia* 46, 322–329, 1980.
7. Rogers, C.S., Suchanek, T.H., and Pecora F.A., Effects of Hurricanes Davis and Frederic (1979) on shallow *Acropora palmata* reef communities: St. Croix, USVI, *Bulletin of Marine Science* 32, 532–548, 1982.

15 Review of Coral Reef Restoration and Mitigation in Hawaii and the U.S.-Affiliated Pacific Islands

Paul L. Jokiel, Steven P. Kolinski, John Naughton, and James E. Maragos

CONTENTS

ABSTRACT

Numerous coral reef mitigation and restoration projects have been conducted in Hawaii and the U.S.-Affiliated Pacific Islands. This chapter reviews the results of these projects and presents a summary of what has been learned. Many of the projects involved transplantation of corals away from proposed construction sites into adjacent areas. Initial transplant mortality was generally low, but long-term mortality often was high due to wave damage and other adverse environmental conditions in the transplant receiving areas. Transplants in wave-sheltered areas showed better long-term success. The terms *mitigation* and *restoration* often are taken to mean reef repair, coral transplantation, or construction of additional habitat (e.g., artificial reefs). However, experience in the Pacific has shown that other feasible options are available. Removal of anthropogenic stress allows natural regeneration processes to occur and often is the most effective approach in remediation. In many situations the natural rates of reef recovery are very rapid, and direct human intervention is unnecessary. Where restoration of a damaged reef is not feasible, a negotiated financial settlement or financial penalties can be used to establish trust funds or undertake other activities that will offset the environmental damage. Managers must develop broad strategic plans and incorporate a wide range of approaches designed to fit each situation on a case-by-case basis. Although protection is the top priority, damage to reefs from various causes will inevitably occur. In these situations direct restoration and mitigation measures must be considered. The cost of reef repair and coral transplantation can be high but effectiveness is generally very low. Protection and conservation, rather than restoration of damaged reefs, is the preferred priority. There is no point in restoring a damaged reef that will continue to be impacted by pollutants. Also, unscrupulous developers or polluters could use a token restoration or mitigation effort as a means of achieving their aims at the expense of the environment; thus, vigilance is required.

15.1 INTRODUCTION

Reef coral communities in the Pacific have been severely impacted by natural events such as storm waves,[1] freshwater floods,[2] and crown-of thorns starfish (*Acanthaster planci*) invasions.[3] Increasingly, reefs are impacted by anthropogenic factors such as ship groundings,[4] dredging and filling,[5] increased sedimentation due to improper land use,[6] and various forms of pollution.[7] In recent years there has been extensive damage to reefs on a worldwide basis due to bleaching and consequent coral death that has been attributed to global warming. Substantial evidence indicates that global warming is being caused by anthropomorphic production of carbon dioxide and other "greenhouse" gasses.[8] Until the past decade, little interest in mitigation and restoration of reefs existed.

Construction and other human activities in Hawaii and the U.S.-Affiliated Pacific Islands have damaged many coral reef communities with little or no associated compensatory mitigation or restoration.[9–11] For example, lengthening of the Moen, Chuuk airport was initiated in 1976 with 16 hectares (40 acres) of reef buried under armor stone, and meaningful mitigation was never achieved. The ability of federal agencies to effectively mitigate unavoidable impacts to Pacific coral reef ecosystems since the passage of the National Environmental Policy Act in 1970 (NEPA) has been described as uncertain due the lack of a comprehensive interagency mitigation strategy and a lack of information on the various options and their effectiveness.[11] However, compensatory mitigation has now become an important management concern, and agencies are working to develop and implement a comprehensive management strategy. Much can be learned from various mitigation and restoration actions in Hawaii and the U.S.-Affiliated Pacific Islands. For the most part the results of recent projects have not been published, although some earlier projects have been documented in the literature.[12,13] Therefore, a large part of the information contained in this chapter was derived from the direct involvement in the projects by the authors, from various unpublished reports, and through personal communication with individuals directly involved in past and current work. This chapter builds upon two previously published summary articles[14,15] and incorporates information from a report in preparation.[16]

The purpose of this chapter is twofold. First, we describe and summarize examples of mitigation and restoration projects that have been conducted in our region. Second, we synthesize and evaluate their effectiveness and list general guidelines for the mitigation and restoration of coral reefs.

15.2 OVERVIEW OF PROJECTS IN HAWAII AND THE U.S.-AFFILIATED PACIFIC ISLANDS

Naughton and Jokiel[14] grouped mitigation/restoration approaches into four main categories: direct action, indirect action, negotiated settlement, and establishment of strategic reserves.

15.2.1 Direct Action

Most of the effective mitigation/restoration projects undertaken in the U.S. Pacific fall into this category. Proactive intervention is directed at the reduction or avoidance of reef damage via project redesign or reestablishment of reef coral populations and/or coral habitats in damaged areas. Techniques for active intervention include reef repair, coral transplantation, reef seeding with coral fragments or larvae, increasing habitat area through placement of artificial reefs, and removal or control of harmful organisms.

15.2.1.1 Reef Repair

In a number of cases action was taken to repair reef damage or remove debris from an impacted site:

15.2.1.1.1 Agana, Guam

During 1992 a large naval vessel dragged its mooring chain across a submerged reef in Agana Harbor, damaging the corals over a wide area. A recovery effort was developed that included righting the overturned corals, stabilizing fragments, and removing debris. Within 5 years considerable recovery of damaged corals and recruitment of new corals occurred, but damage was still evident.[17] A major factor contributing to the recovery was that the site is protected from ocean storm waves and swell, so the broken and dislocated corals remained in place.

15.2.1.1.2 Rose Atoll, American Samoa

In 1993, a 250-ton long-line fishing vessel, *Jin Shiang Fa*, ran aground on a pristine reef at Rose Atoll National Wildlife Refuge.[4,18,19] The vessel released 100,000 gallons of diesel and lubrication oil and broke up rapidly. The spills and crushing action of the grounded ship damaged the reef structure and caused a massive die-off of reef organisms. Impacted areas of the reef were quickly colonized by opportunistic invasive algae and cyanobacteria. Ship debris was spread over several hectares. A salvage tug funded by the ship's insurer removed the ship's bow and other large debris from the reef flat before efforts ceased due to exhaustion of funding (about U.S. $1.2 million). Reef flats deteriorated further when dissolved iron from the corroding wreckage stimulated blooms of invasive algae and cyanobacteria. The U.S. Fish and Wildlife Service (FWS) succeeded in removing 105 tons of metallic debris and fishing gear from the reef during 1999 and 2000. An additional 40 tons of large metallic debris on the fore reef and 10 tons of nonmetallic debris in the lagoon remain at the atoll. Earlier emergency cleanup reduced the extent of damage,[20] but significant damage is still evident. In 2003 FWS succeeded in obtaining funds from the U.S. Coast

Guard's Oil Spill Liability Trust fund to finance remaining cleanup in 2004 and 2005 and monitoring over the next decade.[21]

15.2.1.1.3 Enewetak Atoll, Bikini Atoll, and Johnston Atoll

An incomprehensible scale of reef destruction and contamination resulted from 82 nuclear weapons tests, particularly in the Marshall Islands District of the U.S.-administered Trust Territory of the Pacific Islands, from 1946 to 1962.[22] For example, the "Mike Test" (1952) at Enewetak vaporized the island of Elugelab and left a 70 m deep, 1.9 km wide crater and a deeply fractured reef platform. The subsequent "Koa Test" in 1958 caused the fractured reef next to Mike Crater to break away and plummet to the ocean depths. From 1977 to 1980 the U.S. conducted a partial cleanup and rehabilitation of Enewetak[23] at a cost of U.S. $218 million. Work at Enewetak included removal of debris, derelict ships, piers, and other structures from the reefs in addition to burial of tons of radioactive material produced by 43 atomic and thermonuclear explosions. Cactus Crater on Runit Island, Enewetak, was formed by a nuclear test in 1958. The crater was 30 feet deep and 350 feet across. The crater was filled with thousands of tons of radioactive material. When it became clear that the crater was too small to contain all waste, a mound was created and the top capped with a dome of 18-inch-thick reinforced concrete. Contamination on reef and island ecosystems at Enewetak and several other atolls is still pervasive. For example, food grown in experimental plots still shows high levels of cesium 137.[24] The scale of these "restoration" (i.e., cleanup) efforts has been immense compared to other projects but trivial in view of what was actually achieved to mitigate the extensive damage done to these atolls.

15.2.1.1.4 Northwestern Hawaiian Islands

Derelict fishing gear (marine debris consisting mainly of lines, trawl nets, drift nets, seines and gill nets) accretes into large masses of floating material in the north Pacific that eventually drift into coral reef waters. This material damages corals, entangles wildlife, and can be an agent for the introduction of alien marine species.[25] Drifting clumps of lines and nets can entangle endangered monk seals, sea turtles, and sea birds, causing suffocation or inflicting wounds. Entanglement can prevent these creatures from feeding, and they starve to death. The death of 25 Hawaiian monk seals due to entanglement by derelict fishing gear was documented during 2002; the total population is between 1200 and 1400 animals. Between 1982 and 2002 a total of over 170 Hawaiian monk seals are known to have been entangled in derelict gear. The most effective mitigation effort to date is physical removal of derelict fishing gear. Since 1996 a multiagency effort (National Marine Fisheries Service, Ocean Conservancy, University of Hawaii Sea Grant, U.S. Coast Guard, U.S. Navy and others) has been removing derelict marine debris from Hawaiian reefs. Efforts have been focused on French Frigate Shoals, Lisianski Island, and Pearl and Hermes Reef. Divers pulled behind small boats first locate and map debris. Dive teams cut away the gear, taking care not to harm the coral. The debris is loaded on the small boats and then transferred to large vessels where it is separated into categories, weighed, and documented. As of 2002 over 118 tons of derelict nets had been removed from the reefs and shorelines of the Hawaiian archipelago at a cost in excess of U.S. $3 million.

15.2.1.2 Coral Transplantation

Transplantation of corals has been one of the more common methods used to mitigate damage to coral reefs in Hawaii and the U.S.-Affiliated Pacific Islands. In many cases the focus has been limited to removing corals from areas of future impact and transplanting them into nearby receiving areas. In other instances, corals have been held in reserve and then returned to their original habitats following the impact. More recently, efforts have been made to link mitigation to reef rehabilitation. Corals removed from areas of impending human impact (such as maintenance dredging of harbors and channels) can be used to restore previously impacted areas.[26]

Transplant and culturing techniques with potential application to restoration efforts have been developed for use in reef conservation. The transplanting and subsequent culturing of coral colonies or fragments could allow sustainable production of cultured corals for the aquarium and curio trade, eliminating the need for harvesting of corals from the wild.[27] Transplantation and seeding techniques have been used to protect and propagate rare coral species and thereby maintain biodiversity.[28,29] These methods could be used in the future to restore reefs. Documented examples of mitigation and rehabilitation projects involving coral transplantation in Hawaii and the U.S.-Affiliated Pacific Islands are as follows:

15.2.1.2.1 Kaneohe Bay, Oahu, Hawaii

Major dredging activities in the late 1930s and early 1940s severely impacted Kaneohe Bay.[30] Starting in the early 1960s, raw sewage discharged into the south basin of the bay had a dramatic effect on the reefs.[31] Maragos[32] evaluated coral transplantation as a means of restoration. A number of army surplus bed frames were used as "artificial reefs" at north, central, and south bay locations. Branching colonies of *Porites compressa* and *Montipora capitata* (the two most abundant bay species) were collected and transferred to the experimental sites while submerged in buckets. The corals were attached with rubber-coated wire to the bed frames at the three locations (only *P. compressa* was transplanted at the north bay site). Monitoring of the transplants occurred over an 18-month period. The south bay site had 100% mortality of *P. compressa* and 78% mortality of *M. capitata. Porites compressa* at the north bay site also did poorly, with 83% mortality due to high wave energy and sand scour. *Porites compressa* showed 30% mortality in the central bay, and *M. capitata* showed 61% mortality. However, continual physical removal of the competing bubble alga, *Dictyosphaeria cavernosa*, was required to keep corals from being overgrown at the central bay site.[32] The results were disappointing, but Maragos[32] suggested that successful transplantation would be feasible in areas more favorable to coral survival and growth and protected from excessive wave action and surge. Sewage discharge into the bay was abated in 1979, and indeed subsequent transplant experiments showed much higher success rates.[26]

15.2.1.2.2 Kaneohe Yacht Club Harbor, Kaneohe Bay, Oahu

A coral transplantation project was undertaken in 1996 to 1997 as mitigation for planned maintenance dredging of the Kaneohe Yacht Club Harbor.[26] By this time the reefs had substantially recovered from sewage discharge, which ended in 1979. Luxuriant coral growth in the harbor began to interfere with navigation.[33] Approximately 40 m^2 of branching *M. capitata* and *P. compressa* were transplanted to a nearby dredged area of reef. The receiving reef had been dredged to a depth of 3 m for a seaplane runway circa 1940 and never recovered due to the presence of a thick layer of silt and sand that prevented coral larval settlement. The site was protected from ocean swell and storm-generated waves and appeared similar to the Kaneohe Yacht Club Harbor environment in terms of water motion, depth, and turbidity.[26] Corals were transported underwater in baskets or aboard a boat while submerged in large tubs. Eight coral plots were established and monitored over a 6-year period during which coral coverage in the transplant plots increased approximately 45%. Corals sampled 2 and 3 years after transplantation were fully fecund. Topographic complexity, measured as rugosity, was immediately enhanced by transplantation and increased over time. Over 400 individual fish, including juveniles, were noted to be utilizing the transplant patches after 6 years.[26]

15.2.1.2.3 Marine Base Hawaii, Kaneohe Bay, Oahu

In 1998 approximately 150 colonies of *P. compressa* and *M. capitata* were transplanted away from an area that was to be impacted by extension of a runway drainage culvert at Marine Base Hawaii. Corals growing at 1 to 2 m depths were moved 70 m distant to a new location where they would not be subjected to construction damage and flood discharge from the culvert. The transplant reef is located in an area that is normally calm and protected from strong wave action, except during severe south wind or "Kona" conditions. The corals were placed in baskets and transferred while submerged to the new site. Colonies too large to lift were split with a hammer and chisel for transport. The transplanted corals were placed along two parallel 10-m transects and were photographed in order to determine area estimates for monitoring growth and survival. No evidence of coral distress or mortality was detected over a 6-month period.[34–37] In January and February 2004, unusual strong southerly "Kona" storm wind gusts accelerating to 90 mi/hr caused breaking waves over the transplant site, dislodging and scattering many corals. Approximately 25% of the transplanted corals were lost. However, a majority of the colonies identified as transplants showed evidence of significant posttransplant growth (S. Kolinski, personal observation).

15.2.1.2.4 Kawaihae Small Boat Harbor, Hawaii

In 1994 a large-scale coral transplant pilot study was conducted at Kawaihae. During 1969 to 1970 the entrance channel and turning basin had been blasted from the reef flat with explosives as part of an experimental program named Project Tugboat.[38] Completion of the harbor required extension of an existing breakwater and construction of a new mole and breakwater.[39] The proposed "footprint" covered about 1.8 hectares (4.4 acres) of reef, some of which was occupied by corals and associated organisms.[40] A plan was developed to evaluate the use of coral transplants as a means of mitigating the adverse impacts of harbor construction.[41] Most coastal reefs in the Kawaihae area already supported lush coral communities with cover exceeding 80%, but several sites with low cover were located. These tended to be in suboptimal environments. Massive colonies of *Porites*, *Pocillopora*, and *Montipora* were transferred from the project footprint to seven experimental transplant sites and one "stockpile" area (stockpiled for eventual attachment to the harbor breakwater and mole). The sites were located in a variety of habitats ranging from deep fore-reefs to reef flats, channels, and within the harbor. Most of the corals in the footprint of the new breakwaters were loosely attached to the rubble substratum. The corals to be moved were placed on large mesh wire squares, the corners of which were clipped together to form individual carrying bags when full. Up to four bags were hoisted and tied off under a boat for transport while submerged. At each experimental transplant site, the corals were secured to 6.3 m^2 of wire mesh firmly attached to the substrate with steel stakes. The corals were photographed for identification during monthly monitoring of survivorship.

Approximately 7500 kg of corals were transplanted. After 4 months survival was 100%, which suggested that the process of transplantation was successful. However, the most severe storm swell observed at Kawaihae in over 10 years occurred during the following winter, causing damage to many of the transplants by burial or physical removal. The remaining transplants continued to decline over time, suffering from fish grazing, sedimentation, abrasion, bleaching, and algal overgrowth. Additional corals transplanted into several of the areas that showed the highest survival rate also gradually declined over the course of a year.[41] The study demonstrated that reef corals could be transplanted successfully in large numbers. However, corals transplanted into marginal environments underwent long-term decline.

15.2.1.2.5 Aua, Tutuila, American Samoa

Two fishing vessels grounded in Pago Pago harbor during a typhoon in 1991 were scheduled to be dragged off the reef during November 1999. In order to mitigate damage caused by the removal of the derelicts, a large number of corals was transplanted out of the area that would be impacted by the salvage operation. Approximately 3000 colonies of *Pocillopora meandrina, P. verrucosa, P. eydouxi, Porites lutea,* and other coral species were removed from the area to be impacted. The corals were transferred atop a raft to nearby holding areas. Nearly 1000 of the colonies were tagged for later return to the impact area following removal of the ships. Unfortunately, a storm scattered and damaged the corals prior to final relocation of the tagged colonies. Only 354 of the corals were suitable for final transplantation.[42] These corals were transplanted to rock and hard reef flat substrate in the area damaged by the salvage operation. Corals were reattached using a mixture of Portland cement and molding plaster.[42,43] Approximately 97% of the transplanted corals were located again in a survey 1 year later. Overall tissue survival averaged 66%.[43]

15.2.1.2.6 Tanguisson, Guam

One of the earliest efforts in Guam to restore a degraded reef with transplanted corals occurred in 1979 along a reef within the thermal effluent zone of the Tanguisson Power Plant in Apra Harbor.[28] The intention was to bring in species that are presumably more tolerant of high temperatures to replace those lost when the area was exposed to the heated discharge. Eighteen species in nine genera (*Acropora, Favia, Lobophyllia, Montipora, Pavona, Pocillopora, Porites, Psammocora,* and *Stylophora*) were collected from inside the harbor and from Tumon Bay, transported submerged in buckets, and attached to hard substrates at depths of 0.3 to 3 m within the thermal influence zone and a nearby control area. Additional colonies of what is now called *Porites cylindrica* were collected, fragmented, and scattered at the sites. The attached colonies and fragments were subsequently damaged by typhoon-generated waves. Less than 1% of the corals transplanted into the thermal effluent zone and less than 7% of corals in the control area remained alive after 6 months. None of the scattered fragments were found. The investigators concluded that proper attachment of transplants is important.[28] Further, transplantation is not an option where conditions continue to remain detrimental to coral growth and survival, especially in areas exposed to prevailing wave action, surge, and large periodic storm waves.

15.2.1.2.7 Piti Bay, Guam

In 1990 to 1991, approximately 400 corals were moved to create a 460 by 40 m corridor for transport of prefabricated components and support facilities (a jack-up barge and crane) for construction of the Pacific Underwater Observatory in a large reef sinkhole in Piti Bay, Guam.[44] The Piti reef flat is frequently impacted by typhoon-generated waves and consists largely of unconsolidated sand and gravel resting on a carbonate framework strewn with carbonate boulders that are colonized by corals. The edges of the towpath were marked with buoys. Only carbonate boulders large enough to obstruct movement of the observatory components and support vessels were relocated, generally less than 10 m to an adjacent portion of the reef flat with a similar depth. All corals were kept submerged during transport. Care was taken to avoid coral damage during detachment and movement. There was no transplant mortality, but some slight physical damage was noted. After 6 months these corals had healed, but predation by the starfish *Acanthaster planci* had killed 11 corals, about the same rate as for nontransplant corals. Project success was attributed to the limited disturbance and transfer of colonies within their normal reef flat environment.[44]

15.2.1.2.8 Gun Beach, Tumon Bay, Guam

During 1994 a total of 116 coral colonies (21 species in 10 genera: *Acanthastrea, Acropora, Astreopora, Cyphastrea, Favia, Goniastrea, Pocillopora, Porites, Psammocora,* and *Stylophora*) were removed from obsolete submarine cables and cable supports that were scheduled to be replaced. These corals were moved 16 m distant and attached to reef outcrops at 9 m depth. The receiving area supported few live corals. The corals were detached using hammers and chisels, transported underwater, and attached to receiving substrate with Sea Goin' Poxy Putty®. Colonies greater than 16 cm in diameter were simply placed in natural reef depressions. After 9 weeks, 21% of the transplanted corals had perished and 15% could not be relocated.[45] Much of the mortality was attributed to predation by the coral eating-starfish *Acanthaster planci*, as well as competitive overgrowth by the encrusting sponge *Terpios* sp. The investigators concluded that transplantation of corals was potentially a useful tool in preserving corals, but careful consideration must be given to choice of receiving areas with regards to natural coral predators and competitors.

15.2.1.2.9 Tepungan, Piti, Guam

The installation of a new fiber optics cable in 2001 on a fringing reef flat and slope in the Piti Marine Preserve Area at Tepungan required that corals be transplanted from the path of the cable. Five colonies of *Porites lutea* and 24 *Pocillopora damicornis* were chiseled (with substrate) and/or lifted from the reef flat and shallow areas in the footprint of the intended cable landing and were transported in submerged baskets 60 m to a neighboring reef flat and slope across Tepungan Channel. Colonies were reattached using Sea Going' Poxy Putty and/or Splash Zone® epoxy and were tagged and monitored for a period of 14 weeks.[45–48] An additional survey was conducted by NOAA Fisheries 1 year following transplantation.[49] The 3-month evaluation[48] indicated that 97% of the colonies had survived, including all colonies of *Porites lutea* and all but one of the *Pocillopora damicornis.* One year after transplantation, 68% of colonies remained alive, with most in good condition. Only *P. damicornis* suffered mortality.[49]

15.2.1.2.10 West Rota Harbor, Commonwealth of the Northern Marianas Islands (CNMI)

In 1997, approximately 10,000 corals (mainly *P. damicornis*) were transplanted to mitigate impending damage to nearshore reefs during construction of West Rota Harbor (J. Gourley, R.H. Richmond, S. Burr, personal communication, 1998). Whole colonies and fragments were placed in a submerged cage and transported by boat to a receiving area with depth and substrate characteristics similar to that of the colony source area. The transplants were not attached to receiving substrates. Later in 1997, the region was impacted by high waves and currents caused by a super-typhoon. None of the transplanted corals could be found (J. Gourley, personal communication, 2004).

15.2.1.2.11 Smiling Cove, Saipan, CNMI

Dredging and construction of a marina at Smiling Cove was mitigated by transplanting corals out of the impact area during 1996 and 1997.[50,51] Colonies of *Pocillopora, Porites, Millepora, Fungia, Acropora,* and various coral- and sand-associated macroinvertebrates, were lifted by hand or chiseled away from substratum at 1 to 8 m depths and placed in large wire mesh baskets attached to boats for submerged transport. The organisms were moved 100 m to an area devoid of corals outside of an existing breakwater. Many of the corals were fastened to metal and rock surfaces using Aqua Poxy® epoxy mixed with silica sand. An estimated 12,000 corals were moved in the first phase,[50] and 173 colonies in the second phase.[51] After 7 months, approximately 97% of the transplanted corals survived.[52] In 2004 the transplant area retained a relatively high coral presence that could partially be attributed to the transplantation effort.

15.2.1.2.12 Arakabesan Island, Koror State, Republic of Palau

In 1990 a transplantation effort was undertaken to mitigate the impact of building a jetty on the reef fronting the Palau Pacific Resort. The proposed construction would directly impact 0.18 hectare (0.44 acres) of shallow-water reef community. Coverage by benthic organisms within the footprint of the jetty was estimated at 90% and included 26 species of hermatypic corals, at least three species of octocorals, and various algae, bivalves, and sessile and mobile invertebrates. Fifty-five species of reef fishes were documented in the area.[53] Two methods of coral transplantation were used. A crane used a rope sling to hoist colonies 1 to 2 m in diameter aboard a barge for relocation. Smaller corals were removed by hammer, chisel, or knife, and along with other invertebrates, were transported in nylon bags aboard a small craft to the transplantation site. The receiving site was a nearby sand channel with minimal coral, algae, and fish presence that had been dredged circa 1939. At least 20 coral species in seven genera (*Acropora, Favites, Goniastrea, Leptastrea, Montipora, Pocillopora,* and *Porites*) were moved, along with various invertebrates and epiphytic algae. Less than a week after the transplantation, Typhoon Mike struck Palau. Storm waves scattered, abraded, and buried many of the smaller coral transplants and fragments. The large transplant colonies were less impacted by the storm.[53] Monitoring of the corals continued for 22 months. Rope abrasions and damage that resulted from large colony movement reportedly healed. Fifteen species of corals, nine species of algae, more than 12 species of macroinvertebrates, and more than 20 species of fish reportedly inhabited the transplant reef after 22 months.[53]

15.2.1.3 Seeding Reefs with Larvae, Juveniles, and Fragments

A method that is under development involves seeding a reef with coral larvae. This method may be appropriate when there are insufficient natural sources of larvae to establish colonies and where the substratum is suitable for initial coral settlement. Richmond (personal communication) seeded small areas of sediment-impacted reef with *Acropora* larvae in southern Guam, but settlers succumbed to additional sediment input. Kolinski (in preparation) seeded individual plots of slowly recovering natural reef substrate in Kaneohe Bay with roughly 100,000 larvae of *M. capitata.* No recruits could be located after a 3-year period. Additional seeding of ceramic settlement plates with *M. capitata* at six sites across Kaneohe Bay showed varied levels of settlement but low overall long-term survival.[54]

Stock enhancements using cultured juveniles of certain species have been carried out with success in the U.S. Pacific Islands, but the focus of this work has been on increased harvest and economic return rather than on mitigation or restoration. Juvenile clams reared in hatcheries in Palau and the Marshall Islands have been spread throughout the freely associated states in an effort to establish brood stocks of overexploited populations.[55–57] The gastropod mollusk *Trochus niloticus* and black pearl oyster *Pinctada margaritifera* are cultured and managed in field environments.[58–61] In Hawaii, at least two species of reef-dwelling fish (*Mugil cephalus* and *Polydactylus sexfilis*) have been cultured and released to replenish depleted coastal fisheries.[62,63] The use of cultivated corals to rehabilitate U.S.-affiliated Pacific reefs has not been attempted, although cultured corals have been used as bioindicators in habitat assessments.[64] Introductions of organisms from laboratory facilities and/or from other areas across localities, islands, and archipelagos risks unintended transfer of invasive species, parasites, and pathogens. Consideration must also be given to avoiding inadvertent introductions of deleterious genetic defects to wild populations.[63] Such efforts typically require facilities support, technical expertise, and long-term perspective.

Few efforts to accelerate reef regeneration through seeding of coral fragments are reported for the U.S. Pacific Islands. Birkeland et al.[28] spread buckets of *Porites cylindrica* fragments across exposed reefs in Tanguisson, Guam; however, all were washed away by typhoon-generated waves and currents. Bowden-Kerby[65] reported variable success (2 to 100% survival) in transplanting fragments of four *Acropora* species to shallow sandy back-reef areas in Pohnpei. Kolinski (in preparation) seeded reef areas of Kaneohe Bay, Oahu, with 5- to 10-cm long *Montipora capitata* fragments. Although survival and growth varied between sites, the most degraded reef site experienced exceedingly high levels of fragment survival and growth that resulted in fecund colonies within a 3-year period.

15.2.1.4 Increase Habitat Area

Reef damage can be partially offset by providing additional habitat in the form of artificial reefs or sunken wrecks. Such artificial structures clearly are not natural reefs. However, in some cases such habitats can serve a beneficial and useful purpose as excellent sites for recreational diving and fishing. These areas can provide additional habitat, thereby taking pressure off of natural reefs. Caution is advised because some artificial reef structures may act primarily as benthic fish aggregation devices that can be heavily targeted by fishermen. Without some regulation and oversight, artificial reefs and sunken ships may actually do more harm than good to regional fisheries populations.[66,67]

15.2.1.4.1 Sasanhaya Bay, Rota, CNMI

Extensive damage and loss of a valuable dive site resulted at Sasanhaya Bay when action was taken to eliminate a perceived danger from explosive depth charges aboard a sunken WWII Japanese warship. In May and June of 1996 an explosive ordinance demolition (EOD) team detonated the ordinance in place, which destroyed the historic wreck and caused extensive damage to the surrounding coral reef. Coral cover in the area, which consisted largely of *Porites rus,* was reduced from 60 to 1% in an area within 150 m of the blast. Public outrage by divers, dive tour operators, fishermen, and environmentalists led to the development of a remediation plan. A derelict vessel was cleaned of contaminants and sunk in the area to provide additional dive sites and habitat (J. Naughton and R. Richmond, unpublished observations).

Reef damage similar to that at Sasanhaya Bay, Rota, occurred to the Molokini Marine Life Protected Area in 1984 when an EOD team detonated WWII-era bombs found on the reefs. Sport divers and tour operators were upset about the resulting damage to the corals. When additional bombs were discovered, the EOD teams were not notified. Instead, the tour operators and other volunteer removed the bombs from the reefs at great personal risk in order to prevent further reef damage from detonations (J. Maragos, personal observation). They tied long lines to the bombs and dragged them into deep water where the explosives were cut loose.

15.2.1.4.2 Maalaea Harbor, Maui, Hawaii

A major expansion of Maalaea Harbor was proposed over 20 years ago but was blocked by environmental concerns. Under the most recent proposal, alternative mitigation measures excluded coral transplantation due to lack of suitable receiving environments in the area.[68] The major factors preventing transplantation of corals along the Maalaea coastline are lack of suitable hard substratum in the area and severe wave impact and low tide exposure in the shallows. However, lush coral reef communities have developed on dredged reef faces and basalt riprap.[68] Most of the coral that would be impacted occurs on hard substratum that was created during the original construction of the harbor. A mitigation method to increase habitat area has been proposed for the Maalaea Harbor project.[69] The plan calls for expansion of the proposed sea wall design to include an extension of boulder riprap onto sand flats along the groins to depths of 10 m. This would create an extensive high-rugosity coral reef habitat in areas where only shifting rubble and sand exist today. Engineers involved in planning the project see this option as being cost effective and well within the scope of the engineering plan. Such artificial boulder fields must withstand the largest storm waves experienced at this site. Large interlocking riprap boulders of the same size and set in the same manner as on the sea wall would be suitable. Such high relief boulder riprap areas are rapidly colonized by corals, fish, and invertebrates as shown by observations off the seaward channel at Kawaihae Harbor and on riprap protecting the outfall pipes at Kahe Point, Oahu.[70]

15.2.1.5 Modification of Habitat

In extreme cases, modification of the physical environment may be undertaken in an attempt to correct degradation. Such actions could include dredging to remove accumulated sediments (proposed for Pelekane Bay, Hawaii), modification of shoreline structures to improve flushing and circulation (proposed for Kaunakakai, Molokai), or modification of substrata (increasing relief, rugosity, adding hard substrata as boulders, etc.).

15.2.1.6 Mitigation through Removal of Harmful Organisms

15.2.1.6.1 Molokai, Hawaii

During 1969 to 1970 a large aggregation of over 20,000 crown-of-thorns starfish (*Acanthaster planci*) were studied off south Molokai.[3] They were feeding selectively on the common coral *M. capitata* but not the dominant coral *P. compressa.* Although University of Hawaii marine scientists participating in the evaluation did not believe the reef was in jeopardy, the State of Hawaii Department of Fish and Game undertook extensive eradication efforts over the next few years.[71] Divers killed approximately 26,000 starfish between 1970 and 1975 by injecting them with ammonium hydroxide. Additional surveys were conducted throughout the State of Hawaii, but no other infestations have been detected.

15.2.1.6.2 Waikiki, Hawaii

The red alga *Gracillaria salicornia* was introduced intentionally to two reefs on Oahu, Hawaii, in the 1970s for experimental aquaculture for the agar industry. Some 30 years later, this species has spread from the initial sites of introduction and is now competing with native marine flora and fauna. Large-scale community volunteer efforts were organized to remove *G. salicornia* fragments from the reef area in front of the Waikiki Aquarium.[72] Over 20,000 kg of alien algal fragments were removed from this location in five 4-hr cleanup events. However, based on *G. salicornia* growth rates, ability to fragment, physical tolerance, and low rates of herbivory, it is clear that continued large-scale efforts will be needed to control this invasive alga.

15.2.1.6.3 Kaneohe Bay, Hawaii

Kappaphycus spp., another red alga, was intentionally introduced in small amounts onto reefs in Kaneohe Bay, Oahu for aquaculture experiments by the Hawaii Institute of Marine Biology in 1974.[73,74] The alga has spread and is outcompeting and smothering corals and reducing sessile invertebrate and native algae diversity, leading to a community phase shift across large areas of reef throughout the bay.[74,75] Experiments on methods of control suggest a combination of tactics, including intensive manual removal followed by saline treatments and/or native urchin grazing, may be needed help to control growth, spread, and spatial domination by this genus.[74]

15.2.2 Indirect Action

The most successful and cost-effective means of mitigation and restoration is to reduce or eliminate anthropogenic impact and allow natural processes to restore the reef. In such instances the emphasis is on eliminating the source of the impact, which in any event must be accomplished before any restoration can begin. Once an anthropogenic stress has been removed, natural recovery of a reef system often occurs rapidly without further action. The indirect approach is especially feasible when there is sufficient time to evaluate possible restoration options before the damaging actions are implemented. However, in many cases reef damage occurs without warning (e.g., ship groundings) or when advanced planning and design are inadequate. In these cases "emergency' restoration is often inadequate and hastily organized. Examples include the following:

15.2.2.1 Kaneohe Bay, Hawaii

Removal of sewage outfalls in Kaneohe Bay in 1979 led to dramatic decreases in nutrient levels, turbidity, and phytoplankton abundance and a rapid recovery of coral reef populations.[76–78] By 1983 coral coverage had more than doubled from 12 to 26%.[78] However, proper planning in the early 1960s could have led to initial location of the outfalls outside the bay, avoiding the impact and much of the total cost to relocate them again in the late 1970s.

15.2.2.2 Kahoolawe, Hawaii

The reefs off the former target island of Kahoolawe, Hawaii, were under severe sediment stress due to erosion caused by two centuries of improper land management. Removal of 20,000 feral goats, termination of bombing, and reestablishment of vegetation are reducing erosion on the land with a consequent dramatic impact on the reefs. Sediment on the reefs of Kahoolawe is gradually being winnowed from the shallows faster than it is being delivered from the land. As a result, corals are colonizing the hard substratum that is gradually being uncovered by natural wave processes.[79]

15.2.2.3 Kahe Point, Oahu, Hawaii

An extensive area of reef off Kahe Point was impacted and killed by thermal effluent from a power generation station.[80] When the generating capacity of the plant was increased from 270 to 360 megawatts, the area of dead and damaged corals increased from 0.38 hectare (0.94 acres) to 0.71 hectare (1.76 acres). The requirement for plant expansion and further increases in discharge led to installation of a new outfall pipe in 1976 in deeper offshore waters. This pipe is over 100 m in length, is protected from wave action by heavy rock riprap, and now carries heated effluent offshore and away from the reef. Colonization of the damaged area and the riprap was dramatic, with coral colonization rates among the highest reported in the literature.[70]

15.2.2.4 Hamakua, Hawaii

Discharge of silt-laden water and bagasse from sugar mills along Hawaii's Hamakua coastline over many decades caused extensive damage to coral reefs.[81] Termination of discharges led to a rapid clearing of the sediment and bagasse waste by wave action and subsequent regeneration of coral reefs in the former discharge zones.[82,83]

15.2.3 Negotiated Financial Settlement or "Tradeoffs"

In some cases the primary options discussed above are not available, such as when there is a lack of time for advanced design measures to reduce or avoid impacts. Then, managers must make the best of a bad situation by obtaining some sort of settlement in order to achieve environmental or social benefit as compensation for the damage.

15.2.3.1 Agana Harbor (Guam)

During 1983, the U.S. Department of Defense proposed a project to dredge one of the richest reefs in Agana Harbor (Guam) in order to build a wharf for ammunition ships.[84] This followed other unpopular and nonimplemented Navy proposals for the pier in Guam, as early as 1971 (J. Caperon, R.E. Johannes, and J.E. Maragos, personal observations, 1971). This site was the only location suitable because of the explosive hazard (J. Naughton, notes and reports, unpublished). Environmental managers in the responsible agencies concluded that it would no longer be possible to block the action because of the national defense provision. To oppose the action would be futile so alternative action to mitigate the damage was undertaken. As a mitigation measure, the federal government agreed to create two permanent reef reserves. The Orote and Haputo Ecological Reserve Areas were created in 1984 as part of the U.S. Navy Ammunition Wharf Project.[85,86] The tradeoff could be seen as a net loss, as habitat quality in the reserves is low relative to that destroyed in construction of the wharf, and no active reserve management was required by the agreement.

15.2.3.2 Honolulu, Hawaii

Honolulu Reef Runway, Hawaii, was initiated in 1972 with 308 hectares (763 acres) of reef dredged and filled. Because in-kind mitigation was not possible for this fill project, a tradeoff involving creation of two wetlands in nearby Pearl Harbor was negotiated. This agreement protected nesting habitat for several endangered waterbird species.[87] The two wetlands are now National Wildlife Refuges.

15.2.3.3 Satawal Island, Yap State, Federated States of Micronesia

The bulk carrier *Oceanus* grounded on Satawal on March 18, 1994. The ship cut a large trench in the reef and pulverized the coral. More damage resulted when the ship's coal cargo was transferred to another vessel and when the ship was pulled off the reef. Subsequent shifting of coral rubble created by the grounding destroyed other habitats. Aerial photographs obtained several months after the initial disturbance revealed that sand from the grounding trench had spread to large adjacent reef areas and the island shoreline, magnifying the disturbance.[88] The area impacted was previously the prime fishing and gathering site for the residents of Satawal. Mitigation options were limited due to the remoteness of the island and high wave exposure of the site. However, marine damage assessments, interviews, and aerial photography were organized and accomplished quickly after the grounding by a law firm representing the residents. Evidence compiled by these actions influenced the ship owners to forgo a lawsuit involving "rebuttal" marine surveys. Instead, the defendants opted for an out-of-court settlement,[89] and the residents were eventually awarded approximately U.S. $2 million. A large portion of the award went into a trust fund that is being used to offset the socioeconomic and environmental impact of the grounding (M.A. McCoy, personal communication, 19 July 2001).

15.2.4 Strategic Reserve Network

There is increasing evidence of global reef decline due to global warming, global nutrification, overexploitation, and various other factors.[8] Compelling scientific evidence indicates that marine reserves conserve both biodiversity and fisheries and could help to replenish the seas.[90] As a result, the concept of developing strategic global coral reserves has recently emerged as a means of mitigating and offsetting global decline in reef systems.[91] Meaningful reserves have been and are being established in Palau, Guam, Saipan, and Yap State. Creation of a marine protected area for the Northwestern Hawaiian Islands is under discussion as a means of formally strengthening the protection of the reef areas that resulted as a byproduct of the 1909 Hawaiian Islands National Wildlife Refuge. The Wildlife Refuge protects terrestrial habitats only but has limited human access to the area. The Wildlife Refuge has been a major factor in the preservation of what is now known to be the last major reef system dominated by apex predators.[92]

15.3 MANAGEMENT ACTION

The political, economic, social, and conservation realities dictate that we continue to examine all options of reef restoration and mitigation and apply them in appropriate situations. Jokiel and Naughton[15] found it useful to discuss three categories of management action in relation to reef conservation: prevention, mitigation, and restoration.

15.3.1 Prevention

Prevention includes the management actions of preservation, protection, and avoidance of damage. This management action promotes sustainability primarily through four major activities:

15.3.1.1 Public Awareness

Education can lead to action directly impacting the political process governing management decisions. Effective education can lead to increased awareness and empowerment of the public on issues concerning the protection of coral reefs.

15.3.1.2 Sound Management Practices

Appropriate rules and restrictions designed to avoid the causes of the reef damage must be set.

15.3.1.3 Appropriate Enforcement Practices

Lack of enforcement negates any positive effect accomplished in the first two activities. Without strict enforcement, restrictions on human activity cannot be implemented. Lack of enforcement leads to loss of public support for conservation measures and eventual damage to coral reefs.

15.3.1.4 Assessment and Monitoring

Making intelligent management decisions concerning reef resources requires knowledge of the extent of resources, the ability to detect change, and the ability to identify the cause of change.

15.3.2 Mitigation

The need for mitigation arises when managers must devise a plan to reduce and offset unavoidable damage of an impending negative impact on a coral reef or after an impact for which there was no forewarning. An example of the first would be to negotiate a plan to reduce the impact of a new harbor and provide a means of offsetting habitat loss. An example of the second would be to assess damages from a ship grounding and seek compensation for restoration or mitigation. Actions for proposed project impacts must focus on loss of coral reef habitat, ecological communities, and regional physical and ecological relationships and values. As a general guideline, the following management actions should be undertaken:

15.3.2.1 Eliminate or Reduce Habitat Loss

This is the first line of defense for environmental protection. Search for alternate sites and methods of construction, and develop the best management practice criteria for the project so as to reduce the area of habitat being impacted. If construction must occur, then devise methods to reduce impact. For example, the Kosrae airport and port were initiated after 1980 with 138 hectares (340 acres) of reef and seagrass habitat lost. However, Army Corps permits required the Navy contractor to construct a free-standing rubble-mound revetment and install filter cloth around the entire perimeter of the fill area so that subsequent discharge of dredged slurry would not impact adjacent reefs. Subsequent surveys revealed this mitigation was successful in confining most impacts.[10]

15.3.2.2 Conduct Economic Analysis

Conduct a thorough analysis of the long-term costs of negative impacts to the reef system as part of the economic analysis used to evaluate justification of the project. Numerous valuations have been made for coral reefs.[93–96]

15.3.2.3 Alternative Environmental Actions

If there will be or has been unavoidable loss of habitat, then make the best of less favorable situations by using the loss as leverage to achieve other positive environmental actions. A wide range of actions is available. Work can be undertaken to restore conditions that facilitate natural recovery in degraded reef areas. In some cases the focus might be on establishing and supporting well managed and enforced marine reserves. In other cases it might be feasible to construct well-designed artificial habitats for recruitment of both mobile and sessile reef community members. Another dimension is to secure funding for research and education that leads to improved stewardship of regional reef areas.

15.3.2.4 Install Preventative Measures

Restoration is action taken to correct damage. It is a salvage operation, often an emergency response, with "too little, too late" and it can be very expensive. Measures that reduce or eliminate the need for additional restorative actions should always be considered in mitigation.

15.4 COST-EFFECTIVENESS OF MANAGEMENT ACTIONS

Little information exists on the cost of mitigation and restoration of coral reefs. Estimates available in the literature range from U.S. $13,000 to greater than U.S. $100 million per hectare.[96] Restoration costs can also include remedial action to correct the source of damage. Jokiel and Naughton[15] made a conceptual comparison of cost versus effectiveness of various management actions and concluded that effectiveness of management options decreases rapidly with increasing degradation while cost increases dramatically. Cost is high and effectiveness is low for mitigation efforts. Cost is very high and effectiveness is minuscule for restoration of coral reefs. Given the cost/effectiveness, there will generally be little motivation to restore severely degraded reefs. It is very important to prevent reefs from reaching this state. In many previous cases, resources expended on restoration would have been more cost effective if applied to prevention, preservation, and protection. Limited resources must be directed at more cost-effective measures to protect reefs that are not severely degraded. Scientific research produces information that lowers the cost of management while increasing the effectiveness of management practices. Research increases cost effectiveness of actions across the entire range of management activities.

15.5 SUMMARY

Evaluation of projects to date leads us to the following conclusions:

1. Protection of reefs from environmental degradation must be given highest priority because mitigation and restoration efforts are expensive and often ineffective.[97] Reef protection is the most cost-effective method of achieving sustainability goals for reefs and should be the focus of management activity.
2. Given the documented global decline in coral reefs, restoration and mitigation must be viewed from a broad global strategic perspective rather than from a limited local point of view. Mitigation emphasis is now shifting to the establishment of coral reef reserve networks, which are intended to serve as a primary mitigation tool for reefs throughout the world.
3. Watershed management is inseparable from coral reef management adjacent to human settlements and population centers. An integrated land–ocean plan is necessary, especially in cases involving chronic degradation of reefs due to sedimentation, eutrophication, or shoreline construction activities.
4. Before undertaking any restoration activity on a degraded reef it is critical that the cause of the damage (e.g., sewage, sediment runoff, repeated anchor damage) be eliminated[28] or will be eliminated as an initial phase of the restoration (e.g., Rose Atoll ship metal removal). Efforts at restoration and preservation of reefs near human settlements must consider the condition of the adjacent watersheds and possible future changes on the watershed. Restoration activities on the reefs can take focus off the basic problem. There is no purpose in restoration efforts on a reef that will be subsequently destroyed by poor land management or pollution originating on an adjacent watershed.
5. Mitigation and restoration focus must be on coral reef habitat, the range of community members it supports, and physical and ecological relationships rather than simply transplanting coral colonies.
6. The option of letting nature take its course should be recognized. In many cases, removal of the stress will result in dramatic improvement in the reef communities due to the natural process of reef renewal, especially in areas of good water exchange.
7. If damage does occur, managers have a wide variety of mitigation/restoration tools at their disposal. Reef repair, coral transplant, and artificial reefs are often the first mitigation and restoration techniques that come to mind but can be the least effective in many situations and if chronic anthropogenic stress is not first eliminated. Numerous other tools can serve to meet the objectives. These include elimination of the anthropogenic stresses, enforcement of existing regulations for penalties, establishing new regulations where needed, education of the public, establishment of compensatory environmental trust funds, creation of protected area networks, and establishment of marine reserve networks.
8. Transplantation of coral heads is feasible but has limitations. Initial mortality is low if factors that stress corals are minimized and transplanted corals are secured to the substratum. Transplanting corals into marginal and exposed habitats leads to their eventual demise.

Infrequent wave events along exposed coastlines (intervals of 10 years or more) have major impacts on the structure of coral reefs and are devastating to transplant sites due to the difficulty of securing transplanted corals properly to substrate. The most favorable transplant receiving sites are generally wave-protected lagoon areas.

9. An effective long-term research and monitoring program is necessary to evaluate the success and cost effectiveness of the mitigation/restoration effort.
10. Reef restoration can be a very dangerous concept if used by unscrupulous individuals or organizations or as an alternative to more effective options that eliminate damage to reefs.[13] Token restoration efforts should never be a basis to justify proposed negative environmental actions under the guise of "improving" the environment.
11. A restored reef is not a natural reef unless it is predicted to, or fully recovers to, its natural state. Initially it is an artificially modified community. The loss of large coral heads that are hundreds of years old will take hundreds of years to replace. Restoration can be justified as a means to enhance fisheries production, tourism, recreation, aesthetics, research, conservation, or other activities and may allow natural restoration on otherwise pristine or sparsely inhabited reefs.

ACKNOWLEDGMENTS

Supported in part by USGS-CRAMP co-operative agreement 98RAG1030 and by USEPA Grant CD97918401-0.

REFERENCES

1. Dollar, S.J., Wave stress and coral community structure in Hawai'i. *Coral Reefs* 1982, 71.
2. Jokiel, P.L., Hunter, C.L., Taguchi, S., and Watarai, L., Ecological impact of a fresh-water "reef kill" in Kaneohe Bay, Oahu, Hawaii. *Coral Reefs*, 1993, 177.
3. Branham, J.M., Reed, S.A., Bailey, J.H., and Caperon, J., Coral-eating sea stars *Acanthaster planci* in Hawaii. *Science* 1971, 172.
4. NOAA, Abandoned vessels case history: *Jin Shiang Fa*. NOAA, National Ocean Service, Office of Response and Restoration, Damage Assessment Center, 2001.
5. Brock, V., Van Heukelem, W., and Helfrich, P., An ecological reconnaissance on Johnston Island and the effects of dredging. Hawaii Institute of Marine Biology Technical Report 5, University of Hawaii, Honolulu, 1965.
6. Jokiel, P.L., Hill, E., Farrell, F., Brown, E.K., and Rodgers, K., Reef Coral Communities at Pila'a Reef in Relation to Environmental Factors. Hawaii Coral Reef Assessment and Monitoring Program Report, Kaneohe, HI, 2002.
7. Grigg, R.W. and Dollar, S.J., Natural and anthropogenic disturbance on coral reefs, in *Coral Reefs*, Z. Dubinsky, Ed. Elsvier, Amsterdam, 1990, 453.
8. Hoegh-Guldberg, O., Climate change, coral bleaching, and the future of the world's coral reefs. *Mar. Freshwater Res.*, 1999, 839.
9. Dawson, E.Y., Changes in Palmyra Atoll and its vegetation through the activities of man 1913–1958. *Pacific Naturalist*, 1959, 1.
10. Maragos, J.E., Impact of coastal construction on coral reefs in the U.S.-Affiliated Pacific Islands. *Coastal Management* 1993, 21, 235.
11. Bentivoglio, A., Compensatory Mitigation for Coral Reef Impacts in the Pacific Islands. U.S. Fish and Wildlife Service, Pacific Islands Fish and Wildlife Office, Honolulu, 2003.
12. Carpenter, R.A. and Maragos, J.E., Eds., *How to Assess Environmental Impacts on Tropical Island and Coastal Areas. South Pacific Regional Environment Program Training Manual.* East-West Center Environment and Policy Institute, Honolulu, and Asian Development Bank, Manila, 1989.
13. Maragos, J.E., Restoring coral reefs with emphasis on Pacific reefs, in *Restoring the Nation's Marine Environment*, Thayer, G.W., Ed., Maryland Sea Grant College Pub. UM-SG-TS-92-06, 1992, 141.

14. Naughton, J. and Jokiel, P.L., Coral reef mitigation and restoration techniques employed in the Pacific Islands: I. Overview, *Oceans 2001 Conference Proceedings*, Marine Technological Society/ Institute of Electrical and Electronics Engineers, Inc. Holland Publications, Escondito, CA, 2001, 1, 306.
15. Jokiel, P.L. and Naughton, J., Coral Reef Mitigation and Restoration Techniques Employed in the Pacific Islands: II. Guidelines. *Oceans 2001 Conference Proceedings*, Marine Technological Society/Institute of Electrical and Electronics Engineers, Inc., Holland Publications, Escondito, CA, 2001, 1, 313.
16. Kolinski, S. P., Coral transplantation as a mitigation strategy in Hawaii and the U.S.-Affiliated Pacific Islands: purpose, past success and guidelines for future activities. NOAA report. In prep.
17. Richmond, R.H., Recovering populations and restoring ecosystems: restoration of coral reefs and related marine communities, in *Marine Conservation Biology: The Science of Maintaining the Sea's Biodiversity*, Norse, E., and Crowder, L., Eds., Island Press, Washington, D.C., 2005.
18. Maragos, J.E., Reef and coral observations on the impact of the grounding of the longliner *Jin Shiang Fa* at Rose Atoll, American Samoa. Prepared for the U.S. Fish and Wildlife Service Honolulu. East-West Center, Program on Environment, Honolulu, 1994.
19. Green, A., Burgett, J., Molina, M., Palawski, D., and Gabrielson, P., The impact of a ship grounding and associated fuel spill at Rose Atoll National Wildlife Refuge, American Samoa. U.S. Fish and Wildlife Service Report, Honolulu, HI, 1997.
20. Maragos J. and Burgett, J., Monitoring and partial cleanup at Rose Atoll National Wildlife Refuge after a shipwreck, in *Monitoring Coral Reef Marine Protected Areas, a Practical Guide on How Monitoring Can Support Effective Management of MPAs*, Wilkinson, C., Green, A., Almany, J., and Dionne, S., Eds., Australian Institute of Marine Science, Townsville, and the IUCN Marine Program, Gland, 2003, 40.
21. Helm, R., Final restoration plan for Rose Atoll National Wildlife Refuge. Prepared by the U.S. Fish and Wildlife Service, Portland, and American Samoa Department of Wildlife and Marine Resources, Pago Pago, American Samoa, 2003.
22. Keever, B., Fallout: Enewetak atoll, 50 years ago this week. *Honolulu Weekly*, Oct. 30, 2002.
23. *Honolulu Star Bulletin,* Editorial, Thursday, May 18, 2000.
24. Robison, W.L., Conrado, C.L., Bogen, K.T., and Stoker, A.C., The effective and environmental half-life of 137Cs at Coral Islands at the former U.S. nuclear test site. *J. Environ. Radioact.* 2003, 207.
25. Donohue, M.J., Boland, R.C., Sramek, C.M., and Antonelis, G.A., Derelict fishing gear in the Northwestern Hawaiian Islands: diving surveys and debris removal confirm threat to coral reef ecosystems. *Mar. Poll. Bull.*, 2001, 42, 1301.
26. Kolinski, S.P., Harbors and channels as source areas for materials necessary to rehabilitate degraded coral reef ecosystems: a Kaneohe Bay, Oahu, Hawaii case study, unpublished manuscript.
27. Yates, K.R. and Carlson, B.A., Corals in aquariums: how to use selective collecting and innovative husbandry to promote reef conservation, *Proc. Seventh Int. Coral Reef Symp.,* 1992, 2, 1091.
28. Birkeland, C., Randall, R.H., and Grimm, G., Three methods of coral transplantation for the purpose of reestablishing a coral community in the thermal effluent area at the Tanguisson Power Plant. University of Guam Marine Lab Technical Report 60, 1979.
29. Plucer-Rosario, G. and Randall, R.H., Preservation of rare coral species by transplantation and examination of their recruitment and growth. *Bull. Mar. Sci.*, 1987, 585.
30. Devaney, D., Kelly, M.M., Lee, P.J., and Motteler, L.S., *Kane'ohe a History of Change*, The Bell Press, Honolulu, 1982.
31. Smith, S.V., Kimmerer, W.J., Laws, E.A., Brock, R.E., and Walsh, T.W., Kaneohe Bay sewage diversion experiment: perspectives on ecosystem responses to nutritional perturbation. *Pac. Sci.*, 1981, 279.
32. Maragos, J.E., Coral transplantation, a method to create, preserve and manage coral reefs. University of Hawaii Sea Grant Pub. UNIHI-SEAGRANT AR-74-03, 1974.
33. Kolinski, S.P. and Jokiel, P.L., Coral Transplantation in Conjunction with Dredging of the Kaneohe Bay Yacht Club Harbor, Oahu, Hawaii. Final Report of Feasibility Study, 1996.
34. Marine Research Consultants, Coral transplantation at box drain project under Bracon P-268T at Marine Corps Base Hawaii (MCBH) Kaneohe Bay. Report submitted to Kiewit Pacific Co., 1998.

35. Marine Research Consultants, Coral transplantation at box drain project under Bracon P-268T at Marine Corps Base Hawaii (MCBH) Kaneohe Bay, baseline B. Report submitted to Kiewit Pacific Co., 1999.
36. Marine Research Consultants, Coral transplantation at box drain project under Bracon P-268T at Marine Corps Base Hawaii (MCBH) Kaneohe Bay, post-construction 1. Report submitted to Kiewit Pacific Co., 1999.
37. Marine Research Consultants, Coral transplantation at box drain project under Bracon P-268T at Marine Corps Base Hawaii (MCBH) Kaneohe Bay, post-construction 2. Report submitted to Kiewit Pacific Co., 1999.
38. Day, W.C., Wnuk, W.G., McAneny, C.C., Sakai, K., and Harris, D.C., Project Tugboat: explosive excavation of a harbor in coral. Report no. EERL-TR-E-72-23, U.S. Army Engineer Waterways Experiment Station, Explosive Excavation Research Lab, Livermore, CA, 1975.
39. U.S. Army Engineer District, Honolulu, Final Environmental Assessment for Kawaihae Harbor for Light-Draft Vessels, Honolulu, 1994.
40. U.S. Fish and Wildlife Service, Final Fish and Wildlife Coordination Act Report on the Kawaihae Harbor for Light-Draft Vessels, Kawaihae, Hawaii, Hawaii, in: Final Environmental Assessment for Kawaihae Harbor for Light-Draft Vessels, *Hawaii, Hawaii,* U.S. Army Engineer District, Honolulu, 1993.
41. Jokiel, P.L., Cox, E.F., Te, F.T., and Irons, D., Mitigation of Reef Damage at Kawaihae Harbor Through Transplantation of Reef Corals. Final Report of Cooperative Agreement 14-48-0001-95801, U.S. Fish and Wildlife Service, Pacific Islands Ecoregion, Honolulu, 1999.
42. Hudson, H., Coral restoration project, Pago Pago, American Samoa. Field trip report, NOAA Fisheries, 2000.
43. Jeansonne, J., Coral restoration project, Pago Pago, American Samoa. Draft year one monitoring trip report: July 2001, NOAA Fisheries, 2002.
44. Pacific Basin Environmental Consultants, Inc., Supplemental Coral Transplanting Methodology, 1995.
45. Dueñas and Associates, Inc., Weekly observations of transplanted corals at Gun Beach, North Tumon Bay, Guam. Coral monitoring report No. 2. Prepared for AT&T Submarine Systems, Inc., 1994.
46. Dueñas and Associates, Inc., Department of the Army permit application: trenching of reef flat, installation of conduits and landing of submarine fiber-optic cables at Tepungan, Piti, Guam. Prepared for TyCom Networks (Guam) LLC, 2000.
47. Dueñas and Associates, Inc., Coral transplant and monitoring plan for Tycom Networks Guam LLC fiber optic cable conduit trench in the Tepungan reef flat Piti, Guam. Prepared for Tycom Networks (Guam) LLC., 2001.
48. Dueñas and Associates, Inc., Coral transplant and follow-up monitoring of transplanted corals at Tepungan, Piti, Guam 1 June 2001 to 4 September 2001. Final report prepared for Tycom Networks (Guam) LLC, 2001.
49. Kolinski, S.P., Analysis of year-long success of the transplantation of corals in mitigation of a cable landing at Tepungan, Piti, Guam: 2001–2002. Report prepared for NOAA Fisheries, 2002.
50. Cheenis Pacific Company, Coral transplantation at the outer cove of Smiling Cove, Sadog Tase, Saipan, CNMI. Final report submitted to Marine Revitalization Corporation, 1996.
51. Micronesian Environmental Service, Outer cove coral transplantation project: supplemental report. Report prepared for Marine Revitalization Corporation, 1997.
52. Micronesian Environmental Service, Outer cove coral transplantation project: 7-month assessment. Report prepared for Marine Revitalization Corporation, 1997.
53. MBA International, Coral transplantation, Palau Pacific Resort, a pilot-demonstration project PODCO No. 2156. Final report prepared for the U.S. Army Corps of Engineers, Honolulu Engineer District, Fort Shafter, HI, 1993.
54. Kolinski, S.P., Sexual reproduction and the early life history of *Montipora capitata* in Kaneohe Bay, Oahu, Hawaii, Ph.D. thesis, University of Hawaii, Honolulu, 2004.
55. Heslinga, G.A. and Watson, T.C., Recent advances in giant clam mariculture. *Proc. Fifth Int. Coral Reef Symp.*, 1985, 5, 531.
56. Lindsay, S., Giant clams reseeding programs: do they work and do they use the limited resources wisely? in Dalzell, P. and Adams, T.J.H., Eds., South Pacific Commission and Forum Fisheries Agency workshop on the management of South Pacific inshore fisheries. Manuscript collection of country statements and background papers, Vol. II, SPC, Noumea (New Caledonia), Tech. Doc. Integrated Coastal Fisheries Management Project, No. 11, 1995, 345.

57. Lee, C.S., Ellis, S., and Awaya, K.L., Giant clam farming in the U.S.-Affiliated Pacific Islands. *World Aquaculture* 2001, 32, 21.
58. Heslinga, G.A. and Hillmann, A., Hatchery culture of the commercial top snail *Trochus niloticus* in Palau, Caroline Islands. *Aquaculture,* 1981, 22, 35.
59. Olin, P.G., Aquaculture extension and development in the U.S. Pacific region. *Aquaculture '92: Growing Toward the 21st Century,* 1992, 174.
60. Fassler, C.R. and Walther, M., Mythology, history, and cultivation of Hawaiian pearls. *Aquaculture '98*, World Aquaculture Society, Baton Rouge, LA, 1998, 172.
61. Fassler, C.R. Recent developments in selected Pacific and Indian Ocean black pearl projects. *Sixth Asian Fisheries Forum Book of Abstracts,* Asian Fisheries Society, Quezon, Philippines 2001, 301.
62. Leber, K.M, Arce, S.M., Nishimoto, R.T., and Iwai, T., Developing marine stock enhancement technology in Hawaii: progress and application. *Aquaculture '95 Book of Abstracts,* 1995.
63. Blankenship, H.L. and Leber K.M., A responsible approach to marine stock enhancement. *Amer. Fish. Soc. Symp.* 1995, 15, 167.
64. McKenna, S.A., Richmond, R.A., and Roos, G., Assessing the effects of sewage on coral reefs: developing techniques to detect stress before coral mortality. *Bull. Mar. Sci.,* 2001, 69, 517.
65. Bowden-Kerby, A., Coral transplantation in sheltered habitats using unattached fragments and cultured colonies, *Proc. Eighth Int. Coral Reef Symp.*, 1997, 2063.
66. Grossman, G.D., Jones, G.P., and Seaman, W.J., Do artificial reefs increase regional fish production? A review of existing data. *Fisheries* 1997, 22, 17.
67. Pickering, H. and Whitmarsh, D. Artificial reefs and fisheries exploitation: a review of the "attraction versus production" debate, the influence of design and its significance for policy. *Fish. Res.* 1997, 31, 39.
68. Jokiel, P.L. and Brown, E.K., Coral Baseline Survey of Ma'alea Harbor for Light-Draft Vessels, Island of Maui. Final Report for DACW83-96-P-0216. U.S. Army Engineer District, Honolulu, 1998.
69. Jokiel, P. L., Modification of breakwaters to create enhanced coral reef habitat. Concept paper proposed to U.S. Army Corps of Engineers, Honolulu, 1998.
70. Coles, S.L., Colonization of Hawaiian reef corals on new and denuded substrata in the vicinity of a Hawaiian power station. *Coral Reefs,* 1984, 123.
71. Onizuka, E., Studies on the effects of crown-of-thorns starfish on marine game fish habitat. Final Report of Project F-17-R-2, State of Hawaii Department of Fish and Game, Honolulu, 1979.
72. Smith, J.E., Hunter, C.L., Conklin, E.J., Most, R., Sauvage, T., Squair, C., and Smith, C.M., Ecology of the invasive red alga *Gracilaria salicornia* (Rhodophyta) on O'ahu, Hawai'i. *Pac. Sci.* 2004, 325.
73. Rodgers, S.K. and Cox, E.F., The distributions of the introduced rhodophytes *Kappaphycus alvarezii, Kappaphycus striatum* and *Gracilaria salicornia* in relation to various physical and biological factors in Kane'ohe Bay, O'ahu, Hawai'i. *Pac. Sci.* 1999, 232.
74. Conklin, E.J. and Smith, J.E., Abundance and spread of the invasive red algae, *Kappaphycus* spp., in Kane'ohe Bay, Hawai'i and an experimental assessment of management options. *Biol. Inv.*, 2005, 7, 1029.
75. Smith, J.E., Factors influencing algal blooms on tropical reefs with an emphasis on herbivory, nutrients and invasive species, Ph.D. thesis, University of Hawaii, Honolulu, 2003.
76. Maragos J.E., Evans, C., and Holthus, P., Reef corals in Kaneohe Bay 6 years before and after termination of sewage discharges. *Proc. Fifth Int. Coral Reef Symp.,* 1985, 198.
77. Evans, C.W., Maragos, J.W., and Holthus, P.W., Reef corals in Kaneohe Bay 6 years before and after termination of sewage discharges (Oahu, Hawaiian Archipelago), in *Coral Reef Population Biology,* Jokiel, P.L., Richmond, R.H., and Rogers, R.A., Eds., University of Hawaii, Sea Grant Pub. No.UNIHI-SG-CR-86-01, Honolulu, 1986, 76.
78. Hunter, C.L. and Evans, C.W., Coral reefs in Kaneohe Bay, Hawaii: two centuries of western influence and two decades of data. *Bull. Mar. Sci.,* 1995, 501.
79. Jokiel, P.L., Cox, E.F. and Crosby, M.P., An evaluation of the nearshore coral reef resources of Kahoolawe, Hawaii. Final Report for Co-operative Agreement NA27OM0327, University of Hawaii, Hawaii Institute of Marine Biology, Honolulu, 1993.
80. Jokiel, P.L. and Coles, S.L., Effects of heated effluent on hermatypic corals at Kahe Point, Oahu. *Pac. Sci.* 1974, 28, 1.

81. U.S. Environmental Protection Agency, The Hawaii Sugar Industry Waste Study, U.S. Environmental Protection Agency, Region IX, San Francisco, CA, U.S. Government Printing Office Pub. 981-150, 1971.
82. Grigg, R.W., Hamakua coast sugar mills revisited: an environmental impact analysis in 1983, University of Hawaii, Sea Grant Pub. No. UNIHI-SEAGRANT-TR-85-02, Honolulu, 1985.
83. Grigg, R.W., Hamakua Sugar Company: Haina factories ocean discharges — a comparison analysis of ocean impact from 1971–1991, unpublished manuscript.
84. U.S. Navy, Final Environmental Impacts Statement for an Ammunition Wharf in Outer Apra Harbor, Guam, Mariana Islands. Honolulu, HI, 1983.
85. U.S. Navy, Haputo Ecological Reserve Area Establishment Report, Pacific Division, Naval Facilities Engineering Command. Pearl Harbor, HI, 1984.
86. U.S. Navy, Orote Peninsula Ecological Reserve Area Establishment Report, Pacific Division, Naval Facilities Engineering Command. Pearl Harbor, HI, 1984.
87. Chapman, G.A., Honolulu International Airport reef runway postconstruction environmental impact report. Parsons Hawaii, Honolulu, 1979.
88. Maragos, J.E. and Fagolimul, J.O., Impact of the grounding of the bulk carrier *M/V Oceanus* on the coastal resources of Satawal Island (Yap State, Federated States of Micronesia). Prepared for Paul, Johnson, Park and Niles on behalf of the People of Satawal. East-West Center, Program on Environment, Honolulu, 1996.
89. Kaser, T., $2 million paid for reef damage. *Honolulu Advertiser,* Feb. 9, 1998, B6.
90. Lubchenco, J.S., Palumbi, R., Gaines, S.D., and Andelman, S., Eds., The Science of Marine Reserves. *Ecol. Applications,* 2003, S1.
91. West, J.M. and Salm, R.V., Resistance and resilience to coral bleaching: implications for coral reef conservation and management. *Cons. Biol.* 2003, 956.
92. Friedlander, A.M. and DeMartini, E.E., Contrasts in density, size, and biomass of reef fishes between the northwestern and the main Hawaiian Islands: the effects of fishing down apex predators. *Mar. Ecol. Prog. Ser.* 2002, 230, 291.
93. Spurgeon, J.P.G., The economic valuation of coral reefs. *Mar. Poll. Bull.* 24, 1992, 529.
94. Cesar, H.S.J. and van Beukering, P.J.H., Economic valuation of the coral reefs of Hawaii. *Pac. Sci.* 2004, 58, 231.
95. Van Beukering, P.J.H. and Cesar, H.S.J. Ecological economic modeling of coral reefs: evaluating tourist overuse at Hanauma Bay and algae blooms at the Kihei coast, Hawaii. *Pac. Sci.* 2004, 58, 243.
96. Spurgeon, J.P.G. and Lindahl, U., Economics of coral reef restoration, in *Collected Essays on the Economics of Coral Reefs,* Cesar, H.S.J., Ed., Kalmar University, Kalmar, Sweden, 2000, 125.
97. Edwards, A.J. and Clark, S., Coral transplantation: a useful management tool or misguided meddling? *Mar. Pol. Bull.* 1999, 474.

16 The Coral Gardening Concept and the Use of Underwater Nurseries: Lessons Learned from Silvics and Silviculture

Baruch Rinkevich

CONTENTS

ABSTRACT

The types and numbers of tree species within a forest, as coral colonies in a coral reef, create unique ecosystems by generating biological diversities, formulating the habitats' three-dimensional structures, and changing local biological and environmental conditions. Resulting from intensive anthropogenic pressure, both ecosystems have lost their resilience, their ability to recover and to self-rehabilitate naturally without active human intervention. However, while forestry practices (silviculture) have been developed and tested worldwide for nearly two centuries, the discipline of active reef restoration is less than a decade old. Nevertheless, even though silviculture actions and concepts have long been subjected to rigorous scientific testing, its applications are still elusive. The situation in the field of reef restoration is even more undefined because the concepts and basic protocols have not yet been well studied. Here, one of the novel tools for coral reef restoration, the "gardening coral reefs" concept, where planned underwater nurseries present forestry principles, is discussed, bearing in mind the lessons learned from silvics and silviculture projects. Furthermore, the recently developed approach of a mid-water floating nursery is explained. In the future, as coral reef restoration may become the dominant conservation approach, there would be the need not only to develop improved protocols and defined conceptual bases, but also to adapt ideas, established expertise, and knowledge from silvicultural experience and science.

16.1 RESTORATION THROUGH THE "CORAL REEF GARDENING" CONCEPT

The decline of coral reefs worldwide[1–4] has prompted the need for urgent development of adequate restoration methodologies. Restoration itself has also been drawing increasing attention because most efforts to conserve degrading reefs have failed to yield significant results, whereas traditional rehabilitation measures had not successfully compensated for the fast decline.[1,5–7] Moreover, in many reef areas, the poor state of the reef has reached a critical point (sensu[8]) where management activities could no longer effectively conserve remnants of precious reef populations or prevent further habitat degradation.[7] This situation is further characterized by the lack of state-of-the-art remediation protocols, i.e., established theories and approved restoration techniques, specifically developed for the marine environment, including coral reefs. Existing measures still lag behind those developed for terrestrial habitats.[9–11] As a result, the principles underlying remediation measures have turned out to be just some of the many ill-defined issues of reef restoration.[12,13]

It is also evident that restoring any type of degraded reef area is a complex biological and ecological procedure.[14,15] Until recently, most studies on coral reef restoration were based on small-scale, short-term experimental protocols, testing only some ecological/biological attributes. During the last decade, however, worldwide coral reef restoration operations have been more frequently employed and tested in various reef localities, and the concept of active restoration has been acknowledged as an important approach for reef rehabilitation.[1,2,5,13,14]

One of the most commonly used methodologies for restoration is direct transplantation of coral material (including entire colonies, fragments, nubbins). While the techniques used for removal of coral material and their transportation and reattachment are straightforward and simple, the varying degrees of success that had been reported indicated significant limitations in the direct transplantation methodology. This is caused by the stress imposed on the transplanted coral material, the use of insufficient donor colonies and/or too small fragments, and the disturbances inflicted on the donor coral populations.[12,16,17] In other cases, the failure of corals to recover denuded reef areas also reflects postsettlement mortalities.[18] To alleviate these problems, Rinkevich[5] has suggested the strategy of "gardening coral reefs," a two-step restoration protocol whose central concept is the mariculture of coral recruits (spats, nubbins, coral fragments, and small coral colonies) in nurseries. Firstly, instead of direct transplantation, large *in situ* or *ex situ* pools of farmed corals and spats are constructed. *In situ* nurseries are installed in sheltered zones where the different types of coral recruits are maricultured to sizes suitable for transplantation. This practice also makes use of minute-size coral fragments that would have died in direct transplantation. Secondly, nursery-grown coral colonies are transplanted to degraded reef sites. This approach is associated with theories of silvics and silviculture.[2]

Ex situ and *in situ* coral reef maricultures (the "gardening" concept) are therefore improvements over the common but potentially harmful protocols of direct coral transplantation. *Ex situ* coral mariculture further supports the high survivorship and growth rates of very small coral material (such as settled planulae larvae or delicate nubbins). This concept has already been tested for its applicability in various coral reefs worldwide. Following the suggestion by Rinkevich,[5] several studies tested the applicability, the feasibility, and the detailed developed protocols for holding corals in nurseries under *in situ* conditions,[19–22] in flowing seawater *ex situ* systems,[16] or in closed-seawater *ex situ* facilities (aquaria, holding tanks.[16,23–28] *In situ* coral nurseries can supply the transplanting operations with corals adapted to natural reef conditions,[1] while *ex situ* coral nurseries may facilitate the yield of coral planulae, directly increasing genetic variability of transplanted colonies.[5] Both *ex situ* and *in situ* approaches can also provide ample material for restoration year-round, thus reducing the collection of coral colonies from the wild.[27–28]

16.2 CORALS AND TREES — TWO MAJOR FRAMEWORK BUILDING BLOCKS

Stony corals and trees are the basic framework building blocks of two major biological ecosystems, coral reefs and rainforests. They share a common ecological role by virtue of their biological properties.[29–32] Both building blocks can be propagated by sexual and asexual reproduction (using ramets, nubbins vs. seeds, seedlings, planulae larvae, spat), have direct impact on their ecosystems (trees stabilize soil, enhance litter production, increase soil organic matter, etc.; corals enhance lithification processes, enhance reef rock formation, etc.), and may increase spatial resilience of ecosystems, and have other similarities.[2] By being the primary habitat constructors, these sessile organisms not only bear similarities in their contribution to the ecosystems' structural arrangements but also follow similar basic architectural rules in their growth and characteristics of pattern formation.[33] Reef corals and rainforest trees also harbor numerous inhabitants and provide the ecosystem with the structural design and strength to resist physical disturbances such as storms.[2]

While coral reefs and rainforests exhibit high resilience even to major natural catastrophes, they have been facing dramatic threats from new types of perturbation, as damaging human activities have rapidly increased in scale and intensity. Forest clearing and land conversion to agriculture lead to soil erosion and desertification. Similarly, a major decline in coral coverage due to manmade physical disturbances leads to the collapse of reef communities and the development of algal reefs.[34] Tree logging is recognized as the major destructive agent of forest areas;[35,36] marine-based recreational activities play this role in coral reefs.[1,2] Under regimes of chronic human impacts, both ecosystems undergo dramatic changes in structure and composition of species,[5,36] displaying diminishing capacity to show ecological resilience, to absorb disturbances, to reorganize, and to adapt to changes. In the reef ecosystem, this degradation has been augmented during the last two decades by major bleaching events.[4] Can we utilize the lessons learned from silvics and silviculture?

16.3 SILVICS AND SILVICULTURE — STILL MANY UNRESOLVED ISSUES

One of the most important disciplines in forestry is silviculture. *Silviculture* is the agriculture of trees: how to grow them, how to maximize growth and return, and how to manipulate tree species compositions to meet landowners' objectives. To understand silviculture, one must first understand silvics. *Silvics* involves understanding how trees grow, reproduce, and respond to environmental changes. Silviculture is the applied restoration concept of rehabilitation of terrestrial habitats where natural self-regeneration of forests is not applicable. The failure of forest systems to self-regenerate has stimulated the development of various restoration measures that have proven to be effective (literature cited in Putz et al., 2001[37]). These restoration efforts have succeeded in reversing the trend of deterioration and in creating new habitats for biodiversity.[35]

Forest restoration research is categorized into three major scientific approaches: genetics, nursery, and site preparation. The term "site preparation" includes not only mechanical cultivation but also the use of herbicides, fertilizers, insecticides, and other treatments that are applied when establishing plantations.[38] Nursery researchers are inclined to outplant nursery treatments, geneticists are inclined to outplant progeny tests, and silviculturalists tend to evaluate site preparation methods. Due to these three different routes and the fragmented approaches, even in this well-studied discipline of forestry, trials that comprise all three major categories are scarce; even those that combine two categories are not common. Trials that combine nursery treatments with site preparation treatments are rare. One example is a study by Barber et al.[39] that qualified nursery treatments (fertilization rates) and site preparation treatments (herbicides) under the same experimental design. Nursery/genetic trials were evaluated by Land[40] and genotype/silvicultural interaction were studies discussed by only a few authors (i.e., reference 41).

Although forest restoration has been an important tool for the conservation, preservation, and maintenance of diversity for more than a century, major obstacles in concepts unification still exist. For example, comparing modes of forest protection between regions in Europe is extremely difficult because of wide variation of strategies, procedures, and constraints; the ways forests have been used historically; their present proximity to major urban centers; and even the definition of what constitutes a forest.[42] In this respect, even in silviculture there has yet to be a single, uniform, and universal model and a universally agreed-upon goal for restoration (such as the percentage of forests, which parts should be protected, etc.). It is accepted that the common denominator for a strict forest reserve is not silvicultural management. However, the ideal nonintervention concept of developing appreciable areas of real untouched forests is not realistic for most protected forests.[42] Spatial heterogeneity, enhancing biodiversity of plantations, silviculture systems, structural analyses, and scale of heterogeneity are some of the major unsolved issues of proper restoration measures for degraded forests.[43,44]

One interesting concept is the "close-to-nature silviculture" concept,[45] which was formulated at the end of the 19th century, when new paradigms emerged under the motto of "returning to nature." This concept was examined by numerous studies, and much knowledge was gained from practical experience. However, a closer inspection of the experiences drawn from the past century of applications reveals that the progress made toward the two major original goals, the establishment of mixed stands and the promotion of stand irregularity (reviewed in reference 45) has been insufficient. For that reason, the concept of close-to-nature silviculture is open to various interpretations, which mainly depend on the emphasis given to the terms "culture" and "nature" and the values associated with them. These and other silvicultural ideas and protocols should be considered when establishing the first ideas and concepts in active reef restoration.

16.4 REEF RESTORATION CONCEPTS

Due to biological, structural, and functional analogies between trees and corals, it is natural to suggest that silviculture concepts (well established or even under trial) should assist in establishing the theoretical and practical concepts for coral reef restoration so that a solid restoration framework can be developed.[2] The strategy of coral mariculture through the gardening concept may prove to be the first such sustainable practice for reef restoration, highly comparable to silviculture in its nature. Therefore, the discussion on "gardening the coral reef," in light of the already discussed/established ideas in silviculture, is of great importance. It should also be kept in mind that the concept of gardening denuded reef areas is a superior strategy to the more traditional and widely used measure of "coral transplantation," where the need to harvest coral colonies from existing populations represents a major imperfection.

We[1,2,5,17] consider coral mariculture to be a major component in reef restoration, a concept that has so far been overlooked. This strategy, when established and tested in various localities worldwide, may shape the conceptual and practical platform for reef restoration activities. Pertaining to transplantation, in comparison to the harmful practice of harvesting corals from donor reef areas, the establishment of coral nurseries eliminates the need for the extraction of valuable coral material.[1,5,13,14,17] As aforementioned, a protected nursery phase provides the transplanted material with an acclimation period ensuring better survivorship and growth to a size suitable for transplantation. The transplantation of nursery-grown "propagules" back into their natal reef helps to prevent extinction of genets and species in degrading sites, thus exercising the "rescue effect"[46] on a local scale by preserving genetic heterogeneity. A coral nursery may also serve as a local species pool that supplies reef-managers with coral colonies for sustainable management.[1,2,5,13] Culturing corals in underwater nurseries may also help in structuring the three-dimensional shapes of developing colonies.[14]

Colonial structures emerge as iterative processes of successive layers of material.[47] In branching coral colonies, morphology is established through iteration of two structural units: modules of the first order (the zooids) and modules of the second order (the branches). Variations in the branching morphologies of colonial organisms are frequently correlated with a suite of life-history traits.[48] A significant part of relevant literature examines correlations between environmental qualities and morphometric analyses (either on the zooid, ramet, or genet level, see references 49–55 and literature therein. While there is no doubt that environmental factors may tune phenotypic architecture (this issue is another factor to be studied in underwater coral nurseries), the common high-fidelity of morphological structures that may be exhibited by any species of coral reveals developmental homeostasis, controlled by the genetic background of the species of interest.[56] Adapted maricultural conditions may therefore "secure" the blueprint structures of developing coral colonies.[14,56] Another interesting genetic approach is the application of "tree architectural models" to coral growth forms, see references 33,57,58 and literature therein. Analyses revealed common rules for branching and ramifying axes systems, for the organization of these axes in identical architectural models, for reiteration patterns in the course of growth, and for the physiological variations correlated with environmental parameters.

Much of the literature on ecological restoration pertains to the choice of species to be used.[59] Taxa used for restoration are not selected randomly but often conform to the definition of "key species"; i.e., they perform a function that is more vital for the ecosystem than those of other species. Selected coral species for mariculture are often good representatives of local common coral species and in many cases, are of the branching coral forms.[1,17,19,60] Branching forms are also selected for their high performance in restoration (i.e., survivorship and growth rates). Branching coral colonies potentially provide an increased variety of coral material types that can be used for mariculture including ramets (incorporating a single or several branches); nubbins (fragments the size of a single or several polyps); small whole colonies removed from shallow, frequently disturbed areas, where long-term survival is unlikely; and planula larvae.[1,5,13,61]

16.5 A SPECIAL CASE: MIDWATER FLOATING NURSERY

Only recently, active coral reef restoration has begun to be viewed as a necessary measure,[61] and it has been accepted that methodology should be appropriately adapted to local socioeconomic limitations and subjected to landscape conditions. The nursery component of the gardening concept, however, is a ubiquitous measure that is likely applicable to all locations and could be installed even away from the reef-site itself.[17] Coral nurseries, like tree plantations, are a stark contrast to constructionally complex, multicohort mature and developed reef areas/forests. However, nurseries, as a rehabilitation tool, facilitate complexity and management of the natural habitats by providing flexibility to meet a range of needs, objectives, and specific aims.[62] Although a wide application of reef restoration requires management approaches that are specifically adapted to different operational situations and local conditions, this discipline still needs a ubiquitous rationale to be developed specifically for coral reef rehabilitation.[5,13] Management decisions should always take into consideration the appropriate target coral species and the type of source material most suitable for local cultivation. As such, site-specific considerations and the use of different local coral species as donors require the development of different specific protocols, tailored to the conditions at dissimilar reef areas. The gardening concept, then, could contribute to the formation of a meaningful reef mitigation framework via the conservation of endangered coral species through mariculture of specific genets and rehabilitation through their transplantation. However, dealing with the specifics without formulating the concepts may be a major obstacle for the development of this scientific discipline.

To date, all *in situ* coral nurseries were studied in constructions at or near sea bottom, in shallow-water reef areas.[1,22] These nurseries were subjected to local conditions and therefore, a ubiquitous rationale has not been developed. A novel general approach for an *in situ* coral nursery, the establishment of a midwater nursery (14 m above substrate and in 6 m depth; Figure 16.1) in a protected site, 8 km north of major natural reef areas in the Gulf of Eilat, Red Sea has been tested

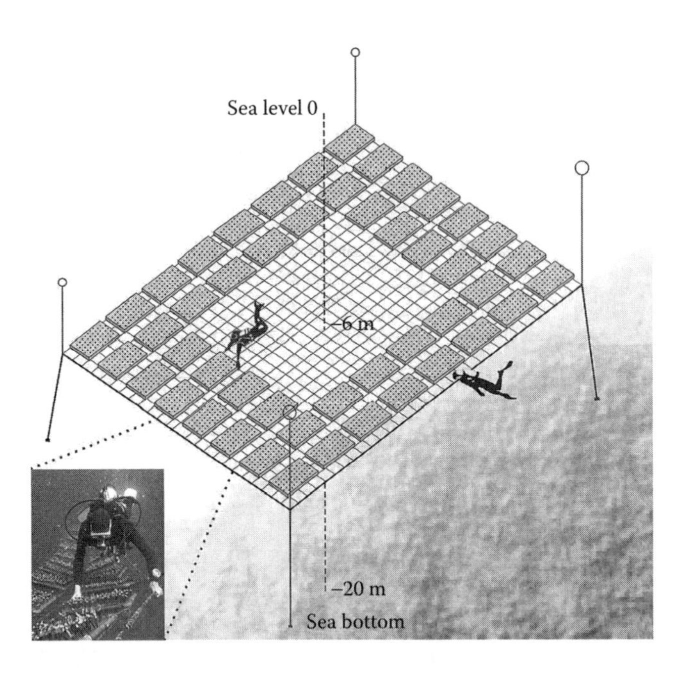

FIGURE 16.1 A mid-water nursery. A prototype that has recently been tested in the Gulf of Eilat (size 10 × 10 m, made of a rope net; adapted from reference 17). The nursery is connected to the sea bottom (20 m depth) by metal cables and is held floating at 6 m depth by several buoys. Coral colonies are maricultured within plastic nets stretched over PVC frames (30 × 50 cm each). Each frame holds up to 110 pins with coral ramets. The frames were situated in an order that allowed divers to approach the developing coral colonies either from the nursery periphery or from its center. Such a nursery may hold several thousands of coral colonies and can be easily maintained by a single pair of divers.

for the first time.[17] This floating nursery, situated away from afflictions of recreational activities and corallivorous organisms, proved to be superior to former versions. More than 7100 coral branches sampled from 11 branching scleractinian species had, after 5 to 10 months, very low (<10%) mortality rates (in some species even zero mortality; Figure 16.2) and very fast growth rates, forming within this limited time frame large colonies originating from small branches (an example case is depicted in Figure 16.3).

Since being afloat in midwater (Figure 16.1), this type of nursery demonstrated that:

1. Water flow in this unique nursery system supplies the developing corals with increased quantities of plankton particles, probably with enhanced rates of dissolved oxygen around coral tissues, resulting in more efficient removal of mucus secreted by the coral tissue.
2. Although sea-bottom nurseries are attached to the reef floor,[1,22] and water movement around the corals results strictly from currents or wave actions, in a midwater floating nursery the entire nursery swings in all directions. This flexibility helps to get rid of debris, sediment particles, and other settling material that might accumulate on developing coral colonies.
3. Since in a midwater nursery the substrate is far below, sedimentation is reduced to a minimum with negligible influence on the developing coral colonies.
4. With proper consideration, a midwater nursery can be positioned at different depths, "tailored" for any coral species–specific needs. Using the midwater nursery also allows the developing corals to gradually acclimatize to conditions of depth and radiation similar to those in their designated transplantation site.

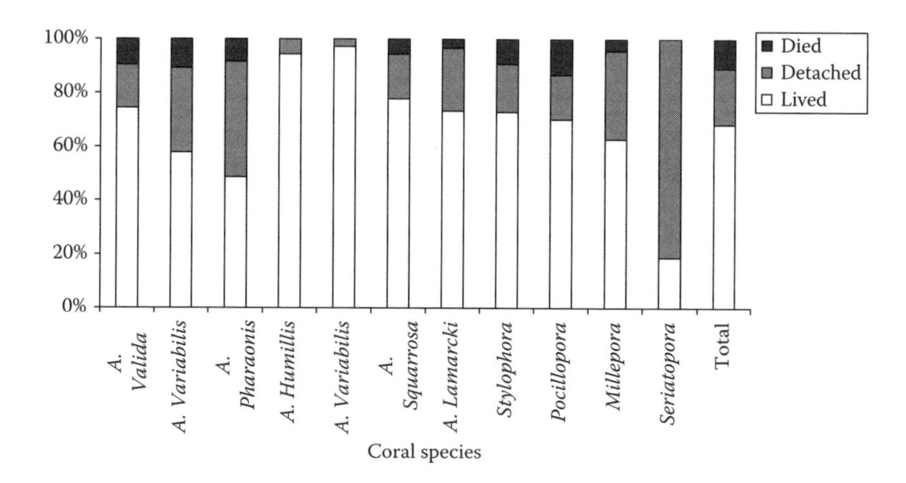

FIGURE 16.2 Status of ramets from 11 branching scleractinian species (*Millepora, Seriatopora, Stylophora, Pocillopora,* and *Acropora* species) maricultured in the midwater nursery after 5 to 10 nursery months (adapted from reference 17.)

5. Installing nurseries at a distance from the coral reef may reduce harmful impacts of corallivorous organisms and recreational activities.
6. Specimens of various commensal and coral-residing species were found to come from the plankton, settling on the newly developed coral colonies (*Trapezia* and *Tetralia* crabs, *Spirobranchus giganteus*, and others). As a result, transplantation may involve not only the cultured coral colonies but also the entire coral infaunal biodiversity.[17] This floating nursery serves, therefore, as a prototype ubiquitous nursery for developing protocols and working rationales applicable worldwide.

16.6 SUMMARY

Ecosystem management in both terrestrial and marine habitats is associated with a blend of social, political, economic, cultural, and ecological themes.[32,63] For example, in silviculture, the choice of methods for thinning forest shading, spacing, and clear-cutting often depend on landowners' and stakeholders' demands. As such, selecting which habitat should be restored may be as important as how much is to be restored. Nonrandom restoration practices such as restoring only habitats

FIGURE 16.3 Typical rapid growth of a small *Acropora pharaonis* branch after 100 nursery days and development into a large colony after 400 nursery days (adapted from reference 17). Scale bar = 1 cm.

adjacent to those inhabited by the target species can dramatically reduce or counteract any restoration act.[64] Moreover, in many cases of forest management, silvicultural practices are designed to promote the regeneration and the stocking of commercial species for timber forest products,[37] as silviculture involves managing and handling the forest in view of its silvics. Not so in coral reefs. In economically important reef sites (such as for the tourist industry), the coral mariculture strategy is being used to develop sustainable material for human and nonhuman stakeholders in a way that all components of reef habitats benefit.

Silviculture is the art and science of producing and tending a forest; application of forest ecology and economics in treatment of forests; and the theory and practice of controlling forest establishment, composition, and growth. Through silviculture, we have healthy growing trees, not only for better wood products but also for the environment. In practice, silviculture consists of the various treatments that could be applied to maintain and enhance utility of trees for any purpose. Through harvesting, cutting, thinning, prescribed burning, and various other methods, the variety and age of tree species within a forest, the density of trees, the arrangement of different layers, and other factors such as lighting and shading can be manipulated. It is important to remember that these management techniques not only affect the present forest but also influence its future characteristics. In coral reefs, however, this rationale of active restoration is not yet well accepted. It is probably best exercised in the gardening concept of reef restoration and in restoration of ship grounding sites. Impacts of ship grounding include dislodging and fracturing of corals, pulverizing of coral skeletons, displacement of sediments from ground, and destruction of the three-dimensional structural complexity of the reef.[65] Salvage operations usually add to reef damages, and damages from fuel and cargo slicking from the ruptured hull may worsen the situation. This causes acute and long-lasting effects on regenerative processes of coral communities. Efforts to restore reef sites damaged by ship grounding include such activities as salvaging coral colonies, coral fragments, and sponges; removing loose debris from the reef floor; reconstructing the three-dimensional structural complexity of the reef; and reattaching detached corals and sponges to cleared reef substrates or specially designed artificial reef structures.[65,66] However, a major snag in this type of reef restoration is that most if not all coral material for restoration comes from the damaged sites and not from adjacent coral reef populations. Another challenge to tackle is the massive amount of rubble.

The task of emulating natural phenomena of reef growth and development with restoration (as with silviculture to forests[62]) is challenging, especially when we consider the socioeconomic constraints and our meager knowledge. At present, even the indicators of ecological sustainability of coral reefs are not well defined; there is still no effort to incorporate biodiversity goals into restoration measures, and operational constraints are not designated. Coral reef biologists may adopt[67] the suggestion that any proposed silviculture system designed to maintain biodiversity and produce timber should be treated as a hypothesis due to the limited number of empirical studies to support or refute the approach.

Like silviculture, reef restoration deals with the three-dimensional structural topography, the creation of structural topography, and increased heterogeneity. So far, the only active reef restoration principle suggested (and applied) with an eye to ecological principles tested in other similar ecosystems is the gardening concept.[1,2,5,13,14,17] In the coral reef as in forest ecosystems, a "total protection" approach would secure and preserve only a certain number of habitats of rare species in any locality. Therefore, silvicultural management or reef restoration is essentially required to maintain large-scale biodiversity in multifunctional production forests and coral reefs.[5,13,42] The restoration of coral reefs, therefore, should become a standard part of conservation practices, and when applied, already tested and approved forest restoration principles may provide important insight into the understanding of the reef ecosystem recovery.[2] The ability to produce and develop many new coral colonies in nurseries[17] may change the way end-users will manage denuded reef areas in the future.

ACKNOWLEDGMENT

This study was supported by BARD (US-Israel Bi-national Agricultural Research and Development Fund), by an INCO-DEV project (REEFRES), by a World Bank/GEF project (Reef Remediation/Restoration Working Group), and by the AID-CDR program (no. C23-004).

REFERENCES

1. Epstein, N., Bak, R.P.M., and Rinkevich, B., Strategies for gardening denuded coral reef areas: the applicability of using different types of coral material for reef restoration, *Restor. Ecol.*, 9, 432, 2001.
2. Epstein, N., Bak, R.P.M., and Rinkevich, B., Applying forest restoration principles to coral reef rehabilitation, *Aquat. Conserv.*, 13, 387, 2003.
3. Wilkinson, C.R., *Status of Coral Reefs of the World: 2002*. Australian Institute of Marine Science, Townsville, Queensland 378 pp, 2002.
4. Bellwood, D.R., Hughes, T.P., Folke, C., and Nyström, M., Confronting the coral reef crisis, *Nature*, 429, 827, 2004.
5. Rinkevich, B., Restoration strategies for coral reefs damaged by recreational activities: the use of sexual and asexual recruits, *Restor. Ecol.* 3, 241, 1995.
6. Risk, M.G., Paradise lost: how marine science failed the world's coral reefs, *Mar. Freshwater Res.*, 50, 831, 1999.
7. Epstein, N., Bak, R.P.M., and Rinkevich, B., Alleviating impacts of anthropogenic activities by traditional conservation measures: can a small reef reserve be sustainedly managed? *Biol. Conserv.*, 121, 243, 2005.
8. Young, T.P., Restoration ecology and conservation biology, *Biol. Conserv.*, 92, 73, 2000.
9. Allison, G.W., Lubchenko, J., and Carr, M.H., Marine reserves are necessary but not sufficient for marine conservation, *Ecol. Appl.*, 8, 79, 1998.
10. Keough, M.J., and Quinn, G.P., Legislative vs. practical protection of an intertidal shoreline in South eastern Australia, *Ecol. Appl.*, 10, 871, 2000.
11. Rose, K.A., Why are quantitative relationships between environmental quality and fish populations so illusive? *Ecol. Appl.*, 10, 367, 2000.
12. Edwards, A.J., and Clark, S., Coral transplantation: a useful management tool or misguided meddling? *Mar. Poll. Bull.*, 37, 8, 1998.
13. Rinkevich, B., Steps towards the evaluation of coral reef restoration by using small branch fragments, *Mar. Biol.*, 136, 807, 2000.
14. Epstein, N., and Rinkevich, B., From isolated ramets to coral colonies: the significance of colony pattern formation in reef restoration practices, *Basic Appl. Ecol.*, 2, 219, 2001.
15. Omori, M., and Fujiwara, S. (Eds.), Manual for restoration and remediation of coral reefs, Nature Conservation Bureau, Ministry of Environment, Japan, 2004.
16. Becker, L.C., and Mueller, E., The culture, transplantation and storage of *Montastraea faveolata, Acropora cervicornis,* and *Acropora palmata*: what we have learned so far, *Bull. Mar. Sci.*, 69, 881, 2001.
17. Shafir, S., van Rijn, J., and Rinkevich, B., A mid water coral nursery, *Proc. 10th Int. Coral Reef Symp.*, in press.
18. Fox, H.E., Coral recruitment in blasted and unblasted sites in Indonesia: assessing rehabilitation potential, *Mar. Ecol. Prog. Ser.*, 269, 131, 2004.
19. Bowden-Kerby, A., Coral transplantation in sheltered habitats using unattached fragments and cultured colonies, *8th Int. Coral Reef Symp.*, Panama, 2, 63, 1997.
20. Oren, U., and Benayahu, Y., Transplantation of juvenile corals: a new approach for enhancing colonization of artificial reefs, *Mar. Biol.*, 127, 499, 1997.
21. Epstein, N. Bak, R.P.M., and Rinkevich, B., Implementation of small scale "no-use zone" policy in a reef ecosystem: Eilat's reef-lagoon 6 years later, *Coral Reefs* 18, 327, 1999.
22. Soong, K., and Chen, T., Coral transplantation: regeneration and growth of *Acropora* fragments in a nursery, *Restor. Ecol.*, 11, 62, 2003.

23. Richmond, R. Coral reef health: concerns, approaches and needs. *Proc. Coral Reef Symp. Practices, Reliable, Low Cost Monitoring Methods for Assessing the Biota and Habitat Conditions of Coral Reefs*, Silver Spring, MD, 22, 1995.
24. Carlson, B.A., Organism response to change: what aquaria tell us about nature, *Am. Zool.*, 39, 44, 1999.
25. Raymundo, L.J., Maypa, A.P., and Luchavez, M.M., Coral seeding as a technology for recovering degraded coral reefs in the Philippines, *Phuket Mar. Biol. Cent. Spec. Publ.* 20, 81, 1999.
26. Szmant, A.M., Coral restoration and water quality monitoring with cultured larvae of *Montastraea "annularis"* and *Acropora palmata. Int. Conf. Sci. Aspects of Coral Reef Assessment, Monitoring and Restoration,* Florida, 188, 1999.
27. Borneman, E.H., and Lowrie, J., Advances in captive husbandry and propagation: an easily utilized reef replenishment means from the private sector? *Bull. Mar. Sci.,* 69, 897, 2001.
28. Petersen, D., and Tollrian, R., Methods to enhance sexual recruitment for restoration of damaged reefs, *Bull. Mar. Sci.,* 69, 989, 2001.
29. Yonge, C.M., The nature of reef-building (Hermatypic) corals, *Bull. Mar. Sci.,* 23, 2, 1973.
30. Connell, J.H., Diversity in tropical rain forests and coral reefs, *Science*, 199, 1302, 1978.
31. Goreau, T.F., Goreau, N.I., and Goreau, T.J., Corals and coral reefs, *Sci. Am.,* 241, 110, 1979.
32. Christensen, N.L., Bartuska, A.M., Brown, J.H., Carpenter, S., D'Antonio, C., Francis, R., Franklin, J.F., MacMahon, J.A., Noss, R.F., Parsons, D.J., Peterson, C.H., Turner, M.G., and Woodmansee, R.G., The report of the Ecological Society of America committee on the scientific basis of ecosystem management, *Ecol. Appl.,* 6, 665, 1996.
33. Dauget, J.-M., Application of tree architecture models to reef-coral growth forms, *Mar. Biol.,* 111, 157, 1991.
34. Lugo, E.L., Caroline, S.R., and Scott, W.N., Hurricanes, coral reefs and rainforests: resistance, ruin, and recovery in the Caribbean, *Ambio*, 29, 106, 2000.
35. Dobson, A.P., Bradshaw, A.D., and Baker, A.J.M., Hopes for the future: restoration ecology and conservation biology, *Science*, 227, 515, 1997.
36. Noble, I.R., and Dirzo, R., Forests as human-dominated ecosystems, *Science* 277, 522, 1977.
37. Putz, F.E., Blate, G.M., Redford, K.H., Fimbel, R., and Robinson, J., Tropical forest management and conservation of biodiversity: an overview, *Conserv. Biol.,* 15, 7, 2001.
38. South, D.B., Rose, R.W., and McNabb, K.L., Nursery and site preparation interaction research in the United States, *New Forests*, 22, 43, 2001.
39. Barber, B.L. Messina, J.S., van Buijtenen, J.P., and Wall, M.M., Influence of nursery fertilization, site quality, and weed control on first-year performance of outplanted loblolly pine, *6th Biennial Southern Silvicultural Res. Conf.,* Memphis, TN, 27, 1991.
40. Land, S.B., Performance and G-E interactions of sycamore established from cuttings and seedlings, *2nd Biennial Southern Silvicultural Res. Conf.,* Atlanta, GA, 431, 1983.
41. McDonald, P.M., Mori, S.R., and Fiddle, G.O., Effects of competition on genetically improved ponderosa pine seedldings, *Can. J. For. Res.,* 29, 940, 1999.
42. Parviainen, J., and Frank, G., Protected forests in Europe approaches — harmonising the definitions for international comparison and forest policy making, *J. Environ. Manag.,* 67, 27, 2003.
43. Kerr, G., The use of silvicultural systems to enhance the biological diversity of plantation forests in Britain, *Forestry*, 72, 191, 1999.
44. Schutz, J.-P., Silvicultural tools to develop irregular and diverse forest structures, *Forestry*, 75, 329, 2002.
45. Schutz, J.-P., Close-to-nature silviculture: is this concept compatible with species diversity? *Forestry*, 72, 359, 1999.
46. Sinsch, U., Postmetamorphic dispersal and recruitment of first breeders in a *Bufo calmamita* metapopulation, *Oecologia,* 112, 42, 1997.
47. Kaandorp, J.A., and de Kluijver, M.J., Verification of fractal growth models of the sponge *Haliclona oculata* (Porifera) with transplantation experiments, *Mar. Biol.,* 113, 133, 1992.
48. Buss, L.W., and Blackstone, N.W., An experimental exploration of Wadington's epigenetic landstacep, *Phil. Trans. R. Soc. Lond.,* 332, 49, 1991.
49. Chamberlain, J.A. Jr. and Graus, R.R., Water flow and hydromechanical adaptations of branched reef corals, *Bull. Mar. Sci.,* 25, 112, 1975.

50. Bottjer, D.J., Branching morphology of the reef coral *Acropora cervicornis* in different hydraulic regimes, *J. Paleontol.*, 54, 1102, 1980.
51. Sebens, K.P., Water flow and coral colony size: interhabitat comparisons of the octocoral *Alcyonium siderium, Proc. Nat. Acad. Sci. USA*, 81, 5473, 1984.
52. Abelson, A., Galil, B.S., and Loya, Y., Skeletal modification in stony corals caused by indwelling crabs: hydrodynamical advantages for crab feeding, *Symbiosis* 10: 233, 1991.
53. Helmuth, B.S.T., Sebens, K.P., and Daniel, T.L., Morphological variation in coral aggregations: branch spacing and mass flux to coral tissues, *J. Exp. Mar. Biol. Ecol.*, 209, 233, 1997.
54. Johnson, A., Flow is genet and ramet blind: consequences of individual group and colony morphology on filter feeding and flow, *Proc. 8th Int. Coral Reef Symp.*, 2, 1093, 1997.
55. Vago, R., Achituv, Y., Vaky, L., Dubinsky, Z., and Kizner, Z., Colony architecture of *Millepora dichotoma* Forskål, *J. Exp. Mar. Biol. Ecol.*, 224, 225, 1998.
56. Rinkevich, B., The branching coral *Stylophora pistillata*: contribution of genetics in shaping colony landscape, *Isr. J. Zool.*, 48, 71, 2002.
57. Dauget, J.-M., La reiteration adaptive, un nouvel aspect de la croissance de certains Scleractiniairies recifaux. Exemple chez *Porites* sp., cf. *Cylindrica* Dana, *C.R. Acad., Sci. Paris*, 313 (III), 45, 1991.
58. Dauget, J.-M., Essai de comparaison entre l'architecture des coraux et celle des vegetaux, *Bull. Mus. Natl. His. Nat. Paris*, (4), 16, 39, 1994.
59. Lessica, P., and Allendorf, F.W., Ecological genetics and the restoration of plant communities: mix or match? *Restor. Ecol.*, 7, 42, 1999.
60. Franklin, H., Christopher, A., Muhando, A., and Lindahl, U., Coral culturing and temporal recruitment in Zanzibar, Tanzania, *Ambio*, 27, 651, 1998.
61. Bowden-Kerby, A., Low-tech coral reef restoration methods after natural transplantation processes, *Bull. Mar. Sci.*, 69, 915, 2001.
62. Palik, B.J., Mitchell, R.J., and Hiers, J.K., Modeling silviculture after natural disturbance to sustain biodiversity in the longleaf pine (*Pinus palustris*) ecosystem: balancing complexity and implementation, *Forest. Ecol. Manag.*, 155, 347, 2002.
63. Thomas, J.W., Forest service perspective on ecosystem management, *Ecol. Appl.*, 6, 703, 1996.
64. Huxell, G.R., and Hastings, A., Habitat loss, fragmentation, and restoration, *Restor. Ecol.*, 7, 309, 1999.
65. Jaap, W.C., Coral reef restoration, *Ecol. Eng.*, 15, 345, 2000.
66. Precht, W.F., Deis, D.R., and Gelber, A.R., Damage assessment protocols and restoration of coral reefs injured by vessel grounding, *Proc. 9th Int. Coral Reef Symp., Indonesia*, 2, 963, 2002.
67. Simberloff, D., The role of science in the preservation of biodiversity, *For. Ecol. Manag.*, 115, 101, 1999.

17 Lessons Learned in the Construction and Operation of Coral Reef Microcosms and Mesocosms

Walter H. Adey

CONTENTS

17.1 INTRODUCTION

Especially since the second World War, considerable effort has been invested in attempts to maintain and display marine organisms in public aquaria and marine laboratories. In addition, in the latter part of the 20th century, numerous hobbyists joined the quest for marine display, with sometimes rather elaborate systems in their homes. The most successful of the professional organism maintenance systems involve the use of sea water, pumped through sand beds or towers, from adjacent, high-quality marine environments (flo-through systems). In some cases, and particularly in the hobby environment, closed aquaculture methods are also employed, mostly using bacterial filtration and/or air-foamed towers (foam fractionation) to remove particulate matter.

A large percentage of these aquaculture and flo-through efforts involve coral reef organisms, although usually as individuals of a relatively few species. Only rarely does current practice involve any semblance of a coral reef ecosystem, either in terms of metabolism or community structure. An introduction to this literature can be found in both research and aquarium science publications (e.g., references 1 through 5) and in numerous books of varying quality that line the bookshelves of high-quality aquarium stores. These same techniques could be employed in a shoreside laboratory to hold less sensitive organisms for timed introduction to a coral reef restoration effort, though loading and organism stress issues limit these techniques. Biodiversity in such systems, even when water quality can be maintained at high levels, is typically low. In large measure, this is due to the filtration of reproductive stages by the water quality–maintaining devices, and to the further destruction of those stages by pumping devices. However, some organisms can be reproduced vegetatively or through budding and fragmentation.

More appropriate to the subject of the present volume, during the last two decades of the 20th century, a number of research and development endeavors were undertaken to model coral reef ecosystems in a semiclosed state. These projects were carried out at a wide variety of scales, ranging from roughly a cubic meter or so up to 3500 m^3. Numerous publications, beginning with reference 6, describe these efforts, and many are cited in this chapter. Two works[7,8] review the coral reef modeling work of the 1980s and the early '90s and compare and contrast the various systems.

The author is aware of nine microcosms and mesocosms specifically constructed to be physical/ecophysiological models of coral reef ecosystems. These systems were built by four separate organizations, and provided (to date) roughly 75 years of summed operation.[7] Mostly, these units were directed to educational display, with research as a secondary feature. A few were constructed primarily to test engineering features of large ecosystem models. Only one of these models, a 1680-l Caribbean coral reef (though with a few Indo-Pacific species) operated long term and was also subjected to considerable biodiversity and ecophysiological analysis. Although a less comprehensive analysis of some of the earlier systems[7] suggested that this system was by no means unique, it is the 1680-l Caribbean system, which operated throughout the 1990s, that provides the significant data on which the conclusions of this chapter are based.

The status of microcosms in restoration ecology prior to the mid-1980s was reviewed in 1987.[9] In that work, the philosophical basis for physical, living systems modeling was extensively discussed and compared to similar engineering endeavors (such as ship model towing basins and hydraulic models of bays and harbors). There is little point in repeating that discussion here, as virtually all of it is just as applicable today as it was 18 years ago. However, about the time that paper was published, research was initiated on the 1680-l Caribbean model that is the primary focus of this chapter. The essential missing ingredient of that report, long-term hard data, is now available. Hereafter, in this chapter, the unit that provided this data is referred to as the Caribbean Model.

The Caribbean Model was specifically modeled after the well-developed, shallow-water, bank barrier reef off the southeast side of the island of St. Croix, U.S. Virgin Islands (for an in-depth discussion of that reef see reference 10). While teaching coral reef ecology and phycology at the West Indies Laboratory of Fairleigh Dickinson University through much of the 1970s, the author had the opportunity to direct and participate in numerous class projects and student papers that related to that very well-developed reef system. At that time, there was little direct tourist or industrial effect on the southeast reef, and fishing was minimal and artisanal. Also, there had not been significant hurricane impact on those reefs for roughly the prior 50 years. Late in the 1970s, based on about 6 years of continuous research on this reef complex, a year-long, upstream/downstream analysis of ecosystem metabolism was carried out.[11] That study correlated to reef structure, and geological development and analyses were carried out at several stations stretched down the 25-km length of the reef.

The extensive coral reef complex after which the Caribbean Model was patterned faces southeast, largely open to the Caribbean trade winds and seas. Yet, it is somewhat protected from the destructive winter "rollers" (large swells out of the North Atlantic) and had developed a roughly 10- to 20-m thick, carbonate reef structure over a period of about 5000 years. Based on extensive coring and carbon 14 (C14) dating, portions of the reef were shown to have upward growth rates greater than 10 m/1000 yr. Generally, growth rates slowed down as reef flat levels were attained and flat widths further developed. Considering the entire pan-tropic coral reef environment, this reef is certainly a strong "performer" (at least it was in the 1970s). Nevertheless, many similarly well-developed coral reefs exist in the Caribbean Sea and West Indies.[12]

At the time when the metabolism and organism cover studies were carried out on the southeast St. Croix reef, all reef/time stages from submature, with very rapid growth rates, to fully mature, with broad reef flats, were present. All major stages were included in the studies that preceded the development of the coral reef models cited in the references for this chapter. In this chapter, "St. Croix reefs" specifically refers to the bank barrier reef on the southeast side of St. Croix during the 1970s. For specific details on the wild-type reefs that relate to the microcosm discussion, consult references 10 and 11. Although physiologically, especially with regard to basic physical–chemical parameters, particularly light and metabolism, the microcosm system was matched to the 5-m-deep fore reef location on a midsection of the St. Croix south reef, community structure was more generally related to a mean of the entire reef.[13]

17.2 CARIBBEAN MODEL (CORAL REEF MICROCOSM)

In this chapter, to provide a ready reference, I briefly describe the Caribbean Model that is our primary source of experimental data. However, for a detailed treatment of background and experimental protocol, readers are referred to references 8, 13, and 14.

A diagram of the physical layout of the Caribbean Model and its basic "closure state" is shown in Figure 17.1. As noted above, the 400-l or metabolic unit is ecophysiologically the microcosm match to a specific point on the St. Croix reef. Water movement, current, and wave action in this primary system are driven by slow-moving, low-shear bellows pumps. These prevent significant damage to swimming and floating reproductive stages. These pumps also supply water to the Algal Turf Scrubber (ATS), which is a nonfiltration, managed algal community that allows the algal turf species resident in the model reef to colonize the ATS, where photosynthesis and growth conditions are optimal. The ATS system is lighted during the dark cycle of the coral reef and thus provides the ability to control the primary chemistry of the water column on a diurnal basis. The 1280-l "refugium," with human operator interactions, provides the broader scale of organism interactions that keep the model from being a constantly changing patch of that reef. This function is particularly important in a nearly closed model. Here, normal population fluctuations could provide a dead end in community structure without the readily available larger reef surface that provides for the immigration and emigration of organisms and/or their reproductive stages.

Table 17.1 provides the primary matching physiological conditions of model to wild reef ecosystem in a more static, long-term mode. The experimentation that determined the more diurnal, dynamic characterization of the unit, as briefly given below, was carried out for 10 months in 1998. At that time, the 400-l microcosm unit had been operating for about 10 years. Both the 400-l microcosm and the 1280-l refugium were in operation and essentially closed (except for a few experimental species) from 1991 to the time of implementation of the experimentation (i.e., about

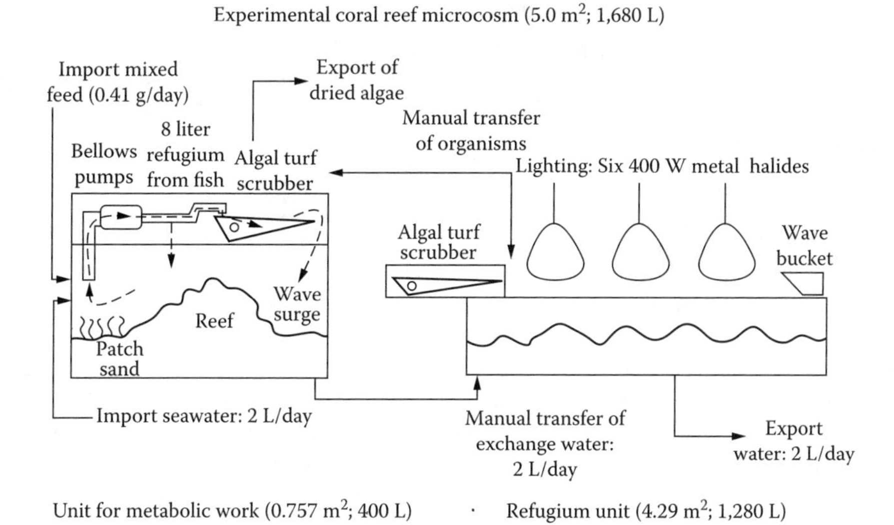

FIGURE 17.1 Diagram of the microcosm. Lights not shown on 400 L metabolic unit; see Adey and Loveland[8] for a detailed mechanical description of this unit. Impeller pumps (not shown) deliver water from the right-hand end of the refugium unit to the Algal Turf Scrubber and from the left side of the refugium unit to the wave bucket. Distilled water is added to compensate for evaporation.

TABLE 17.1
Comparisons between Microcosm and St. Croix Reefs (annual mean or mean daily range with standard error)

	Microcosm	St. Croix Reefs (fore reef)[a]
Temperature (°C) (am-pm)	26.5 ± 0.03 (n = 365) – 27.4 ± 0.02 (n = 362)	24.0 – 28.5
Salinities (ppt)	35.8 ± 0.02 (n = 365)	35.5[a]
pH (am-pm)	7.96 ± 0.01(n = 62) – 8.29 ± 0.02 (n = 39)	8.05 – 8.35[a]
Oxygen concentration (mg l^{-1}) (am-pm)	5.7 ± 0.1 (n = 14) – 8.7 ± 0.2 (n = 11)	5.8 – 8.5
GPP(g O_2 m^{-2} day^{-1}); (mmol O_2 m^{-2} day^{-1})	14.2 ± 1.0 (n = 4); 444 ± 3 (n = 4)	15.7; 491
Daytime NPP (g O_2 m^{-2} day^{-1}); (mmol O_2 m^{-2} day^{-1})	7.3 ± 0.3 (n = 4); 228 ± 9 (n = 4)	8.9; 278
Respiration (g O_2 m^{-2} h^{-1}); (mmol O_2 m^{-2} h^{-1})	0.49 ± 0.04 (n = 4); 15.3 ± 1.3 (n = 4)	0.67; 20.9
N-NO^{-}_{2} + NO^{2}_{3} (μmol)	0.56 ± 0.07(n = 6)	0.28
Calcium (mg l^{-1}); (mmol l^{-1})	491 ± 6 (n = 33); 12.3 ± 0.2 (n = 33)	417.2[a]; 10.4
Alkalinity (meq l^{-1})	2.88 ± 0.04 (n = 59)	2.47[a]
Light[a] (Langleys day^{-1})	220	430 (surface); 220 (5 m deep in fore-reef)

[a] Data from Small and Adey, 2001.[13]

7 years). The model and its refugium had been constructed and fully stocked, from the wild, from about 1988 to 1990.

Figure 17.2, Figure 17.3, and Figure 17.4 provide the most essential diurnal and long-term dynamic characteristics of the model, graphically showing diurnal oxygen cycling, long-term diurnal respiration and primary productivity, and diurnal carbonate cycling. Those diagrams also show the relationship of those basic metabolic characteristics to wild coral reefs, and especially the St. Croix–type reef. The long-term, whole-system (400-l unit) calcification rate of the model was 4.0 ± 0.2 kg $CaCO_3/m^2/yr$. Although the calcification rate follows from the oxygen/carbonate cycling, it provides a more direct and summary measure of the ecophysiological "success" of the model. This matches the "top" 2 to 4% of reefs worldwide.[13] This number cannot be directly related to the St. Croix reefs because those specific data were not collected at that reef. However, the upward growth rates of the St. Croix reef, as determined by core-drilling and C14 analysis, provide equivalent calcification rates, and these are among the higher rates described for coral reefs worldwide. Also, calcification rates of several individual species of stony corals were experimentally obtained by short-term isolation within the model. The highest rates, for a branching *Acropora* species, were 8.1 ± 0.7 kg $CaCO_3/m^2/yr$. This rate is about as high as the highest rates reported from the wild. High calcification rate is among the most important characteristics that allow coral reefs to develop their unique structure and topography and consequently their highly diverse community structure. It is clear from the metabolism and calcification rates briefly described above, for both wild and microcosm coral reefs, that it is possible to match basic reef function in a controllable model.

Although many physical/chemical variables of the St. Croix reefs were considered in the development of the Caribbean Model design (e.g., key nutrients, diurnal oxygen concentrations, diurnal carbonate cycling and carbon import and export, temperatures, wave action, lighting, etc.), in summary, ecophysiological success was determined by whole-system calcification rates. Primary

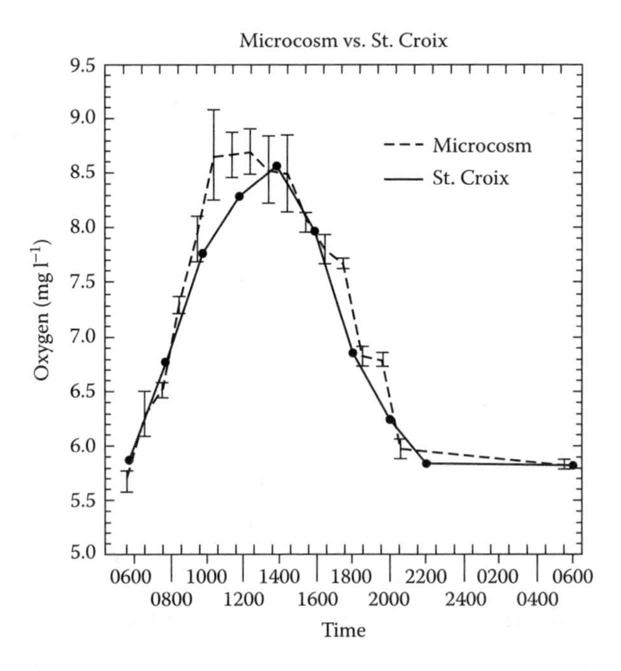

FIGURE 17.2 Comparison between oxygen concentrations of St. Croix and the microcosm. Standard error bars shown where n varies for each point; starting from the left: n = 8, 3, 11, 3, 3, 11, 3, 3, 11, 3, 3, 4, 9, 3, 3. St. Croix data from reference 11. The first set of tank lights comes on at 06:00 h and go off at 18:00 h; the second set comes on at 08:00 h and goes off at 20:00 h. The scrubber lights come on at 19:00 h and go off at 07:00 h.[13]

control of all of these biogeochemical processes was attained through the use of ATS processes. In its essentials, ATS is an algal control system that simulates a large body of water (open-ocean tropical surface water in this case) through manipulation of an externally sited, coral reef algal turf community.[8] ATS has a 25-year history of development and application and is now used in commercial-scale aquaculture and landscape-scale surface water purification.[15] It is therefore applicable to any scale of coral reef restoration, whether for the development of the actual site, as a heuristic tool to help the restorationists understand the dynamics of the project, or simply as an

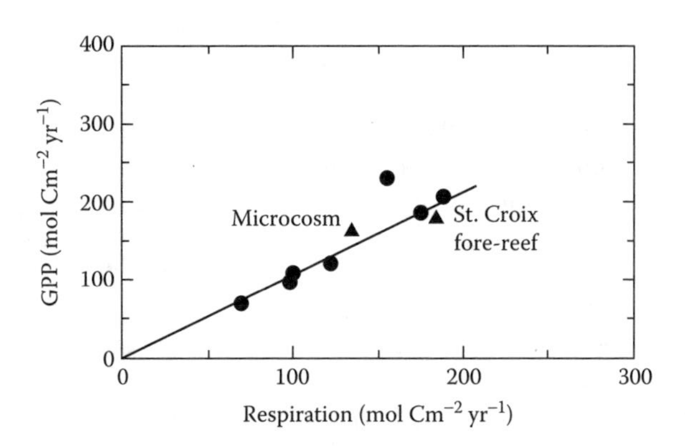

FIGURE 17.3 Gross primary production vs. respiration. Modified after reference 18. Line shows 1:1 relationship.

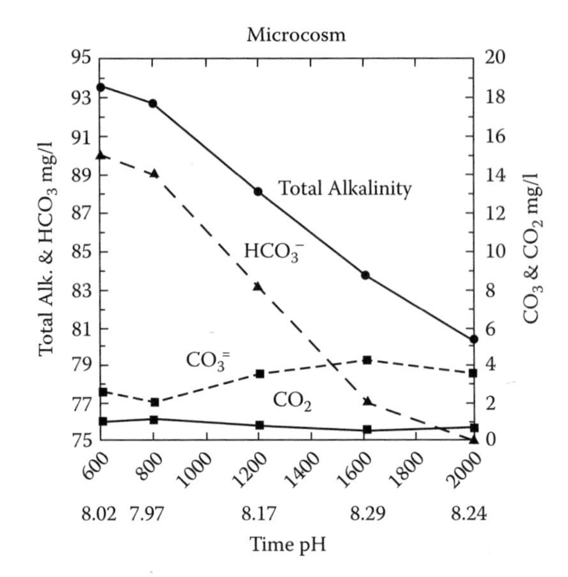

FIGURE 17.4 Daytime carbonate cycle in Caribbean Model, as calculated by nomograph from total alkalinity and pH.[19]

organism maintenance tool. Note that at small scale, the refugium unit in the Caribbean Model was employed as just such a management tool for the smaller "metabolic" reef.

As described above, in the design and construction of the model, considerable efforts have been expended to duplicate flow rates (currents) and wave surge and period (though not wave magnitudes (see Reference 13). In their review of coral reef biogeochemistry, Atkinson and Falter[20] conclude that these are indeed critical factors. In this chapter, as in all precedent studies, the author has emphasized the role of low nutrients, as have most workers in the field. Atkinson and Falter[20] conclude that the nutrient factor in reef biochemistry is poorly understood, and perhaps under some conditions not a significant factor (in the author's view, this may well be so when light levels are very low, either in deep reefs or in poorly lighted models). As earlier demonstrated by Adey[21] and repeated by Atkinson and Falter,[20] C:N:P ratios are high in macroalgae from wild reef systems with oligotrophic nutrient levels, but this does not reduce high primary productivity.[21] In the same environments, physical energy (wave surge and current) allows algal turfs to achieve high uptake rates and normal C:N:P ratios (and thereby achieve the high rates of productivity seen). This same process is repeated in the large-scale ATS water-cleaning systems now in use to achieve oligotrophic water quality in degraded fresh waters.[15]

Carbonate skeletal construction (i.e., framework-building) by coral reef organisms, especially corals, is the key to the development and maintenance of controlled coral reef ecosystems. While it is therefore appropriate to measure the success of coral reef microcosm modeling in part by measuring whole reef calcification rates, it is unlikely that anyone would consider coral reef restoration exclusively for the production of carbonate. The essential secondary consideration would certainly be biodiversity. Some might argue that fish biodiversity alone is the key element. However, except for the specialty artificial reefs built for anglers, where the fish populations are accepted to be based on plankton populations and the reef mostly provides structural habitat, broad-band restored coral reef fish populations can only be maintained with an even greater broad band of algal and invertebrate biodiversity.

Hopefully, considerable efforts are currently underway to define biodiversity in at least a few, scattered wild reefs. However, at this time, the only direct measure of coral reef biodiversity is the Caribbean Model that is the focal point of this chapter.[16] In the 5-m^2, 1680-l volume of that model, after 7 years of semi-closure, 534 species were tallied by a team of 24 specialists (Table 17.2).

TABLE 17.2
Families, Species, and Genera Tallied in the Caribbean Model

Plants, Algae, and Cyanobacteria

Division Cyanophota
Chroococcaceae 6/5
Pleurocapsaceae 4/2
UID family 4/4
Oscillatoriaceae 8/6
Rivulariaceae 4/1
Scytonemataceae 1/1

Phylum Rhodophyta
Goniotrichaceae 2/2
Acrochaetiaceae 2/2
Gelidiaceae 1/1
Wurdemanniaceae 1/1
Peysonneliaceae 3/1
Corallinaceae 11/8
Hypneaceae 1/1
Rhodymeniaceae 3/2
Champiaceae 1/1
Ceramiaceae 3/3
Delesseriaceae 1/1
Rhodomelaceae 7/6

Phylum Chromophycota
Cryptomonadaceae 2/2
Hemidiscaceae 1/1
Diatomaceae 6/4
Naviculaceae 9/4
Cymbellaceae 3/1
Entomoneidaceae 1/1
Nitzchiaceae 6/4
Epithemiaceae 3/1
Mastogloiaceae 1/1
Achnanthaceae 9/3
Gymnodiniaceae 6/4 or 5
Gonyaulacaceae 1/1
Prorocentraceae 2/1
Zooxanthellaceae 1/1
Ectocarpaceae 2/2

Phylum Chlorophycota
Ulvaceae 1/1
Cladophoraceae 4/2
Valoniaceae 2/2
Derbesiaceae 3/1
Caulerpaceae 3/1
Codiaceae 6/2
Colochaetaceae 1/1

Phylum Magnoliophyta
Hydrocharitaceae 1/1

Kingdom Protista

Phylum Percolozoa
Vahlkampfiidae 2/1
UID family 2/2
Stephanopogonidae 2/1

Phylum Euglenozoa
UID family 4/3
Bondonidae 7/1

Phylum Choanozoa
Codosigidae 2/2
Salpingoecidae 1/1

Phylum Rhizopoda
Acanthamoebidae 1/1
Hartmannellidae 1/1
Hyalodiscidae 1/1
Mayorellidae 2/2
Reticulosidae 2/2
Saccamoebidae 1/1
Thecamoebidae 1/1
Trichosphaeridae 1/1
Vampyrellidae 1/1
Allogromiidae 1/1
Ammodiscidae 1/1
Astrorhizidae 1/1
Ataxophragmiidae 1/1
Bolivinitidae 3/1
Cibicidiidae 1/1
Cymbaloporidae 1/1
Discorbidae 5/2
Homotremidae 1/1
Peneroplidae 1/1
Miliolidae 10/2
Planorbulinidae 2/2
Siphonidae 1/1
Soritidae 4/4
Textulariidae 1/1

Phylum Ciliophora
Kentrophoridae 1/1
Blepharismidae 2/2
Condylostomatidae 1/1
Folliculinidae 4/3
Peritromidae 2/1
Protocruziidae 2/1
Aspidiscidae 7/1
Chaetospiridae 1/1
Discocephalidae 1/1
Euplotidae 11/3
Keronidae 7/2
Oxytrichidae 1/1
Psilotrichidae 1/1
Ptycocyclidae 2/1
Spirofilidae 1/1
Strombidiidae 1/1
Uronychiidae 2/1
Urostylidae 4/2
Cinetochilidae 1/1
Cyclidiidae 3/1
Pleuronematidae 3/1
Uronematidae 1/1
Vaginicolidae 1/1
Vorticellidae 2/1
Parameciidae 1/1
Colepidae 2/1
Metacystidae 3/2
Prorodontidae 1/1
Amphileptidae 3/3
Enchelyidae 1/1
Lacrymariidae 4/1

Phylum Heliozoa
Actinophyridae 2/1

Phylum Placozoa
UID Family 5

Phylum Porifera
Plakinidae 2/1
Geodiidae 5/2
Pachastrellidae 1/1
Tetillidae 1/1
Suberitidae 1/1
Spirastrellidae 2/2
Clionidae 4/2
Tethyidae 2/1
Chonrdrosiidae 1/1
Axinellidae 1/1
Agelasidae 1/1
Haliclonidae 4/1
Oceanapiidae 1/1
Mycalidae 1/1
Dexmoxyidae 1/1
Halichondridae 2/1
Clathrinidae 1/1
Leucettidae 1/1
UID family 2/?

Eumetazoa

Phylum Cnidaria
UID family 3/?
Eudendridae 1/1
Olindiiae 1/1
Plexauridae 1/1
Anthothelidae 1/1
Briareidae 1/1
Alcyoniidae 2/2
Actiniidae 3/2
Aiptasiidae 1/1
Stichodactylidae 1/1
Actinodiscidae 4/3
Corallimorphidae 3/2
Acroporidae 2/2
Caryophylliidae 1/1
Faviidae 3/2
Mussidae 1/1
Poritidae 3/1
Zoanthidae 3/2
Cerianthidae 1/1

Phylum Platyhelminthes
UID family 1/1
Anaperidae 3/2
Nemertodermatidae 1/1
Kalyptorychidae 1/1

Phylum Nemertea
UID family 2/2
Micruridae 1/1
Lineidae 1/1

Phylum Gastrotricha
Chaetonotidae 3/1

Phylum Rotifera
UID family 2/?

Phylum Tardigrada
Batillipedidae 1/1

Phylum Nemata
Draconematidae 3/1

Phylum Mollusca
Acanthochitonidae 1/1
Fissurellidae 2/2
Acmaeidae 1/1
Trochidae 1/1
Turbinidae 1/1
Phasianellidae 1/1
Neritidae 1/1
Rissoidae 1/1
Rissoellidae 1/1
Vitrinellidae 1/1
Vermetidae 1/1
Phyramidellidae 1/1
Fasciolariidae 2/2
Olividae 1/1
Marginellidae 1/1
Mitridae 1/1
Bullidae 1/1
UID family 4/?
Mytilidae 2/1
Arcidae 2/1
Glycymerididae 1/1
Isognomonidae 1/1
Limidae 1/1
Pectinidae 1/1
Chamidae 1/1
Lucinidae 2/2
Carditidae 1/1
Tridacnidae 2/1
Tellinidae 1/1

(continued)

TABLE 17.2
Families, Species, and Genera Tallied in the Caribbean Model (Continued)

Phylum Annelida
Syllidae 3/2
Amphinomidae 1/1
Eunicidae 3/1
Lumbrineridae 1/1
Dorvilleidae 1/1
Orbiniidae 1/1
Spionidae 1/1
Chaetopteridae 1/1
Paraonidae 1/1
Cirratulidae 4/3
Ctenodrilidae 4/3
Capitellidae 3/3
Muldanidae 1/1
Oweniidae 1/1
Terebellidae 2/1
Sabellidae 14/4
Serpulidae 6/6
Spirorbidae 2/2
Dinophilidae 1/1

Phylum Sipuncula
Golfingiidae 1/1
Phascolosomatidae 3/2
Phascolionidae 1/1
Aspidosiphonidae 3/2

Phylum Arthropoda
Halacaridae 1/1
UID family 2/?
Cyprididae 2/2
Bairdiidae 1/1
Paradoxostomatidae 1/1
Pseudocyclopidae 1/1
Ridgewayiidae 2/1
Ambunguipedidae 1/1
Argestidae 1/1
Diosaccidae 1/1
Harpacticidae 1/1
Louriniidae 1/1
Thalestridae 1/1
Tisbidae 1/1
Mysidae 1/1
Apseudidae 2/1
Paratanaidae 1/1
Tanaidae 1/1
Paranthuridae 1/1
Sphaeromatidae 1/1
Stenetriidae 1/1
Juniridae 1/1
Lysianassidae 1/1
Gammaridae 4/4
Leucothoidae 1/1
Anamixidae 1/1
Corophiidae 1/1
Amphithoidae 2/2
Alpheridae 2/2
Hippolytidae 2/1
Nephropidae 1/1
Diogenidae 1/1
Xanthidae 2/?

Phylum Echinodermata
Ophiocomidae 1/1
Ophiactidae 1/1
Cidaroidae 1/1
Toxopneustidae 1/1
Holothuriidae 1/1
Chirotidae 1/1

Phylum Chordata
Ascidiidae UID species 1/1
Grammidae 1/1
Chaetodontidae 1/1
Pomacentridae 5/4
Acanthuridae 1/1

Note: Ceramiaceae 3/3 = three species in three genera in the family Ceramiaceae. UID = unidentified.

Since some groups of algae and invertebrates were omitted from the tally due to the lack of a specialist, the authors estimated that the system actually contained about 800 species. Except for a dozen species undergoing manipulation and a few long-lived fish species, this microcosm had been closed to organism import for 7 years prior to analysis; therefore, most of the 500+ species were reproductively maintaining their populations. Based on the biodiversity of this model, and using the relationship $S = kA^z$, it has been estimated that the pantropic biodiversity of wild coral reefs must be at least 3 million species. One must conclude therefore that unless wild coral reefs have a considerably higher biodiversity (the previous maximum estimate being 1 million species[17]), microcosm models can be successful in biodiversity as well as ecophysiological terms.

A glance at the family list in Table 17.2 shows that the biodiversity of this system was very widely based. A few of the reported 230+ families tallied in the Caribbean Model have a half-dozen species in two to five genera. However, most families have only a single species. Thus, while the system is oriented toward the relative success of smaller organisms (e.g., protists and annelids), and at least for the larger organisms, the lower-to-middle levels of food webs, it does not have an ecophysiology that is selecting for a limited range of food web and community structure characteristics. The number of functionally closely related species is likely limited, simply by space (i.e., the small size of the model). There is no room for "pulsing" between adjacent patches in this small model, and species that occupy the same ecological niches cannot coexist because of distance, as many species will be able to do in a large wild coral reef. While species diversity is high by any measure,[16] family diversity for such a small system is likely to be truly extraordinary. The linking of several separate systems through piping and nonstressing pumps would likely proportionally increase species diversity by simulating adjacent patches. In a more sophisticated system, organism cross-access to those patches could be controlled by gates or organism-specific filters; in short, they could function as controlled refugia. This bodes well for the reef restorer who would invest the considerable effort to work with whole coral reef ecosystems and not a limited subset of the more visible or spectacular species.

17.3 THE OPERATIONAL IMPERATIVE

Successful microcosm and mesocosm operation requires the monitoring of a large number of physical and chemical factors. To a large extent, this can be automated with electronic sensors, and the data can be logged and the system computer controlled. Some chemical parameters, such as the low-level nutrients that are characteristic of coral reef systems, still require wet chemistry, though a once-a-week analysis is usually sufficient in a well-run system (even nutrient sensing and control can probably be automated, but the cost could well be prohibitive for restoration programs). Like any piece of complex laboratory equipment (a scanning electron microscope, for example) a dedicated and highly trained technician is needed to manage the monitoring equipment, though in a well-tuned system, considerable time can be available for other duties. ATS management is effectively manual, though this is typically not a large consumer of technician time until a system exceeds hundreds of cubic meters. The ATS system of the Caribbean Model typically required about half an hour per week for physical/chemical maintenance.

An operational feature that is rarely discussed, and in practice is mostly anecdotal in expression, is that of population instability. A microcosm, in effect, is a few-square-meter patch of a larger coral reef ecosystem. In the wild, reef patches of a few square meters can be subject to considerable short-term variability, though stability is achieved to some extent by the smoothing effect of the larger local island or coastal reef that may be measured in square kilometers. On the other hand, even large geographic areas can be subject to population explosions. Coral reef scientists learned this in a very vivid way in the 1970s and 1980s when the *Diadema* populations in the Caribbean slowly built up to very high densities and then suddenly collapsed. Thus, whether the restorer is dealing with an open-water project of many tens of thousands of square meters, a research model, or a holding system for organisms destined for a coral reef being restored, the principles are the same.

The type microcosm of this chapter operated in a closed mode, in essentially the same ecological state, for 7 years (nearly 9 years including the period of intense analysis). Nothing that could remotely be described as a "crash" ever occurred in this coral reef model, in spite of occasional physical or electronic failures. However, on the scale of several months, single populations of the system (e.g., *Caulerpa* spp., various dinoflagellate species, and a few species of polychaete worms) would undergo a population explosion. Usually, this tendency, for which the observant operator typically had weeks of warning time, would last a few months, and then the subject population would reduce to "background" levels. Sometimes, the explosion would recur years later; in other cases it was a one-time experience.

It has been my experience that microcosms and mesocosms require an ecologist, fully acquainted with "normal" community structure of the "wild-type" system. Effectively, that ecologist/operator performs as the highest, and most omnivorous, predator. This "variable" predator (if not on vacation or otherwise detained) is instantly available at "full population" to limit short-term population inbalance. In a 10-year study of scaling effects on Chesapeake Bay mesocosms, researchers often were able to solve scaling problems by just such ecological manipulation.[22] In the case of *Caulerpa* and *Prorocentrum* "explosions" in the Caribbean Model, the operator's function was obvious, a once-a-week "grazing" (i.e., hand harvest) of the *Caulerpa* or similarly a once-a-week scraping and collection of mostly glass surfaces for *Prorocentrum* until the explosion tendency subsided. In other cases, the short-term introduction of a fish or invertebrate predator (such as angelfish for fire worms) could be quite successful. In some cases, it has even been valuable to maintain such "managed predators" in the refugium unit where they are readily available for such service. It is the view of some coral reef ecologists, struggling to understand why wild reefs are now apparently rapidly degrading, that extensive (albeit local) fishing (effectively large-scale species manipulation) is the primary factor responsible for reef decline. This demonstrates that human interaction with coral reefs can occur at any scale. This aspect of microcosm operation has considerable application to coral reef restoration, as will be discussed further below.

17.4 IMPLICATIONS OF MICROCOSM MODELING FOR CORAL REEF RESTORATION

An extensive ecophysiological as well as biodiversity and community structure understanding of the wild system to be modeled is the primary key to the restoration of coral reef ecosystems. The success achieved with the microcosm briefly described herein was based, more than any other single factor, on having an existing, functioning wild ecosystem (and not an elusive ideal) that had to be matched. Most critically, if negative factors of water quality, such as turbidity and nutrients, cannot be returned to the state of the wild system to be emulated, success is unlikely. For example, although the direct role of elevated nutrients has not been fully defined (indeed, it is probably quite complex), it appears certain that most coral reef ecosystems require that the primary nutrients, phosphorus and nitrogen, be measured in single digits of parts per billion. In the most successful of the coral reef microcosms discussed in this chapter, primary nutrients were maintained at extremely low levels for many years at a time. Nevertheless, if the coral reef restorationist felt that a particular reef ecosystem could maintain itself in the face of moderate nutrient levels, once its community structure was restored during a period of high water quality, nutrient scrubbing, very large volumes of water (tens of millions of gallons/day) can be achieved if adjacent land surface is available.[15] Alternatively, if coral reefs of restoration interest have been degraded by the low water quality of incoming fresh waters from human activities in nearby terrestrial environments, ATS fresh water systems can scrub rivers, streams, and sewage outfalls of nutrients, sediments, and toxic compounds. Also, *in situ* culture methods have been devised that can re-establish reefs destroyed by hurricanes, low water quality, or excessive temperatures.[23] However, if return of low water quality or high temperatures cannot be avoided, considerable effort can be expended that will be of little value to long-term reef restoration.

Development and operation of the kind of laboratory ecosystem model described in this chapter can be an extremely valuable heuristic tool and can provide ecological experience that could be achieved only over long periods of time on a wild reef. Effectively, a microcosm or mesocosm coral reef can be a pilot project carried out prior to the initiation of a full restoration operation. While the system described above was operated for a decade, such long-term operation is not essential to success. Short-term stability in coral reef models is usually attainable on the scale of months. Also, such microcosms are useful in the developing and testing of many experimental protocols. Microcosms are often capable of producing experimental results in a shorter time frame than in the wild because variables are more controllable.

Finally, restoration of a wild reef will generally require considerable organism manipulation by a wide range of specialists working as a team. Particularly if a high-quality source reef and its organisms are at a considerable distance from the reef to be restored, a moderately sized microcosm (probably a mesocosm in most cases) would be an extremely valuable resource to hold the organisms being manipulated. As described above, for the Caribbean Model, success in achieving organism stability in the metabolism unit was assured by having a "refugium" that allowed more-or-less instant manipulation of many organisms. Another aspect of ATS-controlled units is the ability to modify such systems, either in the main microcosm or mesocosm, or perhaps more appropriate in a restoration effort, in a side loop, for aquaculture purposes. This has been demonstrated in a large-scale commercial aquaculture for fin fish, and could be modified relatively easily to produce large numbers of a particular species, algal, invertebrate, or vertebrate, that are needed for introduction but are not readily available at reasonable shipping distances.

17.5 SUMMARY

Twenty-five years of intensive and repeated ecosystem modeling of coral reefs, as microcosms and mesocosms, has demonstrated this approach to be a viable experimental tool. Such a system or systems can provide prior, experimental understanding of the problems to be overcome in a specific coral reef restoration project. Also, in a situation where healthy source reefs lie at a considerable distance from the reef to be restored or would be depleted by the required removal of one or several species, local model systems can be an invaluable organism culture and manipulative asset during the restoration process. Control of global degrading factors may be critical to long-term coral reef restoration. We now have the tools to achieve global water quality control, although whether socio-political factors will allow that to happen remains to be seen. A system of mesocosms could provide the biodiversity bridge to a time of better ocean water quality.

REFERENCES

1. Carlson, B. 1987. Aquarium systems for living corals. *Int. Zool. Yb.* 26: 1–9.
2. Moe, M. 1989. *The Marine Aquarium Reference.* Green Turtle Publications. Plantation, FL.
3. Spotte, S. 1995. *Captive Seawater Fishes, Science and Technology.* Wiley. New York.
4. Atkinson, M., B. Carlson, and G. Crow. 1995. Coral growth in high nutrient, low pH sea water: a case study of corals cultured at the Waikiki Aquarium, Honolulu, Hawaii. *Coral Reefs* 14: 215–223.
5. Borneman, E. 2001. *Aquarium Corals: Selection, Husbandry, and Natural Selection.* TF Neptune Publications, Charlotte, VT.
6. Adey, W. 1983. The microcosm, a new tool for reef research. *Coral Reefs* 1: 193–201.
7. Luckett, C., W. Adey, J. Morrissey, and D. Spoon. 1996. Coral reef mesocosms and microcosms—successes, problems, and the future of laboratory models. *Ecol. Eng.* 6: 57–72.
8. Adey, W. and K. Loveland. 1998. *Dynamic Aquaria: Building Living Ecosystems.* 2nd ed. Academic Press. New York.
9. Adey, W. 1987. Marine microcosms. In Jordan, W., M. Gilpin, and J. Alber, Eds. *Restoration Ecology.* Cambridge Univ. Press, Cambridge, U.K. pp. 134–149.
10. Adey, W. 1975. The algal ridges and coral reefs of St. Croix. *Atoll Res. Bull.* 187: 1– 67.
11. Adey, W. and R. Steneck. 1985. Highly productive eastern Caribbean reefs: synergistic effects of biological, chemical and geological factors. NOAA Symposium Series for Undersea Research 3: 163–187.
12. Adey, W. 1978. Coral reef morphogenesis: a multidimensional model. *Science* 202: 831–837.
13. Small, A. and Adey, W. 2001. Reef corals, zooxanthellae, and free-living algae: a microcosm study that demonstrates synergy between calcification and primary production. *Ecol. Eng.* 16: 443–457.
14. McConnaughey, T., W. Adey, and A. Small. 2000. Community and environmental influences on coral reef calcification. *Limnol. Oceanogr.* 45: 1667–1671.
15. www.hydromentia.com.
16. Small, M., W. Adey, and D. Spoon. 1998. Are current estimates of coral reef biodiversity too low? The view through the window of a microcosm. *Atoll Res. Bull.* 458: 1–18.
17. Reaka-Kudla, M. 1997. The global biodiversity of coral reefs: a comparison with rain forests. In Reaka-Kudla, M., D. Wilson, and E.O. Wilson, Eds. *Biodiversity II: Understanding and Protecting our Biological Resources.* Joseph Henry Press, Washington, D.C.
18. Gattuso, J.-P., M. Frankignoulle, and R. Wollast. 1998. Carbon and carbonate metabolism in coastal aquatic ecosystems. *Annu. Rev. Ecol. Syst.* 29: 405–433.
19. Cleseri, L., A. Greenburg, and R. Tressel. 1989. *Standard Methods for the Examination of Water and Wastewater.* 19th ed. American Public Health Association, Washington, D.C.
20. Atkinson, M. and J. Falter. 2003. Coral reefs. In Black K. and G. Shimmield. *Biogeochemistry of Marine Systems.* CRC Press, Oxford, U.K. 372 pp.
21. Adey,W. 1987. Food production in nutrient poor seas: bringing tropical ocean deserts to life. *Bioscience* 37: 340–348.
22. Petersen, J. E. et al. 2003. Multiscale experiments in coastal ecology: improving realism and advancing theory. *Bioscience* 53: 1181–1197.
23. www.reefball.com.

18 Ethical Dilemmas in Coral Reef Restoration

Rebecca L. Vidra

CONTENTS

18.1 INTRODUCTION

Coral reef restoration projects provide unique and important opportunities to apply the science of coral reef ecology to improve ecologically fragile but degraded ecosystems. Restoration represents an intersection of objective-based science and policy-based practice, involving scientists, ecosystem managers, public agencies, nongovernmental organizations, and the public. People with different training, objectives, values, languages, and cultures need to work together to develop and implement restoration plans and monitor restoration outcomes. Those involved with these projects often are faced with many challenges of logistics such as prioritizing projects and determining what constitutes restoration success. A tremendous amount of energy has been devoted to these challenges. Yet, these questions also lead to ethical dilemmas that are more difficult to address with systematic or prescriptive approaches.

Within the restoration community, philosophers, scientists, and practitioners have sustained a dialogue addressing these ethical dilemmas. In this chapter, I draw from that dialogue and my own discussions with restoration ecologists to outline some of the potential ethical dilemmas that may arise when developing coral reef restoration projects and programs. This certainly is not an exhaustive list or a prescription for avoiding dilemmas. Instead, these issues suggest that everyone involved in restoration consider their own values and recognize the potential for ethical dilemmas to arise throughout the restoration process. Two fictitious case studies illustrate how these dilemmas may emerge during the course of restoration. Acknowledging ethical issues and maintaining an open dialogue among restorationists is critically important to achieve successful restoration results.

18.2 IS RESTORATION THE SOLUTION OR A "BIG LIE"?

Ecological restoration is increasingly becoming an important management tool for healing degraded ecosystems. Responding to a growing need for scientists and practitioners to collaborate, the Society for Ecological Restoration was formed in 1987. That same year, Jordan et al.[1] published a seminal book *Restoration Ecology: A Synthetic Approach to Ecological Restoration,* which outlined the opportunities and challenges in this growing field. As restoration ecology has matured, the ethics of attempting restoration has been vigorously debated both within the discipline and in the context of larger questions of environmental ethics. This debate, involving scientists, practitioners, and philosophers, has resulted in a unique reflection on the ability of humans to recreate nature.

A goal of many restoration projects is to restore or create an ecosystem equal in value to a natural ecosystem. For example, commenting on the consensus built by coral reef ecologists, Precht[2] has stated that "the most appropriate course of action is to replace damaged and disturbed reefs with fully functional, restored ecosystems at a rate resulting in no net loss of ecosystem value (i.e., rate of reef destruction offset by rate of reef repair)." Philosophers have challenged this notion, arguing that ecosystems have an intrinsic value that can never be restored by humans. Philosopher Robert Elliott[3] compares restoration projects to "fake art," arguing that reproductions can never attain the same value as the original. Because this value is generated through the natural genesis of ecosystems, philosopher Eric Katz[4] suggests that restoration projects are "big lies," mere representations of natural ecosystems.

Many coral reef enthusiasts would consider the waterscape of Atlantis in Paradise Island, Bahamas to be an example of a fake human-created ecosystem. In fact, the promotional material touts the 11 exhibit lagoons as "the largest marine habitat in the world, second only to Mother Nature."[5] Yet, there is a surreal aura of such a place, a "fake" feeling that cannot be completely disguised. The challenge for the coral reef restoration community is to avoid designing restoration projects that have this feeling.

In addition, Katz[6] has argued that indeed restoration is one more way humans dominate nature, stating "Once and for all, humanity will demonstrate its mastery of nature by "restoring" and repairing the degraded ecosystems of the biosphere. Cloaked in an environmental consciousness, human power will reign supreme." Others argue that restoration can serve as the ultimate healing process of humans with nature.[7–10] Again, we must balance our confidence in restoration with an acknowledgment that we may never know enough to create a fully natural ecosystem.

Another criticism of restoration is that our confidence in our ability to restore nature may indeed serve as excuse for further degradation. If we can create an ecosystem with equal value to one that was destroyed or degraded, why should we preserve natural systems? Are ship groundings really disastrous if we have a prescription and the financial resources to recreate the reefs? Do we compromise our ability to advocate for reef protection when we have the tools to recreate or "fix" reefs? As we become increasingly confident about the restoration process, we may also need to acknowledge that "natural" systems are preferable, in some way, to restored ecosystems.

Restoration may give us the opportunity to save threatened ecosystems or it may give us the excuse to destroy or dominate natural landscapes. This debate and discussion about the value of human-created ecosystems from philosophical as well as scientific viewpoints continue today.[11,12] This discussion should not be limited to philosophers but should be taking place on docks and conference rooms, among scientists and volunteers. As restorationists, we should continue to contemplate our notions of the value of restored ecosystems when considering the motivations behind restoration projects.

18.3 ETHICAL DILEMMAS IN THE PRACTICE OF RESTORATION

The larger questions about ecological restoration referenced above have dominated discussions about ethics in the field. Yet, ethical dilemmas arise during all phases of restoration, from translating results from a scientific study into real-world recommendations to determining whether the project is ultimately successful. The following categories of ethical dilemmas represent the wide range of potential challenges that restoration ecologists face.

18.3.1 ADDRESSING THE VALUES OF STAKEHOLDERS: ECOLOGICAL, ECONOMIC, AND EDUCATIONAL GOALS

Designing a restoration project involves goals that are influenced by what the involved stakeholders value as success.[13–15] Measuring success may mean counting fish species, calculating nutrient fluxes, or evaluating the educational appeal of a reef. Most restoration projects involve a diverse set of stakeholders, and project goals often reflect scientific approaches, human values, and management concerns.[16] For example, integrating the economic goals of the client with the recommendations from ecologists can provide some challenging decisions for the project manager. Often, the public becomes involved in restoration efforts, either as affected parties or as motivated volunteers. Their input can also affect the restoration process. Restoration projects can also serve as educational opportunities both for scientists interested in testing the efficacy of restoration methods and for the public to learn about ecosystem creation. Thus, these projects likely need to encompass a wide range of goals and values of the involved stakeholders (e.g., Anderson et al.[17]).

Integrating these sometimes disparate values can present logistical challenges in the planning, implementation, and/or monitoring phases of the restoration project. Ethical dilemmas arise when these values conflict, leading to choices being made about how the restoration is done and how to define the goals of the restoration project. For example, consider a coral reef restoration program funded by a nonprofit organization. The mission of this organization could be to repair damage from ship groundings with the help of volunteers. Multiple goals could be part of these projects: to enhance the ecological value of these sites, to educate the public about coral reef recovery, or to stimulate the local economy by providing tourist opportunities. These goals, all of which are reasonable, may lead to very different methods for coral reef restoration and criteria for judging success. Not all of these projects may lead to the same ecologically desirable outcome.[18]

Once the goal is established, the criteria used to evaluate the success of the restoration project depend on the perspective and bias of the evaluator. One of the major challenges for restoration ecologists is the development of success criteria by identifying ecologically sustainable endpoints.[19] For example, a community ecologist may take species composition and species interactions into account when evaluating a restoration project.[20,21] Because of the high variability at the community level and the potential for multiple stable states, the ecologist may identify several potentially successful communities or choose the most complex as the restoration criterion. Restoration projects may have broad and vague goals established, such as to restore the ecological health or integrity of a site. However, there is no real consensus on what a healthy ecosystem is, and making that determination may certainly involve societal values.[22,23]

Ecosystem managers may use very different criteria based on productivity, recreational or aesthetic value, or some other economic measure for evaluating the success of a restoration project. While management objectives often differ on a site-by-site basis, they may not always coincide with ecological criteria for success. In order to address these sometimes disparate goals, we need to integrate science into management decisions throughout all phases of the restoration process.

18.3.2 Integrating Science into Management Decisions

Restoration ecology represents a crossroads between basic and applied science, providing an opportunity for integrating the results of scientific research into management decisions.[24,25] Fruitful partnerships between scientists, consultants, practitioners, government agencies, and nonprofit organizations are necessary to ensure that the science is timely and research questions well-formed to meet the needs of restorationists. While scientists and managers may have similar restoration goals, they may be motivated by different career objectives, work under different timeframes, and have different sets of logistical constraints. These differences, along with the intense pressure on academic researchers to publish, often make it difficult to develop true research partnerships.[26,27]

Managers must base their plans and decisions upon the goal(s) of the project. The challenge for scientists is to provide information that improves ability to predict outcomes of these decisions.[16] Yet, how much do we need to know before we suggest restoration methods, demonstrate alternative evaluation tools, or argue for more monitoring? Sometimes, a value conflict may arise when choosing between an immediate action and a careful study of alternatives, particularly given the time constraints of most restoration projects.[28,29] For these reasons, we should consider how to best integrate the best available science into management decisions.

Restoration ecologist and philosopher Eric Higgs has previously suggested that engaged scientists respond to ethical, political, and social issues in their work. He argues for "good restoration ecology [that] must operate expressly to benefit the restoration of ecosystems."[30] In fact, restoration ecologists are asked to operate at the intersection of science, practice, and policy.[18] Yet, the foundation of the scientific process is objectivity, which can arguably be difficult to maintain while investigating questions related to restoration ecology. Translating the needs of restoration practice into scientifically testable and value-free hypotheses is difficult but necessary.[31]

Restoration ecologists are frustrated with communicating uncertainty to policy makers and the public.[32] For example, what if the results of your study show that one particular restoration method results in higher species diversity but is not significantly different from other methods? Intuition and experience as a restoration ecologist may lead you to believe that this method is inherently better and will eventually yield significantly better results. Communicating the uncertainty to managers is difficult, yet perhaps it is not the role of the scientist to make judgments about the acceptable level of uncertainty.[23,30]

The "precautionary principle" has often been suggested as a necessary attitude when dealing with environmental science and other problems with direct implications for the well-being of society. Simply put, this principle cautions us to avoid environmentally damaging actions even when there is not enough scientific evidence to prove a link between the action and the damage.[28] In the context of restoration, the precautionary principle serves as a guide for choosing and evaluating methods. Even in the absence of significant scientific results at the $\alpha = 0.05$ level, there may be enough evidence that the restoration is working. We are challenged then to expand our acceptable criteria beyond traditional scientific norms when the implications of the results are needed now.

18.3.3 Conflicts of Interest: Are You a Scientist or an Advocate?

Restoration ecologists have the opportunity to be involved in and advocate for the application of their research to restoration projects. They may indeed have an obligation to provide research that will benefit society and the environment.[33] Restoration research is almost always management oriented, and it could be argued that scientists have a moral duty to ensure that their work serves some greater societal good, especially when the research is publicly funded.[34] Ecologists and conservation biologists have recently debated the role of the scientist as an advocate, challenging the notion that scientists should remain objective and policy-neutral (see papers in *Conservation Biology* 10(3) and *Bioscience* 51(6)).

Scientists involved in restoration research may feel conflicted by their dual roles: to gather knowledge objectively and to serve the environment and society.[28,32] Yet, those who are perceived to have a political agenda may compromise their credibility among peers.[35] This intersection of science and advocacy has been discussed in many circles,[36,37] with no concrete suggestions for separating these dual and often conflicting roles.

Scientists may be influenced by their advocacy for restoration in many different ways. Consider a fundamental research question related to coral reef restoration: Do manmade structures facilitate similar recruitment of benthic organisms as natural reef structures? When designing an appropriate study to test this question, a scientist may be influenced by his/her support of restoration in various ways. Designing the experiment, analyzing the data, and interpreting and presenting the results all pose potential ethical dilemmas. In most restoration studies, we are looking for similarities between restored and natural sites. Yet, perhaps the more interesting question is whether there are significant differences between these sites and what can be learned from these differences.

18.3.4 Learning from Failures and Communicating Uncertainty

Perhaps the greatest challenge for restoration ecologists today is to press for the establishment of a *Journal of Failed Results.* Countless studies are done that yield no significant results or that can be considered failures because of an inability to reject the null hypothesis. Yet, this valuable information is not communicated to fellow scientists in any public forum. In restoration, this lack of reporting is especially detrimental; the negative results may tell as much of a story as positive results do.[38] The most successful results are published, but it may be the less successful that provide the most information.[28,39]

There are many cases of restoration projects that did not achieve expected results. Whether these projects failed to provide important habitat (e.g., Zedler and Callaway[40]), did not create appropriate physical conditions (e.g., Kusler and Kentula[41]), or resulted in exotic species invasion (e.g., D'Antonio and Meyerson[42]), restoration projects are often seen as experiments in practice. On the other hand, it is just as important to not exaggerate the importance of restoration studies. A crucial element of restoration research needs to remain the evaluation of restoration approaches, whether or not they are ultimately successful. The restoration community needs to encourage and provide an outlet for communication of all results of restoration research and practice.

18.4 SPECIAL CHALLENGES FOR CORAL REEF RESTORATION

Those involved in coral reef restoration efforts face many of the ethical dilemmas mentioned above. We need to continue to challenge our ideas about restoration, always returning to the fundamental questions of the value of restored ecosystems. One way of exploring the intricacies of potential ethical dilemmas in coral reef restoration is to evaluate case studies. Case studies are often used to briefly present scenarios that involve ethical dilemmas.[43] The following studies are completely fabricated but illustrate some of the potential challenges that restorationists may face.

18.4.1 Case Study 1: To Restore or Not to Restore?

The ABC reef off the coast of Island Y is a popular reef for both snorkeling and diving. Its close proximity to shore renders it highly accessible to tourists who bring in hundreds of thousands of dollars annually to the local economy. There are several other reefs near Island Y, some of which feature endemic bryozoan species and diverse fish communities. While not the most diverse reef in this particular region, the ABC reef certainly attracts the most use.

In 2003, Hurricane X hit the island region, destroying much of the structure of coral reefs, including ABC reef. Limited funds are available from the island's government to restore these reefs. Conservation groups in the area have lobbied for restoration of an isolated reef which was previously rich in both coral and fish species and hosts the endemic species. However, there is intense pressure from the tourism industry to restore ABC reef, thereby restoring the tourist draw to the island.

When considering the options for this restoration, consider the following questions:

1. How are priorities set for coral reef restoration?

Coral reef restoration, like any other type of restoration, involves tradeoffs. Given the tremendous cost and effort of restoration, not all degraded sites will be restored. Choosing potential restoration sites may involve ecological considerations. For example, sites that will provide habitat for a wide range of species or protection for a threatened species or those that connect natural patches of habitat may be favored.

Yet, the economic realities of restoration cannot be overstated: coral reef restoration is expensive. In this case, there is a real economic motivation for restoring ABC reef even though it is not the most diverse or perhaps ecologically valuable site. There may also be educational value to restoring ABC reef, providing an opportunity to educate visitors about the reef recovery process.

Setting priorities for restoration efforts certainly involves ethical dilemmas as it requires those involved to acknowledge their values (or the values of the organization they represent), evaluate potential solutions, consider the best chances for success, and determine how to measure that success. Certainly, all of these aspects of the decision-making process involve logistical constraints. However, we would be doing ourselves and the ecosystem a disservice by ignoring how the values of stakeholders influence this process.

2. Should reefs disturbed by hurricanes or other natural disturbances be restored?

Natural disturbances are an integral part of ecosystem development and maintenance of biodiversity. In fact, monitoring recovery of coral reefs from these disturbances will continue to inform our understanding of the potential recovery of restored reefs (e.g., Connell et al.[44]). Restoration, particularly in the United States, has typically focused on anthropogenically degraded sites such as ship grounding sites. Yet restoring reefs that have been hit by hurricanes is also a priority in other regions and for nongovernmental organizations.

Given the tremendous pressure on coral reefs and their current threatened status, the argument could be made that we should restore as many reefs as possible. Those reefs that are potentially the most diverse or provide an important ecosystem function may have priority over others. On the other hand, how much do we really know about actually restoring the coral reef ecosystem? Can we create ecosystems of equal value to natural systems? Should we attempt to do this for reefs that have been naturally disturbed? One of the arguments for restoration generating "fake" reproductions is that the intrinsic value of the system is created through the natural genesis of that system. Should we then focus our efforts on manmade disturbances and allow for the natural recovery on those reefs disturbed by nature?

18.4.2 Case Study 2: Translating Conflicting Results into Management Recommendations

Julie is a marine ecologist studying the recovery of corals on restoration sites. With growing concern for the health of the world's reefs, she recently became interested in using restoration to enhance the diversity and function of these diverse ecosystems. Currently, Julie is working on a government-funded project to compare natural rates of recovery on unrestored coral reef to coral transplant survival on manmade structures. She is primarily interested in understanding how natural recovery can be sped up on these reefs to achieve a diverse benthic community.

After two seasons of data collection, Julie finds some interesting patterns. On the restored sites, fleshy algal growth has rapidly colonized the limestone structure, precluding establishment of a diverse benthic community. While many of the coral transplants on these reefs have survived, there are no significant differences between the transplanted and natural recovery reefs in terms of the number of coral heads or the percentage cover of corals.

When considering what to do with these results, Julie contemplates the following questions:

1. Do ecologists have a duty to publish "negative" or nonsignificant results?

Julie's results suggest that the restored reefs may not achieve the desired restoration objective, to create a diverse benthic community, at least not initially. There are many interesting implications of these results. Perhaps the manmade structures are inhibiting benthic recruitment of species or facilitating the establishment and spread of fleshy algae. Perhaps the "control" reefs are more indicative of natural recovery and should be studied further. The way that Julie interprets her results could influence her recommendations for future restoration efforts.

While many of us would agree that all results of restoration projects should be published, there may also be political implications to consider. These particular results challenge the long-held notion of restoration ecologists that "if you build it, they will come." What if Julie's results are used by a group or agency to suggest that coral reef restoration does not work or is a waste of money?

2. How can coral reef scientists balance their roles as scientists and their interest in advocacy for coral reef restoration?

Julie may face an ethical dilemma when considering where to publish these results and how to interpret their implications. She is a strong believer in the potential for restoration to save coral reefs, yet her results suggest that natural recovery may be just as good, if not better, approach. Should she suppress her results until more time goes by or more data are collected? Should she try to put a positive spin on the restoration site by emphasizing other aspects of the restoration, even though they were not directly part of her study? It may be impossible for a scientist to remove all personal values and policy preferences from a scientific study. This case study illustrates how easy (and how tempting) it is for a scientist to blur the lines between science and advocacy.

18.5 CONCLUSIONS AND RECOMMENDATIONS

I have outlined only a few of the dilemmas that those involved in coral reef restoration efforts may face. At first glance, many of these may seem like logistical challenges. Every project will involve diverse stakeholders with opinions, values, and hopes for the restoration project. Restorationists will always be forced to choose between restoration options or potential restoration sites. The point that I wish to emphasize here is that these decisions, along with many others, involve the values, personal preferences, and the enthusiasm of everyone engaged in the restoration process. Identifying and addressing ethical dilemmas should therefore be part of any restoration project.

Stepping back from the logistics of the restoration (i.e., type of structure, coral species transplants, monitoring), one must evaluate the reasons for the restoration and whether the endpoint really will be worth all the effort. Particularly when dealing with the endangered coral reef ecosystem, some will continue to argue that money and effort would be better spent protecting pristine or less disturbed ecosystems instead of trying to recreate new reefs. Throughout this book, authors have recommended an adaptive management approach to coral reef restoration in order to continually evaluate and adjust restoration strategies. Part of this approach should address these ethical challenges as well.

The most important and relevant way to address these ethical dilemmas and others is to maintain an open dialogue among those involved in coral reef restoration. For example, the journal *Ecological Restoration* has initiated a forum for addressing relevant ethics issues. Restorationists are invited to submit their ethical dilemmas, to which a panel of philosophers, practitioners, and scientists respond with comments.[45] The Ecological Society of America has also recently addressed ethical dilemmas. The journal *Frontiers in Ecology and the Environment* has embraced the case study approach by publishing case studies that address common dilemmas faced by ecologists, along with guidance for discussing these issues in small groups.[46]

Creating special sessions at professional conferences can be a productive way to find out the ethical dilemmas that restoration ecologists encounter. Through my own work with the Society for Ecological Restoration International, I have found that people want to talk about these issues and will take advantage of the opportunity to do so.[31] Not only should ecologists and managers talk openly about these ethical dilemmas, but the public should be invited to contribute as well. While coral reef restoration efforts provide important opportunities for people to effect positive change, the challenges of doing so should be made transparent.

Coral reef restoration provides an exciting opportunity for people who are passionate about conserving these threatened ecosystems to get involved. We would be well served to consider our ethical dilemmas, challenge our notions about "good" restoration, and be honest about our motives in order to create fruitful partnerships and successful projects.

ACKNOWLEDGMENTS

This work was inspired by conversations with and comments by Ted Shear, Bill Precht, and members of the Society for Ecological Restoration International and is an outgrowth of graduate work completed in the Curriculum in Ecology at the University of North Carolina, Chapel Hill and the Department of Forestry at North Carolina State University.

REFERENCES

1. Jordan, W.R., M.E. Gilpin, and J.D. Aber. 1987. *Restoration Ecology: A Synthetic Approach to Ecological Restoration.* Cambridge, U.K.: University Press.
2. Precht, W.F. 2001. Improving decision-making in coral reef restoration. *Bulletin of Marine Science* 69(2): 329–330.
3. Elliott, R.1982. Faking nature. *Inquiry* 25: 81–93.
4. Katz, E. 1992. The big lie: human restoration of nature. *Research in Philosophy and Technology* 12: 231–241.
5. Atlantis Marine Habitats, Paradise Island. Retrieved on June 2, 2005 from http://www.atlantis.com/atlantis_layers1024.asp.
6. Katz, E. 1992. The call of the wild: the struggle against domination and the technological fix of nature. *Environmental Ethics* 14: 265–273.
7. Kane, G.S. 1993. Restoration or preservation: reflections on a clash of environmental philosophies. *The Humanist* 53(6): 27–31.
8. McGinnis, M.V. 1996. Deep ecology and the foundations of restoration. *Inquiry* 39: 203–217.
9. Light, A. 2004. Restorative relationships. Forthcoming in *Healing Nature, Repairing Relationships: Landscape Architecture and the Restoration of Ecological Spaces*, R. France, ed. Cambridge, MA: MIT Press.
10. Shapiro, E. 1995. Restoring habitats, communities, and souls. Pages 224–239 in A.D. Kanner, T. Roszak, and M.E. Gomes, eds. *Ecopsychology: Restoring the Earth, Healing the Mind.* New York, NY: Sierra Club Books.
11. Throop, W. ed. 2000. *Environmental Restoration: Ethics, Theory, and Practice.* Amherst, NY: Humanity Books.
12. Higgs, E. 2004. *Nature by Design: People, Natural Process, and Ecological Restoration.* Cambridge, MA: MIT Press.
13. Diamond, J. 1987. Reflections on goals and on the relationship between theory and practice. Pages 329–336 in W.R. Jordan III et al., eds. *Restoration Ecology: A Synthetic Approach to Ecological Restoration.* Cambridge: University Press.

14. Davis, M.A. and L.B. Slobodkin. 2004. The science and values of restoration ecology. *Restoration Ecology* 12(1): 1–3.
15. Winterhalder, K., A.F. Clewell, and J. Aronson. 2004. Values and science in ecological restoration—A reply to Davis and Slobodkin. *Restoration Ecology* 12(1): 4–7.
16. Westman,W.E. 1991. Ecological restoration projects: measuring their performance. *Environmental Professional* 13: 207–215.
17. Anderson, J.L., R.W. Hilborn, R.T. Lackey, and D. Ludwig. 2003. Watershed restoration—adaptive decision-making in the face of uncertainty. Pages 203–232 in R.C. Wissmar and P.A. Bisson, eds. *Strategies for Restoring River Ecosystems: Sources of Variability and Uncertainty in Natural and Managed Ecosystems.* Bethesda, MD: American Fisheries Society.
18. Precht, W.F., R.B. Aronson, S.L. Miller, B.D. Keller, and B. Causey. 2005. The folly of coral restoration programs following natural disturbances in the Florida Keys National Marine Sanctuary. *Ecological Restoration* 23(1): 24–28.
19. Hobbs, R.J. and J.A. Harris. 2001. Restoration ecology: repairing the Earth's ecosystems in the new millennium. *Restoration Ecology* 9: 239–246.
20. Gilpin, M.E. 1987. Experimental community assembly: competition, community structure, and the order of species introductions. Pages 151–161 in W.R. Jordan III et al., eds. *Restoration Ecology: A Synthetic Approach to Ecological Restoration.* Cambridge, U.K.
21. Palmer, M.A., R.F. Ambrose, and N.L. Poff. 1997. Ecological theory and community restoration ecology. *Restoration Ecology* 5: 291–300.
22. Lackey, R.T. 2001. Values, policy, and ecosystem health. *Bioscience* 51(6): 437–443.
23. Cairns, J. Jr., 2003. Ethical issues in ecological restoration. *Ethics in Science and Environmental Politics* 2003: 50–61.
24. Aber, J.D. and W.R. Jordan. 1985. Restoration ecology: an environmental middle ground. *Bioscience* 35: 399.
25. Allen, E.B., W.W. Covington, and D.A. Falk. 1997. Developing the conceptual basis for restoration ecology. *Restoration Ecology* 5(4): 275–276.
26. Lach, D., P. List, B. Steel, and B. Shindler. 1993. Advocacy and credibility of ecological scientists in resource decisionmaking: a regional study. *Bioscience* 53: 170–179.
27. Huenneke, L.F. 1995. Involving academic scientists in conservation research—perspectives of a plant ecologist. *Ecological Applications* 5: 209–214.
28. Coblentz, B.E. 1990. Exotic organisms: a dilemma for conservation biology. *Conservation Biology* 4: 261–265.
29. Buhl-Mortensen, L. and S. Welin. 1998. The ethics of doing policy-relevant science: the precautionary principle and the significance of non-significant results. *Science and Engineering Ethics* 4: 410–412.
30. Higgs, E.S. 1997. What is good ecological restoration? *Conservation Biology* 11: 338–348.
31. Mills, T.J. and R.N. Clark. 2001. Roles of research scientists in natural resource decision-making. *Forest Ecology and Management* 153: 189–198.
32. Vidra, R.L. 2003. What are your ethical challenges? *Ecological Restoration* 21: 120–121.
33. Lubchenco, J. 1998. Entering a century of the environment: a new social contract for science. *Science* 279: 491–497.
34. Schrader-Frechette, K. 1994. *Ethics of Scientific Research.* Landham, MD: Bowman and Littlefield.
35. Rykiel, E.J. 2001. Scientific objectivity, value systems and policymaking. *Bioscience* 51: 433–436.
36. Salzman, L. 1995. Scientists and advocacy. *Conservation Biology* 9: 709–710.
37. Ehrlich, P.R. 2002. Human natures, nature conservation, and environmental ethics. *Bioscience* 52: 31–43.
38. Bradshaw, A.D. 1987. Restoration: an acid test for ecology. Pages 23–29 in W.R. Jordan III et al., eds. *Restoration Ecology: A Synthetic Approach to Ecological Restoration.* Cambridge, U.K.: University Press.
39. Friedman, P.J. 1996. An introduction to research ethics. *Science and Engineering Ethics* 2: 443–456.
40. Zedler, J.B. and J.C. Callaway. 1999. Tracking wetland restoration: do mitigation sites follow desired trajectories? *Restoration Ecology* 7: 69–73.
41. Kusler, J.A. and M.E. Kentula. 1990. *Wetland Creation and Restoration: The Status of the Science.* Washington, DC: Island Press.

42. D'Antonio, C. and L.A. Meyerson. 2003. Exotic plant species as problems and solutions in ecological restoration: a synthesis. *Restoration Ecology* 10: 703–713.
43. Penslar, R.L., ed. 1995. *Research Ethics: Cases and Materials.* Bloomington, IN: Indiana University Press.
44. Connell, J.H., T.P. Hughes, and C.C. Wallace. 1997. A 30-year study of coral abundance, recruitment, and disturbance at several scales in space and time. *Ecological Monographs* 67: 461–488.
45. Egan, D. 2004. Looking for the ethical restorationist. *Ecological Restoration* 22(1):1.
46. Dudycha, J.L. and C.K. Geedey. 2003. Adventure of the mad scientist: fostering science ethics in ecology with case studies. *Frontiers in Ecology and the Environment* 1: 330–333.

19 The Volunteer Movement in Coral Reef Restoration

Robin J. Bruckner

CONTENTS

19.1 INTRODUCTION

The concept of coral reef restoration often conjures up images of disastrous ship grounding incidents of the magnitude reported in the news. One might picture a huge oceangoing vessel running too close to shore, grinding to a halt on a reef where it lies stranded, maybe for days, causing significant structural damage from the grounding and subsequent removal of the vessel. Unfortunately, incidents of this scale do occasionally occur, and there is a specific process known as Natural Resource Damage Assessment (NRDA) for emergency response, damage assessment, and restoration. Often, this process results in a multimillion dollar settlement that may take years to resolve and requires major structural reconstruction to repair a reef and return it, in appearance at least, back to something resembling its original form. Restoration at sites requiring significant reef reconstruction may be supplemented with additional biological restoration techniques. For instance, corals and other benthic organisms that would have been found in the area prior to the grounding incident might be transplanted or relocated to the structurally restored site in an effort to return the ecosystem to a more functional state, in a shorter timeframe, than could be achieved if recovery were left solely to nature.

As federal resource trustees for coral reef environments, the National Oceanic and Atmospheric Administration (NOAA) has been involved in a number of significant ship grounding restoration efforts, many spanning years and involving activities ranging from initial response to damage assessment, emergency repairs, monetary settlement for damages from those responsible, restoration planning, and compensatory restoration activities. In the past, reef restoration projects primarily focused on stabilization and reconstruction of the reef structure, with a lesser emphasis on reintroduction of benthic invertebrates. One weakness of NRDA settlements is that monitoring is not required by law to be part of the settlement agreement and is therefore not typically included. Responsible parties tend to be apprehensive that additional compensation could be sought based on monitoring results; thus, the effectiveness of these types of efforts is often unknown or inadequately studied.[1] As a result, large-scale reef restoration efforts in the United States have been conducted on a site-by-site basis using a variety of different approaches, and have only infrequently included limited science-based monitoring to evaluate the benefits of a particular technique.

While restoration at significant ship grounding sites does occasionally occur, the majority of incidents that damage reefs and kill important reef-building corals receive less publicity but may have catastrophic long-term consequences. In the Florida Keys, over 600 small pleasure and fishing boats are reported to ground on shallow nearshore reefs and grassbed communities each year; many more such incidents go unreported.[2] In addition to direct physical damage from human actions, tropical storms, predator outbreaks, and disease epizootics can also have significant localized and regional impacts. In areas like the Florida Keys National Marine Sanctuary (FKNMS), where thousands of tourists visit the reefs each year, the cumulative effects on reef ecosystems may be more insidious than the large ship groundings. The type of large-scale structural restoration that occurs for major ship groundings is often not reasonable or appropriate for small-boat groundings and periodic natural events. Given the rapid loss of live coral cover documented in recent years,[3] however, it is critically important to develop methods to restore, preserve, and increase the quantity of reef-building corals in locally impacted areas.

To gain valuable insight into the success of various restoration approaches and to determine optimal strategies to maximize survivorship of coral transplants, many projects are beginning to place greater emphasis on experimental design and follow-up monitoring activities. Pilot research projects are also seeking to develop and evaluate new restoration techniques by targeting important ecological processes. For instance, in the Florida Keys, efforts are focused on enhancement of sexual recruitment of reef-building coral species by larval culture and seeding, reestablishment of important reef herbivores, and evaluation and control of coral predators and disease. One factor hampering rapid progress in this new direction is the considerable commitment of both financial and human resources required for implementation and evaluation of outcomes. Most public agencies are competing for increasingly limited resources and more often than not are faced with shrinking budgets for programs and services coupled with a higher demand.

19.2 THE IMPORTANCE OF VOLUNTEERS IN REEF RESTORATION EFFORTS

Conservationists and resource managers are fighting a growing battle to protect coral reefs through such strategies as the installation of navigational aids and mooring buoys, establishment of no-take reserves, and "ridge-to-reef" approaches to reduce land-based sources of pollution. Managers have learned that these efforts are most successful when supported by stakeholders and supplemented by voluntary efforts to improve the health of reef ecosystems. Volunteers allow restoration programs to expand the scope of their efforts and supplement available federal or state funding through in-kind services. The public has become aware of the growing coral reef crisis, and many concerned individuals are reaching out to support conservation activities and become involved in restoration efforts. Perhaps one of the most underutilized benefits of working with volunteers is their capacity to collect monitoring information that can ultimately help determine the effectiveness of restoration projects and provide researchers with valuable information to guide adaptive management and improve methodologies for future restorations. The involvement of volunteers also provides the foundation for a deeply committed constituency that is critical to ensure the conservation of coral reefs and that will assist local, regional, and federal agencies in the ongoing stewardship of these fragile and diverse marine ecosystems.

Recent studies reveal a wealth of interesting and useful information that can help volunteer programs recruit and maintain involvement in community-based coral reef restoration.[4] Federal agencies and academic institutions have conducted a great deal of social science research on the use of volunteers in the restoration of land-based systems, and this information is directly applicable to those working to restore coastal habitats, including coral reefs. Herbert Schroeder, an environmental psychologist with the U.S. Forest Service, examined the "inner psyche" of volunteer restorationists, looking at their motives, their values, and their perception of nature. He observed volunteers to be very motivated and to derive a great deal of satisfaction from their efforts, despite long hours of physically demanding tasks, sometimes under less-than-ideal weather conditions. His research identified several main themes surrounding the high level of motivation and enthusiasm of volunteers,[5] which can be condensed into three major interacting factors. The most important motivation, or the primary purpose behind the work of volunteers, was to preserve, protect, and restore nature, with an emphasis on preserving and restoring biodiversity. Volunteers feel a sense of urgency and immediacy about the fragility of coastal habitats in general and about coral reef habitats in particular, and the impending changes to and loss of familiar sites and native species. The global message emphasizing the decline of coral reef habitats substantiates their perception that these precious, historically resilient resources will be irretrievably lost unless immediate action is taken. Working locally in small groups, volunteers can contribute to a much larger effort that they believe will benefit future generations, thereby preserving nature for others to enjoy. Second is their belief that they can make an important and real difference in preventing this loss. By getting actively involved, volunteers see the possibility of actually changing the course of reef decline and achieving a better outcome for the future. Third is the ability to see tangible progress from their efforts in a fairly short time span. Combined, these factors create a powerful incentive to reinforce restoration efforts through the use of volunteers.

Successful coral conservation efforts are dependent upon strong community support. The sense of impending and irretrievable, but preventable, loss of a precious resource motivates even those who live far from coastal systems to seek out reef-related activities such as research and restoration. For example, through groups such as Earthwatch Institute, volunteers use their often-scarce leisure time to help researchers collect important data that can provide the basis for species and habitat protection measures. Through active participation, volunteers learn a great deal about the ecosystem they enjoy and care about and can see the results of their work through journal publications and policy changes. NOAA is helping to maintain and further the motivation and enthusiasm of volunteers by highlighting the importance of restoration to the future of the local, regional, and global environment and by providing tools, technical expertise, information, and financial assistance to increase these efforts.

One program in particular, NOAA's Community-based Restoration Program, emphasizes the involvement of the general public in the restoration of coral reef and other marine and coastal habitats.[6] Despite the challenges of working in an underwater environment, including variable environmental conditions, limited communication, and the need for specialized training and a certain comfort level underwater, volunteers have made significant contributions. Their efforts have helped to reintroduce keystone herbivores, remove invasive, exotic algae (Figure 19.1) and waste tires (Figure 19.2, Figure 19.3), collect "corals of opportunity" resulting from storms or other disturbances and stabilize them for future use in local restoration efforts, remove abandoned and lost crab traps (Figure 19.4) that pose a physical threat to corals and that continue to ghost fish for reef species, and provide volunteer services for a multitude of projects to restore ecologically related habitats such as seagrass meadows and mangrove forests. Volunteers have also played a critical role in data collection for long-term monitoring efforts. Through the use of volunteers in monitoring the federally conducted reef restoration efforts at the *Fortuna Reefer* grounding site in Puerto Rico, information was gathered that subsequently led to midcourse correction efforts at the site, resulting in a retention of more restored coral fragments onsite than would have otherwise occurred.[7] These NOAA-supported projects, conducted with a variety of national, regional, and local partners, have increased public awareness of the cumulative impacts of coastal development and human activities on these ancient coral ecosystems, and are beginning to provide additional opportunities and

FIGURE 19.1 "Human chain" of volunteers passing burlap bags containing exotic algae removed from a reef in the Waikiki, Hawaii, area to the beach for sorting and disposal. Photo credit: Bruce Casler.

FIGURE 19.2 Volunteer diver attaching polypropylene line to remove waste tires off Broward County, Florida. Photo credit: Matthew Hoelscher.

FIGURE 19.3 Deployed in bundles as a certified reef in the 1960s and 1970s, waste tires are causing considerable impact to reefs off Broward County, FL as they move about the seafloor freely. Photo credit: Mathew Hoelscher.

FIGURE 19.4 Volunteer boaters unloading derelict crab pots for disposal. Photo credit: Leslie Craig.

improved methodologies for coral reef restoration. This chapter provides information on community-based coral restoration efforts, describes the results of working with volunteers, and highlights the potential for continued volunteer involvement in future projects.

19.3 INGREDIENTS FOR SUCCESS: THE IMPORTANCE OF TRAINING AND COMMUNICATION

Restoring and monitoring submerged habitats such as seagrass meadows, kelp forests, and coral reefs presents a unique set of challenges and requires skill sets beyond those necessary for terrestrial restoration. Restoration practitioners work to train volunteers so that they understand the end goals of a project and can carry out restoration efforts according to established protocols. More extensive training is crucial to ensure underwater restoration activities proceed smoothly, result in intended outcomes, and do not compromise the safety of volunteers. Special permitting requirements also must be met since many activities take place on coral reefs that are generally under active management or protection. While training volunteers to work around major structural engineering projects may be impractical, training volunteers to assist in biological restorations has proven to be both practical and invaluable.

Coral reefs support a diverse array of user groups, from commercial and recreational fisherman, to tourism-related businesses, to scientists and academics, to waterfront property owners who feel a strong connection to the adjacent resources. In order for coral restoration efforts to be successful, practitioners must take into account the various ways that stakeholders envision a natural and healthy reef system. For instance, a longtime resident, a tourist, a coral reef ecologist, a layman, a fisherman, and a SCUBA dive operator may all have different ideas about coral health and integrity that have been shaped by the nature and extent of their interaction with the environment (i.e., hands-on learning vs. academic learning; business vs. pleasure). Often, those without a scientific background equate coral reef health with certain visual cues such as large and colorful populations of reef fish, for example. "Experts," on the other hand, tend to evaluate ecological health and reef integrity quite differently, as they can more readily distinguish live and dead coral from algae, visually assess the apparent health of particular coral reef species, and conduct rapid field assessments in a defined area to estimate the abundance of coral versus other species. Garnering public support, particularly among residents who live near reef environments and who are often directly impacted by their perception of reef health and well-being, is best achieved by providing hands-on opportunities for local participation. This allows for an exchange of information between various stakeholders that can lead to a proactive approach to identification of restoration areas. Public participation in the process, starting with early outreach and education, goes a long way toward securing support for community-based restoration activities and can, to some degree, eliminate process- and context-based concerns that relate to how and where restoration is being carried out.

One of the best examples of a volunteer training program in coral reef restoration stemmed from a highly successful community-based pilot project, the Reef Medics Program. Reef Medics began as a collaborative effort between Mote Marine Laboratory (MML), The Nature Conservancy (TNC), and the FKNMS, and included participation by both federal (NOAA) and state agencies. The program was specifically designed to reduce the cumulative impacts of vessel groundings by recruiting and engaging local volunteers to report, assess, restore, and monitor small-vessel grounding sites. Reef Medics activities focused on development of program training materials and implementation of training workshops. Workshops were offered at two training intensities, designated "Reef Observer" and "Reef Partner."

Participants were recruited via a local radio station, TNC volunteer newsletters, Mote's volunteer network e-mail list, and press releases in the *Key West Citizen* newspaper. Reef Observers learned about the impacts of vessel groundings and received training in reporting ship grounding incidents. Participants were divided into teams of two to four people, each working with a Reef Medics trainer in a land-based field situation to obtain skills needed to "Know Where You Are" using a global positioning system (GPS) to provide standardized coordinates and triangulation to report vessel grounding sites. Participants also were taught boating and navigational skills designed to increase their ability to avoid damaging coral reef and seagrass resources.

Reef Partner training consisted of intensive workshops designed to create a knowledgeable volunteer force to respond to, assess, restore, and monitor reefs after small-vessel groundings, and involved land- and water-based training in damage assessment and coral restoration activities. Initial efforts focused on assessment of simulated groundings on dry land. A marked transect line was run down the middle of the site, and various objects, such as rocks, concrete blocks, and buckets, were placed within the "damaged area" to simulate features to be mapped at a real site. Assessment kits consisting of a slate with underwater compass, 30-m tape, and gear bag were made up for each team of two participants. Damage Assessment/Restoration Worksheets were provided for teams to record measurements and observations, and 1-m^2 quadrats were provided for a detailed mapping exercise. Handheld GPS units were available for participants to determine coordinates and gain more practice. At a nearby shallow patch reef with prior vessel injuries, a baseline outlining the damaged area was deployed by instructors a day or two prior to the workshop, and sites for attaching corals were identified and marked with floats.

The in-water training for the Reef Partner workshop was conducted at carefully chosen sites that included areas of two previous groundings (*Voyager*, April 1997, and *Bateau Due*, November 1999). Corals that had been stabilized using formed concrete "Reef Crowns" by a contractor and Sanctuary staff prior to the Reef Medics training exercises provided features for mapping and an example of one approach to reef restoration. Participants could also see restored corals that had been attached at the *Voyager* site during a pilot project. During the second Reef Partner workshop, the grounding of a motor cruiser on shallow Admiral Reef provided volunteers with a unique opportunity to become part of the response and assessment effort and to help restore a recently impacted site. Volunteers received an overview orientation of the site, turned over disturbed corals to upright positions where appropriate, and practiced repair skills involving reattachment of dislodged corals. While the Reef Medics-in-training were only able to resecure a few coral fragments, the site proved extremely valuable as a training exercise for mapping grounding sites and recognizing recent vessel damage to coral reef habitat.

The Reef Medics program has created a volunteer constituency in the Florida Keys with the skills to respond to small vessel grounding events and is an example of a training program with far-reaching benefits that include reducing the likelihood of a grounding incident, and simple and practical steps volunteers can take to mitigate damage from groundings. The program is ongoing thanks to a core group of volunteers that has persisted despite some limiting drawbacks to the program. Chief among these are the strict federal SCUBA diving regulations that prohibit most volunteers from participating deeper than snorkeling depths due to rigorous dive certification standards.[8] This has meant that most volunteer assistance is needed above the water on the deck of the boat, mixing cement for divers reattaching corals, handling gear and heavy, water-filled buckets of coral for holding and transplanting, and performing other activities that do not have the glamour or appeal of underwater reattachment of corals. Nonetheless, experienced snorkelers are still able to assist with field assessments and mapping, emergency stabilization of overturned corals, and removal of coral rubble from small-vessel grounding sites, which tend to be located at relatively shallow depths.

19.4 VOLUNTEER RESTORATION AND MONITORING CASE STUDIES

19.4.1 Removal of Alien Algae from Hawaii's Reefs

Healthy coral reef systems are typically dominated by reef-building corals, with nearly all algal production consumed by grazers.[9] Native algae communities are often associated with coral reefs in adjacent shallow waters, sand flats, and reef flat areas. However, over the last several decades,

Hawaiian reefs have experienced increased growth of nonindigenous or "alien" marine algae. Several species of alien algae were introduced for commercial aquaculture and subsequently abandoned and now pose a severe and immediate threat to near-shore coral reef habitats.[10] As alien alga spread, they smother areas dominated by corals and native algae, and reef habitats rapidly shift from diverse coral communities to monotypic stands of alien algae. As the system shifts, it experiences decreases in fish diversity and abundance, and in time, a once-healthy reef is converted to a highly degraded area with few surviving corals and fish.[11]

Some of Hawaii's most famous beaches periodically become inundated with thousands of pounds of algae that wash ashore and decompose, costing local municipalities hundreds of thousands of dollars each year for removal in a constant effort to preserve property values and retain tourism critical to local economies. Areas most heavily impacted include Kaneohe Bay and the southern Oahu shore, including the world-famous Waikiki area, the southeastern shore of Molokai, and West Maui, each of which harbors some of Hawaii's most expansive coral reef ecosystems.[12,13]

A broad-reaching partnership has been working since 2001 to restore and protect key areas of coral reef and native algae ecosystems in windward and southern Oahu by controlling the spread of invasive, nonnative marine algae. The long-term goal of the project is to remove and control alien marine algae in Kaneohe Bay and the Waikiki area, educate the local community on coral reef ecology and the impacts of alien algae species, generate stewardship for local resources through hands-on involvement in the project, and expand the effort to other locations through interisland education and outreach. This partnership recognizes that the opportunity for volunteers to play a key role in the restoration may be a crucial factor in the long-term success of the effort. Pilot projects have involved federal and state agencies, citizen groups, local communities, and nongovernmental organizations, including the Waikiki Aquarium, The Nature Conservancy, NOAA's Community-based Restoration Program, the Hawaii Coral Reef Research Initiative, the University of Hawaii, Hawaii's Department of Land and Natural Resources, Reef Check, the U.S. Fish and Wildlife Service, the National Park Service, private dive operators, and numerous schools. This extensive partnership represents perhaps the largest grassroots effort in the state to address restoration of coral reef resources in Hawaii. Activities have focused on Alien Algae Cleanup Events in the Waikiki area and Kaneohe Bay where, despite 30 years of unchecked growth, the distribution of "gorilla ogo" (*Gracilaria salicornia*) algae is still relatively confined. Controlled and monitored volunteer protocols have been developed to remove algae in ways that do not make the situation worse, since most algae can regenerate from tiny, broken fragments. Algae collected by divers is placed in burlap bags underwater, before bringing it to the surface and hauling it to shore on surfboards, to reduce the chances of liberating algae fragments into the water column (Figure 19.1). Although volunteer efforts are considered relatively small-scale with respect to the scope of the problem, involvement of more than 1600 volunteers during 13 events resulted in the removal of more than 80 tons of algae. In addition to a noticeable reduction in alien algae biomass on affected reefs, outreach and publicity generated from these events have raised awareness of the invasive species issue inside and outside Hawaii.

While the immediate goal of this reef restoration project was to assess the feasibility of alien algae control in Hawaii, new methodologies are now being tested that might meet the challenge of restoring larger areas, more quickly. A device known as a Venturi system, whose suction is created by a bladeless water vacuum, is proving to be more effective at removal, with estimates of up to 10 tons of algae removed in a 2-hr period. Trained experts operate this device while volunteers examine and document the algae collected and remove native algae and invertebrates, which are quickly returned to the reef.

Partners are also working to encourage the regrowth of native algae in areas where it once was dominant. Similar to land-based revegetation practices after invasive vegetation removal, volunteers can cultivate and plant large quantities of native algae quickly and inexpensively. Assessments guide the selection of the most appropriate native species to be cultivated and out-planted in particular areas. Planting sites are chosen by considering where traditional distributions of specific native species existed prior to the invasion of alien algae and where planting may serve the dual purpose of reseeding the native species and controlling the recolonization and spread of alien algae along the fringe of alien algae blooms. Community members play a critical role in this effort, as their knowledge of historic native algal species distribution is invaluable. Once historic ranges are identified, native species are collected from nearby areas to ensure genetic consistency. Samples of native algae are collected and cultured in large tubs or bins. Within these containers, native algae are given fertilizer and aeration to allow them to grow rapidly and reproduce. Native algae are then strung on a piece of raffia or some other organic twine much like flowers on a lei, wrapped around a piece of rubble to facilitate their attachment to the substrate, out-planted, and monitored through Post-Removal Community Surveys. This approach is inexpensive, is easy to learn and teach, and does not require large facilities to implement.

It is interesting to note that native algae out-planting using raffia lei is a traditional Hawaiian practice that has been used for centuries.[14] Traditional cultures are a wealth of institutional knowledge and insight that might not be available anywhere else. Contemporary managers able to think laterally as to what kinds of expertise may be available to most effectively manage local resources may rejuvenate traditional management practices, as has occurred in Hawaii. The application of traditional practices can be a very effective mechanism to foster support for management efforts, as it opens a channel for community members to lend insight and thus feel like they can contribute in a significant way.

While large sections of reef have not yet been cleared due to the magnitude of the invasive algae problem, these restoration techniques offer great promise and lend themselves well to volunteer participation. Project proponents are optimistic that these preliminary efforts will lead to cumulative efforts that will demonstrate significant progress. Additionally, much of the support services (food for volunteers, surfboards and dive equipment, and hauling away waste algae for use by a green-waste recycling company to produce compost) has been donated to the project by local businesses. Community groups on other islands have requested training in how to conduct their own cleanup events, and TNC has developed a protocol to guide implementation of community-led alien algae control programs. Through the alien algae cleanups, community and political awareness of the ecological threat of alien algae to Hawaiian reefs has been heightened. More and more residents are becoming directly involved in conservation and marine resource management as these events foster community support for protecting coral reef resources.

19.4.2 Removal of Waste Tires from Reefs off Broward County, Florida

One of the preferred structures used to create artificial reefs involves the placement of derelict ships in sandy substrates. These artificial reefs develop sizeable and diverse fish assemblages, making the sites popular with commercial and recreational fishermen and SCUBA divers. Before the use of derelict vessels as artificial reefs gained popularity however, other materials were used that were more readily available and appeared particularly attractive at the time, since they also offered a solution to solid waste disposal. One such project initiated in Broward County, Florida, in 1967 involved building an artificial reef using waste car and truck tires that were widely available and otherwise filling up local landfills. The used tires were bundled in groups of eight and bound with strapping that was designed not to rust or corrode. By the end of 1973 over a million tires, and closer to 2 million by some estimates, had been deployed within a permitted area. Over time, storm events and ocean currents began to break up this tire reef, and tires now routinely wash ashore. Tires that remain in the water continue to move with wave action and currents, and many are forced up against the middle of three parallel reefs off Broward County, with accumulations at individual sites estimated in the hundred thousands (Figure 19.3). Repeated impact of tires on the reef has severely damaged many corals and reduced structural relief available for reef fish habitat.

In 2001, a scientist with Nova Southeastern University recruited volunteer divers to clean one area of a reef by removing these tires and bringing them back to shore for recycling. Volunteers were recruited through presentations at dive club meetings and environmental organization e-mail lists, and included groups like Oceanwatch, Under Seas Adventures dive club, Hammerheads dive club, Palm Beach Reef Research Team, South Florida Reef Research Team, Miami Beach Fire Rescue dive team, VONE Research, Nova Southeastern University Oceanographic Center/National Coral Reef Institute students and staff, and the Broward County Department of Planning and Environmental Protection. Volunteer divers were briefed on protocols and safety issues, historical information about the placement of tires in the water, and the rationale for removing them. In turn, volunteers enthusiastically offered suggestions to improve or refine the methods for tire removal. Divers worked in teams of three and had specific tasks such as handling lines, lifting and moving tires, removing any corals for subsequent transplantation, and cleaning tires of algal growth with wire brushes as required by the recycling company. Volunteer divers were placed on site with 25-foot-long polypropylene lines, each of which had a loop spliced in one end. Divers lifted tires, one at a time, and threaded the line through each tire, placing 10 tires on each of 20 lines during a dive (Figure 19.2). The end of the line was threaded through the loop and tied for retrieval by a commercial dive company that towed them back to Port Everglades just below the surface using lift bags. Tires were lifted by crane from the dock into a recycling truck and removed for chipping and reuse as asphalt patch material.

To evaluate the benefits of the volunteer tire removal effort and the extent of reaccumulation of tires, cleaned 20 × 20 m areas were monitored in December 2001 and July 2002. Monitoring dives showed that the tires were even more mobile than initially proposed. By December 2001, within 4 months of the removal activities, the site was recovered by a single layer of tires. Seven months later, the site was fully covered by multiple layers of tires. While the removal of 1600 tires from a field of several hundred thousand may appear futile, the overwhelming size of the problem did not deter these groups of newly informed, dedicated volunteers from participating. Three important purposes were served by the project. Tire removal efforts have raised local awareness about the potential to conduct community-based conservation and restoration of valuable coral reef resources that directly support the economy of southeast Florida. The organization and training of a group of dedicated volunteers was accomplished; their energy and enthusiasm could be translated into other habitat restoration efforts. Finally, some of the material damaging the reefs was removed, and through education and outreach, the importance of finding a long-term solution with guaranteed results was emphasized to the broader public.

19.4.3 Herbivore Reintroduction and Enhancement of Coral Recruitment

Reef-building corals, with their diverse growth forms, are responsible for much of the structural relief of shallow-water coral reefs and support a high diversity of ecologically and economically important fishes and invertebrates. Because corals coexist in a dynamic balance with reef algae and because algae grow more rapidly, high levels of herbivory are critical to allowing corals to maintain their competitive dominance. On Western Atlantic coral reefs, there has been a shift away from dominance by corals since the die-off of the herbivorous, long-spined black sea urchin *Diadema antillarum* in the early 1980s, and many reefs have become overgrown by thick turf and fleshy algae.[15] Densities of these urchins today are still too low on many reefs to achieve major recruitment over any spatial scale. Their dramatic die-off and lack of significant recovery has resulted in changes in the characteristics and trophic structure of coral reef substrate, including declines in cover of certain crustose coralline algae that release chemical cues triggering coral larvae settlement and increases in fleshy algae that directly impact coral recruitment and survival of juvenile corals.[16] One potentially effective ecological restoration approach involves reintroduction of herbivores known to control algal cover, such as *Diadema antillarum*.[17]

In one restoration effort, volunteers with The Nature Conservancy examined the role of urchins in enhancing coral recruitment. This project built upon work conducted by researchers at the University of North Carolina Wilmington, University of Miami Rosenstiel School of Marine and

Atmospheric Science, and NOAA Fisheries Southeast Fisheries Science Center, described elsewhere in this book. Volunteers collected wild *Diadema* and confined them in corrals on reefs with low coral cover to improve the substrate, such that cultured coral larvae would be more likely to settle and grow. Corrals made of nylon mesh were built around large mounds of reef and stocked with adult urchins at the historic, pre–die off density of approximately 5 urchins/m^2. The survivorship of the urchins within the corrals was assessed, and algal abundance inside the corrals was compared with that of surrounding, untreated coral heads. A drastic reduction of turf algae was evident inside the corrals within 3 months. Coral larvae were subsequently introduced, and settlement and survivorship of the juvenile corals were monitored. After 2 weeks, pieces of coral rubble within the seeded areas were examined under a dissecting microscope and showed scores of settlers. Three months later, the corralled areas were surveyed, and many more juvenile corals were observed compared to the initial survey. Over 500 volunteer hours were logged over the duration of the project, with volunteers assisting with small-boat operation, literature searches, collection and placement of urchins in the enclosures, and construction, maintenance, and cleaning of fouling organisms from the corrals. Work is continuing to create colonies of adult urchins by concentrating wild adults and by introducing lab-raised urchins to build pockets of higher population density. *Diadema* are broadcast spawners, so researchers are trying to help them spawn in aggregations that improve the rates of fertilization, which should lead to greater abundance of urchin larvae and more successful recruitment of urchins, and eventually, corals.

19.4.4 "Corals of Opportunity" — Establishing Coral Nurseries for Use in Restoration

In Broward County, Florida, volunteers are working with scientists to rescue damaged or dislodged corals after ship groundings, tropical storms, or similar incidents and relocate the detached corals and coral fragments to a coral "nursery" — an artificial reef holding area. Nova Southeastern University and Broward County scientists and managers developed a protocol to survey local reefs and to identify overturned and dislodged corals in imminent danger of mortality, and they identified suitable transplantation sites to receive the donor corals. Members of nonprofit environmental and diving groups were trained on coral search, identification, collection, and transplantation procedures. Volunteer and scientific divers conducted searches to collect and move dislodged colonies to a nursery site, and salvaged colonies were attached to the artificial reefs using underwater cement. The colonies at the nursery site and naturally detached control colonies of the same species on a nearby reef were tagged and monitored quarterly to learn which types of coral were most likely to live on after experiencing the effects of detachment on their growth and apparent health. Through the establishment of a coral nursery, a ready supply of coral colonies was made available to replenish areas damaged by ship groundings. In addition, these efforts will enhance future transplantation by focusing on species known to recover best.

The coral nursery project has been highly visible and has offered numerous opportunities for public outreach and participation. Outreach efforts have included several seminars for volunteers, newspaper articles and editorials, oral presentations at scientific meetings, and research publications. Partners in the effort include Broward County's Department of Planning and Environmental Protection, the Ocean Watch Foundation, the Florida Marine Research Institute, Nova Southeastern University Oceanographic Center/National Coral Reef Institute, and local volunteers. Funding to kick off the effort was provided through a partnership between NOAA's Community-Based Restoration Program and the National Fish and Wildlife Foundation.

19.4.5 Volunteer Participation in Long-Term Coral Reef Restoration Monitoring Efforts

In 1997, the 325-foot vessel *Fortuna Reefer* ran aground on a coral reef off Mona Island, Puerto Rico in an area dominated by *Acropora palmata.* The initial restoration, described elsewhere in this book, involved securing fragments of *A. palmata* (elkhorn) coral to the reef and to dead standing elkhorn skeletons as part of an emergency restoration that took place within months of the incident. The subsequent monitoring undertaken by volunteers was critical in identifying a major weakness of the restoration that manifested over time — chronic failure of the stainless steel wire used to initially secure the fragments. It also highlighted an area of improvement for future restorations to increase survivorship through more strategic placement of fragments to minimize their contact with a common bioeroding sponge. Data collection and analysis led to a midcourse correction to restabilize remaining fragments and will contribute to the development of new and improved techniques for restoring coral reef habitats.

Volunteer participation in this research was critical to maximize the amount of data that could be collected on each monitoring trip. The primary site location, a small offshore island west of mainland Puerto Rico, was logistically difficult to get to and work from (volunteers camped on the island), and most work required the use of SCUBA, limiting the data that could be collected by individual divers. Volunteers were essential not only to collect certain data themselves but, by setting transect lines and quadrats, taking photographs, and performing assessments for fish and algae, they maximized the time researchers could spend collecting detailed and specific coral reef data. Volunteers recruited through Earthwatch Institute participated in 12-day missions, and were provided with a list of background reading materials in advance. Research scientists delivered a series of informational training lectures that covered coral reef ecology and focused on the ability of volunteers to positively identify 20 species of stony corals and the major groups of fish. Lectures were followed by land-based skill-building activities so volunteers would fully understand what was required of them underwater. Shallow-water training sessions allowed volunteers to practice their skills and alternate tasks so they could determine the most effective working teams and activities that best suited each individual. Team development was enhanced, and individual responsibilities identified, by conducting control standardized transects run by volunteers and scientists, analyzed to determine variability between observers and match underwater skill levels with appropriate tasks. The project demonstrates the importance of volunteers in a science-based monitoring program. Monitoring information was used to guide an adaptive management response for a federal NRDA coral reef restoration case, resulting in maintenance of restored coral fragments onsite that most likely would have otherwise been lost.[7]

19.5 CONCLUSION

Criticism about the use of volunteers to conduct meaningful restoration and collect research data that will provide the basis for crucial management and policy decisions has been negated by studies dating back as far as the 1960s. A great deal of literature has been produced that highlights the value of volunteer support, ideas, enthusiasm, and resources that ordinary members of the public can bring to the conservation world.[18] Much of this type of restoration research is labor intensive but technically straightforward. Finding the appropriate level of detail for volunteer data collection is critical to the integrity of the work, however. So is following task identification with quality control of test data and ensuring that the volunteers with an aptitude for specific research tasks are utilized in a consistent fashion so that data generated by volunteers is nearly identical to that collected by experienced scientists. Research studies are often expensive and time consuming, with large volumes of data needed to make reliable predictions about current trends and to guide management and policy decisions that will lead to expected outcomes. As with restoration of terrestrial environments, there is a rising demand for information on coral reef restoration results and an increasing gap between the resources that are necessary and those that are available for this purpose. Volunteers help fill that resource gap, and in doing so, also become groomed as the future stewards of the environment they work in through the increased knowledge and understanding they gain through hands-on activities. People from all walks of life, with little to thorough knowledge of coral reef ecosystems, are becoming more aware of environmental issues in general and are looking for opportunities to do their part to offset destructive, largely manmade environmental problems.

Community-based restoration is a widely applicable model for undertaking the challenge of environmental protection and restoration that may not be within the reach of statutory mandates. Civic environmentalism is a supplement to traditional regulation policy and is a proactive, rather than reactive, approach that is far more palatable to the general public. Involving nonexperts in reef restoration activities can further serve to help convince policy makers to formulate better policies and make the case in understandable and meaningful terms to the public at large to support these policies.[19]

The physical act of hands-on involvement in projects designed to nurse coastal and marine ecosystems back to health serves as a powerful instrument for transforming environmental attitudes and subsequent actions. Hands-on restoration provides the strong foundation necessary to fully understand relationships between humans and coral reef environments, to create values related to these relationships, and to generate emotional commitment to them. It provides a basis for the creation of the large, deeply committed constituency that is critical to ensure the conservation of natural areas like coral reefs.[20] Grassroots involvement is the key that will assist local, regional, and federal agencies in the ongoing stewardship of the nation's marine and coastal resources.

The trend toward using volunteers is making marine habitat restoration more effective, more plausible, and more widespread. In the past, restoration activities were accomplished almost solely through traditional funding mechanisms using paid technical staff. Now volunteers are making a huge difference. In the United States and around the world, volunteers are involved in a multitude of activities to restore habitats important to fish and wildlife. Students, scout troops, faith-based organizations, nongovernmental organizations, civic groups, and ordinary community residents give freely of their time to remove invasive plants and replant native vegetation; grow oyster seed and construct oyster reefs; plant mangroves, saltmarsh, and seagrass; and conduct other activities, including coral reef restoration, which might otherwise not be possible or practical without their involvement. As the case studies above demonstrate, volunteers have proven their effectiveness in restoring and monitoring coral reef habitats, and volunteer-based efforts in coral reef ecosystems should continue to be encouraged.

ACKNOWLEDGMENTS

This manuscript would not have been possible without the detailed information provided by grantees to the NOAA Community-Based Restoration Program through progress reports and phone interviews, and by project oversight and follow-up from dedicated NOAA Restoration Center staff. This manuscript was improved by critical reviews from Louise Kane, Melanie Gange, Andrew Bruckner, Leslie Craig, Eric Co, and other project proponents, and anonymous reviewers. The opinions and views expressed in this document are those of the author and do not necessarily reflect the policy or opinion of the National Marine Fisheries Service, the National Oceanic and Atmospheric Administration, or the U.S. government.

REFERENCES

1. Miller, M.W., The importance of evaluation, experimentation, and ecological process in advancing reef restoration success, *Proc. 9th Intl. Coral Reef Symp.,* 2, 977, 2000.
2. Precht, W.F., Ders, D.R., and Gelba, A.R., Damage assessment protocol and restoration of coral reefs injured by vessel grounding, *Proc. 9th Intl. Coral Reef Symp.,* 2, 973, 2000.
3. Bruckner, A.W., Potential application of the U.S. Endangered Species Act as a conservation strategy, *Proc. Carib. Acropora Workshop: NOAA Tech Memo NMFS-OPR-24,* 199, 2003.
4. Gobster, P.H., Human actions, interactions, and reactions, in *Restoring Nature,* Gobster, P.H. and Hull, R.B., Eds., Island Press, Washington DC, 2000, intro.

5. Schroeder, H.W., The restoration experience: volunteers' motives, values, and concepts of nature, in *Restoring Nature,* Gobster, P.H. and Hull, R.B., Eds., Island Press, Washington DC, 2000, chap. 12.
6. http://www.nmfs.noaa.gov/habitat/restoration.
7. Bruckner, A.W. and Bruckner, R.J., Survivorship of restored *Acropora palmata* fragments over 6 years at the *M/V Fortuna Reefer* ship grounding site, Mona Island, Puerto Rico, *Proc. 10th Intl. Coral Reef Symp.,* in press.
8. Goodwin, B., personal communication, 2004.
9. Carpenter, R.C., Partitioning herbivory and its effects on coral reef algal communities, *Ecol. Monogr.,* 56, 345, 1986.
10. Squair, C.A, An introduction to invasive alien algae in Hawaii: ecological and economic impacts, *Marine Bioinvasions: 3rd Intl. Conf. on Marine Bioinvasions,* 115, 2003.
11. Stimson, J., Larned, S., and Conklin, E., Effects of herbivory, nutrient levels, and introduced algae on the distribution and abundance of the invasive macroalgae *Dictyosphaera cavernosa* in Kaneohe Bay, Hawaii, *Coral Reefs,* 19, 343, 2001.
12. Russell, D.J., The ecological invasion of Hawaiian reefs by two marine red algae: *Acanthophora spicifera* and *Hypnea musciformis* and their association with two native species, *Laurencia nidifica* and *Hypnea cervicornis, ICES Mar. Sci Symp.,* 194, 110, 1992.
13. Smith, J.E., Hunter, C.L., and Smith, M.C., Distribution and reproductive characteristics of nonindigenous and invasive marine algae in the Hawaiian Islands, *Pac. Sci.,* 56, 299, 2002.
14. Co, E., personal communication, 2004.
15. Lessios, H.A., Mass mortality of *Diadema antillarum* in the Caribbean: what have we learned? *Annu. Rev. Ecol. Syst.,* 19, 371, 1988.
16. Morse, D.E., et al., Control of larval metamorphosis and recruitment in sympatric agariciid corals, *J. Exper. Mar. Biol. Ecol.,* 116, 193, 1998.
17. Edmunds, P.J. and Carpenter, R.C., Recovery of *Diadema antillarum* reduces macroalgal cover and increases abundance of juvenile corals on a Caribbean reef, *PNAS,* 98, 5067, 2001.
18. Foster-Smith, J. and Evans, S.M., The value of marine ecological data collected by volunteers, *Biol. Conserv.,* 113, 119, 2003.
19. Hull, R.B. and Robertson, D.P., The language of nature matters: we need a more public ecology, in *Restoring Nature,* Gobster, P.H. and Hull, R.B., Eds., Island Press, Washington DC, 2000, chap. 5.
20. Jordan III, W.R., Restoration, community and wilderness, in *Restoring Nature,* Gobster, P.H. and Hull, R.B., Eds., Island Press, Washington DC, 2000, chap. 1.

20 Monitoring the Efficacy of Reef Restoration Projects: Where Are We and Where Do We Need to Go?

Cheryl Wapnick and Anne McCarthy

CONTENTS

The last few decades have led to an increased awareness of problems regarding the conservation of coral reef systems and the natural and anthropogenic causes of damage to these resources. Natural events that contribute to coral reef destruction primarily include hurricanes and typhoons, earthquakes, and lava flows while human-induced damages generally stem from, but are not limited to, fishing, dredging activities for harbors and channels, beach renourishment projects, vessel groundings, water quality degradation, and anchorings. All of these activities have the potential to fundamentally change the physical, biological, and geomorphological structure of coral reefs. Understanding the long-lasting effects, such as ecological shifts, is paramount in determining whether human intervention can act to rehabilitate coral reef impacts.

Even under the most favorable conditions, coral growth rates and coral reef recovery following disturbances are slow. For example in the Florida Keys, the coral growth rate has been reported as 0.65 to 4.85 m per 1,000 years,[1] and most evaluations of the natural recovery process following a moderate disturbance event have reported at least a decade for recovery to predisturbance conditions.[2–5] Coral reef restoration efforts have the potential to accelerate the natural recovery process or in some cases, where onsite restoration is impossible, to mitigate for the loss of resource services provided. To better understand the success of these restoration efforts and evaluate the procession of recovery, long-term monitoring and evaluation is essential. According to Jaap,[6] "[i]f a project is worthy of executing a restoration, it is worthy of monitoring the progress of the recovery."

20.1 WHY DO WE MONITOR?

Monitoring is a vital component of coral reef restoration projects for several reasons. First, it is a necessary means in evaluating and further developing indicators of functional success of the restoration efforts. To accomplish this, the evaluators must initially identify and define what are acceptable "indicators of function" as well as the expected levels of long-term results. Because changes to a coral reef system may be imperceptibly slow over short periods of time and ecological processes may vary greatly spatially and temporally, the examination of long-term trends is essential.[7] A monitoring plan should be developed as such to address the questions concerning the anticipated levels of long-term success using the predetermined indicators as the gauge of the restoration effort (e.g., structural integrity, topographic complexity, return of fish assemblages, coral recruitment).

One of the most important aspects of monitoring is its role in adaptive management. Adaptive management is based on the provision of feedback loops between scientists and managers allowing for the identification of need and implementation of corrective action (i.e., midcourse correction). By evaluating the success of a restoration project in progress, researchers and managers can determine whether the proposed objectives are indicating signs of success and whether the program warrants the expenditure of additional community or corporate resources.[8] As a result, the need to modify, revise, or discontinue the project can be identified. If necessary, the study design can be adjusted or enhanced to further evaluate emerging questions as the project progresses. In fact, without periodic monitoring and subsequent adjustments, it may be impossible to determine whether a restoration project is heading in the right direction.[9] Precht et al.[5] indicated that this view of restoration management considers three important criteria:

1. Allowing for acceptance and adaptation in management decisions based on the results of monitoring studies
2. Allowing for direction and guidance of the restoration and monitoring studies by a multidisciplinary team of experts
3. Allowing for integration of feedback into restoration management, specifically decisions that may require timely implementation of incremental changes

Another application of monitoring and adaptive management is the identification of ways to improve future restoration efforts, as well as monitoring programs such as increasing the frequency or temporal extent of monitoring events, expanding the spatial scale of the monitoring studies, or dropping or adding an indicator of functional success. For example, monitoring of the damaged area that resulted from the cruise ship *Maasdam*, that struck Soto's Reef, George Town, Grand Cayman Islands, British West Indies on January 12, 1996, consisted of baseline,[10] 6-month,[11] 1-year,[12] and 2-year events.[13] Although short-term monitoring indicated that the restoration measures were relatively successful, considering the magnitude of the damage, monitoring should have been continued for at least 10 years, including additional monitoring events at 4, 6, 8, and 10 years after completion of restoration.[6] The use of monitoring programs as a basis for improving future designs and for evaluating prospective projects will ultimately lead to more efficient restoration and restoration monitoring.[9]

20.2 WHAT DOES MONITORING PROVE?

20.2.1 Functional versus Compliance Success

The evaluation of a reef restoration project's "success" is one of the most vital, yet highly controversial, aspects of coral reef monitoring. There are two fundamental ways to define the term "success" and how it relates to restoration. Compliance success is often used to evaluate restoration projects and only considers whether permit requirements were met or simply whether the restoration project and/or the monitoring component were implemented.[5] Functional success, in contrast, is based on whether the ecological functions of the systems have been restored.[14] A much more difficult task, the evaluation of functional success should include the processes that take place within and between habitats as a result of the physical, chemical, and biological characteristics of that habitat.[9] In order to effectively determine the success of a restoration effort (compliance and functional), it is necessary to establish both the goals and the objectives of the restoration project as well as the success criteria (and possibly the methods that will be used to evaluate the criteria) prior to full implementation of the restoration plan.

20.2.2 Examples of Long-Term Monitoring Programs

Vessel groundings are a common cause of reef damage in the Florida Keys, and depending on the size of the ship and the prevailing sea conditions, damage can be severe and extensive.[6,15] Secondary impacts include the destruction of previously undamaged resources via the movement of disturbance-formed rubble, increased sedimentation related to newly exposed reef framework, and additional damage caused by vessel salvation and towing efforts, all of which compound the devastation caused by the grounding itself.[5] Several ship-grounding restoration projects, including those devised for the motor vessels *Maitland, Elpis, Wellwood*, and *Houston* incidents, can be used as examples of long-term monitoring programs demonstrating how success is being evaluated. The timing and location of the *Maitland* and *Elpis* grounding events and subsequent restoration efforts, for example, provided an opportunity to assess restoration success in terms of the *in situ* recruitment of coral populations.[16]

20.2.2.1 *M/V Alec Owen Maitland*

In October 1989, the *Maitland*, a 43-m oil field supply vessel, went hard aground on living coral reefs in approximately 2.5 m of water, south of Carysfort Reef in the Key Largo National Marine Sanctuary, now part of the Florida Keys National Marine Sanctuary (FKNMS).[17] The resulting damage included the breakage of underlying coral substrate, the formation of deep craters in the coralline seabed, and the creation of substantial amounts of coral rubble that were ejected from the craters.[18] Under the auspices of the National Marine Sanctuaries Act, funds were collected from the vessel's owner to be used for site rehabilitation. Structural restoration began in 1995 and involved practices to stabilize loose rubble, fill in areas of lost reef framework, and accelerate the recovery rate by providing stable substrate for coral recruitment and fish habitation.[16,19] The objective of this restoration project was to "recreate a stable foundation which closely emulates the adjacent natural seabed and which would foster future recruitment of the local biota."[18] After 5 years, monitoring efforts indicated the coral assemblage at this site was dominated by one species, *Porites astreoides*.[16] The long-term monitoring plan calls for an evaluation of community structure based on the abundance and coverage of functional groups (corals, algae, octocorals, etc.), coral recruitment and survivorship, and determining spatial and temporal trends between the restoration site and control sites.

20.2.2.2 *M/V Elpis*

Within 3 weeks of the *Maitland* grounding, the *Elpis*, a 143-m cargo freighter, ran aground on the same reef system at nearby Elbow Reef in 8.5 to 10 m of water.[17,19] Similar damage and restoration efforts to the *Maitland* grounding site, with the exception of structure design, were applied in 1995 to this deeper location with the hopes of fulfilling the same goals. As part of a joint monitoring effort with the *Maitland* site, 5 years after the restoration efforts, scientists observed that the juvenile coral assemblage at this site was dominated by *P. astreoides, Favia fragum,* and *Agaricia* sp. with an estimated juvenile density of 50% greater than the *Maitland* site.[16] The long-term monitoring plans adopted by the FKNMS evaluate the same criteria as set for the *Maitland* restoration site.

20.2.2.3 *M/V Wellwood*

In 1984, the 122-m *M/V Wellwood* grounded on Molasses Reef, a bank barrier reef within the then-designated Key Largo National Marine Sanctuary in approximately 6 m of water.[20] Unlike the *Maitland* and *Elpis* grounding sites, the *Wellwood* site did not undergo coral reef restoration until 2002. After 5 years following the grounding, juvenile coral abundance and diversity were not observed to increase.[21] This indicates that significant restoration efforts, including the reestablishment of topographic complexity, are required to return damaged spur-and-groove habitat to its former state on a time scale of decades.[5,22] Studies looking at abundance and diversity of fish and coral populations indicated the need to conduct re-creation of three-dimensional structure to enhance recovery of the injured site.[20,23] The FKNMS will be looking at changes in the community structure in comparison to a reference site following the recent completion of structural restoration at the *Wellwood* site.

20.2.2.4 *M/V Houston*

The *Contship Houston*, a 187-m vessel, ran aground between American Shoal and Maryland Shoal within the FKNMS in February 1997. The restoration efforts, funded by the vessel operator and insurance companies, included the reattachment or repair of 3220 displaced corals to the seafloor as well as the stabilization of large rubble berms with epoxy cement;[6] injured areas flattened by the grounding were stabilized and three-dimensional structure was added using flexible concrete mats and limestone boulders. Monitoring of this restoration project has concentrated and continues to concentrate on the physical stability of the concrete mats, boulders, and rubble berms as well as the survivorship of transplanted corals and the recruitment of stony and soft corals and sponges.[24,25]

Restoration and monitoring programs that focus on changes in community composition and structural stability of coral reefs commonly caused by large ship groundings provide the opportunity to understand the natural recovery process that often follows physical disturbances and how this process can be enhanced. Scientific monitoring of the various groundings throughout a relatively small area within the Florida Keys has supplied vital information regarding the impacts of a range of restoration efforts and how the definition of "success" varies among programs. The above examples can be used to assist researchers and managers in designing, assessing, and implementing similar programs in the future.

20.3 SUGGESTED MONITORING PROTOCOL

20.3.1 Science-Based Monitoring Programs

Several factors must be included in a long-term monitoring plan to effectively establish whether the objectives of the restoration program have been achieved. On the grand scale, the use of ecologically based, hypothesis-driven, long-term monitoring programs is the only approach to answering essential questions regarding:

1. The length of time necessary for natural recovery to occur at any particular site without manipulation
2. Whether natural recovery will result in an alternate community state
3. Whether anthropogenically disturbed reefs respond differently than do naturally disturbed reefs[5]
4. Whether restoration activities accelerate natural recovery

Thayer et al.[9] included 12 steps in the process of developing a monitoring plan. The first steps involved the identification of the project goals and the establishment of a basic understanding of the project habitat and characteristics. At this stage, it is recommended that information be collected on the monitoring of comparable projects and that experts be consulted. If the restoration program involves multiple objectives, it is imperative that the primary goal be identified and that the remaining goals be ranked by importance, in order to better make planning decisions that may involve tradeoffs.[26] For example, time and financial constraints may prevent a restoration program from incorporating both the monitoring of transplanted corals and the recruitment of juvenile corals to an artificial reef structure. As a result, the scientists and managers must decide which of these objectives is more essential in determining success of the program. Testable hypotheses should then be formulated to address the project objectives. If the recruitment of juvenile corals is chosen as the primary objective in the above example, than the testable hypothesis may be that recreating the lost three-dimensional structure will result in a coral recruitment rate equal to or greater than that of the reference or control site.

Although experimentation is oftentimes not included in monitoring programs, one of the most crucial aspects of any monitoring protocol is that it be science based and hypothesis driven to the greatest extent possible. Although this component of monitoring may be difficult to achieve, the formulation and testing of hypotheses for each project goal is the underlying component in our ability to quantify success of the restoration project. As a result, the use of hypotheses allows for the adaptation of management decisions (future and corrective actions) to meet restoration goals.[5] Once the null and alternative hypotheses have been formed, historical data should be collected to help detect signs of long-term trends and reasons for declines in habitat.[9] For example, historical data may demonstrate that coral growth rates naturally vary in a specific area on a seasonal scale. As a result, these data may improve the overall quality of the results by establishing a baseline of how the habitat functioned prior to the disturbance.

If one considers the definition of ecosystem restoration to be "the return of an ecosystem to a close approximation of its conditions prior to disturbance,"[27] then it is logical to base the assessment and evaluation of restoration programs on those conditions encountered at natural, undisturbed reference sites.[5] Disturbed reference sites that are similar to the project site, however, may also be helpful. In this case, the disturbed reference sites, sampled in conjunction with the project site, would be used to represent the rate of recovery in the project area if restoration had not been performed.[9] Precht et al.[5] used both methods to test the hypothesis that following a ship grounding, areas of high relief would converge to an alternate community state found in natural hardground communities if restoration was not used to accelerate or enhance recovery. The comparison of the benthic assemblages of the 1984 *Wellwood* grounding site (disturbed spur-and-groove habitat), a natural hardground reference site, an undisturbed spur-and-groove habitat, and a hardground site that had been flattened over one century ago indicated that ship grounding sites in former spur-and-groove habitat will not recover to their original state without a significant restoration effort. The implications of this study were further strengthened when contrasted against the 1989 *Elpis* grounding site, which showed recovery of topographic complexity to surrounding reference levels within a decadal time frame of restoration activities performed in 1995.

Whenever possible, monitoring programs should utilize experimentally based designs and incorporate statistical analysis that would enable the adequacy of the resultant data to be determined. Therefore, prior to the onset of the monitoring program, the investigator should decide upon a method for collecting data in order to obtain statistically valid, quantitative information that addresses the designated objectives. For example, if the objective includes the restoration of coral population densities to background levels, the data collection and analysis would include detailed examination of species diversity and coral species densities per square meter. In contrast, if the objective involves the restoration of overall coral cover, general data regarding coral density per square meter would be sufficient, and species diversity and coral density by species would be excluded from the data collection and analysis. Each method would likely produce different results. If necessary, a statistician or a field scientist experienced in experimental design should be consulted. Pilot studies can be useful in determining the number of samples necessary to meet the monitoring goal.[7]

The adequacy of the resultant data will be reflected in the degree, independence, and suitability of the chosen replication; the measurement techniques; the distribution of samples in time and space; and the sufficiency of the resources used to execute the sampling design; and is what establishes the success and validity of the assessment.[26] The restoration effort is acknowledged to be successful once the results lie within the predetermined levels or indicate that the prescribed criteria have been met.[28]

Techniques such as classification analysis, multi-dimensional scaling, principal component analysis, and ordination can all be applied to discern patterns of change or stability in a system[7] and can be rigorously applied to provide the valid data the evaluator is seeking. Nonparametric, multivariate analyses and qualitative information (e.g., photographs or video) can also be useful tools for evaluating ecological data that do not conform to the assumptions of parametric statistics. Despite the type of statistical analysis employed, reporting the resultant data in a format compatible with other studies increases the utility of the assessment for the adaptive management of the project as well as related programs. In the same respect, the results should be reported to project managers and engineers as soon as possible.[9]

20.3.2 Specific Design for Each Restoration Site

No two disturbance events, restoration sites, or circumstances are identical. Thus, the monitoring design and analysis must be specific to each incident and location. For example, discrete, high-magnitude stressors like ship groundings produce results that are immediately evident, while more chronic, gradual disturbances, such as overfishing, may not demonstrate visible effects for several years.[7] In the latter case, the monitoring program must incorporate an extensive enough time scale to capture any potential effects. Techniques utilized by monitoring programs also need be chosen wisely with regard for potential socioeconomic or financial constraints of the site location. For instance, though independent sampling designs make it more difficult to detect significant effects and offer less statistical power to detect signals, they are less costly and time consuming than repeated measures designs and are therefore more feasible for reef monitoring studies in developing countries.[29] Although practicality is an important factor to consider, the investigator must ensure that the legal obligations from the compliance standpoint are satisfied. This point stresses the importance of prioritizing the objectives of the restoration plan as described earlier, as well as differentiating between management and research goals and the expectations of each.[26]

Injury classifications and restoration project types vary among sites and are based on several factors that all play a critical role in the design of the monitoring plan. These factors may include, but are not limited to, the type, extent, and severity of the damage as well as the type of restoration applied. Classification schemes can be useful tools but are not currently widely used among coral reef trustees. Some workers, however, have already begun using fairly straightforward methods to classify injury types specific either to the cause of injury[30] or a physical description of the injury.[31] The intent of the classifications is to provide resource managers with an understanding of injury types and identify restoration alternatives for the injury classes. For example, a vessel grounding that destroys the underlying reef framework and where artificial structure is added to recreate the topographic complexity will demand different types of restoration and monitoring objectives than a grounding that scrapes the surface of an area and where a "no restoration" alternative is chosen (natural recovery). These classification systems are a first step in developing some unified approaches that should be considered by all trustees in South Florida in order to begin "speaking the same language" with regard to coral reef injuries.

Given the above factors, the investigator must specify the objectives of the restoration project and define the success criteria for each objective. The parameters for evaluation would then be selected based upon the injury classification, the criterion chosen, the specific restoration tasks performed, and what each could prove or disprove. For example, if the type of restoration included the transplantation of displaced coral heads to the seafloor, one of the objectives would be the long-term survivorship or vitality of the restored coral heads. The success criterion for this objective could be defined as a statistically equivalent or greater level of survivorship exhibited by the restored corals as compared to reference (undisturbed) corals over an identified time span. Parameters that could be used during monitoring to evaluate the success of this restoration method may include continued adherence to the substrate; competition with benthic algae; the percent damage due to algae, sponges, or other invertebrates; and the percent damage due to disease or bleaching.[6,32] Regardless of the injury classification, success criteria, or parameters used in the monitoring plan, strong statistical analysis is required to determine reef performance or success.

Certain restoration and monitoring programs have relatively straightforward parameter requirements. For instance, disturbances that result in severe framework damage, such as major ship groundings or intense storms, would likely necessitate that structural integrity be monitored over time, while disturbances such as a major coral disease epidemic resulting in areas barren of live coral would entail monitoring of coral recruitment or coral colony recovery. Choosing the right parameters to monitor and how to measure those parameters, however, is not always an easy task. Coral reef systems are extremely complex and involve numerous interactions among both plants and animals. The abundance of possible causes and the mystery around the ecological repercussions of community change make it difficult to adequately determine which variables to monitor.[29] Under all circumstances, however, it is best to be able to justify the need and to find supporting literature for monitoring each parameter. Thayer et al.[9] provide matrices of parameters that should be considered for monitoring based on the structural and functional characteristics of the habitat in question.

The chosen monitoring design, including the timing of sampling, must also consider the type of characteristics that are being evaluated. Thayer et al.[9] suggest that structural attributes be frequently monitored for numerous years immediately following restoration. In contrast, the monitoring of functional characteristics should commence once the system has begun to mature and the function of concern has had the opportunity to become sufficiently established. Investigations regarding the grounding sites in the FKNMS can again be used as a resource for recommendations. Miller and Barimo[16] used juvenile coral recruit densities as the principal basis for comparison of successful recovery among the *Maitland, Elpis*, and *Wellwood* grounding sites. Although the study established that this one ecological aspect of coral reefs is necessary to attaining a restored reef community, other factors such as the survival, growth, and reproduction of these recruits also need to be considered to ensure a functional habitat. For example, the mere settlement of coral recruits at a damaged site does not directly translate into the growth and survivorship of these corals on a decadal scale.[33,34] As a result, comparisons made among the three Florida grounding sites after only 3 to 4 years of restoration are not sufficient to draw long-term conclusions regarding the success of these projects. Functional success can only be determined by adjusting the variables monitored to consider the growth and survival of the coral recruits on larger time scales.[16]

There is no discrete recipe that can be followed for conducting a monitoring program; however, the recognition and application of the above guidelines would prove helpful to the implementation of a valid and practical long-term monitoring plan. In summary, it is imperative to gear the monitoring program to be science based and hypothesis driven to the greatest extent possible. This includes the potential use of experimentation, the statement of prioritized objectives preceding the onset of monitoring, and the utilization of experimentally based designs. These designs should be specific to each restoration site and must consider the legal obligations from the compliance perspective. Each monitoring design should also reflect the injury classification and restoration type as well as which factors need to be examined in order to maximize efficiency of the program. One of the most important aspects to developing a monitoring plan is to establish and justify the success criteria for the restoration program prior to commencement of the monitoring program. Although the monitoring techniques utilized for each restoration site will continue to vary, the procedures described in this chapter can assist in standardizing the basic methods used on a regional scale. This standardization, in turn, will enable researchers and managers to gain confidence in monitoring programs and the resulting data they have to offer.

20.4 PROBLEMS TO OVERCOME AND SUGGESTED SOLUTIONS

The fundamental need to define the objectives and success criteria of restoration projects prior to their implementation has been stressed throughout this chapter. Despite the multitude of reef restoration efforts, the ability to evaluate the functional success of many projects has been compromised by vague or nonexistent objectives or the lack of explicitly established success criteria.[5] For example, the general goal to "restore scleractinian coral populations" cannot be properly evaluated without first stating that successful restoration would entail the survival, recruitment, growth, and reproduction of the recruited corals. Therefore, without defined success criteria, it is impossible to determine whether the restoration project has achieved its proposed function.

One suggestion for defining success criteria is for researchers to learn to describe "success" over time using the information provided by past monitoring programs. In other words, scientists should avoid trying to reinvent the wheel. Identifying past mistakes and achievements and understanding where and why individual programs were effectively executed can help define success for future programs. Monitoring is a relatively recent movement. As a result, we are at the stage in which the dissemination of information regarding successful procedures and techniques is vital for the effective progression of these programs. By incorporating what we learn from past accomplishments and failures, we can more effectively set new goals that will result in improved management decisions and better protection of the resource.[5]

Although the use of past restoration results can be useful, it is undeniable that the science of coral reef monitoring is comparatively new and experimental in nature. This results in a relatively low abundance of relevant literature, making it particularly difficult to determine which parameters to monitor for each restoration site. In the future, once sufficient information has been accumulated on this topic, we propose the development of a region-wide (e.g., South Florida, Caribbean) monitoring guidebook to help determine restoration and monitoring options for different types of disaster sites. This guidebook would include a flowchart or matrix of templates outlining the parameters available for monitoring each habitat according to disturbance type as well as guidelines to help choose which parameters to examine. The creation of this book would involve researchers and managers working in conjunction to ensure that an accurate and realistic set of guidelines is provided and that the correct type of information for making management decisions is included.

Despite good intentions, another significant problem of most restoration project designs is the failure to be experimentally based and hypothesis driven. One reason is that the legislative language and funding (recovered damages) supporting these programs limits the efforts to only those methods necessary to evaluate the repaired site. For example, the language of the National Marine Sanctuaries Act[35] requires that any remaining recovered funds, after use by the Secretary or any other federal or state agency for reimbursement of response costs, first be utilized "to restore, replace, or acquire the equivalent of the sanctuary resources that were the subject of the action including the costs of monitoring." Residual monies can then be applied toward the restoration of a degraded sanctuary resource of the corresponding affected sanctuary or the restoration of disturbed resources of other national marine sanctuaries. Although the NSMA is typical of natural resource damage assessment legislation, the language in the Act leaves little room for experimentation in restoration and monitoring programs.

The State of Florida, however, has a unique provision in its statutes related to coral reef injuries and the trust fund established to hold monetary damages for the benefit of restoring the degraded resources. The State of Florida's Ecosystem Management and Restoration Trust Fund (EMRTF) is a large fund used by various divisions and administered by Florida's Department of Environmental Protection. The purpose of the trust fund is management and restoration of Florida's ecosystems; it is used for beach renourishment, surface water improvements, and coral reef restoration projects. In addition to damages from vessel groundings, the EMRTF's main sources of funding included violations under Section 373, F.S., \$30 M in documentary stamp taxes, and 0.2% sales tax. Much like the NMSA, in Section 380.0558, F.S., it states the primary purpose is to reimburse the state for any costs associated with response, assessment, restoration, and monitoring but goes further to define that excess funds be "dedicated to the research, protection, restoration of, or substitution for, the coral reefs and other natural resources injured or destroyed." The provision gives the Board of Trustees of the Internal Improvement Trust Fund (i.e., the governor and cabinet) the authority to fund alternative projects on the basis of anticipated benefits. In the Florida Keys, the EMRTF has paid for the emergency removal of a sunken vessel on coral reef habitat, and the rehabilitation of a patch reef that had been repeatedly hit by small recreational vessels.

NOAA collects monetary damages for injuries to natural resources under a variety of statutes and deposits them into the Damage Assessment and Restoration Revolving Fund (DARRF). The DARRF was statutorily created to retain these recoveries and to allow their use, without further appropriation, for costs associated with injury assessment, response, restoration, and monitoring. Funds are distributed as required and approved by NOAA's Office of General Counsel, which reviews each request in relation to the legislative language. In the Florida Keys, this fund has supported limited, yet costly and necessary, projects that were either of an emergency nature or that would further degradation of the resources. One recent project funded by the DARRF was the removal of over 66 tons of illegally placed materials on the seafloor of the FKNMS in order to allow for natural recovery of hardbottom and seagrass habitats.

Neither trust fund has to date and to the authors' knowledge been specifically set aside for projects that are of a restoration research nature or for experimentation in restoration or monitoring. We suggest enhancing the use of these trust funds by levying an additional percentage on top of the settlement amount of a significant type of violation (e.g., ship groundings, renourishment activities, and dredging) and placing it into an earmarked category within the trust funds specifically for restoration and monitoring research. When human-induced damages occur due to one of these violations, a predetermined percentage of the settlement cost would be directed into the relevant trust fund. For example, a vessel grounding research levy may be 5% of the total settlement. Another proposal is to increase flexibility in the language of the statutes such that the recovery of this additional percentage and any excess funds may be used toward restoration methods development and monitoring research or methods development efforts. The long-term benefits of developing new restoration techniques through experimentation will outweigh the costs of conducting the studies when the results should ultimately determine more effective, efficient, and cost-effective approaches to rehabilitating coral reef habitats.

In an attempt toward making the system of monitoring more relevant to decision-makers, as requested by the National Research Council,[36] monitoring plans must include the collection of evaluative data regarding the benefits of the restoration program in addition to general ecological information. Obviously, this abundance of information oftentimes requires additional funding sources. Researchers and managers alike are aware of the relatively low budgets generally provided for research and assessment activities.[8] The previous recommendations could alleviate some of these funding shortfalls and provide for greater evaluative data. Unfortunately, even when monies are available for sufficient monitoring, they are often spent collecting incomplete or irrelevant data that do not assist in making the necessary decisions regarding a variety of human activities.[36] The primary focus is often put on data collection with insufficient attention to analysis, synthesis, and the interpretation of data needed for adaptive management. This could be easily remedied by developing a statistically sound monitoring program with the capacity to adequately answer predetermined hypotheses and evaluate predefined success criteria. The collection of monitoring data using the standard format for that specific variable and technique would also serve to increase the usefulness and value of the data to managers or other researchers.[9] Once the data have been synthesized and interpreted, the conclusions would need to be compiled, explained using language appropriate for a general audience, and made available to the public.

The supervising agency should be responsible for ensuring that the design of the monitoring project is sufficient for the associated restoration plan as well as that the implementation of the monitoring portion is completed in accordance with the terms of the approved proposal. The agency may assume several additional roles to ensure that these matters are suitably handled including, but not limited to, the following[37]:

1. Assistance with the identification, selection and hiring of contractor(s)
2. Supervision of field work
3. Evaluation of field work in terms of related contract or permitting requirements
4. Review and assessment of monitoring reports
5. Adaptation of procedures (midcourse) so as to comply with the authorized standards

The inherent conflicts of interest regarding restoration and monitoring projects are another common problem that needs to be resolved. Researchers, by nature, commonly view restoration and monitoring as a scientific endeavor and thus often aspire to examine every aspect of each project. Managers, in contrast, are constrained by the function of their role as decision-makers and are obligated to focus on the specific question at hand. Collaboration and unification between researchers and managers is one step toward the promotion and better implementation of adaptive management strategies. The provision of an evaluation and update by managers regarding the scientific requirements of the agencies is another step toward a resolution.

Legal counsel for responsible parties is often involved in coral cases. It is in the best interests of all to keep this a cooperative process and avoid litigation. Thus far, this has been the norm with only few exceptions, including the *Windspirit*, a 130-m sailing cruise ship that anchored on coral in Francis Bay in the Virgin Islands National Park in 1990.[6] Past settlements of coral claims have often been insufficient to fund comprehensive monitoring programs. This is often detrimental to the scientific aspect of the program in that researchers and managers are forced to base results and future management decisions on limited data (or no data if the entire plan is negotiated out of the claim). By immediately assuming responsibility for both the hiring of contractors to implement the required restoration and the funding of a monitoring component, the responsible party can reduce lengthy legal debates and accelerate recovery while reducing costs and maintaining all components of the restoration program.[6]

20.5 CONCLUSION

The information and guidelines described in this chapter are provided to help effectively monitor the harmful impacts that anthropogenic activities frequently impart on coral reef systems. Restoration programs are vital to preventing the continual loss of this declining habitat; however, improvements in the management and design of these pursuits cannot be realized without the implementation of carefully designed restoration monitoring programs. This chapter describes the purposes for monitoring and what monitoring programs can achieve, and it offers several examples of monitoring programs that have been used in the past. We provide suggestions of various fundamental concepts and techniques that should be considered when designing a restoration monitoring program and describe common problems to be overcome as well as potential solutions for future programs. The widespread application of protocols discussed in this chapter can lead to a consistent and prudently managed coral reef restoration monitoring system on a regional scale.

REFERENCES

1. Shinn, E.A., J.H. Hudson, R.B. Halley, and B.H. Lidz. 1977. Topographic control and accumulation rate of some Holocene coral reefs, south Florida and Dry Tortugas. Pages 1–7 in *Proceedings of the 3rd International Coral Reef Symposium,* Miami, FL.
2. Pearson, R.G. 1981. Recovery and recolonization of coral reefs. *Marine Ecology Progress Series* 4:105–122.

3. Sheppard, C. 1982. Coral populations on reef slopes and their major controls. *Marine Ecology Progress Series* 7:83–115.
4. Connell, J.H., T.P. Hughes, and C.C. Wallace. 1997. A 30-year study of coral abundance, recruitment, and disturbance at several scales in space and time. *Ecological Monographs* 67:461–488.
5. Precht, W.F., R.B. Aronson, and D.W. Swanson. 2001. Improving scientific decision-making in the restoration of ship-grounding sites on coral reefs. *Bulletin of Marine Science* 69:1001–1012.
6. Jaap, W.C. 2000. Coral reef restoration. *Ecological Engineering* 15:345–364.
7. Rogers, C.S., G. Garrison, R. Grober, Z.-M. Hillis, and M.A. Franke. 2001. *Coral Reef Monitoring Manual for the Caribbean and Western Atlantic.* National Park Service, Virgin Islands National Park, St. John, USVI.
8. Seaman, W.J. and A.C. Jensen. 2000. Purposes and practices of artificial reef evaluation. Pages 2–19 in W.J. Seaman, ed. *Artificial Reef Evaluation with Application to Natural and Marine Habitats.* CRC Press, Boca Raton, FL.
9. Thayer, G.W., T.A. McTigue, R.J. Bellmer, F.A. Burrows, D.H. Merkey, A.D. Nickens, S.J. Lozano, P.F. Gayaldo, P.J. Polmateer, and P.T. Pinit. 2003. *Science-Based Restoration Monitoring of Coastal Habitats, Volume One: A Framework for Monitoring Plans under the Estuaries and Clean Waters Act of 2000 (Public Law 160-457).* NOAA Coastal Ocean Program Decision Analysis Series No. 23, NOAA National Centers for Coastal Ocean Science, Silver Spring, MD.
10. Jaap, W.C. and J. Morelock. 1996. Baseline monitoring report, restoration project, Soto's Reef, George Town, Grand Cayman Island, British West Indies. Technical Report. Holland America-Westours and Cayman Islands Department of the Environment, Seattle and George Town.
11. Jaap, W.C. and J. Morelock. 1997. Six-month monitoring report, restoration project, Soto's Reef, George Town, Grand Cayman Island, British West Indies. Technical Report. Holland America-Westours and Cayman Islands Department of the Environment, Seattle and George Town.
12. Jaap, W.C. and J. Morelock. 1997. One-year monitoring report, restoration project, Soto's Reef, George Town, Grand Cayman Island, British West Indies. Technical Report. Holland America-Westours and Cayman Islands Department of the Environment, Seattle and George Town.
13. Jaap, W.C. and J. Morelock. 1998. Two-year monitoring report, restoration project, Soto's Reef, George Town, Grand Cayman Island, British West Indies. Technical Report. Holland America-Westours and Cayman Islands Department of the Environment, Seattle and George Town.
14. Quammen, M.L. 1986. Measuring the success of wetlands mitigation. *National Wetland Newsletter* 8:6–8.
15. Hanisak, M.D. 2004. Recolonization of a coral reef following the grounding of the freighter *Wellwood* on Molasses Reef, Florida Keys, in Harbor Branch Oceanographic Institution. http://www.hboi.edu/marinesci/resprojects/wellwood/wellwood1.html.
16. Miller, M.W. and J. Barimo. 2001. Assessment of juvenile coral populations at two reef restoration sites in the Florida Keys National Marine Sanctuary: indicators of success? *Bulletin of Marine Science.*
17. NOAA. 1995. Coral reef restoration to begin in Florida Keys National Marine Sanctuary at *Maitland* and *Elpis* grounding sites. http://www.publicaffairs.noaa.gov/press95.html.
18. Bodge, K.R. 1996. Structural restoration of the *M/B Alec Owen Maitland* and *M/V Elpis* Vessel Grounding Sites; Florida Keys National Marine Sanctuary. Engineering Summary Report for the National Oceanic and Atmospheric Administration, Olsen Associates, Inc., Silver Spring, Maryland.
19. McKain, D.W. and V.E. McKain. 1996. Unique reef replication. *Ocean News and Technology,* May/June, p. 32.
20. Gittings, S.R. and T.J. Bright. 1988. The *M/V Wellwood* grounding: a sanctuary case study, the science. *Oceanus,* 31:36–41.
21. Smith, S.R., D.C. Hellin, and S.A. McKenna. 1998. Patterns of juvenile coral abundance, mortality, and recruitment at the *M/V Wellwood* and *M/V Elpis* grounding sites and their comparison to undisturbed reefs in the Florida Keys. Final Report. NOAA Sanctuary and Reserves Division and the National Undersea Research Program/University of North Carolina at Wilmington.
22. Aronson, R.B., and D.W. Swanson. 1997. Video surveys of coral reefs: uni- and multivariate applications in *8th International Coral Reef Symposium,* 2:1441–1446.
23. Dennis, G.D. and T.J. Bright 1998. The impact of a ship grounding on the reef assemblage at Molasses Reef, Key Largo National Marine Sanctuary, Florida. *Proceedings of the 6th International Coral Reef Symposium,* 2:213–218.

24. Deis, D.R., G.P. Schmahl, and S.K. Shutler. 2003. *Contship Houston* Cooperative Assessment and Restoration "The Good, the Bad, and the Ugly." Invited speaker at a Workshop on Cooperative Damage Assessment and Restoration Projects, National Oceanic and Atmospheric Administration, Damage Assessment and Restoration Program, Washington, D.C.
25. Waxman, J., R. Shaul, G.P. Schmahl, and B. Julius. 1999. Innovative tools for reef restoration: the *Contship Houston* grounding. Abstract in *International Conference on Scientific Aspects of Coral Reef Assessment, Monitoring and Restoration, Program and Abstracts.* p. 200.
26. Lindberg, W.J., and G. Relini. 1997. Integrating evaluation into reef project planning. Pages 196–212 in W.J. Seaman, ed. *Artificial Reef Evaluation with Application to Natural Marine Habitats.* CRC Press, Boca Raton, FL.
27. Cairns, J.J. 1995. *Rehabilitating Damaged Ecosystems.* Lewis Publishers, Boca Raton, FL.
28. Portier, K.M., G. Fabi, and P.H. Darius. 2000. Study design and data analysis issues. Pages 21–50 in W.J. Seaman, ed. *Artificial Reef Evaluation with Application to Natural Marine Habitats.* CRC Press, Boca Raton, FL.
29. Aronson, R.B., P.J. Edmunds, W.J. Precht, D.W. Swanson, and D.R. Levitan. 1994. Large-scale, long-term monitoring of Caribbean coral reefs: simple, quick, inexpensive techniques. *Atoll Research Bulletin* 421:1–19.
30. Deis, D.R. 2000. The use of natural resource damage assessment techniques in the assessment of impacts of telecommunication cable installation on hard corals off Hollywood, Florida. In *Overcoming Barriers to Environmental Improvement, 25th Annual Conference of the National Association of Environmental Professionals.* June 25–29, 2000. Portland, ME.
31. Graham, B.D., and R. Mulcahy, 2003. Coral reef injuries and relevant case studies: management tools for directing restoration, in *Proceedings of the 10th International Coral Reef Symposium.*
32. PBS&J. 2003. AT&T Hollywood, Florida Station Year Four Monitoring Report for Repaired Corals and Artificial Reef Modules. A Report for Paul Shorb, AT&T, Bedminster, NJ, Jacksonville, FL.
33. Rogers, C.S. 1993. Hurricanes and anchors: preliminary results from the National Park Service regional reef assessment program. Page 420 in R.N. Ginsburg, ed. *Proceedings of the colloquium on Global Aspects of Coral Reefs: Health, Hazards, and History.* University of Miami, FL.
34. Rogers, C.S., and V.H. Garrison. 2001. Ten years after the crime: lasting effects of damage from a cruise ship anchor on a coral reef in St. John, USVI. *Bulletin of Marine Science* 69:793–803.
35. National Marine Sanctuaries Act 16 U.S.C. 1431 ET. SEQ., as amended by Public Law 106-513 Sec. 312. Destruction or Loss of, or Injury to, Sanctuary Resources, (d) Use of Recovered Amounts, 1972.
36. National Research Council, U.S. 1996. *Linking Science and Technology to Society's Environmental Goals.* National Academy of Sciences, Washington, D.C.
37. Fonesca, M.S., W.J. Kenworthy, and G.W. Thayer, 1998. Example Propeller and Mooring Scar Restoration Plan. in NOAA's. C. O. P. D. A. S. No. 12, ed. Guidelines for the Conservation and Restoration of Seagrasses in the United States and Adjacent Waters. U.S. Department of Commerce, National Oceanic and Atmospheric Administration, Coastal Ocean Office, Silver Spring, MD.

Index

D

E

F

G

H

I

J

K

L

N

O

P

Q

R

S

T

U

V

W

X

Y

Z